Kelvin's Baltimore Lectures and Modern Theoretical Physics

Studies from the Johns Hopkins Center for the History and Philosophy of Science

Observation, Experiment, and Hypothesis in Modern Physical Science, edited by Peter Achinstein and Owen Hannaway, 1985

Kelvin's Baltimore Lectures and Modern Theoretical Physics: Historical and Philosophical Perspectives, edited by Robert Kargon and Peter Achinstein, 1987

Kelvin's Baltimore Lectures and Modern Theoretical Physics

Historical and Philosophical Perspectives

edited by
Robert Kargon
and
Peter Achinstein

The MIT Press
Cambridge, Massachusetts
London, England

This book was set in Baskerville and Optima by Asco Trade Typesetting Ltd. in Hong Kong and printed and bound by Halliday Lithograph in the United States of America.

Library of Congress Cataloging-in-Publication Data

Kelvin's Baltimore lectures and modern theoretical physics.

(Studies from the Johns Hopkins Center for the History and Philosophy of Science)
Bibliography: p.
Includes index.
1. Mathematical physics—History. 2. Physics—Philosophy—History. 3. Kelvin, William Thomson, Baron, 1824–1907. I. Kargon, Robert. II. Achinstein, Peter. III. Kelvin, William Thomson, Baron, 1824–1907. IV. Series.
QC19.6.K45 1987   530.1′5   86-27368
ISBN 0-262-11117-9

Contents

The Lectures
William Thomson, Lord Kelvin

Essays

William Thomson, Baron Kelvin (1824–1907), in 1898.

Preface

This book, sponsored by the Johns Hopkins Center for the History and Philosophy of Science, commemorates the centenary of Lord Kelvin's Baltimore lectures on wave theory and molecular dynamics. The twenty lectures, delivered at the Johns Hopkins University in the fall of 1884, were recorded stenographically by A. S. Hathaway and reproduced from his handwritten transcriptions by the "papyrograph" method, a process akin to that of the mimeograph. This volume presents the lectures in essentially their original form, without the great alterations Kelvin made for the version published in 1904 by the Cambridge University Press as *The Baltimore Lectures on Molecular Dynamics and the Wave Theory of Light*. The lectures are followed by ten essays on historical and philosophical themes relating to Kelvin and to theoretical physics from Kelvin's day to the present.

The lectures themselves illuminate the state of classical physics before the onset of the twentieth-century revolution in physics involving radioactivity, x rays, the electron, relativity, and quantum mechanics. As the detailed record of a "master class," they also reveal the mind of the nineteenth century's most celebrated physicist at work, developing ideas at the blackboard and responding to his distinguished audience of professors and advanced students.

The essays, written from historical and philosophical perspectives, deal with Kelvin's differences with James Clerk Maxwell, with George Francis FitzGerald, and with Lord Rutherford, and cover a range of issues in theoretical physics since Kelvin, including quantum measurement, space and time, action at a distance, and parts and wholes.

The editors would like to thank Julia Driver, Takehiko Hashimoto, Andrew Lang, Gail Schley, Nancy Thompson, and David White for their invaluable help in preparing this volume. Publication was made possible by grants from several sources at the Johns Hopkins University, including the Center for the History and Philosophy of Science, the Dean's Office, and the Philosophy Department. We are grateful to George Fisher, Dean of Arts and Sciences, to Susan Martin, Director of the Milton S. Eisenhower Library, and to Jerome

Schneewind, Chairman of the Philosophy Department, for their support of this project.

The editors are very grateful to the National Endowment for the Humanities for its support of the research project (RO-20836) during which some of the work for this volume was done.

Kelvin's Baltimore Lectures and Modern Theoretical Physics

Introduction

Beginning October 1, 1884, on the invitation of President Daniel Coit Gilman, Sir William Thomson (later Lord Kelvin) delivered a series of lectures on molecular dynamics at the Johns Hopkins University in Baltimore. Henry Rowland, professor of physics at Johns Hopkins and the leading producer of physics Ph.D.s in nineteenth-century America, predicted that the occasion would have "a very great influence upon American science."[1]

Indeed, despite the fact that in these lectures Kelvin argued against the electromagnetic theory of light and proposed a model for the ether that few Americans were willing to take up, his enthusiasm for the discipline, his personal magnetism, and his performance gave a tremendous boost to the morale of the fledgling physics profession in the United States, and provided a focus for interest in mathematical physics, a subject not at the time noticeably strong in the western hemisphere.

Although Gilman left the choice of lecture topics to Kelvin, he passed on the views of his consultants, Rowland (of Hopkins) and Oliver Wolcott Gibbs (of Harvard). Gibbs hoped that the lectures would concentrate on "the obscure and difficult points in our modern physics," for instance "the difficulties we meet in the wave theory of light, in the atomic and molecular theory of matter, in electricity as regards the want of any physical theory whatever."[2] Kelvin responded with a proposal that he lecture on two or three of five subjects: vortex motion, gyrostatics, molecular dynamics, "Fourier-mathematics," and "the equilibrium of an elastic solid."[3] Rowland encouraged him to concentrate on one. "It seems to me," he wrote, "that the subject of 'Molecular Dynamics' would interest the greatest number of persons and indeed be of the most profound scientific interest."[4] Apparently convinced, Kelvin settled on molecular dynamics, and noted: "I shall bear in mind as much as possible the suggestions of Profs. Rowland and Wolcott Gibbs in endeavouring to call attention rather to difficulties and deficiencies in Mathematical Physics, than in attempting to go over well-cultivated ground."[5]

The attendees were expected to have advanced knowledge of physics and mathematics. Among them were the British physicists Lord Rayleigh and George

Forbes; Professors Kikuchi and Fujioka of Japan; American instructors in physics from eastern and western colleges, including Albert Michelson and Edward Morley; attendees from Canada, Germany, and Russia; and, of course, Hopkins faculty and students, including Rowland, Thomas Craig, Fabian Franklin, Henry Crew, Gustav Liebig, Joseph Sweetman Ames, and Christine Ladd Franklin.[6]

The sessions, which were held in a small lecture hall, were conducted as "master classes." The tone was conversational and informal; Kelvin made almost no use of notes. Lord Rayleigh told his son: "What an extraordinary performance that was! I often recognized the morning's lecture was founded on the questions which had cropped up when we were talking at breakfast."[7] The regulars (dubbed "the twenty-one coefficients" by George Forbes, in an allusion to the coefficients defining an elastic solid) took the class out of the lecture hall, plying Kelvin with questions, searching for books and references in the library, and tabulating solutions.

The main point of the series of lectures was to take up, in part, one of Gibbs's points. Kelvin wanted to place the wave theory of light upon a manifestly physical basis. He began, in the first lecture, by pointing to four major areas of difficulty: dispersion, the ether, reflection and refraction, and double refraction. Kelvin addressed each of these difficulties from three standpoints: the propagation of a disturbance through an elastic medium taken as a whole, molecular vibrations, and the influence of molecules on the propagation of waves. Usually, Kelvin lectured from one of these standpoints, then engaged his audience in a discussion of the details,[8] and then shifted to another of the standpoints for the second part of the lecture.

Concerning molecules, Kelvin expressed the conviction that many of the problems of the propagation of light could be treated successfully by supposing "a rude mechanical model": a rigid spherical shell with another infinitely rigid shell inside it, and so on *ad infinitum*, with ether on the outside and dense matter at the center, a set of "zig-zag" springs connecting the outermost shell with its interior neighbor, and spiral springs connecting the others. At the center of this system was the molecule. The number of shells was not important—they could be infinite in number, thus constituting a continuous "atmosphere." In the lectures, an actual model was used to illustrate this conception—the "coefficients" called it "the wiggler." Composed of steel wires and wood laths, it was driven by a pendulum system. Another model employed a series of weights attached to vertical springs.[9] It was in connection with these models, both theoretical and concrete, that Kelvin made his famous pronouncements about the use of models and analogies in physics. The following passage in Lecture XI lays out the terrain:

Although the molecular constitution of solids supposed in these remarks and mechanically illustrated in our model is not to be accepted as true in nature, still the construction of a mechanical model of this kind is undoubtedly very instructive, and we could not be satisfied unless we could see our way clear to make a model.... My object is to show how to make a mechanical model which shall fulfil the conditions required in the physical phenomena that we are considering, whatever they may be. At the time when we are considering the phenomena of elasticity in solids, I want to show a model of that. At another time, when we have vibrations of light to consider, I want to show a model of the action exhibited in that phenomenon. We want to understand the whole about it; we only understand a part. It seems to me that the test of "Do we or do we not understand a particular point in physics?" is, "Can we make a mechanical model of it?"

This leads precisely to Kelvin's rejection of the electromagnetic theory of light. In Lecture I he complained:

If I knew what the magnetic theory of light is, I might be able to think of it in relation to the fundamental principles of the wave theory of light. But it seems to me that it is rather a backward step from an absolutely definite mechanical motion that is put before us by Fresnel and his followers to take up the so-called electro-magnetic theory of light.

Later, in Lecture XX, he maintained:

I never satisfy myself until I can make a mechanical model of a thing. If I can make a mechanical model I can understand it. As long as I cannot make a mechanical model all the way through I cannot understand; and that is why I cannot get the electro-magnetic theory. . . . I want to understand light as well as I can, without introducing things that we understand even less of. . . . But so soon as we have rotators to take the part of magnetism, and realize by experiment Maxwell's beautiful ideas of electric displacements and so on, then we shall see electricity, magnetism, and light closely united and grounded in the same system.

The use of mechanical models, even to describe partially the understanding of a phenomenon, avoids the pitfall of "explaining" something with ideas even less well understood. It also avoids another of Kelvin's nemeses: the "aphasia" of mathematics. A second theme to emerge from the lectures is the necessity of giving physical meaning to mathematical formulas. George Forbes, one of the more faithful attendees, reported that "rather than make a meaningless mathematical assumption he would prefer to burden his formulas with undetermined quantities, and even if unable to reach the final solution, would rejoice in the richness of the formulas, which showed a potentiality of overcoming many difficulties." [10] In Lecture XV, Kelvin commended Rankine and Faraday as warriors against the temptation to substitute mathematical manipulation for physical understanding: "Rankine did a great deal to cure the mathematical disease of aphasia from which we suffered so long; Faraday did most. The old mathematicians used neither diagrams to help people to understand their work nor words to express their ideas. It was formulas and formulas alone. Faraday was a great reformer in that respect with his language of 'lines of force' etc."

Kelvin's classes, which became known as the Baltimore Lectures, were generally regarded as a success.[11] In 1904 Joseph Sweetman Ames, who had attended as a student, wrote in *Physical Review* that it was "impossible to express in words the enthusiasm that was felt by Lord Kelvin's audience as they listened to his daily lectures."[12]

Albert Michelson and Edward Morley, who were later to collaborate on the famous experiment on ether drag that bears their names, followed the proceedings very closely. During the first lecture Kelvin remarked: "We do not know at this moment whether the earth moves dragging the luminiferous ether with it, or whether it moves more nearly as if it were through a frictionless fluid." Michelson discussed such problems with Kelvin and with Rayleigh. He reported: "They both seemed to think that the first step should be to repeat Fizeau's experiment. I have accordingly ordered the required apparatus and hope to be able to settle the question within a few months."[13] Michelson and Morley left Baltimore together by train. One of the topics they are known to have discussed was the suggestion by the British physicists for further work on ether drag.[14]

In general, the Baltimore Lectures fulfilled Rowland's hopes for them. They encouraged the American physicists, brought together a strong group with mutual interests, and drew attention to Johns Hopkins as a center for physical research. Despite the fact that many in the audience were committed to the electromagnetic theory whether or not a mechanical basis for it could be found, the lectures did, as Rowland had predicted, stimulate interest in molecular dynamics and ether physics among American physicists. By stating the case for the molecular-mechanical world view, by forcefully demonstrating its methods, and by fearlessly pointing out its difficulties, Kelvin prepared the way for the thoroughgoing critique of that world view that was to come within two decades.

Editor's Note

Alterations to the original text were kept to a minimum. Trivial misspellings (e.g. *analagous, infinitessimal*) were corrected silently, but most archaic and British spellings were retained. Minor stylistic quirks were left alone, but idiosyncrasies that were felt likely to baffle modern readers were fixed unobtrusively. All added words and all replacements for misused words are bracketed. Brackets around entire sentences or paragraphs, rather than around single words or short phrases, are present in the original text and indicate alterations made by the transcriber, A. S. Hathaway, or by Kelvin (William Thomson); this is made plain by the initials (H. or W.T.) just inside the closing bracket.

As typesetting is less flexible than handwriting, mathematical material had to be styled in a more conventional manner than in the original volume. Where necessary, long or complex expressions were broken out of running text and displayed. However, each expression is essentially faithful to the original;

the order of symbols was not changed in any equation, and only where typographic considerations required it was an expression altered (e.g., by "breaking down" a derivative or an *a*-over-*b* term with a virgule so that it could be kept in running text).

The diagrams have been reproduced. Although they are not located exactly as they were in the original volume (where, again, Mr. Hathaway did not have to deal with the vagaries of typesetting), they are placed between sentences in a manner that comes close to the original treatment.

The use of commas was a good deal less standardized and more extravagant in the 1880s than it is today. Superfluous, misplaced, and omitted (to the modern eye) commas present what may be the greatest impediment to the present-day reader of the lectures. It is hoped that the reader will not mind the limited number of silent corrections in this area; they are restricted to sentences that were felt to be truly difficult to follow as originally punctuated.

Notes

1. Henry Rowland to William Thomson, January 8, 1884 (Rowland Papers, Johns Hopkins University).

2. Oliver Wolcott Gibbs to Daniel Coit Gilman, enclosure in letter from Gilman to Thomson, February 19, 1883 (Kelvin Papers, Cambridge University Library).

3. Thomson to Gilman, October 2, 1883 (Gilman Papers, Johns Hopkins University).

4. Rowland to Thomson, January 8, 1884 (Kelvin Papers).

5. Thomson to Gilman, February 5, 1884 (Gilman Papers).

6. List of Attendees (Gilman Papers).

7. R. J. Strutt, *John William Strutt, 3rd Baron Rayleigh* (London: Edward Arnold, 1924), p. 145.

8. George Forbes, "Molecular Dynamics," *Nature* 31 (1885): 461–463, 508–510, 601–603.

9. Ibid., p. 508.

10. Ibid., p. 461.

11. J. J. Sylvester to Thomson, November 25, 1884 (Kelvin Papers).

12. J. S. Ames, "Review of *Baltimore Lectures*," *Physical Review* 19 (1904): 62–64.
     Another student attendee, Henry Crew, noted: "Rowland presided, but was frequently seen to be nodding with sleep. My guess is that Rowland was not seriously interested in Kelvin's approach to the subject, since he (Rowland) had some years before decided that the electromagnetic theory of light was the only one worthy of serious consideration." (Henry Crew to R. Woodbury, December 4, 1935 [Henry Crew Papers, Northwestern University]; the first sentence is crossed out and marked "Please

forget this sentence," but see also R. S. Shankland, "The Michelson-Morley Experiment," *American Journal of Physics* 32 [June 1964]: 16–35, at p. 25).

13. Michelson to J. Willard Gibbs, December 15, 1884 (quoted in Dorothy Livingston, *Master of Light* [New York: Scribner, 1973], p. 109).

14. Livingston, p. 108.

Notes of Lectures
on
Molecular Dynamics.
and
The Wave Theory of Light.

Delivered at the Johns Hopkins University Baltimore.

By
Sir William Thomson,
Professor in the University of Glasgow.

Stenographically Reported by
A. S. Hathaway,
Lately Fellow in Mathematics of the Johns Hopkins University.
1884.

Lecture I

The most important branch of physics which at present makes demands upon molecular dynamics seems to me to be the wave theory of light. When I say this, I do not forget the one great branch of physics which at present is reduced to molecular dynamics, the kinetic theory of gases. In saying that the wave theory of light seems to be that branch of physics which is most in want, which most inevitably demands applications, of molecular dynamics just now, I mean that as the kinetic theory of gases is a part of molecular dynamics, is founded upon molecular dynamics, works wholly within molecular dynamics, to it molecular dynamics is everything, and it must be advanced by molecular dynamics, so the wave theory of light is only beginning to demand imperatively applications of that kind of dynamical science.

The wave theory of light began very much in the hands of Fresnel, afterwards of Cauchy, and to some degree, though not perhaps to so great a degree, in the hands of Green. It was wholly molecular dynamics, but of an imperfect kind in the hands of Fresnel. Cauchy attempted to found his mathematical investigations on a molecular treatment of the subject. Green almost wholly shook off the molecular treatment, and worked out all that was to be worked out in that way for the wave theory of light, by the dynamics of continuous matter. Indeed, I do not know that it is possible to add substantially to what Green has done in this subject. Substantial additions are scarcely to be made to a thing that is applied as Green's work is, on the explanation of the propagation of light, the refraction and the reflection of light at the bounding surface of two different mediums, and the propagation of light through crystals, by a strict mathematical treatment, founded on the consideration of homogeneous, elastic matter. Green's treatment is really complete in this respect, and there is nothing substantial to be added to it. But there is a great deal of exposition wanting to let us make it our own. We must study it; we must try to see what there is in the very concise and sharp treatment, with some very long formulae, which we find in Green's papers.

The wave theory of light, treated on the assumption that the medium through which the light is propagated is continuous and homogeneous, except where distinctly separated by a bounding inter-face between two different mediums, is really completed by Green. But there is a great deal to be learned from that kind of treatment that perhaps scarcely has yet been learned, because the subject has not been much studied and reduced to a very popular form hitherto.

Cauchy seemed unable to help beginning with the consideration of discrete particles mutually acting upon one another. But except in his theory of dispersion he virtually came to the same thing somewhat soon in his treatment, everytime he began it afresh, as if he had commenced right away with the consideration of a homogeneous, elastic solid. Green preceded him, I believe, in this subject. I read a statement of Lord Rayleigh that there seems to have been a matter of fact attributing to Cauchy of that which Green had actually done before. Green had exhausted the subject; but there is no doubt that Cauchy worked in an independent way.

What I propose in this first Lecture—we must have a little mathematics, and I must not be too long, with any kind of preliminary remarks—is to call your attention to the outstanding difficulties. The first difficulty that meets us in the dynamics of light is the explanation of dispersion, that is to say, of the fact that the velocity of propagation of light is different for different wave lengths or for light of different periods in one and the same medium. Treat it as we will, vary the fundamental suppositions as much as we can, as much as the very fundamental idea allows us to vary them, and we cannot force from the dynamics of a homogeneous elastic solid a difference of velocity of wave propagation for different periods.

Cauchy pointed out that if the spheres of action of individual molecules be comparable with the wave lengths, the fact of the difference of velocities for different periods or for different wave lengths in the same medium is explained. The best way, perhaps of putting Cauchy's fundamental explanation is to say that there is heterogeneousness through space, comparable with the wave lengths in the medium—that is, if we are to explain dispersion by Cauchy's unmodified supposition. We shall consider that a little later. I have no doubt it is perfectly familiar already to many of you that it is essentially insufficient to explain the facts.

Another idea for explaining dispersion has come forward more recently, and that is the assumption of molecules loading the luminiferous ether and somehow or other elastically connected with it. The first distinct statement that I have seen of this view is in Helmholtz's little paper on anomalous dispersion. I shall have occasion to speak of that a good deal and to mention other names whom Helmholtz quotes in this respect, so that I shall say nothing about it historically, except that there we have in Helmholtz's paper and by some German mathematicians who preceded him quite another departure in respect to the explanation of dispersion. The Cauchy hypothesis gives us something comparable with the wave length in the geometrical dimensions of the body. Or, to take a crude matter of fact view of it, let us say the ratio of the distance from molecule to molecule (from the center of one molecule to the center of the next-nearest molecule) to the wave length of light is the fundamental characteristic, as it were, to which we must look for the explanation of dispersion upon Cauchy's theory.

We may take this fundamental idea in connection with the two hypotheses for accounting for dispersion: that we must have some relation either to wave length or to period, and it seems (altho' this is a proposition that would require modification) at first sight that with very long waves the velocity of propagation should be independent of the period or wave length. That, at all events, seems to be the case when the subject is only looked upon according to Cauchy's view. We are led to say then that it seems that for very long waves there should be a constant velocity of propagation. Experiment and observation now seems to be falling in very distinctly to affirm the conclusions that follow from the second hypothesis that I alluded to to account for dispersion. In this second hypothesis, instead of having a geometrical dimension in the solid which is comparable with the wave length we have a fundamental time relation—a certain definite interval of time somehow ingrained in the constitution of the solid with a definite relation to the period. So that, instead of a relation of length to length, we have a relation of time to time.

Now, how are we to get our time element ingrained in the constitution of matter? We can scarcely put that question now-a-days. We are all familiar with the time of vibration of the sodium atom, and the great wonders revealed by the spectroscope are all full of indications showing a relation to absolute intervals of time in the properties of matter. This is now [so] well understood, that it is no new idea to propose to adopt as our unit of time one of the fundamental periods—for instance, the period of vibration of light in one or other of the sodium D lines. You all have a dynamical idea of this already. You all know something about the time of vibration of a molecule, and how the time of vibration of light in passing through any substance in supposing it nearly the same as the natural time of vibration of the molecules of the substance, gives rise to the absorption. We all know of course, according to this idea, the old dynamical explanation, first proposed by Stokes, of the dark lines of the solar spectrum.

We have now this interesting point to consider, that if we would work out the idea of dispersion at all, we must look definitely to times of vibration, in connection with solid [ponderable matter] itself. To get a firm hypothesis that will allow us to work at the subject, let us imagine the luminiferous ether occupied by something different from the luminiferous ether itself. That something might be a portion of denser ether, or a portion of more rigid ether; or we might suppose a portion of ether to have greater density and greater rigidity, or different density and different rigidity from the surrounding ether. We will come back to that subject in connection with the explanation of the blue sky, and, particularly, Lord Rayleigh's dynamics of the blue sky. In the meantime, I want to give something that will allow us to bring out a very crude mechanical model of dispersion.

In the first place, we must not listen to any suggestion that we must look upon the luminiferous ether as an ideal way of putting the thing. A real matter between us and the remotest stars I believe there is, and that light consists of real motions of that matter—motions just such as are described by Fresnel and Young, motions in the way of transverse vibrations. If I knew what the magnetic theory of light is, I might be able to think of it in relation to the fundamental principles of the wave theory of light. But it seems to me that it is rather a backward step from an absolutely definite mechanical motion that is put before us by Fresnel and his followers to take up the so-called electro-magnetic theory of light in the way it has been taken up by several writers of late. In passing I may say that the one thing about it that seems intelligible to me, I scarcely think is admissible. What I mean is, that there should be an electric displacement perpendicular to the line of propagation and a magnetic disturbance perpendicular to both. It seems to me that when we have an electro-magnetic theory of light, we shall see electric displacement as in the direction of propagation—simple vibrations as described by Fresnel with lines of vibration perpendicular to the line of propagation—for the motion actually constituting light. I merely say that in passing, as perhaps some apology is necessary for my insisting upon the plain matter of fact dynamics and the true elastic solid as giving what seems to me the only tenable foundation for the wave theory of light in the present state of our knowledge.

The luminiferous ether we must imagine to be a substance which so far as luminiferous vibrations are concerned moves as if it were an elastic solid. I do not say it is an elastic solid. That it moves as if it were an elastic solid in respect to the luminiferous vibrations is the fundamental assumption of the wave theory of light.

An initial difficulty that might be considered [insuperable] is, how can we have an elastic solid, with a certain degree of rigidity pervading all space, and the earth moving through it at the rate the earth moves around the sun, and the sun and solar system moving through it at the rate in which they move through space, at all events relatively to the other stars?

That difficulty does not seem to me so very insuperable. Suppose you take a piece of Burgundy pitch, or Trinidad pitch, or what I know best for this particular subject, Scotch shoemaker's wax. That is the substance I used in the illustration I intend to refer to. I do not know how far the others would succeed in the experiment. Suppose you take one of these substances, the shoemaker's wax, for instance. It is brittle, but you can bend it into the shape of a tuning fork and make it vibrate. Take a long rod of it, and you can make it vibrate as if it were a piece of glass. But leave it lying upon its side for a night and it will flatten down gradually. The weight of a letter will flatten it. Experiments have not been made as to the fluidity or non-fluidity of such a substance as shoemaker's wax; but that time is all that is necessary to allow it to yield absolutely as a fluid, is not an improbable supposition with reference to any one of the substances I have mentioned. Scottish shoemaker's wax, I have used in this way: I took a large slab of it, perhaps a couple of inches thick, fitting in a glass jar ten or twelve inches in diameter. I filled the glass jar with water and laid the slab of wax in it with a quantity of corks underneath and two or three lead bullets on the upper side. This was at the beginning of an Academic year. Six months passed away and the lead bullets had all disappeared, and I suppose the corks were half way through. Before the year had passed, on looking at the slab, I found that the corks were floating in the water at the top, and the bullets of lead were tumbling about in the bottom of the jar.

Now, if a piece of cork, in virtue of the greater specific gravity of the shoemaker's wax, would float upwards through that solid material and a piece of lead, in virtue of its greater specific gravity would move downwards through the same material, though only at the rate of an inch per six months, we have an illustration, it seems to me, quite sufficient to do away with the fundamental difficulty from the wave theory of light. Let the luminiferous ether be looked upon as a wax which is elastic and I was going to say brittle (we will think of that yet of what the meaning of brittle would be) and capable of executing vibrations like a tuning fork when times and forces are suitable—when the times in which the forces tending to produce distortion act, are very small indeed, and the forces are not too great to produce rupture. When the forces are long continued then very small forces suffice to produce change of shape. Whether infinitesimally small forces produce change of shape or not we do not know; but very small forces suffice to produce change of shape. All we have got with respect to the luminiferous ether is that the exceedingly small forces required to be brought with play in the luminiferous vibrations do not, in the times during which they act suffice to produce any sensibly permanent distortion. The come and go effects taking place in the period of the luminiferous vibrations do not give rise to the consumption of any large amount of energy, not large enough an amount to cause the light to be wholly absorbed in say its propagation from the remotest visible star to the earth.

If we have time, we shall try a little later to think of some of the magnitudes concerned, and think of, in the first place, the magnitude of the shearing force in luminiferous vibrations of some assumed amplitude on the one hand, and the magnitude of the shearing force concerned, when the earth, say, moves through the luminiferous ether on the other hand. The subject has not been gone into very fully; so that we do not know at this moment whether the earth moves dragging the luminiferous ether altogether with it, or whether it moves more nearly as if it were through a frictionless fluid. It is conceivable that it is not impossible that the earth moves through the luminiferous ether almost as if it were moving through a frictionless fluid and yet that the luminiferous ether has the rigidity necessary for the performance of the luminiferous vibration [in] the period from the four hundred million millionth of a second to the eight hundred million millionth of a second corresponding to the visible rays, or from the periods which we now know in the low rays of radiant heat as recently experimented on and measured for the wave length by Abney, to the high ultra-violet rays of light, known chiefly by their chemical actions. If we consider the exceeding smallness of the period from the 100 million millionth of a second to the 1600 million millionth of a second through the known range of radiant heat and light, we need not fully despair of understanding the property of the luminiferous ether. It is no greater mystery at all events than the shoemaker's wax. That is a mystery, as all matter is; the luminiferous ether is no greater mystery.

We know the luminiferous ether better than we know any other kind of matter in some particulars. We know it for its elasticity; we know it in respect to the constancy of the velocity of propagation of light for different periods. Take the eclipses of Jupiter's satellites or something far more telling yet, the bursting of luminous stars and so on, as referred to by Professor Newcomb in a recent discussion at Montreal on the subject of the velocity of propagation of light in the luminiferous ether. These phenomena prove to us with tremendously searching test, to an excessively minute degree of accuracy, the constancy of the velocity of propagation of all the rays of visible light through the luminiferous ether.

Luminiferous ether must be a body of most extreme simplicity. It may be perhaps soft. We might imagine it to be a body whose ultimate property is to be incompressible; to have a definite rigidity for vibrations in times less than a certain limit, and yet to have the absolutely yielding character that we recognize in wax-like bodies when the force is continued for a sufficient time.

It seems to me that we must know a great deal more of the luminiferous ether than we do. But instead of beginning with saying that we know nothing about it, I say that we know more about it than we do about air or water, glass or iron—it is far simpler, there is far less to know. That is to say, the natural history of the luminiferous ether is an infinitely simpler subject than the natural history of any other body. It seems probable that the molecular theory of matter may be so far advanced sometime or other that we can understand an excessively fine-grained structure and understand the luminiferous ether as differing from glass and water and metals in being very much more finely grained in its structure. We must not attempt, however, to jump too far in the inquiry, but take it as it is, and take the great facts of the wave theory of light as giving us strong foundations for our convictions as to the luminiferous ether.

Imagine for a moment that we make a rude mechanical model. Let this be an infinitely rigid spherical shell; let there be another absolutely rigid shell inside of that, and so on, as many as you please.

Naturally, we might think of something more continuous than that, but I only wish to call your attention to a crude mechanical explanation, possibly, of the effects of dispersion. Suppose we had luminiferous ether outside, and that this hollow space is of very small diameter in comparison with the wave length. Let zig-zag springs connect the outer rigid boundary with boundary number two.

I use a zig-zag, not a spiral spring, which observes the helical properties which we are not ready for yet, such properties as sugar and quartz have in disturbing the luminiferous vibrations. Suppose we have shells 2 and 3 also connected by a sufficient number of zig-zag springs and so on; and let there be a solid enclosed in the center with spring connections between it and the shell outside of it. If there is only one of these interior shells, you will have one definite period of vibrations. Suppose you take away everything except that one interior shell; displace that shell and let it vibrate. The period of its vibration is perfectly definite. If you have an immense number of such shells, with movable molecules inside of them distributed through some portion of the luminiferous ether, you will put it into a condition in which the velocity of the propagation of the wave will be different from what it is in the homogeneous luminiferous ether. You have what is called for, viz., a definite period; and the relation between the period of vibration in the light considered and the period of the free vibration of the shell will be fundamental in respect to the attempt of a mechanism of that kind to represent the phenomena of dispersion.

If you take away everything except the one shell, you will have almost exactly, I think, the view of Helmholtz's paper—a crude model, as it were, of what Helmholtz makes [the subject of] his paper on anomalous dispersion. Helmholtz, besides that, supposes a certain degree or coefficient of viscous resistance against the vibration of the inner shell, relatively to the outer one. Helmholtz does not reduce it to a gross mechanical form like this, but merely assumes particles connected with the luminiferous ether and assumes a viscous motion to operate against the motion of the particles.

If we had only absorption and dispersion to deal with, there would be no difficulty whatever in accounting for all that is necessary. When the period of luminiferous vibration is smaller than the natural vibration of the first shell, we have a certain state of things; when it is the same we have what is prettiest, the mathematical conditions of absorption and the infinite vibrations are wanting. What is meant by absorption in the interior? The conversion of luminiferous vibrations into heat or some other mode of action or the dynamics must be such that when the motion of vibration of this inner shell is through a greater and greater range the period ceases to fulfill the conditions of exactness, and so, without absorption the infinity vibration is not met with. This part of the subject will occupy us more fully a little later.

If we had only dispersion to deal with there would be no difficulty in getting a full explanation by putting this not in a rude mechanical model form, but in a form which would commend itself to our judgement as presenting the actual mode of action of the particles, whatever they may be, upon the particles of luminiferous ether. We except the heavier matter; but oxygen, hydrogen and such as those must somehow or other act in the luminiferous ether, have some sort of elastic connection with it; and I cannot imagine anything that commends itself to our ideas better than this sort of thing. By taking enough of these interior shells, and by neglecting the idea of absolute continuity with no limit whatever to the period, we may come as it were to the kind of mutual action that exists between any particular atom and the luminiferous ether. It seems to me that there must be something in this, that this, as a symbol, is certainly not an hypothesis, but a certainty.

But alas for the difficulties of the undulatory theory of light, refraction and reflection at plane surfaces worked out by Green differ in the most irreducible way from the facts. They correspond in some degree to the facts, but there are differences that we have no way of explaining at all. A great many hypotheses have been presented but none of them seems at all tenable.

First of all is the question, are the vibrations of light perpendicular to, or are they in, the plane of polarization—defining the plane of polarization as the plane through the incident and refracted rays, for light polarized by reflection. Think of light polarized by reflection at a plane surface and the question is, are the vibrations in the reflected ray perpendicular to the plane of incidence and reflection, or are they in the plane of incidence and reflection. I merely speak of this subject in the way of index. We shall consider very fully Green's theory and Lord Rayleigh's work upon it. I come to the conclusion with absolute certainty, it seems to me, that the vibrations must be perpendicular to the plane of incidence and reflection of the light that is polarized by reflection.

Now there is this difficulty—outstanding—the theory which gives this result does not give it rigorously, but only approximately. We have by no means so good an approach in the theory to complete extinction of the vibrations in the reflected ray (when we have the light in the incident ray vibrating in the plane of incidence and reflection) as observation gives. I shall say no more about that difficulty, because it will occupy us a good deal later on, except to say that the theoretical explanation of reflection and refraction is not satisfactory. It is not complete, and it is unsatisfactory in this, that we do not see any way of mending it.

But suppose for a moment that it might be mended and there is a question connected with it which is this: Is the difference between two mediums a difference corresponding to difference of rigidity, or does it correspond to difference of density? That is an interesting question, and some of the work that was done upon it seemed most tempting in respect to the supposition that the difference between two mediums is a difference of rigidity and not a difference of density. When fully examined, however, the seemingly plausible way of explaining the facts of refraction and reflection by difference of rigidity and no difference of density I found to be delusive, and we are forced to the view that there is difference of density and very little difference of rigidity.

It seemed to me, in working out this subject very carefully, and endeavoring to understand Lord Rayleigh's work upon it, and learn what had been done by others, for a time, to be too much of an assumption that the rigidity was exactly the same and that the whole effect was due to difference of density. Might it not be, it seemed to me, that the luminiferous ether on the two sides of interface at which the refraction and reflection takes place, might differ both in rigidity and in density. It seemed to me then by a piece of work (which I must verify, however, before I state quite confidently about it) that, by supposing the luminiferous ether in the commonly called denser medium to be considerably denser than it would be where the rigidity is equal, and the rigidity to be greater than in the other mediums, we might get a better explanation of the polarization by reflection than Green's result gives. Green's work ends with the supposition of equal rigidities and unequal densities. He puts the whole in his formulae to begin with, but he ends with this supposition and his result depends upon it.

Not to deal in generalities, let us take the case of glass and a vacuum, say. It seemed to me that by supposing the rigidity of luminiferous ether in glass to be greater than in vacuum and the density to be greater, but greater in a greater proportion than the rigidity so that the velocity of propagation is less in glass than in vacuum, we should get a better explanation of the details of polarization by reflection than Green's result gives.

It is only since I have left the other side of the Atlantic that I have worked at this thing, and going at it with considerable interest. I inquired of everybody I met whether there were any observations that would help me. At last I was told that Professor Rood had done what I desired to know, and on looking at his paper, I found that it settled the matter.

My question was this: Has there been any measurement of the intensity of light reflected at nearly normal incidence from glass or water considerably greater than Fresnel's formula gives. Fresnel gives

$$\left(\frac{\mu - 1}{\mu + 1}\right)^2$$

for the ratio of the intensity of the reflected ray to the intensity of the incident ray in the case of normal incidence, or incidence nearly normal. I wanted to find out whether that had been verified. It seems that nobody had done it at all until Professor Rood, of Columbia College, New York, took it up. His experiments showed to a rather minute degree of accuracy an agreement with Fresnel's formulae, so that the explanation I was inclined to make was disproved by it. I myself had worked with the reflection of a candle from a window glass and had come to the same conclusion, even through such very crude and rough approximate results. At all events, I satisfied myself that there was not so great a deviation from Fresnel's law as would allow me to explain the difficulties of refraction and reflection by assuming greater rigidity, for example, in glass than in air. We are now forced very much to the conclusion from several results, but directly from Professor Rood's photometrical experiments, that the rigidity must be very nearly equal in the two.

There is quite another supposition that might be made that would give us the same law, the supposition that the reflection depends wholly upon difference of rigidity and that the densities are equal in the two. That gives rise to the same intensity of reflected light, so that the photometric measurement does not discriminate between the two extremes, but it does prevent us from pushing in on the other side of [a] generally accepted result in the manner that I had thought of.

We may look upon the explanation of polarization by reflection and refraction as not altogether unsatisfactory, although not quite satisfactory, and you may see that this kind of modification of the luminiferous ether is just what would give us the virtually greater density. How this gives us precisely the same effect as a greater density I shall show when we work the thing out mathematically. We shall see that this supposition is equivalent to giving the luminiferous ether a greater density, without making the addition to the density—according to the idea of vibration.

I am approaching an end I had hoped to get [to] sooner. We have the subject of double refraction in crystals, and here is the great hopeless difficulty. I do not find it quite correctly stated even in places in which it is referred to. For instance, even Lord Rayleigh says that Fresnel's view requires us to suppose the rigidity of the luminiferous ether to depend on the direction of the vibration—which is not quite true. The rigidity cannot depend on the direction of the vibration.

If we look into the matter of the distortion of the elastic solid, we may consider, possibly, that that is not wonderful; but Fresnel's supposition as to the direction of the vibration of light is that the conclusion that the plane of vibration is perpendicular to the plane of polarization proves, if it is true, that the velocity of propagation of light in uniaxial crystals depends on the direction of vibration and not on the plane of the distortion. In the vibrations of light, we have to consider the medium as being distorted and tending to recover its shape.

Let this be a piece of uniaxial crystal iceland spar, for instance, a round or square column, with its length in the direction of the optic axis, which I will represent on the board by a dotted line.

Fig. 1

Now the relation between light polarized by passing through iceland spar on the one hand and light polarized by reflection on the other hand shows us that if the line of vibration is perpendicular to the plane of polarization, then the velocity of propagation of light in different directions through iceland spar depends solely on the line of vibration and not at all on the plane of distortion.

There is no way in which that can be explained by the rigidity of an elastic solid. Look upon it in this way, in the first place. Take a cube of iceland spar, keeping the same direction of the axis as before.

Fig. 2

Let the light be passing downwards, as indicated by the dotted arrowheads. What would be the mode of vibration, with such a direction of propagation? Let us suppose, in the first place, the vibrations to be in the plane of the diagram. Then the distortion of that portion of matter will be in the direction indicated; a portion which was rectangular swings into [the] shape represented by the dotted lines. The force tending to cause a piece of matter which has been displaced to resume its original shape depends on this kind of distortion. The mathematical expression of it would be *n* a constant of rigidity, multiplied into *a*, the amount of the distortion. How that is to be reckoned is familiar to many of you, and we will not enter into the details just now. But just consider this other case, where the direction of propagation of the light is horizontal, as indicated by the dotted arrowheads, that is to say, propagated perpendicular to the axis of the crystal (Fig. 3).

Fig. 3

What would be the nature of the distortion here [,] the vibration being still on the plane of the diagram? The distortion will be in this way in which I move my two hands. A portion which was rectangular will swing into this shape indicated by the dotted lines. The return force will then depend upon a distortion of that kind. But a distortion of that kind (Fig. 3) is identical with a distortion of this kind (Fig. 2) and the result must be, if the effect depends upon the return force in an elastic solid, that we must have the same velocity of propagation in this case and in this case (Figs. 2 and 3).

But observe this is the case of the extraordinary ray; and you know that we have greater velocity of propagation in the first case, and less in the second. There is an outstanding difficulty that is absolutely inexplicable on the bare theory of an elastic solid.

The question now occurs, may we not explain it by loading the elastic solid. But the difficulty is to load it unequally in different directions. Lord Rayleigh thought that he had got an explanation of it in his paper to which I have referred. He was not aware that Rankine had exactly the same idea. Lord Rayleigh at the end of this paper puts forward the supposition that difference of effective inertias in different directions may be adduced to explain the difference of velocity of propagation in iceland spar. But if that were the case, the wave propagation would not follow Huygens' law. It would follow the law according to which the velocity of propagation would be inversely proportional to what it is according to Huygens' law. Huygens' geometrical construction for the extraordinary ray in iceland spar gives us an ellipsoid of revolution according to which the velocity of propagation of light will be found by drawing from the centre of the ellipsoid a perpendicular to the tangent plane. For example (Fig. 4) CN will correspond to the velocity of propagation of the light when the front is in the direction of this (tangent) line.

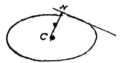

Fig. 4

If the velocity is different in different directions in virtue of an effective inertia, as your Instructor will humbly hold, Lord Rayleigh's idea is that the vibrating molecules might be like oblate spheroids vibrating in a frictionless fluid. It will have greater effective inertia when vibrating in the direction of its axis perpendicular to its flat side, less effective inertia when vibrating in its equatorial plane. That is a very beautiful idea, and we absolutely want it to explain the difficulty if the pushing forward of the conclusions from it were verified by experiment. Stokes has made the experiment. He did not know of Rankine's paper. Rankine made the first suggestion in the matter but did not push the question further than to give it as a mode of getting over the difficulty in double refraction. Stokes took away the poetry of it. He experimented on the refracting index of iceland spar for a variety of incidences, and found with minute accuracy indeed that Huygens' construction was verified and that therefore it was impossible to account for the unequal velocity of propagation in different directions by the beautiful suggestion of Rankine and Lord Rayleigh.

I have not been able to make a suggestion, but I have great hopes that these spring arrangements are going to work us out of the difficulty. I will, just in conclusion, give you the idea of how it might.

We can easily suppose these spring arrangements to have different strengths in different directions; and their law will suit exactly. Their law will give the fundamental thing we want [,] which is that the velocity of propagation of light shall depend on the direction of vibration, and not on the distortion. Besides that, this will obviously verify Huygens' law—it gives us exactly the same law as the elastic theory gives.

But alas, alas, we have one difficulty which seems still insuperable and prevents my putting this forward as the explanation, and that is, that I cannot get the requisite difference of effective inertia in different directions for the different wave lengths to suit. If we take this theory, we should have, instead of the very nearly equal difference of refractive index for the different rays in such a body as iceland spar, with dispersion merely a small thing in comparison with those differences, that the difference of refractive index in different directions would be comparable with dispersion and modified by dispersion to a prodigious degree, and in fact we should have anomalous dispersion coming in between the velocity of propagation in one direction and the velocity of propagation in another. The impossibility of getting a sufficiently constant difference of wave velocity in different directions for the different periods in those directions seems to me to be a puzzler.

So now, I have given you one hour and seven minutes and brought you face to face with a difficulty which I will not say is insuperable, but something in which nothing ever has been done from the beginning of the world to the present time that will give us the slightest explanation.

I shall do to-morrow, what I had hoped to do to-day, give you a little mathematics, knowing that it is not going to explain everything, but I think we will have an interest in working out the motions of an elastic solid and obtaining a few solutions that depend on the equations of motion of an elastic solid. I shall first take the case of zero rigidity; that will give us sound. We shall take the most elementary sounds possible, namely a spherical body alternately expanding and contracting. We shall pass from that to the case of a single globe vibrating to and fro in air. We shall pass from that to the case of a tuning fork, and endeavor to explain the cones of silence which you all know in the neighborhood of a vibrating tuning fork. I hope we shall be able to get through that in a short time and pass on our way to the corresponding solutions of the motions of a wave proceeding from a center in respect to the wave theory of light.

Lecture II

In the first place, I will take up the equations of motion of an elastic solid. I assume that the fundamental principles are familiar. At the same time, I should be very glad if any person present would, without the slightest hesitation, ask for explanations, if anything is not understood. I want to be at once on a professional footing with you, so that the work shall be rather something between you and me than something in which I shall be making a performance before you in a matter in which many of you may be quite as competent as I am, if not more so.

I want, if we can get something done in half an hour on these problems of *molar* dynamics, as we may call it to distinguish [it] from molecular dynamics, to come among you for a few moments and then go on to a problem of molecular dynamics to prepare the way for motions of mutual interference among particles under varying circumstances that may perhaps have applications in physical science and particularly to the theory of light.

The fundamental equations of equilibrium of elastic solids are, of course, included in D'Alembert's form of the equations of motion. I shall keep to the notation that is employed in Thomson and Tait's *Natural Philosophy*, which is substantially the same notation as is employed by other writers.

Let *a*, *b*, *c*, denote distortion, viz.: *a* is a distortion in the plane perpendicular to *OX* produced by slippings of the two planes which intersect in *OX*.

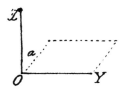

Let us consider this state of strain in which, without other change, a portion of the solid in the plane *yz* which was a square section becomes a rhombic figure. The measurement of that state of strain is given very fully in Thomson and Tait's geometrical preliminary for the theory of elastic solids. It is called a simple shear. It may be measured either by the rate of shifting of parallel planes per unit distance perpendicular to them, or, which comes to exactly the same thing, the change of the angle, measured in radians. Then I shall put down inside this small angular space the letter a, to denote the angle measured in radians.

I use the word "radians"; it is not a very common word; I suppose you know what I mean. In Cambridge in the olden time we used to have a very illogical nomenclature, viz.: "the unit angle"—a very absurd use of the article "the". It is illogical to talk of an angle being measured in "the" unit angle; there is no such thing as measuring anything, except in terms of "a" unit. The unit in which it is convenient to measure angles in analytical mechanics is the angle whose arc is radius. That used to be called the unit angle. My brother James Thomson proposed to call it the radian.

There are three principal distortions, a, b, c, relative to the axes of OX, OY, OZ; and again, three principal dilatations—condensations of course if any one is negative—e, f, g, which are the ratios of the augmentation of length to the length.

The general equation of energy will of course be an equation in which we have a quadratic function of e, f, g, a, b, c, the expression for which will be

$$\tfrac{1}{2}(11e^2 + 12ef + 13eg + 14ea + 15eb + 16ec + 21ef + 22f^2 + 23fg + \cdots).$$

We do not deal with 11, 12, ... etc. as numbers but as representing the twenty-one coefficients of this quadratic subject to the conditions $12 = 21$, etc. If we denote this quadratic function by E, then

$$\frac{dE}{de} = 11e + 12f + 13g + 14a + 15b + 16c.$$

This is a component of the normal force required to produce this compound strain e, f, g, a, b, c. According to the notation of Thomson and Tait, let $P = dE/de$, $Q = dE/df$, $R = dE/dg$, $S = dE/da$, $T = dE/db$, $U = dE/dc$. We have, then, the relation $Pe + Qf + Rg + Sa + Tb + Uc = 2E$, the well known dynamical interpretation of which you are of course familiar with. A little later we shall consider these 21 coefficients, first, in respect to the relations among them which must be imposed to produce a certain kind of symmetry relative to the three rectangular axes; and then see what further conditions must be imposed to fit the elastic solid for performing the functions of the luminiferous ether in a crystal.

Before going on to that we shall take the case of a perfectly isotropic material. We can perhaps best put it down in tabular form in this way:

In the first place in this square which has to do with the distortions a, b, c, alone, if we let n represent the rigidity-modulus the three diagonal terms will each be n, and those outside the diagonal will be zero. Six of the coefficients are thus determined. With reference to the upper right and lower left-hand corner squares, let us consider what possible relations there can be for an isotropic body between longitudinal strains and distortions. Clearly none. No one of the longitudinal strains can call into play a tangential force in any of the faces; and conversely, if the medium be isotropic, no distortion produced by slipping in the faces parallel to the principal planes can introduce a longitudinal stress—a stress parallel to any of the lines OX, OY, OZ. Therefore we have zeros in those squares. We know that $11 = 22 = 33$ and each of these will be represented by $[\mathfrak{A}]$.* Now consider the effect of a longitudinal pull in the direction of OX. If the body be only allowed to yield longitudinally, that clearly will give rise to a negative pull in the directions parallel to OY, OZ. We have then a cross connection between pulls in the directions OX, OY, OZ. Isotrophy requires that the mutual relations must be all equal, so that we have just one coefficient to express these relations. That coefficient is denoted by $[\mathfrak{B}]$; and that fills up our 36 squares, which represent but 21 coefficients in virtue of the relations $12 = 21$, etc. We can now write down our quadratic expression for the energy,

$$E = \tfrac{1}{2}[\mathfrak{A}(e^2 + f^2 + g^2) + 2\mathfrak{B}(fg + ge + ef) + n(a^2 + b^2 + c^2)].$$

Instead of these letters $\mathfrak{A}$, $\mathfrak{B}$, which have very distinct and obvious interpretations, we may introduce the resistance of the solid to compression, the reciprocal of what is commonly called the compressibility, or perhaps, to avoid ambiguities, what we may call the bulk modulus, k. Then it is proved in Thomson & Tait, and in an article in the *Encyclopaedia Britannica* which perhaps some of you may have, that $\mathfrak{A} = k + \tfrac{4}{3}n$, $\mathfrak{B} = k - \tfrac{2}{3}n$. The considerations which show the above relations with the bulk modulus also show us that we must have $n = \tfrac{1}{2}(\mathfrak{A} - \mathfrak{B})$. This is most important. Take a solid cube with its three axes parallel to OX, OY, OZ. Apply a pull along two

* *The original text reads "Saxon A" and shows the letter that appears in the preceding table. German $\mathfrak{A}$ and $\mathfrak{B}$ are used here because "Saxon" type is not available.—Editor*

faces perpendicular to OX, and an equal pressure on two faces perpendicular to OY; that will give a distortion in the plane xy. Find the value of that simple shear; it is done in a moment. Find the shearing force required to produce it calculated from $\mathfrak{A}$ and $\mathfrak{B}$ and equate that to the force calculated from the rigidity modulus n and then you find this relation. The relations for complete isotrophy are exhibited here in this quadratic expression for the energy, with the equation $n = \frac{1}{2}(\mathfrak{A} - \mathfrak{B})$.

We shall pass on to the formation of the equations of motion. For equilibrium, the force applied at any point x,y,z of the solid, reckoned per unit of bulk at that point, must be equal to $(dP/dx + dU/dy + dT/dz)$. If the body be held distorted in any way, by bodily forces applied all through the interior, the resultant of the elastic force on an infinitesimal portion of matter at the point x,y,z is obviously $(dP/dx + dU/dy + dT/dz)\delta x\delta y\delta z$. If the pull P augments as you go forward in the direction OX, there will, in virtue of that, be a resultant forward pull dP/dx upon the infinitesimal element. On the other hand U is the stress corresponding to rotation around the axis OZ.

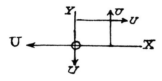

The two tangential forces, U perpendicular to OY, and U perpendicular to OX on the one pair of forces and the pair of forces equal and in opposite directions on the other faces constitute two balancing couples, as it were. If the force parallel to OX increases as we proceed in the direction y positive, there will be a resultant positive force on this element, because it is pulled to left by the smaller and to right by the larger, and there will be an augmentation to the force in the direction of OX by dU/dy. Quite similarly, dT/dz is a contribution to the force parallel to OX.

Now, let there be no bodily forces acting through the material, but let the inertia of the moving part and the reaction against acceleration in virtue of inertia constitute the equiliberating reaction against elasticity. The result is, that we have the equation

$$\frac{dP}{dx} + \frac{dU}{dy} + \frac{dT}{dz} = \rho\frac{d^2\xi}{dt^2},$$

if by ρ we denote the density and by ξ we denote the displacement from equilibrium in the direction OX of that portion of matter having x,y,z for coordinates of its mean position.

I said I would use the notation of Thomson and Tait, who employ α, β, γ to denote the displacements; but errors are too common when α and a are mixed up, especially in print, so we will take ξ, η, ζ instead. I have had volumes of trouble in reading Helmholtz's paper on anomalous dispersion, on this account very frequently not being able to distinguish with a glass whether a certain letter was a or α.

The values S, T, U we had better write out in full, although the others may be obtained from the value of any one by symmetry. The expenditure of chalk is often a saving of brains. They are:

$$S = n\left(\frac{d\eta}{dz} + \frac{d\zeta}{dy}\right), \quad T = n\left(\frac{d\xi}{dz} + \frac{d\zeta}{dx}\right), \quad U = n\left(\frac{d\xi}{dy} + \frac{d\eta}{dx}\right).$$

We have $P = \mathfrak{A}e + \mathfrak{B}(f + g)$. There are two or three other forms which are convenient in some cases and I will put them down (writing m for $k + \frac{1}{3}n$):

$$P = (m + n)\frac{d\xi}{dx} + (m - n)\left(\frac{d\eta}{dy} + \frac{d\zeta}{dz}\right)$$

$$= (m - n)\left(\frac{d\xi}{dx} + \frac{d\eta}{dy} + \frac{d\zeta}{dz}\right) + 2n\frac{d\xi}{dx}$$

$$= m\left(\frac{d\xi}{dx} + \frac{d\eta}{dy} + \frac{d\zeta}{dz}\right) + n\left(\frac{d\xi}{dx} - \frac{d\eta}{dy} - \frac{d\zeta}{dz}\right).$$

We shall denote very frequently by δ the expression $d\xi/dx + d\eta/dy + d\zeta/dz$, so that, for example, the second of these expressions is

$$P = (m - n)\delta + 2n\frac{d\xi}{dx}.$$

If we want to write down the equations of a heterogeneous medium, as will sometimes be the case, especially in following Lord Rayleigh's work on the blue sky, we must keep these symbols m, n inside of the symbols of differentiation; but for homogeneous solids, we treat m and n as constant. I forgot to say that δ is the cubic dilatation or the augmentation of volume per unit volume in the neighborhood of the point x, y, z, which is pretty well known, and helps us to see the relations to rigidity, and so on. If we suppose zero rigidity, $P = m\delta$ is the relation between pressure and volume. In order to verify this, take the second equation in P and make $n = 0$ and we obtain $P = m\delta$, the equation for the compression of a compressible fluid, in which m has become the bulk modulus.

This sort of work is called molar dynamics. It is the dynamics of continuous matter; there are no molecules, no heterogeneousnesses at all. We are preparing the way for dealing with heterogeneousnesses in the most analytical manner by supposing m and n to be functions of x, y, z. Lord Rayleigh studied the blue sky in that way, and very beautifully. The treatment is quite perfect of its kind. He considers an imbedded point to represent the particle of water, of dust, or unknown material, whatever it is, that causes the blue sky. He supposes a sudden change of rigidity and of density in the luminiferous ether; not an absolutely sudden change, but a change not homogeneous all around, and confined to a space which is small in comparison with the wave length.

I want to take up another subject which will prepare the way to what we shall be doing afterward, which is the particular dynamical problem of the movement of a system of connected particles. I suppose most of you know the linear equations of motion of a connected system—the cycloidal motion; the equations whose integral always leads to the same formula as the cycloidal pendulum, viz.: a determinant equaled to zero, whose roots are essentially real.

As an example, take three weights, one of 7 pounds, another of 14 pounds, and another of 28 pounds, say. The lowest weight is hung upon the middle weight by a spiral spring; the middle is hung upon the upper by a spiral spring, and the upper is attached to a fixed point by a spiral spring.

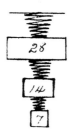

It is a pretty illustration, and I find it very useful to myself. I am speaking, so to say, to professors who sympathize with me, and might like to know an experiment which will be instructive to their pupils.

Just apply your finger to any one of the weights—the upper weight, for example. You soon learn the periods. Move it up and down gently in the period which you find to be that of the three all moving in the same direction. You will get a very pretty oscillation, the lowest weight moving through the greatest amplitude, the second through a less, and the upper weight through the smallest. That is No. 1 motion, corresponding to the greatest root of the cubic equation which expresses the solution of the mathematical problem. No. 2 motion will come after a little practice. You soon learn to give an oscillation a good deal further than before, in which the lowest weight moves downward while the two upper move upwards, or the two lower move downwards while the upper moves upwards, or it might be that the middle weight does not move at all in this second mode, in which case the excitation must be by putting the finger on the upper or the lower weight. These periods depend upon the magnitude of the weights, and the strength of the springs that we use, and are soon learned in any particular set of weights and springs. It might be a good problem for junior laboratory students to find the weights and springs which will insure a case of the nodal point lying between the upper and middle weights, or at the middle weight, or between the middle and lower weights. The next mode of vibration, corresponding to the smallest root of the cubic equation, is one in which you always have one node between the upper and middle weights, and one node between the middle and lowest (the first and third weight vibrating in the same direction, and the middle weight in an opposite direction to the first and third).

It is assumed that there is no mass in the springs. If you want to vary your laboratory exercises, take smaller masses for the weights, and more massive springs, and you pass on again to a very beautiful illustration of the velocity of sound. For that purpose a long spiral spring of steel wire 20 feet long, hung up, will answer. You can get the gravest fundamental modes without any attached weights at all. In this problem which we have been considering we have three separate weights and not a continuous spring; and we have three, and only three modes of vibration, when the springs are massless. We have an infinite number of modes when the mass of the springs is taken into account. In any convenient arrangement of heavy weights, the stiffness of the springs is so great and their masses so small that the period of vibration of one of the springs will be very short; but take a long spring, a spiral of best pianoforte steel wire, perhaps, and hang it up and you will find it a nice illustration for getting the gravest fundamental modes.

I want to put down the dynamics of our problem for any number of masses. You will see at once that that is just the case that I spoke of yesterday of extending Helmholtz's singly vibrating particle connected with the luminiferous ether to a multiple vibrating heavy elastic atom imbedded in the luminiferous ether, which I think must be the true state of the case. A solid mass must act relatively to the luminiferous ether as an elastic body imbedded in it of enormous mass compared with the mass of the luminiferous ether that it displaces. In order that the vibrations of luminiferous ether may not be absolutely stopped by the mass, there must be an elastic connection. It is easier to say what must be than to say that we can understand the result. The result is almost infinitely difficult to understand in the case of ether in glass or water or carbon disulphide, but the luminiferous ether in air is very easily understood. We just think of the molecules of oxygen and nitrogen as if they were groups of jelly relative to the luminiferous ether; and you do not, in the slightest degree, need to take into account the motions of the particles of oxygen, nitrogen and carbon dioxide in our atmosphere relatively to the propagation of waves through the air. Think of it in this way: the period of vibration is from the 100 million millionth of a second to the 1600 million millionth of a second. Now think how far a particle of oxygen or nitrogen moves in the course of that exceedingly small time. You will find that it moves through an exceedingly small fraction of the wave length. Inasmuch as each particle moves through a very small fraction of the wave length during its period, I am fully confident that the wave motion takes place independently of the translatory motion of the particles of oxygen and nitrogen in performing their functions according to the kinetic theory of gases. You may therefore really look upon the motion of light waves through our atmosphere as being solved by a dynamical problem such as this, applied to a case in which there is so little effective inertia that the velocity of light is not altered, perhaps, more than one-third per cent by it. More difficulties surround the subject when you come to impact on solid bodies.

In this case, let the particles of the bodies be represented by $m_1, m_2 \cdots m_j$. I am going to suppose the several particles to be acted upon by connecting springs. I do not want to use spiral springs here. The spiral of the spring in these experiments has no effect; but I want to introduce a spiral for investigating the dynamics of the helical properties, as shown by sugar. It is usually called the rotary property, although a misnomer. The magneto-optical property which was discovered by Faraday is rotational; the property exhibited by quartz and sugar and such things has not the essential elements of rotation in it, but has the characteristic of a spiral spring in the constitution of the matter that exhibits it. We apply the word helical to the one and the word rotational to the other.

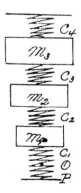

I am going to suppose one more connecting particle P—a particle of the elastic solid—which is moved to and fro with a given motion whose displacement downwards from a fixed point O we shall call ξ. Let C be the coefficient of elasticity of the first spring, connecting the particle P with the particle m_1; C_2 the coefficient of elasticity of the next spring connecting m_1 and m_2; $C_j + 1$, the coefficient of elasticity of the spring connecting m_j to a fixed point. We are not taking gravity into account; we have nothing to do with it. Although in the experiment it is convenient to use gravity, it would be still better if we could go to the centre of the earth and perform the experiment. The only difference would be, these springs would not be pulled out by the weights hung upon them. In all other respects the problem would be the same, and the same symbols would apply.

We are reckoning displacements downward as positive, the displacement of the particle m_i being x_i. The force acting upon m_1, in virtue of the spring connection between it and P, is $C_1(\xi - x_1)$; and in virtue of the spring connection between it and m_2, is the opposing pull—$C_2(x_1 - x_2)$; so that the equation of motion of the first particle is

$$m_1 \frac{d^2 x_1}{dt^2} = C_1(\xi - x_1) - C_2(x_1 - x_2).$$

For No. 2 particle we have

$$m_2 \frac{d^2 x_2}{dt^2} = C_2(x_1 - x_2) - C_3(x_2 - x_3);$$

and so on.

Now suppose P to be arbitrarily kept in some simple harmonic motion in time or period T. I might introduce a fresh set of letters and say, let $\xi = \text{const} \times \cos\omega t$, ω be the angular velocity; but we take the formula $\xi = \text{const} \times \cos(2\pi t/T)$. We assume that every part of the apparatus is moving with a simple harmonic motion, as will be the case if there be infinitesimal resistance and the simple harmonic motion of P is kept up long enough; so that we can write $x_1 = \text{const} \cdot \cos(2\pi t/T)$, etc. I am going to alter the m's so as to do away with the $4\pi^2$ which comes in from differentiation. I will let $m_1/4\pi^2$ denote the mass of the first particle, and $m_2/4\pi^2$ the mass of the second particle, etc. The result will be that the equations of motion become

$$-\frac{m_1}{T^2} x_1 = C_1(\xi - x_1) - C_2(x_1 - x_2) \text{ etc.}$$

Our problem is reduced now to one of algebra. There are some interesting considerations connected with the determinant which we shall obtain by elimination from these equations. To find the number of terms is easy enough; and it will lead to some remarkable expressions. But I wish particularly to treat it with a view to obtaining by very short arithmetic the result which can be obtained from the determinant in the regular way only by enormous calculation. We shall obtain an approximation, to the accuracy of which there is no limit if you push it far enough, that will be exceedingly convenient in performing the calculations.

In the next lecture we shall begin with the solution of the equations that are on the board for sound. We shall then try to go on a step further with this dynamical problem.

Lecture III

We will now go on with the problem of molar dynamics, the propagation of sound or of light, from a source. I advise you all who are engaged in teaching, or in thinking of these things for yourselves, to make little models. If you want to imagine the strains that were spoken of yesterday, get such a box as this covered with white paper and mark upon it the directions of the forces S, T, U. I always take the directions of the axes in a certain order so that the direction of positive rotation shall be from y to z, from z to x, from x to y. What we call positive is the same direction as the revolution of a planet seen from the northern hemisphere, or opposite to the motion of the hands of a watch. I have got this box for another purpose, as a mechanical model of an elastic solid with 21 independent moduluses, the possibility of which used to be disproved, and after having been proved, the result has been doubted for a long time.

Let us take our equations.

$$\rho \frac{d^2 \xi}{dt^2} = \frac{dP}{dx} + \frac{dU}{dy} + \frac{dT}{dz},$$

where

$$P = (m - n)\delta + 2n\frac{d\xi}{dx},$$

$$U = n\left(\frac{d\xi}{dy} + \frac{d\eta}{dx}\right),$$

$$T = n\left(\frac{d\zeta}{dx} + \frac{d\xi}{dz}\right)$$

$$\left\{\delta = \frac{d\xi}{dx} + \frac{d\eta}{dy} + \frac{d\zeta}{dz}\right\}.$$

We shall not suppose that m and n are variables, but take them constant.

If we do not take them constant, we shall be ready for Lord Rayleigh's paper, already referred to. I will do the work upon the board in full, as it is a case in which the expenditure of chalk saves brain; but it would be a waste to print such calculations, for the reason that a reader of mathematics should have pencil and paper beside him to work the thing out. * * * The result is that

$$\rho\frac{d^2\xi}{dt^2} = m\frac{d\delta}{dx} + n\nabla^2\xi. \tag{1}$$

We take the symbol $\nabla^2 = d^2/dx^2 + d^2/dy^2 + d^2/dz^2$. In the case of no rigidity, or $n = 0$, the last term goes out. We shall take solutions of these equations, irrespectively of the question of whether we are going to make $n = 0$ or not, and we shall find that one standard solution for an elastic solid is independent of n and is therefore a proper solution for an elastic fluid.

I have, in this Royal Institute lecture of mine in Feb. 1883, on the Size of Atoms, inserted a note on some mathematical problems which I set when I was examiner for the Smith's Prize at Cambridge, Jan. 30, 1883. One was to show that the equations of motion of an isotropic elastic solid are what we have here obtained, and another to show that so and so was a solution. We will just take that, which is: Show that every possible solution of these three equations [(1) etc.] is included in the following: $\xi = d\varphi/dx + u$, $\eta = d\varphi/dy + v$, $\zeta = d\varphi/dz + w$, where φ, u, v, w are some functions of x, y, z, t. Of course every possible solution is included in these formulae because u, v, w may be any functions, but the condition is added that u, v, w are such that $du/dx + dv/dy + dw/dz = 0$.

If we calculate the value of the cubic dilatation, we find

$$\delta = \nabla^2\varphi + \frac{du}{dx} + \frac{dv}{dy} + \frac{dw}{dz} = \varphi.$$

Again, by substituting $\xi = d\varphi/dx + u$ in (1), we find (bearing in mind $\delta = \nabla^2\varphi$)

$$\rho\left(\frac{d^2}{dt^2}\frac{d\varphi}{dx} + \frac{d^2u}{dt^2}\right) = (m+n)\nabla^2\frac{d\varphi}{dx} + n\nabla^2u.$$

Now, we have

$$\rho\frac{d^2}{dt^2}\frac{d\varphi}{dx} = (m+n)\nabla^2\frac{d\varphi}{dx}.$$

This is not proved as yet; the proof is reserved. Multiply this by dx, and the similar equations by dy, dz, and add. We thus get a complete differential; in other words, the relation which φ must satisfy is

$$\rho \frac{d^2 \varphi}{dt^2} = (m + n)\nabla^2 \varphi$$

[in addition to the relation $\delta = \nabla^2 \varphi$, which, as will be seen in the next lecture, determines φ as the potential corresponding to the density, $\delta/4\pi$]. Therefore, if φ satisfies this, we have u, v, w satisfying equations of the same form:

$$\rho \frac{d^2 u}{dt^2} = n\nabla^2 u, \quad \rho \frac{d^2 v}{dt^2} = n\nabla^2 v, \quad \rho \frac{d^2 w}{dt^2} = n\nabla^2 w.$$

By solving these four similar equations, one involving $(m + n)$, and three involving n, we can get solutions of (1), that is certain. That we get every possible solution, I shall hope to prove to-morrow. The velocity of the sound wave, or condensational wave, is $\sqrt{(m + n)/\rho}$. The velocity of the wave of distortion in the elastic solid is $\sqrt{n/\rho}$. I shall not take this up because I am very anxious to get on with the molecular problem; but you see brought out perfectly well the two modes of waves in an isotropic homogeneous solid, the condensational wave and the distortional wave. The condensational wave follows the equations of motion of sound, which is the same as if n were null; and this gives the solution of the propagation of sound in a homogeneous medium, like air, etc. The solution is worked out ready at hand for the distortional wave because the same forms of equations give us separate components u, v, w, the same solution that gives us the velocity potential for the condensational waves, gives us the separate components of displacement for the distortional waves.

What I am going to give you to-morrow will include a solution which is alluded to by Lord Rayleigh. There is nothing new in it; Lord Rayleigh knew it perfectly well. I am going to pass over the parts of the solution which [, as] interpreted by Stokes [,] explains that beautiful and curious experiment of Leslie's. Lord Rayleigh quotes from Stokes, ending his quotation of 8 pages with "The importance of the subject and the masterly manner in which it has been treated by Prof. Stokes will probably be thought sufficient to justify this long quotation." I would just like to read two or three things in it. Lord Rayleigh says (*Theory of Sound*, Vol. II, p. 207) "Prof. Stokes has applied this solution to the explanation of a remarkable experiment by Leslie, according to which it appeared that the sound of a bell vibrating in a partially exhausted receiver is diminished by the introduction of hydrogen. This paradoxical phenomenon has its origin in the augmented wave length due to the addition of hydrogen in consequence of which the bell loses its hold (so to speak) on the surrounding gas." I do not like the words "paradoxical phenomenon"; "curious phenomenon" or "interesting phenomenon" would be better. There are no paradoxes in science. We may call it a dynamox, but not a paradox. Lord Rayleigh goes on to say, "The general explanation cannot be better given than in the words of Prof. Stokes: 'Suppose a person to move his hand to and fro through a small space. The motion which is occasioned in the air is almost exactly the same as it would have been if the air had been an incompressible fluid. There is a mere local reciprocating motion in which the air immediately in front is pushed forward and that immediately behind impelled after the moving body, while in the anterior space generally the air recedes from the encroachment of the moving body, and in the posterior space generally flows in from all sides to supply the vacuum that tends to be created; so that in lateral directions, the flow of the fluid is backwards, a portion of the excess of the fluid in front going to supply the deficiency behind.'" It will take some careful thought to follow it. I wish I had Green here to read a sentence of his. Green says, "I have no faith in speculations of this kind unless they can be reduced to regular analysis." Stokes speculates in a way, but is not satisfied without reducing it to regular analysis. He gives here some very elaborate calculations that are also important and interesting in themselves, partly in connection with spherical harmonics, and partly from their exceeding instructiveness in respect to many problems regarding sound. Passing by all that 5 or 6 pages of mathematics—I will not tax your brains with trying to understand the dynamics of it in the course of a few minutes; I am rather calling your attention to a thing to be read than reading it—Stokes comes more particularly to Leslie's experiments. Instead of a bell vibrating, Stokes considers the vibrations of a sphere becoming alternately prolate and oblate; and he shows that the principles are the same.

I have intended merely to arouse an interest in the subject. I proposed the springs as offering a solution. For any one of the springs let there be a certain change of pull C, per unit change of length. It is not the slightest matter whether a spring is long or short, only, if it is long, let it be so much the stiffer; but long or short, thick or thin, it must be massless. I mean that it shall have no inertia. I am going to put a little memorandum on the board to keep this proposed explanation by the springs in mind. I hope we will reach it today. I think it has its applications straight away to anomalous dispersion and possibly elsewhere, though we are getting into the almost hopeless problem of explaining double refraction in crystals, and so on, by the wave theory of light.

To return to the consideration of these springs, we will suppose a good fixing at the top, so firm and stiff that the changing pull of the spring does not give it any sensible motion. The masses may be equal or unequal, and are connected by springs. Let us attach here a bell pull or something or other, that you can pull by, and call that P. This, in our application to the luminiferous ether, will be the rigid shell lining between the luminiferous ether and the first moving mass.

The equation of motion for the first mass becomes, on bringing ξ to the left hand side,

$$-C_1\xi = \left(\frac{m_1}{T^2} - C_1 - C_2\right)x_1 + C_2x_2;$$

and similarly for the second mass. I shall use i to denote any integer. I find the letter i too useful for that purpose to give it up, and when I want to write the imaginary $\sqrt{-1}$, I use $\underline{1}$. Let us call the first coefficient on the right a_1, the similar coefficient in the next equation a_2, and so on, so that $a_i = m_i/T^2 - C_i - C_{i+1}$. The ith equation will thus be

$$-C_ix_{i-1} = a_ix_i + C_{i+1}x_{i+1}.$$

Now write down all these j equations; form the determinant by which you find all of the others in terms of ξ, and the problem is solved.

If we had a little more time I would like to determine the number of terms in this determinant. We will come back to that because it is exceedingly interesting; but I want at once to put the equations in an interesting form, borrowing a suggestion from Laplace's treatment of the celebrated Diophantine problem. What we want is really the ratios of the displacements, and we shall therefore write

$$\frac{C_ix_{i-1}}{-x_i} = u_i,$$

introducing the minus sign, so that when the displacements are alternately positive and negative the successive ratios will be all positive. We have then:

$$\frac{C_1\xi}{-x_1} = u_1 = a_1 - \frac{C_2{}^2}{u_2},$$

$$u_2 = a_2 - \frac{C_3{}^2}{u_3}, \ldots u_i = a_i - \frac{C_{i+1}{}^2}{u_{i+1}}, \ldots u_j = a_j \quad (u_{j+1} = \infty).$$

We can now form a continued fraction which, for the case that we want, is rapidly convergent. If this be differentiated with respect to T^{-2}, we find a very curious law, but I am afraid we must leave it for the present. The solution is

$$u_1 = \frac{C_1 \xi}{-x_1} = a_1 - \cfrac{C_2{}^2}{a_2 - \cfrac{C_3{}^2}{a_3 - \cdots \cfrac{}{a_{j-1} - \cfrac{C_j{}^2}{a_j{}^2}}}}.$$

Thus if we are given the spring connections and the masses, everything is known when the period is known. If you develop this, you simply form the determinant; but the fractional form has the advantage that in the case when the masses are larger and larger and the spring connections are not larger in proportion we get an exceedingly rapid approximation to its value by taking the successive convergents. The differential coefficient of this continued fraction with respect to the period is essentially negative, and thus we are led beautifully from root to root, and see the following conditions: First suppose we move with very great rapidity; then when the whole has come to a periodic movement, it is necessary that P and the first particle move in opposite directions. The vibrations of the first particle is hurried up when the motion of P is of a shorter period than the shortest of the possible independent motions of the system, and if you want to hurry up a particle you shove it at the end of one range and pull it at the end of the other. You meet this principle quite often; it is well known in the construction of clock escapements. To hurry up the vibratory motion we must add to the return force of particle No. 1 by the action of the spring connected to the handle P. From looking at the thing, and learning to understand it by making the experiment, if you do not understand it by brains alone, you will see that everything that I am saying is obvious. It is not satisfactory to speak of these things in general terms unless we can submit them to a rigorous analysis.

That is the configuration in which the motion of P is of a shorter period than the shortest that will give us any of the critical periods. Suppose now the motion of P to be less rapid and less rapid; a state of things will come in which, the motion of P being slower and slower, the motion of the first particle will be greater and greater. That is to say, if we go on diminishing and diminishing the motion we shall find, for some range of motion of P, that the motion of m and each of the other particles will be greatly increased relatively to the motion of P. In analytical words, if we begin with a configuration of values corresponding to T very small, and then if we increase T, making it greater and greater, we shall find an infinity will appear; we shall find x_1/ξ will become infinite. In the first place, we begin with $u_1, u_2, \ldots u_j$ all positive—T small will cause them all to be positive as you will see. Take the differential coefficient of u_i with respect to T and it will be found to be essentially negative. In other words, if we increase T, we shall diminish $u_1, u_2 \ldots$. In some classes of cases, not necessarily in all, u_1 will first become zero; then we get the first infinity $x_1/\xi = \infty$. If we diminish T a little further u_2 will become zero; diminish T a little further and u_3 will become zero. We shall go into this to-morrow but I should like to have you know beforehand what is going to come from this kind of treatment of the subject.

Lecture IV

We found yesterday

$$\rho\frac{d^2\xi}{dt^2} = (k + \tfrac{1}{3}n)\frac{d\delta}{dx} + n\nabla^2\xi \quad (m = k + \tfrac{1}{3}n);$$

and we saw that we get two solutions, which when fully interpreted correspond to two different velocities of propagation, on the assumptions that were put before you as to a condensational or a distortional wave. We will approach the subject again from the beginning, and you will see at once that the sum of these solutions expresses every possible solution.

In one of our solutions of yesterday, we took, instead of ξ, η, ζ, other symbols u, v, w, which satisfied the condition $du/dx + dv/dy + dw/dz = 0$. In other words, the u, v, w of yesterday express the displacements in a case in which the dilatation or condensation is zero. Now, just try for the dilatation in any case whatever, without such restriction. That we can do as follows: Differentiate (1) with respect to x (taking account of the constancy of m and n), and the corresponding equations with respect to y and z, and add. We thus find

$$\rho\frac{d^2\delta}{dt^2} = (m + n)\nabla^2\delta = (k + \tfrac{4}{3}n)\nabla^2\delta.$$

This equation, you will remember, is the same as we had yesterday for φ. We shall consider solutions of this equation presently; but now remark that, whatever be the displacements, we have a dilatation corresponding to some solution of this equation. When we pass on from this equation to find ξ, η, ζ, subject to other conditions, we can look upon it in this way. Suppose for the moment $d\varphi/dx$, $d\varphi/dy$, $d\varphi/dz$ to be three displacements which we may compound with the actual displacements ξ, η, ζ, if you please. I have made no supposition, as yet, as to what φ may be. I say, let these three differential coefficients denote merely three displacements at any point x,y,z. Let us now determine φ so that δ is the dilatation corresponding to them. That is to say, let us take $\nabla^2\varphi = \delta$. We know how to find φ from this equation. It is the problem of attraction, viz.: $\nabla^2\varphi = -4\pi\delta/4\pi$. Therefore $-\delta/4\pi$ will be, in the familiar case of attraction, the density of the distribution of matter of which the potential is φ; so that we shall have

$$-\varphi = \iiint \frac{\delta'\,dx'\,dy'\,dz'}{4\pi\sqrt{(x-x')^2 + (y-y')^2 + (z-z')^2}};$$

where δ' denotes the value of δ at the point x',y',z'; and we may put in the limits of integration $-\infty$ to $+\infty$. This is the familiar expression for the potential of matter of the density δ, distributed through all space. If we have other boundary conditions, we must put those in.

For any possible solutions of equations (1) etc., we have a value of δ which is a function of x, y, z; take the above volume integral corresponding to this value of δ through all points of space x',y',z', and we obtain the corresponding φ function which fulfills the condition $\nabla^2\varphi = \delta$. Now, let us compound as follows the displacements $d\varphi/dx$, etc., with the actual displacements:

$$\xi - \frac{d\varphi}{dx} = u,\ \eta - \frac{d\varphi}{dy} = v,\ \zeta - \frac{d\varphi}{dz} = w;$$

and remarking that we have $du/dx + dv/dy + dw/dz = 0$, the proposition that we proposed yesterday is established.

To obtain a solution of (1) etc., we have simply to find δ from the equation

$$\rho\frac{d^2\delta}{dt^2} = (m+n)\nabla^2\delta;$$

and u, v, w from the similar equations which we found yesterday with n in place of $(m+n)$ * subject to the conditions $du/dx + dv/dy + dw/dz = 0$.

We shall take our φ solution and see how we can vary that and obtain different forms of φ solutions. We can do that for the purpose of illustrating

---

* This requires that φ should satisfy the same equation as δ, which is the proposition (left undemonstrated in the last lecture) by which the equations for u, v, w were obtained. Later on, in response to a question raised by Dr. Franklin, [an] indirect proof of the proposition $\rho\,d^2\varphi/dt^2 = (m+n)\nabla^2\varphi$ is given. A direct demonstration may be obtained from the value of

$$\varphi = -\iiint \frac{\delta'\,dx'\,dy'\,dz'}{4\pi r},$$

remembering that

$$\rho\frac{d^2\delta'}{dt^2} = (m+n)\nabla'^2\delta'$$

and that

$$-\iiint \frac{\nabla'^2\delta'}{4\pi r}dx'\,dy'\,dz' = \delta = \nabla^2\varphi.$$

H.

different problems in sound, and in order to familiarize you with the wave that may exist along with the wave of distortion in any true elastic solid which is incompressible. We ignore this condensational wave in the theory of light. We are sure that its energy, at all events if it is not null, is very small in comparison with the luminiferous vibrations we are dealing with. But to say that it is absolutely null would be an assumption that we have no right to make. When we look through the little universe that we know, and think of the transmission of electrical force and of the transmission of magnetic force and of the transmission of light, we have no right to assume that there may not be something else that our philosophy does not dream of. We have no right to assume that there may not be condensational vibration in the luminiferous ether. We only do know that any vibrations of this kind which are excited by the reflection and refraction of light are certainly of very small energy compared with the energy of the light from which they proceed. The fact of the case as regards reflection and refraction is this, that unless the luminiferous ether is absolutely incompressible, the reflection and refraction of light must generally give rise to waves of condensation. Waves of distortion may exist without waves of condensation, but waves of distortion cannot be reflected at the bounding surface between two mediums without exciting in each medium a wave of condensation. When we come to the subject of reflection and refraction, we shall see how to deal with these condensational waves and find how easy it is to get quit of them by supposing the medium to be incompressible. But it is always to be kept in mind to be examined into, are there or are there not very small amounts of condensational waves generated in reflection and refraction, and may, after all, the law of electric force not depend on the waves of condensation?

Suppose that we have at any place in air or in luminiferous ether (I cannot distinguish now between the two ideas) a body that through some action we need not describe, but which is conceivable, is alternately positively and negatively electrified; may it not be that this will be the cause of condensational waves? Suppose this, that we have two spherical conductors united by a fine wire, and that an alternating electromotive force is produced in that fine wire, for instance with an alternating dynamo-electric machine; and suppose that sort of thing goes on away from disturbance—at a great distance up in the air, for example. The result of the work of that dynamo-electric machine will be that one conductor will be alternately positively and negatively electrified and the other conductor, negatively and positively electrified. It is perfectly certain, if we turn the machine slowly, that in the neighborhood of the conductors we will have alternately positively and negatively electrified elements with reversals, perhaps two or three hundred per second of time without a gradual transition from negative through to zero to positive, and so on; and the same thing all through space; and we can tell exactly what the potential is at each point. Now, does any one believe that if that revolution was made fast enough the electro-static law would follow? Every one believes that if that process be conducted fast enough, several million times or millions of million times per second, we should be far from fulfilling the electrostatic law in the electrification of the air in the neighborhood. It is absolutely certain that such an action as that going on would give rise to electrical waves. Now it does seem to me probable that those electrical waves are condensational waves in luminiferous ether; and probably it would be that the propagation of these waves would be enormously faster than the propagation of ordinary light waves.

I am quite conscious, when speaking of this, of what has been done in the so-called electro-magnetic theory of light. I know the propagation of electric impulse along an insulated wire surrounded by gutta percha, which I worked out myself about the year 1854 and in which I found a velocity comparable with the velocity of light. We then did not know the relation between electrostatic and electro-magnetic units. If we had, that might have been obtained in the way that Maxwell has brought out so beautifully from the proper coefficients of capacity for the gutta percha. If we work that out for the case of air instead of gutta percha, we get practically the same v, I think, for the velocity of propagation of the impulse. That is a very different case from this and I have waited in vain to see how we can get any justification of the way of putting it in the so-called electro-magnetic theory of light. Simplify it down to the uttermost and take that case; there is a case of excitation of a kind that we know; we know the a, b, c, of it, and the laws of it, and feel certain that if this operation be performed but fast enough there will be waves. It seems to me that there are exceedingly strong probabilities that there will be waves of condensation and rarefaction of the luminiferous ether. I may refer to a little article of mine in which I gave a sort of mechanical representation of electric, magnetic, and galvanic forces—galvanic force I called it then, a very badly chosen name. It is published in the first volume of the reprint of my papers. It is shown in that paper that the static displacement of an elastic solid follows exactly the laws of the electro-static force, and that rotary displacement of the medium follows exactly the laws of magnetic force.

It seems to me that an incorporation of the theory of the propagation of electric and magnetic disturbances with the wave theory of light is most probably to be arrived at by this view that I am now indicating. In the wave theory of light, however, we shall simply suppose the resistance to compression of the luminiferous ether and the velocity of propagation of the condensational wave in it to be infinite. We shall sometimes use the words "practically infinite" to guard against supposing these quantities to be absolutely infinite.

I will now take two or three illustrations of this solution for condensational waves. Part of the problem that I referred to yesterday says, prove that the following is a solution of these equations:

$$\varphi = \frac{1}{r} \sin \frac{2\pi}{\lambda}\left(r - t\sqrt{\frac{m+n}{\rho}}\right) \quad \left[= \frac{1}{r}\sin q \right].$$

We might put this in a more analytical form, but the analysis consists in the verification of the thing. For that purpose, let us take the Laplacian of φ. We use this theorem,

$$\frac{d^2}{dx^2}(uv) = v\frac{d^2u}{dx^2} + 2\frac{du}{dx}\frac{dv}{dx} + u\frac{d^2v}{dx^2},$$

and find

$$\nabla^2\varphi = \frac{1}{r}\left\{\frac{2\pi}{\lambda}\cdot\frac{2}{r}\cos q - \frac{4\pi^2}{\lambda^2}\sin q\right\} - 2\frac{2\pi}{\lambda}\cdot\frac{1}{r^2}\cos q + 0 = -\frac{4\pi^2}{\lambda^2}\varphi.$$

Our equation for φ is therefore

$$\rho\frac{d^2\varphi}{dt^2} = (m+n)\nabla^2\varphi = -(m+n)\frac{4\pi^2}{\lambda^2}\varphi.$$

We will now make it a little more analytical and say the thing to be proved is that which is written down, letting the assumption be $\varphi = 1/r \sin 2\pi(r/\lambda - t/T)$, where T is the period of vibration and λ the wave length. Substitute and the equation becomes

$$\rho\frac{4\pi^2}{T^2}\varphi = -\frac{4\pi^2}{\lambda^2}(m+n)\varphi;$$

or the velocity of propagation

$$\frac{\lambda}{T} = \sqrt{\frac{m+n}{\rho}} = \sqrt{\frac{k+\frac{4}{3}n}{\rho}}.$$

Here then is the determination of a form of motion which is possible for an elastic solid. We shall consider the nature of this motion presently. The presence of $1/r$ prevents it from being a pure wave motion. Passing over that consideration for the present, we note that it is less and less effective, relatively to the motion considered, the farther we go from the center.

In the meantime, we remark that the velocity of propagation in an elastic solid is little greater than in a fluid with the same resistance to compression. k is the bulk modulus and measures resistance to compression, n is the rigidity modulus. I may hereafter consider relations between k and n for real solids. k is generally several times n, so that $\frac{1}{3}n$* is very small in comparison

* [The lecturer used throughout this investigation $m = k + \frac{1}{3}n$, instead of $m + n = k + \frac{4}{3}n$. The fact that this should be $\frac{4}{3}n$ is the occasion for a further consideration of the subject in a subsequent lecture.—H.]

with k, and therefore in ordinary solids the velocity of propagation of the condensational wave is exceedingly little greater than if the solid were deprived of rigidity and we had an elastic fluid of the same bulk modulus.

I shall want to look at this motion in the neighborhood of the source. That beautiful investigation of Stokes quoted by Lord Rayleigh has to do entirely with the region in which the change of value of this coefficient $(1/r)$ from point to point is considerable. Without looking at that now, let us find the displacement and see what it will be. $d\varphi/dx$, $d\varphi/dy$, $d\varphi/dz$ are the three components of the displacements. Clearly, the displacement will be in the direction of the radius, because everything is symmetrical; and its magnitude will be $d\varphi/dr$.

Dr. Franklin: The equation at the top of the board puzzles me. It is the same equation we have had before for δ, with φ written in the place of δ. I do not understand how the φ got there.

Sir Wm. Thomson: Let us see how this is. It is quite correct, but we will just look at that question. Our equation was $\rho \, d^2\delta/dt^2 = (m + n)\nabla^2\delta$; and then we had $\nabla^2\varphi = \delta$. δ must fulfill the first condition, and if φ fulfills that condition δ does certainly fulfill it in virtue of the second condition. That ought to prove that when δ fulfills the first condition φ also fulfills it. That is not quite rigorous perhaps, but I think it is obvious from the finding of φ from δ. If we take the Laplacean of $\rho \, d^2\varphi/dt^2 = (m + n)\nabla^2\varphi$, we find (since $\nabla^2\varphi = \delta$) $\rho \, d^2\delta/dt^2 = (m + n)\nabla^2\delta$. All we have to do is to find δ to fulfill this condition, and having found it, we will find φ to fulfill it.

Having obtained a solution of our equations, let us see what we can make in interpreting it. The component of the displacement in the direction of x is

$$\frac{d\varphi}{dx} = \frac{2\pi x}{\lambda r^2}\left(\cos q - \frac{\lambda}{2\pi r}\sin q\right).$$

When r is great in comparison with $\lambda/2\pi$, the second term becomes very small in comparison with the first and we have

$$\frac{d\varphi}{dx} = \frac{2\pi}{\lambda}\frac{x}{r^2}\cos q.$$

Also

$$\frac{d\varphi}{dr} = -\frac{1}{r^2}\sin q + \frac{2\pi}{\lambda}\frac{2}{r}\cos q.$$

Therefore, when the distance from the origin is large in comparison with $\lambda/2\pi$, the displacement is sensibly equal to $(2\pi/\lambda)(1/r)\cos q$, and is therefore approximately in the inverse proportion to the distance; and the intensity of the sound, if it were to be applied to sound, would be inversely as the square of the distance. At a considerable distance from the place in which there is circulation around the source is the permanent term which I have written down.

I want to get a second and a third solution. Take

$$\psi = \frac{\lambda}{2\pi} \, {}^* \frac{d\varphi}{dx} = \frac{x}{r^2}\left(\cos q - \frac{\lambda}{2\pi r}\sin q\right)$$

as the velocity potential for a fresh solution. I take it that you all know that if we have one solution φ, for the velocity potential, we can get any other solution by φ any linear function of $d\varphi/dx$, $d\varphi/dy$, $d\varphi/dz$. Now let us find the displacements $d\psi/dx$, $d\psi/dy$, $d\psi/dz$. Here I want to prove that though this solution is no longer symmetrical with respect to r, so that there will be motions other than radial in the neighborhood of the source, the motion is approximately radial at a distance from the source. Work it out, and you will find that

$$\frac{d\psi}{dx} = \frac{2\pi}{\lambda}\frac{x^2}{r^3}\left[-\sin q + \frac{\lambda}{2\pi r}\cdot\frac{(r^2 - 3x^2)}{x^2}\cos q\right].$$

The principal term is then $-(2\pi/\lambda)\,(x^2/r^3)\sin q$. We might go on to the third and fourth terms, increasing the multiplicity. That splendid work of Stokes in which this multiplicity is dealt with to show the effect of hydrogen in killing sound is one of the finest things written in physical mathematics. But we will drop those terms and think only of the principal terms.

The principal term in the expression for the displacement is

$$\xi \doteqdot -\frac{2\pi}{\lambda}\frac{x^2}{r^3}\sin 2\pi\left(\frac{r}{\lambda} - \frac{t}{T}\right).$$

I use $\doteqdot$ as the sign for approximate equality. This approximate equality is true for assigned distances from the centre great in comparison with the wave length. Let me remark, it is the differentiation of $\sin q$ that gives the effective terms of the displacement; and in differentiating ψ with respect to y, you have simply to take the differential coefficient of r in $\sin q$ with respect to y, instead of x. So that we may just write down the principle terms of the y and z displacements

$$\eta \doteqdot -\frac{2\pi}{\lambda}\frac{xy}{r^3}\sin q, \; \zeta \doteqdot -\frac{2\pi}{\lambda}\frac{xz}{r^3}\sin q.$$

These component displacements, being proportional to x, y, z, show that the resultant displacement is in the direction of the radius, and that its magnitude is $-(2\pi/\lambda)\,(x/r^2)\sin q$. If we write $x = r\cos i$, this becomes

---

* $\lambda/2\pi$ is introduced merely for convenience. The solution is not essentially different from $\psi = d\varphi/dx$.

$- (2\pi/\lambda) \, (\cos i/r) \sin q$; or the displacement is inversely proportional to the distance. If $i = 0$, we have a maximum; if $i = \pi/2$, we have zero. The upshot of it is that the displacement is a maximum in the axis OX, zero in the axes OY, OZ, and symmetrical with respect to the axes.

The third solution is to take $d^2\varphi/dx^2$ as our velocity potential. At a distance from the origin, great in comparison with the wave length, the displacement is in the direction of the radius, and its magnitude is

$$\frac{d}{dr} \frac{d^2\varphi}{dx^2}.$$

Now the interpretation of these cases is as follows: The first solution, a globe alternately becoming larger and smaller; the second solution, a globe vibrating to and fro in a straight line; the third solution, two globes vibrating to and fro meeting one another, or the disturbance in the neighborhood of the prongs of a tuning fork.

That, however, requires a little nice consideration, and we shall take it up in a subsequent lecture. The third mode does not quite represent the motion in the neighborhood of the prongs of a tuning fork; there must be an unknown amount of the first mode compounded with the third mode for this purpose. The expression for the vibration in the neighborhood of a tuning fork, going so far from the ends of it that we will be undisturbed by the shape of the thing, will be given by the velocity potential $A\varphi + d^2\varphi/dx^2$. That will be the velocity potential for the chief terms, the terms which alone have an effect at a distance. The differentiation will be performed simply with reference to the r in the term $\sin q$ or $\cos q$; and will be the same as if the coefficient of $\sin q$ or $\cos q$ were constant. A differentiation of this velocity potential will show that the displacement is in the direction of the radius from the centre of the system, and the magnitude of the displacement will be

$$\frac{d}{dr}\left(A\varphi + \frac{d^2\varphi}{dx^2} \right).$$

A is an unknown quantity depending upon the tuning fork. I want to suggest this as a junior exercise, to try tuning forks with different breadths of prongs. When you take tuning forks with prongs a considerable distance asunder you have much less of the φ to take. Try a tuning fork with flat prongs, pretty close together, and you will have much more of the φ to take. The φ part of the velocity potential corresponds to the swelling of the tuning fork, the air between the prongs becoming larger and smaller. The larger and flatter the prongs are, the greater is the proportion of the φ solution, and the larger the value of A in that formula.

The experiment that I suggest is this: that you take tuning forks and turn them around until you find the cone of silence, or find the angle between the line joining the prongs and the line going to the place where you hear no sound. The suddenness of transition from sound to no sound is startling. Having the tuning fork in the hand, turn it slowly around near one ear until you find the place of silence; a very small angle of turning around the vertical axis from that place gives you a loud sound. I think it is very likely that the place of no sound will depend on the angle of vibration. If you excite it very powerfully, you will find greater inclination; less powerfully, less inclination. It will certainly vary with the tuning fork.

Lecture V

I stated in the last lecture that the second solution corresponding to the velocity potential $d\varphi/dx$ would represent the effect, at a great distance from the mean position, of a body vibrating to and fro in a straight line. I said a sphere, but we may take a body of any shape vibrating to and fro in a straight line, and at a very great distance from the mean position, the motion produced will be represented by the velocity potential $d\varphi/dx$. Then the velocity potential $d^2\varphi/dx^2$, in the third solution, would, I believe, represent (without an additional term $A\varphi$) the motion at a distance, when the origin of the sound consists in two globes, let us say, for fixing the ideas, placed at a distance from one another very great in comparison with their diameters and set to vibrating to and fro. Suppose this is a globe in one hand, and this is one in the other. I now move my hands towards and from each other— that sort of motion produced by the exciting bodies would, at a very great distance, be expressed exactly by the velocity potential $d^2\varphi/dx^2$.

But when you have two globes, or two flat bodies, near one another, you need an unknown amount of the φ vibration to represent the actual state of the case. That unknown amount might be determined theoretically for the case of two spheres. The problem is analogous to Poisson's problem of the distribution of electricity upon two spheres, and it has been solved by Stokes for the case of fluid motion [see *Mem. de l'Inst. Paris*, 1811, pp. 1, 163; *Stokes' Papers*, Vol. I, p. 230—"On the resistance of a fluid to two oscillating spheres"]. You can thus tell the motion exactly in the neighborhood of two spheres vibrating to and fro provided the amplitudes of their vibrations are small in comparison with the distance between them; and you can find the value of A for two spheres of any given radii and any given distance between them. For such a thing as a tuning fork, you could not, of course, work it out theoretically; but I think it would be an interesting experiment for junior laboratory work.

I suppose you are all familiar with the zero of sound in a tuning fork; but I have never seen it described correctly anywhere. I shall take that up on Monday. We shall see that we have no theoretical means of determining the inclination of the line going to the mean position of the area for silence to the line joining the prongs; but that this is dependent upon the proportions of the body. In turning the tuning fork around, you can get with great nicety the position for silence; and a surprisingly small turning of the tuning fork from the position of silence causes the motion to be heard. It would be very curious to find whether the position of zero sound varies relatively to the fork as the amplitude of the vibrations increases. I doubt whether any perceptible difference will be found in any ordinary case however we vary the amplitude of the vibrations. But I am quite sure you will find considerable difference, according as you take tuning forks of cylindrical proportions or tuning forks like the more modern ones that Koenig makes, with very broad flat ends.

Now for our molecular problem.

I want to see how the quantities vary, when we vary the period. Remember that $a_i = m_i/T^2 - C_i - C_{i+1}$, so that $da_i/dT^{-2} = m_i$. Write for the moment δ for d/dT^{-2}, and differentiate the equation for u_i; we have

$$\delta u_i = m_i + \left(\frac{C_{i+1}}{u_{i+1}}\right)^2 \delta u_{i+1}, \; \delta u_{i+1} = m_{i+1} + \left(\frac{C_{i+2}}{U_{i+2}}\right)^2 \delta u_{i+2}, \cdots \delta u_j = m_j.$$

Substitute successively, and we find

$$\delta u_i = m_i + \left(\frac{C_{i+1}}{u_{i+1}}\right)^2 m_{i+1} + \left(\frac{C_{i+1}C_{i+2}}{u_{i+1}u_{i+2}}\right)^2 m_{i+2} + \cdots \left(\frac{C_{i+1}\cdots C_j}{u_{i+1}\cdots u_j}\right)^2 m_j.$$

This is our expression, and remark the exceedingly important property of it that it is essentially positive, i.e., the variation of u_i with respect to T^{-2} is essentially positive. Also $du_i/dT = 2T^{-3}\delta u_i$. Now,

$$u_i = \frac{C_i x_{i+1}}{-x_i} \text{ or } \frac{C_i}{u_i} = -\frac{x_i}{x_{i+1}}, \; \frac{C_iC_{i+1}}{u_iu_{i+1}} = \frac{x_ix_{i+1}}{x_{i-1}x_i}, \text{ etc.}$$

The result therefore is this remarkable expression for the differential coefficient of u_i with respect to the period:

$$\frac{du_i}{dT} = -\frac{2}{T^3}\frac{1}{x_i^2}(m_ix_i^2 + m_{i+1}x_{i+1}^2 + \cdots m_jx_j^2)\cdots. \tag{2}$$

This is certainly a very remarkable theorem, and one of great importance with reference to the interpretation of the solution of our problem. Remember that x_i is the displacement of m_i at any part of the motion. You may habitually think of the maximum values of the displacements, but it is not necessary to confine yourselves to the maximum values. Instead of x_1,

$x_2, \cdots x_j$ we may take constants equal to the maximum values of the x's, multiplied into $\sin(2\pi t/T)$—remembering that each of them varies with a simple harmonic motion. The masses are positive, and we have squares of the displacements, so that the second member of (2) is essentially negative. Hence, as we augment the period, the functions u_i, etc., each one decreases, and as we decrease the period, each one increases.

Let us now consider this spring arrangement. I am going to suppose, in the first place, that the period of vibration is very small, and is then gradually increased. As you increase the period, the values of each one of the quantities $u_1, u_2, \cdots$ decreases. It is interesting to remark that since du_i/dT is always negative, every one of the u's decreases throughout every variety of configuration, as T increases. In the first place, T may be taken so small that the u's are all very large positive quantities; for

$$u_i = \frac{m_i}{T^2} - C_i - C_{i+1} - \frac{C_{i+1}{}^2}{u_{i+1}}$$

may be certainly made very large positive by taking T small enough if at the same time the succeeding quantity, u_{i+1}, is large (a condition which is fulfilled since we always have $u_{j+1} = \infty$).

Observe that the u's all positive implies that ξ, x_i, x_j are alternately positive and negative. In other words the handle P and the several particles m_1, $m_2, \cdots m_j$, are each moving in a direction opposite to its neighbor. Since the magnitudes of the ratios $u_1, u_2, \cdots u_j$ of the several amplitudes decrease with the increase of the period, the amplitude of particle m_i is becoming smaller in proportion to the amplitude of the succeeding particle m_{i+1}; that is to say, the handle P is hurrying up the system. I am going to show you that as every one of these quantities u_i decreases, the first that passes through zero is u_1—corresponding to infinite motion of the particles of the system, in comparison with the motion of the handle P. This is the first critical case; after that u_1 becomes negative, and the motion of P is in the direction of the motion of the first particle. Also, if we have further negative values, the order of procedure always is that the negative value passes along the line from particle m_1 towards the fixed end. In other words, as we go on increasing the period we shall find that the next critical case that takes place is that particle m_1 has zero motion, or $u_1 = C_1 \xi / - x_i = -\infty$.

Let us look at the state of things when u_i has approached very near to zero. We shall have $u_{i-1} = a_{i-1} - C_i{}^2/u_i$ a very large negative quantity. This fact alone shows that u_{j-1} must have preceded u_j in becoming zero, since it must have passed through zero before becoming large negative.* Therefore, as we augment T the first of the u's to become zero is $u_1 = C_1 \xi / - x_1$; or the

---

* [This does not show, however, that u_{i-1} may not have passed through zero more than once before u_i.—H.]

motion of particle m_1, and also of each of the other particles, is infinite in comparison with the motion of P. Just before this state of things, all the particles $P, m_1, \cdots m_j$ are moving each opposite to its neighbor; just after it, P has reversed its motion with reference to the first particle, and is moving in the same direction with it.

That is also the configuration, just before the second critical case, in which we have u_1 large negative, u_2 small positive, [and] $u_3, \cdots u_j$ all positive. At this critical case, we have $u_1 = C_1 \xi / x_1 = -\infty$ or $x_1 = 0$. The period of motion of P that will produce this state of things is equal to the period of the free vibration of the system of particles, with mass m_1 held at rest and each of the other masses moving in an opposite direction to its neighbor. When the period of P is equal to the period of motion of the system with the first particle held at rest, then the only motion of the system that fulfills the condition of being a simple harmonic motion is that in which the amplitude of vibration of the second particle in one direction is such as to produce a pull in that direction equal to the pull exercised on the spring by P in the opposite direction; which keeps the first particle at rest. Immediately after this critical case, u_1 has changed from large negative to large positive and u_2 from small positive to small negative; or the first particle has reversed the direction of its motion with respect to P and the second particle.

The third critical case is that of the second particle coming to rest, and reversing its motion; but I shall not go further with these critical cases. I am only giving you an indication of how to perceive the thing. There is a great deal more to think of as to the a's becoming negative, etc. My object was simply to indicate the state of things merely, and I will just jump over the remaining critical cases and take up T very great.

It would be curious to find the solution when the period is infinitely great out of these equations. When T is infinite, m_i / T^2 vanishes, and $a_i = -C_i - C_{i+1}$. That applied to the equations for the u's ought to find the solutions quite readily. The solutions which you find are very curious, but it is like the case of so many problems which all the great mathematicians used to be fond of proposing and of putting their heads together to solve. If you were successful in finding out the right way of doing them the solutions were easy, otherwise they were hard.

You know, when you think of this case, that when T is infinitely great, P is moving infinitely slowly, so that the inertia of each particle has no sensible effect; and all the particles are in equilibrium. Let F be the force, then, on the spring; that is to say, pull P down with a force F and hold it at rest. What will be the displacements of the different particles? Answer, $x_j = F / C_{j+1}, x_{j-1} = F / C_{j+1} + F / C_j$ and so on. The number j'th particle is displaced to a distance equal to the force, divided by the coefficient of elongation of the spring. To obtain the displacement of particle $j - 1$, we have to add the displacement resulting from the elongation of the next spring C_j, and so on. The general equation then is

$$x_i = \left(\frac{1}{C_{j+1}} + \frac{1}{C_j} + \cdots \frac{1}{C_{i+1}} \right) F.$$

$$\therefore u_i = -C_i \left(\frac{1}{C_{j+1}} + \cdots \frac{1}{C_i} \right) \bigg/ \left(\frac{1}{C_{j+1}} + \cdots \frac{1}{C_{i+1}} \right).$$

It is a curious problem to substitute the value of $a_i = -C_i - C_{i+1}$ in the continued fraction which gives u_i, and verify this solution.

I just want to call your attention a little bit to magnitudes; for the problem we really care for is not this. It is like fiddling while Rome is burning to be explaining fluorescence when the explanation of refraction of light in crystals is waiting. The difficulty is not to explain phosphorescence and fluorescence but to explain why there is so little sensible fluorescence and phosphorescence. This thing brings everything to fluorescence and phosphorescence. The state of things as regards our system would be this: Suppose we have this handle moved backwards and forwards until everything is in a perfectly periodic state. Then suddenly stop moving P. The system will continue vibrating for a definite time with a complex vibration which will really embody something of all the modes. That I believe is fluorescence.

But now comes Mr. Michelson's question, and Mr. Newcomb's question, and Lord Rayleigh's question, as to velocity of groups. There again we are all afloat with vibrations of this kind. Suppose a succession of luminiferous vibrations commences. In the commencement of the luminiferous vibrations the attached molecules imbedded in the luminiferous ether do not immediately get into the state of a simple harmonic vibration which will create a regular light. It seems quite certain that there must be an initial fluorescence. Let light begin shining on uranium glass; for the thousandth of a second, perhaps, after the light has begun shining on it, you should find an initial state of things which differs from the permanent state of things exactly the same as fluorescence differs from no light at all.

There is still another question, which is of profound interest, and seems to present many difficulties, and that is, the actual condition of the light which is a succession of groups. Lord Rayleigh has told us in his printed paper in respect to the agitated question of the velocity of light, and then again at the meeting of the British Association at Montreal, he repeated very peremptorily and clearly the fact that the velocity of a group of waves must not be confounded with the wave velocity of an infinite succession of waves. He seems to be quite certain that what he said is true. But here is a difficulty which has only occurred to me since I began speaking to you on the subject; and I hope, before we separate, we shall see our way through it. All light consists in a succession of groups. Why is light not polarized? We are going to work our way slowly on until we get expressions for sequences of vibrations of existing light. Take any conceivable supposition as to the origin of light, in a flame, or a wire made incandescent by an electric current, or any other source of light; we shall work our way up from these sound equations to the kinds of expression that light must have, from any conceivable source. Now, in a source consisting of a motion that kept going on in exactly the same way, the light from that source would be plane polarized, or circularly polarized, or elliptically polarized, and would be absolutely constant. In reality, there is a multiplicity of succession of groups of waves. One molecule, of enormous mass in comparison with the luminiferous ether that it displaces, gets a shock, and it performs a set of vibrations until it comes to rest or gets a shock in some other direction; and it is sending forth vibrations with the same want of regularity that is exhibited in a group of sounding bodies consisting of bells, tuning forks, organs, etc., every one of which is sending forth its strain and each of which is propagated, some distance away from the source, as if there were no others. We thus see that light is entirely compounded of groups of waves; and if the velocity of a group of waves, or even the center of gravity of a group, differs from the velocity of absolutely continuous sequences of waves, we have all ground cut from under us in respect to the velocity of waves of light.

I mean to say, that all light consists of groups following one another, in that way, and that there is a difficulty to see what to make of the beginning and end of the vibrations of a group; and that then there is the question which was talked over a little in Section A at Montreal—will the mean effect of the group be the same as that of an infinite sequence of uniform waves, and will the deviation from regular periodicity at the beginning and end of the group have but a small influence in comparison with the whole? It seems almost certain that it must have but a small influence from the known facts regarding the velocity of light and the approximate regularity that we have—but I am leading you into a muddle because I am far into the difficulty and have not understood it. Still you can all think a good deal with me about the connections of this subject.

Lecture VI

I want to ask you to note that when I spoke of $k + \frac{4}{3}n$ not differing scarcely from k for most solids, I was rather under the impression for the moment that the ratio of n to k was smaller than it is; and also you will remember that we had $k + \frac{1}{3}n$ on the board. The square of the velocity of a condensational wave in an elastic solid is $k + \frac{4}{3}n$. For solids fulfilling the supposed relation of Navier and Poisson between compressibility and rigidity we have $n = \frac{3}{5}k$; and for such cases the numerator becomes $\frac{9}{5}k$. It would be k if there were no rigidity; it is $\frac{9}{5}k$ if the rigidity is that of a solid for which Poisson's ratio has its supposed value.

Metals are not enormously far from fulfilling this condition but it seems that for elastic solids generally, n bears a less proportion to k than this. It is by [no] means certain that it fulfills it even approximately for metals; and for india rubber, on the other hand, and for jellies, n is an exceedingly small fraction of k, so that in these cases the velocity of the condensational wave is $\sqrt{k/\rho}$. The velocity of propagation of a distortional wave is $\sqrt{n/\rho}$; so that for jellies, the velocity of propagation of condensational waves is enormously greater than that of distortional waves.

I am asked to define velocity potential. Those who have read German writers on hydrodynamics already know the meaning of it perfectly well. It is purely a technical expression which has nothing to do with potential or force. "Velocity potential" is a function of the coordinates such that its rate of variation per unit distance in any direction is equal to the component of velocity in that direction. A velocity potential exists when the distributions of velocity are expressible in this way; in other words when the motion is an irrotational one. The most convenient definition of irrotational motion is, the motion such that the velocity components are expressed by the differential coefficients of a function. That function is the velocity potential. When the motion is rotational there is no velocity potential.

This is the strict application of the words "velocity potential" which I have used. A corresponding language may be used for displacement potential.

It is not good language, but it is convenient, it is rough and ready. So that when we are speaking of component displacements in any case, whether of static displacement in an elastic solid or of vibrations, in which the components of displacement are expressible as the differential coefficients of a function, we may say that it is an irrotational displacement. If from the differentiation of a function we obtain components of velocity, we have velocity potential; whereas, if we get components of displacement, we have displacement potential. The functions φ that we used are not then, strictly speaking, velocity potentials but displacement potentials.

I want you in the first place to remark what is perfectly well known to all who are familiar with differential equations, that taking the solution $\varphi = (1/r) \cdot \sin q$ as a primary (where $q = (2\pi/\lambda) \, (r - t\sqrt{n/\rho})$ if we are considering the [condensational] wave) we may derive other solutions by differentiations with respect to the rectangular coordinates. The first thing I am going to call attention to is that at a distance from the origin, whatever be the solution derived from this primary by differentiation, the corresponding displacement is nearly in the direction through the origin of coordinates.

Take any differential coefficient whatever,

$$\frac{d^{i+j+k}\varphi}{dx^i \, dy^j \, dz^k};$$

the term of this which alone is sensible at an infinitely great distance is that which is obtained by successive differentiation of $\sin q$. That distance term in every case is as follows:

$$\left(\frac{2\pi}{\lambda}\right)^{i+j+k} \cdot \left(\frac{dr}{dx}\right)^i \cdot \left(\frac{dr}{dy}\right)^j \cdot \left(\frac{dr}{dz}\right)^k \cdot \frac{1}{r} \, {\sin \atop \cos} q.$$

It will be $\sin q$ or $\cos q$, according as $i + j + k$ is even or odd. We do not need to trouble ourselves about the algebraic sign, because we shall make it positive, whether the differential coefficient is positive or negative. Now

$$\frac{dr}{dx} = \frac{x}{r}, \quad \frac{dr}{dy} = \frac{y}{r}, \quad \frac{dr}{dz} = \frac{z}{r}.$$

Thus, our type solution becomes, omitting the constant factor,

$$\frac{x^i y^j z^k}{r^{i+j+k+1}} \, {\sin \atop \cos} q.$$

This expresses the most general type of displacement potential for a condensational wave proceeding from a center. I have not formally proved that this is the most general type, but it is very easy to do so. I am rather going into the thing synthetically. It is so thoroughly treated analytically by many writers that it would be a waste of your time to go into anything more, at present, than a sketch of the manner of treatment, and to give some illustrations.

But now to prove that the displacement at a distance from the origin of the disturbance is always in the direction of the radius vector. Once more, the differential coefficient of this displacement potential, which has several terms depending upon the differentiation of the r's, x's etc. has one term of paramount importance, and that is the one in which you get $2\pi/\lambda$ as a factor. The smallness of λ in proportion to the other quantities makes the factor $2\pi/\lambda$ give importance to the term in which it is found. The distance terms, then, for the components of the displacement are

$$\xi \doteqdot \frac{x^i y^j z^k}{r^{i+j+k+1}} \cdot \frac{2\pi}{\lambda} \cdot \frac{x}{r} \frac{\cos}{\sin} q = R\frac{x}{r},$$

$$\eta \doteqdot R\frac{y}{r},$$

$$\zeta \doteqdot R\frac{z}{r}.$$

These are the components of a displacement which is radial; and the expression for the radial displacement is

$$R = \frac{2\pi}{\lambda} \cdot \frac{x^i y^j z^k}{r^{i+j+k+1}} \frac{\cos}{\sin} q.$$

The sum of any number of such expressions will express the distance effect of sound proceeding from a source. It is interesting to see how, simply by making up an algebraic function in the numerator out of the x's, y's, and z's, we can get a formula that will express any amount of nodal subdivision where silence is felt. The most general result for the radial displacement is

$$R = \sum \frac{C x^i y^j z^k}{r^{i+j+k+1}} \frac{\cos}{\sin} q.$$

Remark that $x/r, y/r, z/r$ are merely angular functions and may be expressed at once as $\sin\theta \cos\psi$, $\sin\theta \sin\psi$, $\cos\theta$; and that therefore R is an integral algebraic function of $\sin\theta \cos\psi$, $\sin\theta \sin\psi$, $\cos\theta$.* It is thus easy to see that you can vary indefinitely the expressions for sound proceeding from a source

---

* [The lecturer had not the factor $\frac{\cos}{\sin} q$ upon the board in his expression for R, and so overlooked that the factor $(1/r) \cos q$ must enter into terms of even order, and $(1/r) \sin q$ into terms of odd order. Thus the most general function gives $R = (1/r) (R_0 \cos q + R_1 \sin q)$; and we have merely lines of silence radiating from the source, viz., the intersections of the cones $R_0 = 0$, $R_1 = 0$. For cones of silence, either R_0, R_1 may have a common factor or one of these angular functions may be wanting in the expression for R.—H.]

with cones of silence and corresponding nodes or lines in which those cones cut the spherical wave surface. It is interesting to see that even in the neighborhood of the nodes the vibration is still perpendicular to the wave surface; so that we have realized, in any case, a gradual falling off of the intensity of the wave to zero and a passing through zero, which would be equivalent to a change of phase, without any motion perpendicular to the radius vector.

The more complicated terms that I have passed over are those that are sensible in the neighborhood of the source. Suppose, for instance, that you have a bell vibrating. The sound slipping out and in over the sides of the bell and around the opening gives rise to a very complicated state of motion close to the bell; and similarly with respect to a tuning fork. If you take a spherical body, you may somewhat nearly express the motion in terms of spherical harmonics, and so on. You can see that in the neighborhood of the sounding body there will be a great deal of slipping in directions perpendicular to the radius vector, the displacements along the radius vector being compounded with motions out and in; but it is interesting to notice that all these motions become insensible at distances from the center large in comparison with the wave length. It is the consideration of these motions at distances that are moderate in comparison with the wave length that Stokes has made the basis of that very interesting investigation with reference to Leslie's experiment of a bell vibrating in a vacuum, to which I have already referred.

We may just notice, before I pass away from the subject, two or three points of the case, with reference to a tuning fork, a bell, and so on. Suppose the sounding body to be a circular bell. In that case clearly, if the bell be held with its lip horizontal, and if it be kept vibrating steadily in its gravest ordinary mode, the kind of vibration will be this: a vibration from a circular figure

into an elliptical figure

along one diameter, and a swinging back through the circular figure

into an elliptical figure

along the other diameter at right angles to the first. Clearly there would be practically a plane of silence here and another at right angles to it here (represented on the diagrams by dotted lines). Hence the solution for the radial component corresponding to this case, at a considerable distance from the bell, is

$$R = (\tfrac{1}{2} - \cos^2\theta)\frac{\cos q}{r},$$

in order that the component may vanish when $\cos^2\theta = \tfrac{1}{2}$, or $\theta = \pm 45°$.

On the other hand a tuning fork vibrating to and fro or an elongated elliptical bell (shaped like that which I have, which was obtained from that fine old Frenchman, Koenig's predecessor, [Marloye] that makes an exceedingly loud sound), has an advantage in acoustic experiments over a circular bell. If you set a circular bell to vibrating and leave it to itself you always hear a beating sound, because the bell is not quite symmetrical. Excite it with a bow, and take your finger off and leave it to itself, and if you do not choose the proper place to touch it, so that no vibration will occur there when you take your finger off, it will execute the resultant of two fundamental modes.

I do not know whether that experiment with plates is familiar to all of you. I would be glad to know whether it is. I make it always before my own classes, in illustrating the subject. Take a circular plate—just one of the ordinary circular plates that are prepared. Excite it in that way, putting the finger on to make the quadrantal vibration

That would be a case to which this solution obtained for the bell would apply. According to that notation the axis of x would be in this direction

for a circular plate with two lines of silence at right angles to one another. If sand be sprinkled upon the plate, and I take my finger off, the sand at that point begins to oscillate, and I hear a beating sound. But by a little trial, I find one place where if I touch the finger, and excite it so as to make a quadrantal vibration, if I then take off my finger the sand remains undisturbed and there will be no beat. Then having found one place, I know there will be another place which is got by touching it here 180° from the first place, and that I can get another pair of nodes perpendicular to the first pair where there is also silence. Put your finger in between those places, force the plate to vibrate and take off your finger and you will have very loud beats, because the vibrations of the plate are not equal—the two sounds always differ from one another. Try it in that way, and you will find it a most interesting thing with reference to circular plates. I have never seen it in any book, and I have done it every year for 45 years.

Take a division of the circumference into six equal parts by three diameters, and you find the same thing over again. Go on by trial touching the plate at two points 60° asunder, and bowing it 30° from either, and you will get a sound resting on the three diameters determined by your fingers. Take off your fingers and you will in general get a beat. Follow your way around, little by little—it is very pretty when you come near a place of no beat. The moment you take off your fingers, you see the lines of nodes going backwards and forwards with a very slow oscillation. Get exactly the position, bow it, take off your finger, and the lines remain absolutely still. Take a point midway between those two and another 60° from that and you have a beat from loud sound to silence. If you try for it until you get exactly the intermediate position you will have the strongest beat possible, which is a beat from loud sound to silence. Advance your fingers another 30° and you will find the nodes remain absolutely still when you remove your fingers. You may go on in this way with eight and ten subdivisions, and so on; but you must not expect that the places for the octantal subdivisions correspond to the places for the quadrantal subdivisions. The places for quadrantal subdivision will not in general be places for octantal subdivision. You must experiment separately for the octantal places, and you will find generally that their diameters are oblique to the quadrantal.

The reason for all this is quite obvious. In each case, the plate being only approximately circular and symmetrical, the general equation for the motion has two approximately equal roots corresponding to the nodes or divisions by one, two, three, or four diameters, and so on. Those two roots always correspond to sounds differing a little from one another. The effect of putting the finger down at random is to cause the plate, as long as your finger is on it, to vibrate forcibly in a simple harmonic vibration of a period greater than the one root and less than the other. But as soon as you take your finger off, it follows the law of superposition of fundamental modes; each fundamental mode being a simple harmonic vibration. I have often tried musicians with two notes which were very nearly equal, and said to them, "Now, which of the two notes is the graver?" Sometimes they could tell, if the difference was not too small; but they are not accustomed to listen to the thing with physical ears, and do not always say which is the graver note. A person can tell at once, after having made a few experiments of that kind, that this is the graver and that the less grave note, even though he may have what musicians call an uncultivated ear, or a very bad ear for music, not good enough in fact to guide him in singing or make him sing in tune. It is very curious, when you have two notes which you thoroughly know are different, that if you sound first one and then the other, most people will say they are about the same. But sound them both together, and then you hear the discord of the two notes in approximate unison.

We need not go further into these divisions of disturbance in air. In every case there is a plane of silence. If you take a square plate or bell vibrating in a quadrantal mode, for instance, then you have two vertical planes of silence at right angles to one another. If you make it vibrate with six or more subdivisions, you will have a corresponding number of planes of silence. I may go more into the case of the tuning fork. We have in general for quadrangular vibrations

$$R = (A - \cos^2\theta)\frac{\cos q}{r}$$

where A is an unknown constant. In the particular case we have been considering, that constant is essentially $\frac{1}{2}$.

With reference to the motions in the neighborhood of the tuning fork, you get this beautiful idea, that we have essentially harmonic functions to express them. Essentially algebraic functions of the coordinates appear in these distant terms, but in the other terms which Professor Stokes has worked out, and which has been worked out in Professor Rowland's paper on electro-magnetic disturbances* in a very full way, quite that kind of analysis appears, and it is most important. I have not given you that part but only called your attention to the part with reference to the distance equation, partly because I think it is interesting for sound and partly because it prepares us for our special subject, waves of light.

Tomorrow I think we shall begin and try to get sources of waves of light. I want to lead you up to the idea of what the simplest element of light is. It must be polarized, and it must consist of a single sequence of vibrations. A body gets a shock so as to vibrate; that body of itself then constitutes the very simplest source of light that we can have; it produces an element of light. An element of light consists essentially in a sequence of vibrations. It is very easy to show that, and to prove that the velocity of propagation of sequences in the luminiferous ether is constant. It goes on, only varying with

* *Phil. Mag.*, XVII, 1884, p. 413. *Am. Jour. Math.*, VI, 1884, p. 359.

the variation of the source. As the force gradually subsides in giving out its energy, the amplitude evidently decreases; but there will be no [throwing] off of waves forward, no spreading out or lagging in the rear, no ambiguity as to the velocity. But when that comes into collision with other bodies, what is the result? According to the discussion to which I have referred, the velocity should be quite uncertain, depending upon the number of waves in the sequence, and all this seems to present a complicated problem.

But I am anticipating a little. We shall speak of this hereafter. Somebody asked me if I was going to get rid of the subject of groups of waves. I do not see how we can ever get rid of it in the wave theory of light. We must try to make the best of it, however.

This question of the vibration of particles is a peculiarly interesting and important problem. I hope you are not tired of it yet. You see that it is going to have many applications. In the first place it is at the base of the theory of the propagation of waves. When we take our particles uniformly distributed and connected by constant springs we may pass from the solution of the problems for the mutual influence of a group of particles to the theory, say, of the longitudinal vibrations of an elastic rod, or, by the same analysis, to the theory of the transverse vibrations of a cord.

I am going to refer you to Lagrange's *Mécanique Analytique* [Part 2, p. 339]. The problem that I put before you here is given in that work under the title of vibrations of a linear system of bodies. Lagrange applies what he calls the algorithm of finite differences to the solutions. The problem which I put before you is of a much more comprehensive kind; but it is of some little interest to know that cases of it may be found, ramifying into each other.

I want to put before you some properties of the solution which are of very great importance. I want you to note first the number of terms. We have

$$C_j x_{j-1} = -a_j x_j,$$

$$C_{j-1} x_{j-2} = a_{j-1} x_{j-1} - C_j x_j = \frac{a_j a_{j-1}}{C_j} x_j - C_j x_j,$$

etc. All the x's being expressed in this way in terms of x_j, let N_j be the number of terms in x_{j-i}. These terms are obtained by substituting the values of x_{j-i+1}, x_{j-i+2} in the formula

$$-C_{j-i+1} x_{j-i} = a_{j-i+1} x_{j-i+1} + C_{j-i+2} x_{j-i+2}.$$

None of the terms can destroy one another except for special values, and the conclusion is that we have the following formula for obtaining the number of terms:

$$\mathcal{N}_i = \mathcal{N}_{i-1} + \mathcal{N}_{i-2}.$$

This is an equation of finite differences. Apply the algorithm of finite differences, as Lagrange says; or we may try for solutions of this equation by the following formula: $\mathcal{N}_i = \zeta\mathcal{N}_{i-1}$. We thus find $\zeta^2 = \zeta + 1$, or $\zeta = (1 \pm \sqrt{5})/2$. We can satisfy our equation by taking either the upper or the lower sign. The general solution is, of course,

$$\mathcal{N}_i = C\left(\frac{1 + \sqrt{5}}{2}\right)^i + C'\left(\frac{1 - \sqrt{5}}{2}\right)^i$$

where C, C' are to be determined by the equation $\mathcal{N}_0 = 1, \mathcal{N}_1 = 1$. It is rather curious to see an expression of this kind for the number of terms in a determinant. You will find that the following is a solution of the more general equation $\mathcal{N}_i = a\mathcal{N}_{i-1} + b\mathcal{N}_{i-2}$:

$$\mathcal{N}_i = \mathcal{N}_1\frac{(r^i - s^i)}{r - s} + b\mathcal{N}_0\frac{(r^{i-1} - s^{i-1})}{r - s},$$

where r, s are the two roots of the equation $x^2 = ax + b$. The coefficients of $\mathcal{N}_0, \mathcal{N}_1$ must of course be integral functions of $r + s = a$ and $rs = -b$. If one of the roots be a proper fraction, s for example, we may omit the large powers of s, and therefore for large values of i we may be sure of obtaining $\mathcal{N}_i$ to within a unit by calculating the integral part of

$$\frac{\mathcal{N}_i r^i + \mathcal{N}_0 b r^{i-1}}{r + b/r}.$$

The values of $\mathcal{N}_i$ up to $i = 12$ for the case of our problem $(a = b = \mathcal{N}_0 = \mathcal{N}_1 = 1)$ are

$$i = 2, 3, 4, 5, 6, \quad 7, \quad 8, \quad 9, \quad 10, 11, \quad 12.$$

$$\mathcal{N}_i = 2, 3, 5, 8, 13, 21, 34, 55, 89, 144, 233.$$

Lecture VII

Lagrange, in the second section of the second part of his *Mecanique Analytique* on the oscillation of a linear system of bodies, has worked out very fully the motion in the first place for disjoined bodies, and secondly for bodies forming a continuous cord. The case that we are working upon is not restricted to equal masses and equal connecting springs, but includes this particular linear system of Lagrange, in which the masses and springs are equal. I hope to take up that particular case, as it is of great interest. We shall take up this subject first to-day, and the propagation of disturbances in an elastic solid second.

It was pointed out by Dr. Franklin that the formula for

$$N_i = aN_0 \frac{r^i - s^i}{r - s} + bN_0 \frac{r^{i-1} - s^{i-1}}{r - s}$$

(which is equivalent to assuming $N_{-1} = 0$ so that $N_1 = aN_0$) may be thus simplified.

We have

$$r^2 = ar + b,$$

or, multiplying by r^{i-1},

$$ar^i + br^{i-1}.$$

So that the expression simplifies down to

$$N_i = N_0 \frac{r^{i+1} - s^{i+1}}{r - s}. \qquad *$$

We have, for example, $r - s = \sqrt{5}$, $r = (1 + \sqrt{5})/2 = 1.618$. If we work this out by very moderate logarithms for the case $N_{12} = r^{13}/(r - s)$, dropping s^{13}, we find

---

* This may be obtained directly, by determining C, C' in terms of N_0 and $N_{-1} = 0$.

$$13 \log 1.618 - \log\sqrt{5} = 13 \times .209 - .3495 = 2.3675 = \log 233,$$

which comes out exact. This working with only 4 place logarithms.

I want to call your attention to something far more important than this. The dynamical problem, quite of itself, is very interesting and important, connected as it is with the whole theory of modes and sequences of vibration; but the application to the theory of light, for which we have taken this subject up, gives to it more interest than we could have for it as a mere dynamical problem. I want to justify a fundamental form into which we can put our solution, which is of importance in connection with the application we wish to make.

Algebra shows that we must be able to throw $-x_1/\xi$ into the form

$$\frac{q_1}{K_1{}^2/T^2 - 1} + \frac{q_2}{K_2{}^2/T^2 - 1} + \cdots \frac{q_j}{K_j{}^2/T^2 - 1}$$

where $q_1, q_2, \cdots q_j$ are some constants, and $K_1, K_2, \cdots K_j$ are the values of the period T for which $-x_1/\xi$ becomes infinite. We can put it into this form certainly, for if x_1, ξ be expressed in terms of x_j, they will be functions of the $(j-1)$st and j'th degrees, respectively, in $1/T^2$. This is easily seen if we notice that $x_{j-1} = -(a_j/C_j)x_j$ is of the first degree in $1/T^2$, and that the degree of each x is raised a unit above that of the succeeding x by the factor $a_j = m_i/T^2 - C_i - C_{i+1}$ in the equation

$$-C_i x_{i-1} = a_i x_i + C_{i+1} x_{i+1}.$$

Therefore, writing z for $1/T^2$, we have

$$\frac{-x_1}{C_1 \xi} = \frac{A z^{j-1} + A' z^{j-2} + \cdots}{B z^j + B' z^{j-1} + \cdots},$$

which on being expanded into partial fractions becomes

$$\frac{C_1}{1/T^2 - 1/K_1{}^2} + \frac{C_2}{1/T^2 - 1/K_2{}^2} + \cdots \frac{C_j}{1/T^2 - 1/K_j{}^2}$$

which takes the required form on putting $C_i K_i{}^2 = q_i$.

We know that the roots of the equation of i'th degree in z which makes $-x_1/C_1\xi$ become infinite are all real; they are the periods of vibration of a system of connected bodies. We have formal proof of it in the work which we have gone through in connection with such a system. I am putting our solution in this form, because it is convenient to look upon the characteristic feature of the ratio of T to one or other of the fundamental periods. In the first place it is obvious that if we know the roots $K_1, K_2, \cdots$ the determination of $q_1, q_2, \cdots$ is algebraic. Another form which I shall give you is an answer to that algebraic question, what are the values of $q_1, q_2 \cdots$? It is an answer in a form that is particularly appropriate for our considerations because it introduces the energy of the vibrations of the several fundamental modes in a remarkable manner. We will just get that form down distinctly.

Take the differential coefficient of $C_1 \xi / - x_1$ with respect to $1/T^2$, writing this form for the moment

$$\frac{q_1}{D_1} + \frac{q_2}{D_2} + \cdots .$$

Thus

$$\frac{d}{dT^{-2}} \frac{C_1 \xi}{-x_1} = \frac{K_1^2 q_1 / D_1^2 + K_2^2 q_2 / D_2^2 \cdots}{(q_1 / D_1 + q_2 / D_2 \cdots)^2} .$$

For the case $T = K_1$, our differential coefficient becomes K_1^2 / q_1, which determines

$$q_1 = K_1^2 \left| \frac{d}{dT^{-2}} \frac{C_1 \xi}{-x_1} \right. .$$

Now you will remember that we had

$$\frac{d}{dT^{-2}} \frac{C_1 \xi}{-x_1} = m_1 + m_2 \left(\frac{x_2}{x_1}\right)^2 + \cdots m_j \left(\frac{x_j}{x_1}\right)^2 .$$

For the moment, take the expression for the simple harmonic motion, and you see at once that that comes out in terms of the energy. Adopt the temporary notation of representing the maximum value by an accented letter. Then we have at any time of the motion

$$x_1 = x_1' \sin \frac{2\pi t}{T} ,$$

if we reckon our time from the time of each particle passing through its middle position, remembering that all the particles pass the middle position at the same instant. We have therefore for the velocity of particle No. 1

$$\dot{x}_1 = \frac{2\pi}{T} x_1' \cos \frac{2\pi t}{T} .$$

The energy, which at any time is partly kinetic and partly potential, will be all kinetic at the moment of passing through the middle position. Take then the energy at that moment. For $t = 0$, we have $x_1 = 0$, $\dot{x}_1 = (2\pi/T) x_1'$. Denoting the whole energy by E (and remembering that the mass $= m_1 / 4\pi^2$) we have

$$E = \tfrac{1}{2}(m_1 x_1'^2 + m_2 x_2'^2 + \cdots m_j x_j'^2) \frac{1}{T^2} .$$

Thus, the ratio of the whole energy to the energy of the first particle $(E/\frac{1}{2}(m_1 x_1'^2/T_2))$ being denoted by R^{-1}, we have

$$m_1 R^{-1} = \frac{d}{dT^{-2}} \frac{C_1 \xi}{-x_1}.$$

This is true for any value of T whatever. From this equation, find then the ratios of the whole energy to the energy of the first particle when $T = K_1, K_2, \cdots$. Denoting these several ratios by $R_1^{-1}, R_2^{-1} \cdots$, we find

$$q_1 = \frac{K_1^2 R_1}{m_1}, q_2 = \frac{K_2^2 R_2}{m_1}, \cdots.$$

Our solution becomes then

$$\frac{-x_1}{C_1 \xi} = \frac{T^2}{m_1} \left(\frac{K_1^2 R_1}{K_1^2 - T^2} + \frac{K_2^2 R_2}{K_2^2 - T^2} + \cdots \right).$$

This is the much more convenient form, as it shows us everything in terms of quantities whose determinations are suitable, viz.: the periods and [the] energy ratios.

It remains, lastly, to show how, from our process without calculating the determinants, we can get everything that is here concerned. Our process of calculating gives us the u's in order, beginning with u_{j-1}. That gives us the x's in order, and thus we have all that is embraced in the differential coefficient with respect to $1/T^2$. Everything is done, if we can find the roots. I will show how you can find the roots from the continued fraction, without working out the determinant at all. The calculation in the neighborhood of a root gives us the axiom of x's corresponding to that root, and then by multiplying the squares of the ratios of the x's to x_1 by the masses and adding, we have the corresponding energy.

The case that will interest us most will be the successive masses greater and greater, and the successive springs stronger and stronger, but not in proportion to the masses so that the periods of vibration of limited portions of the higher numbered particles of the linear system shall be very large. For example, so that if we hold at rest particles 4 and 6, the natural time of vibration of particle 5 will be longer than No. 2's would be if we held Nos. 1 and 3 at rest and set No. 2 to vibrating.

We will just put down once more two or three of our equations:

$$\frac{C_1 \xi}{-x_1} = a_1 - \frac{C_2^2}{u_2}, \cdots u_i = a_i - \frac{C_{i+1}^2}{u_{i+1}}; \quad a_i = \frac{m_i}{T^2} - C_i - C_{i+1}.$$

Without considering whether u_{i+1} is absolutely large or small, let us suppose that it is large in comparison with C_{i+1}. u_i will then be of the order a_i, u_{i-1} of the order a_{i-1}, and so on. We are to suppose that $a_1, a_2, \cdots a_i$ are in ascending order of magnitude. Now,

$$u_1 \cdot u_2 \cdots u_i = (-1)^i C_1 \cdots C_i \frac{\xi}{x_i}.$$

We thus have this important proposition: that the magnitudes of the vibrations of the successive particles decrease from particle No. 1 towards No. j; and x_i is exceedingly small in comparison with ξ, even though there is only a moderate proportion of smallness with respect to the ratios C_1/u_1, $C_2/u_2, \cdots C_i/u_i$. Thus see how small is the motion at a considerable distance from the point at which the excitation takes place, under the suppositions that we have been making.

Now, as to the calculations. I do not suppose anybody is going to make these calculations, but I always feel in respect to arithmetic somewhat as Green has expressed in reference to analysis. I have no satisfaction in formulas unless I feel their arithmetical magnitude—at all events, when formulas are intended for operations of that kind. So that if I do not exactly calculate the formulas, I would like to know how I would calculate them and express the order of the magnitudes. It might not be worth while to go into the number of terms *per se*, but the number of terms is closely related to the order of the magnitudes we have been dealing with. We are not going to make the calculations, but you will remark that we have every facility for doing so. In the first place, it is the exceeding rapidity of convergence of the formulas. The question is to find $C_1\xi/-x_1$; everything, you will find, depends upon that. The exceeding rapidity of the convergence is manifest. Since u_2 is large, u_1 is equal to a_1 with a small correction; similarly $u_2 = a_2$ with a small correction, and so on; so that two or three terms of the continued fraction will be sufficient for calculating the ratio denoted by u_1. The continued fractions converge with enormous rapidity upon the suppositions we have been making. We thus know the value of the differential coefficient du_1/dT^{-2}. We can in this way obtain several values of u_1 and begin to find it coming near to zero. Then take the usual process. Knowing the value of the differential coefficient allows you to diminish very much the number of trials that you must make for calculating a root. The process of finding the roots of this continued fraction will be quite analogous to Newton's process for finding the roots of an algebraic equation; and I tell any of you who may intend to work at it, that if you choose any particular case you will find that you will get at the roots very quickly.

I should think something like an arithmetical laboratory would be good in connection with class work in which students might be set at work upon problems of this kind, both for results, and in order to obtain facility in calculation. I think we will not say anything more about this problem just now, and we will leave it as we have it.

I hinted to you in the beginning about the kind of view that I wanted to take of molecules connected with the luminiferous ether, and affecting by their inertia its motions. I find since then that Lord Rayleigh really gave in a very distinct way the first indication of the explanation of anomalous dispersion. I will just read a little of the paper on the Reflection and Refraction of Light by Intensely Opaque Matter [*Philosophical Mag.*, May, 1872]. He commences, "It is, I believe, the common opinion, that a satisfactory mechanical theory of the reflection of light from metallic surfaces has been given by Cauchy, and that his formulae agree very well with observation. The result, however, of a recent examination of the subject has been to convince me that, at least in the case of vibrations performed in the plane of incidence, his theory is erroneous, and that the correspondence with fact claimed for it is illusory, and rests on the assumption of inadmissible values for the arbitrary constants. Cauchy, after his manner, never published any investigation of his formulae, but contented himself with a statement of the results and of the principles from which he started. The intermediate steps, however, have been given very concisely and with a command of analysis by Eisenlohr (*Pogg. Ann.*, vol. CIV, p. 368), who has also endeavored to determine the constants by a comparison with measurements made by Jamin. I propose in the present communication to examine the theory of reflection from thick metallic plates, and then to make some remarks on the action on light of a *thin* metallic layer, a subject which has been treated experimentally by Quincke.

"The peculiarity in the behavior of metals towards light is supposed by Cauchy to lie in their *opacity*, which has the effect of stopping a train of waves before they can proceed for more than a few wave-lengths within the medium. There can be little doubt that in this Cauchy was perfectly right; for it has been found that bodies which, like many of the dyes, exercise a very intense selective absorption on light, reflect from their surfaces in excessive proportion just those rays to which they are most opaque. Permanganate of potash is a beautiful example of this given by Prof. Stokes. He found (*Phil. Mag.*, Vol. VI, p. 293) that when the light reflected from a crystal at the polarizing angle is examined through a Nicol held so as to extinguish the rays polarized in the plane of incidence, the residual light is green, and that, when analyzed by the prism, it shows bright bands just where the absorption spectrum shows dark ones. This very instructive experiment can be repeated with ease by using sunlight, and instead of a crystal a piece of ground glass sprinkled with a little of the powdered salt, which is then well rubbed in and burnished with a glass stopper or otherwise. It can without difficulty be so arranged that the two spectra are seen from the same slit one over the other, and compared with accuracy.

"With regard to the chromatic variations it would have seemed most natural to suppose that the opacity may vary in an arbitrary manner with the wave-length, while the optical density (on which alone in ordinary cases the refraction depends) remains constant, or is subject only to the same sort of variations as occur in transparent media. But the aspect of the question has been materially changed by the observations of Christiansen and Kundt (*Pogg. Ann.*, vols. CXLI, CXLIII, CXLIV) on anomalous dispersion in Fuchsin and other colouring-matters, which show that on either side of an absorption-band there is an abnormal change in the refrangibility (as determined by prismatic deviation) of such a kind that the refraction is *increased* below (that is, on the red side of) the band and *diminished* above it. An analogy may be traced here with the repulsion between two periods which frequently occurs in vibrating systems. The effect of a pendulum suspended from a body subject to horizontal vibration is to increase or diminish the virtual inertia of the mass according as the natural period of the pendulum is shorter or longer than that of its point of suspension. This may be expressed by saying that if the point of support tends to vibrate more rapidly than the pendulum, *it* is made to go faster still, and *vice versa*"—I cannot understand the meaning of that sentence, at all. There is a terrible difficulty with writers in abstruse subjects to make sentences that are intelligible. It is impossible to find out from the words what they mean; it is only from knowing the thing that you can do so—"Below the absorption-band the material vibration is naturally the higher, and hence the effect of the associated matter is to increase (abnormally) the virtual inertia of the aether, and therefore the refrangibility. On the other side the effect is the reverse." Then follows a note, "See Sellmeier, *Pogg. Ann.* vol. CXLIII p. 272." Thus Lord Rayleigh goes back to Sellmeier, and I suppose he is the originator of all this. "It would be difficult to exaggerate the importance of these facts from the point of view of theoretical optics, but it lies beside the object of the present paper to go further into the question here."

There is the first clear statement that I have seen. Prof. Rowland has been kind enough to get these papers of Lord Rayleigh for me, with an immense deal of trouble. An interminable number of books have been brought to me, and in every one of them I have found something very important.

Sellmeier, Lord Rayleigh, Helmholtz and Lommel seems to be about the order. Lommel does not quote Helmholtz. I am rather surprised at this, because Lommel comes three or four years after Helmholtz; 1874 and 1878 are the respective dates. Lommel's paper is published in Helmholtz's Journal [*Ann. der Physik und Chemie* 1878, vol. 3, p. 339] so I suppose Helmholtz has no objection. Helmholtz's paper is excellent. Lommel goes into it still further, and has worked out the vibrations of associated matter to explain ordinary dispersion.

I only found this forenoon that Lommel [*Ann. der Ph. und Chem.* 1878, Vol. 7, p. 55] also goes on to double refraction of light in crystals—the very problem I am breaking my head against. He is satisfied with his solution, but I do not think it at all satisfactory. It is the kind of thing that I have seen for a long time, but could not see that it was satisfactory; and I do see reason for its not being satisfactory. He goes on from that and obtains an equation which would approximately give Huygens' surface. I have not had time to determine how far it may be correct. The exceedingly close agreement of Huygens' surface with the facts of the case which Stokes has found absolutely cuts the ground from under a large number of very tempting modes of explaining double refraction.

Lecture VIII

We shall take some fundamental solutions for wave motion such as we have already had cosiderable to do with, only we shall consider them as now applicable to distortional waves, instead of condensational waves. That is, we can take our primary solution in the form

$$\varphi = \frac{1}{r} \sin \frac{2\pi}{\lambda} (r - Ct),$$

where $C = \sqrt{(k + \frac{4}{3}n)/\rho}$ if the wave is condensational, and $= \sqrt{n/\rho}$ if the wave is distortional. But for a distortional wave, we must also have what is denoted by $\delta = 0$.

In the first place, we know that φ satisfies

$$\rho \frac{d^2 \varphi}{dt^2} = n \nabla^2 \varphi,$$

our value of C being $\sqrt{n/\rho}$. (I want, very much, a name for that function ∇—delta turned upside down. I do not know whether Prof. Ball has any name for it or not, Sir Wm. Hamilton uses it a great deal, and I think perhaps Prof. Ball may know of a name for it.) The conditions to be fulfilled by the three components of displacement, ξ, η, ζ, of a distortional wave are, in the first place,

$$\rho \frac{d^2 \xi}{dt^2} = n \nabla^2 \xi, \quad \rho \frac{d^2 \eta}{dt^2} = n \nabla^2 \eta, \quad \rho \frac{d^2 \zeta}{dt^2} = n \nabla^2 \zeta;$$

and we must have besides

$$\frac{d\xi}{dx} + \frac{d\eta}{dy} + \frac{d\zeta}{dz} = 0.$$

Thus ξ, η, ζ must be three functions, each fulfilling the same equation. There is a fulfilment of this equation by the functions φ; and as we have one solution, we can derive other solutions from that by differentiation. Let us see, then, if we can derive three solutions from this value of φ which shall fulfil the remaining condition. It is not my purpose here to go into an analytical investigation of solutions, it is rather to show solutions which are of fundamental interest. Without further preface, then, I will show you one, and another, and then I will interpret them both.

Take for example the following, which obviously fulfils the equation $d\xi/dx + d\eta/dy + d\zeta/dz = 0$:

$$\xi = 0, \quad \eta = -\frac{d\varphi}{dz}, \quad \zeta = \frac{d\varphi}{dy}.$$

In each case the distance terms only of our solution are what we wish. Thus,

$$\eta = -\frac{d\varphi}{dz} \doteqdot -\frac{2\pi}{\lambda} \cdot \frac{z}{r^2} \cos q,$$

$$\zeta = \frac{d\varphi}{dy} \doteqdot \frac{2\pi}{\lambda} \cdot \frac{y}{r^2} \cos q.$$

Remark that in this solution the displacement at a distance from the source is perpendicular to the radius vector; i.e., we have

$$x\xi + y\eta + z\zeta = -y\frac{d\varphi}{dz} + z\frac{d\varphi}{dy} \doteqdot 0.$$

Before going further, it will be convenient to get the rotation. It is an exceedingly convenient way of finding the direction of vibration in distortional displacements. The rotations about the axes of x, y, z, will be

$$\frac{d\zeta}{dx} - \frac{d\eta}{dy} \doteqdot -\frac{4\pi^2}{\lambda^2} \frac{y^2 + z^2}{r^3} \sin q,$$

$$\frac{d\xi}{dz} - \frac{d\zeta}{dx} \doteqdot \frac{4\pi^2}{\lambda^2} \frac{xy}{r^3} \sin q,$$

$$\frac{d\eta}{dx} - \frac{d\xi}{dy} \doteqdot \frac{4\pi^2}{\lambda^2} \frac{xz}{r^3} \sin q.$$

These rotations are proportional to

$$\frac{x^2}{r^3} - \frac{1}{r}, \quad \frac{xy}{r^3}, \quad \frac{xz}{r^3};$$

that is to say, besides the x component $-(1/r)$, we have an r component x/r^2. We have a rotation around the radius vector r, and a rotation around the axis of x, whose magnitudes are proportional to x/r^2 and $1/r$.

If you think out the nature of the thing, you will see that it is this: a globe, or a small body at the origin, set to oscillate about Ox as axis. You will have turning vibrations everywhere; and the light will be everywhere polarized in planes through Ox. The vibrations will be everywhere perpendicular to the radial plane through Ox.

In the first place we have (omitting the constant factor $2\pi/\lambda$)

$$\xi = 0, \quad \eta = -\frac{z}{r^2}\cos q, \quad \zeta = \frac{y}{r^2}\cos q.$$

That presents a wave spreading out in all directions from the axis of x. For if $y = 0$, $z = 0$, the displacements are zero, or we have the case of zero vibration in the axis of x. Again, the displacements are everywhere perpendicular to Ox (since we always have $\xi = 0$), and being perpendicular also to the radius vector, they are perpendicular to the radial plane through the axis of x.

Let us consider the state of things in the planes yz. Suppose we have a small body here at the origin or center of disturbance, and that it is made to turn in this way:

(indicating a twisting motion about an axis perpendicular to the plane of the paper) in a given period T. What is the result? Waves will proceed out in all directions and the intersections of the wave fronts with the plane (yz) of the paper will be circles. We shall have vibrations perpendicular to the radius vector of magnitude $\cos q/r$, which is the same in all directions. The rotation, which is simply the polar rotation about the axis of x in the plane yz, is

$$\frac{2\pi}{\lambda}\frac{\sin q}{r}$$

(also the same in all directions). There is therefore a minimum displacement where there is a maximum distortion, and *vice versa*. At a point of maximum distortion (positive or negative) there is zero rotation; at a point of maximum rotation there is zero distortion. We have polarized light consisting of vibrations in the plane and perpendicular to the radius vector, and therefore the plane of polarization is the radial plane through Ox.

Here we have a simple source of polarized light; it is the simplest form of polarization and the simplest source that we can have. Every possible light consists of sequences of light from simple sources. Is it probable that the shocks to which the particles are subjected in the electric light, or in fire, or in any ordinary source of light, would give rise to a sequence of this kind? No; because there is nothing to make a body vibrate by itself. We can arbitrarily do it, for we can do what we will with the particle. That privilege occurred to me in Philadelphia last week, and I showed the vibrations by having a large bowl of jelly made with a ball placed in the middle of it. I really think you will find it interesting enough to try it for yourselves. It allows you to see the vibrations we are speaking of. I wish I had it to show you just now, so that you might see the thing in force. It saves brain very much.

I had a large glass bowl quite filled with some spirit jelly, and a wooden ball floating in the middle of it.

Try it, and you will find it a very pretty illustration. Apply your hand to the ball, and give it a twisting motion thus, and you have exactly the kind of motion here expressed in the plane yz. The motion in any oblique direction, such as at this point (x,y,z) you will find to consist of polarized light vibrating perpendicular to the radial section. The amplitude of the vibration here (in the vertical axis) is zero; here at the surface (in the plane yz) it is $(1/r)\cos q$; and if you use polar coordinates, calling this angle θ (indicating on the diagram), then the amplitude here (at xyz) is $(1/r)\cos q\sin\theta$, giving, when θ is a right angle, the previous expression.

I say that this is the simplest source and the simplest strain of polarized light that we can imagine. But it cannot be induced naturally, because no natural vibrator could do it. The next simplest is a globe or small body vibrating to and fro in one line. We will take the solution for that presently. Still we have not got up to the essential complexity of the natural vibrator. I may take my hand and give torsional oscillations to the globe; I can take my hand (and that makes a very pretty modification of the experiment) and shove out on the globe, making it vibrate, and people cannot help saying, "0, there is the natural time of the vibration, you find it, if you only leave it alone to itself." But it is only proper for an *illustration* of vibrations spreading out from a center. We are bothered here also by reflection back, as it were, from the containing bowl, just as in suspending a rope to show waves running along it we are bothered in the experiment by the rope not being infinitely long. You can always see a set of vibrations running along the rope, beginning at the lower end and reflected back from the upper end where it is fastened to the ceiling. But in this experiment, you do not see the waves travelling out at all because you get it in a certain set of vibrations, depending on this finite material. But just imagine the bowl to be infinitely large, and that you commence making torsional oscillations; what will take place? A spreading outwards of this kind of vibrations, the beginning being, as we shall see, abrupt. We shall scarcely reach that to-day, but we shall consider the abruptness of the beginnings and endings of the vibrations in an elastic solid; and in every case in which the velocity of propagation is independent of the wave length we have no end at all, but waves travelling outwards, with a gradual falling off of intensity.

When you apply your hands and force the ball to perform those torsional vibrations, you have waves proceeding from it; but if you then leave it to itself, there is no vibrating energy in it at all, except the slight angular velocity that you leave it with. A vibrator which can send out a succession of impulses, independently of being forced to vibrate from without, must be a vibrator with the means of conversion of potential into kinetic energy in itself. A tuning fork, and a bell, are sample vibrators [for] sound. The simplest sample vibrator that we can get to represent the origin of the simplest sequence of light is just like a tuning fork. Two bodies, joined by a spring would be more symmetrical than a tuning fork. Two globes joined by [a] spring—that will give you the idea; or (which will be a vibration of the same type still) one spherical body vibrating backwards and forwards from having been drawn so

into an oval shape, and let go.

I will look, immediately, at a set of vibrations produced in an elastic solid by a sample vibrator. But suppose you produce vibrations in your jelly solid by taking hold of this ball and shoving it to and fro horizontally, or again shoving it up and down vertically, and think of the kinds of vibrations it will make all around. Think of that, in connection with the formulae, and it will help us to interpret them. But it will take a higher order of vibrator to get the kind of vibration that comes from the natural source. We might have those torsional vibrations; but among all the possible vibrations of atoms in the clang and clash of atoms that there is in a flame, or other source of light, a not very rare case I think would be that which I am going to speak of now. It consists of opposite torsional vibrations at the two ends of an elongated mass; or, to simplify our conception for a moment, imagine two globes connected by a columnar spring; twist them in opposite directions, and let them go. There might be a source of vibrations; and if the potential energy of the spring is very large in comparison with the energy that has been carried off in a thousand or a hundred thousand vibrations, you will have a set of vibrations following the same law that we get in the case already considered.

Before passing on to the to and fro vibrator we will think of this motion for a moment, but we will not work it out, because it is not so interesting. To suit our drawing we shall suppose one globe here, and another upon the opposite side on a level with the first so that the line of the two is perpendicular to the boards. Give these globes opposite torsional vibrations about their common axis, and what will the result be? A single one produces zero light in the axis and maximum light in the equatorial plane. The two going in opposite directions will produce zero light in the equatorial plane and zero light in the axis; so that you will proceed from zero in the equatorial plane to a maximum between the equatorial plane and the poles, and zero at the poles; and you will have opposite vibrations in each hemisphere. That constitutes a possible case of vibrations of polarized light, proceeding from a possible independent vibrator. If you had, among all the elements concerned in the production of the light, some such action, or configuration as that if a shock took place at one end of a molecule, another should simultaneously take place in an opposite direction upon the other end; that might set the thing to vibrating in that way; and that is one of the possible sets of vibrations constituting light.

But by far the most simple and natural supposition in respect to an independent vibrator is afforded by the illustration of a bell, or a tuning fork, or an elastic body deformed from its natural shape and left to vibrate. In all these cases, you remark, the center of gravity of the vibrator is at rest; and you can not have any thing else from an independent action. The vibrator must have potential energy in itself, and its center of gravity must be at rest, except insofar as the reaction of the medium upon it causes a slight motion of the center of gravity.

I will put down the solution which corresponds to a to and fro vibration in the axis of x, viz.,

$$\xi = \frac{4\pi^2}{\lambda^2}\varphi + \frac{d^2\varphi}{dx^2}, \quad \eta = \frac{d^2\varphi}{dydx}, \quad \zeta = \frac{d^2\varphi}{dzdx}.$$

φ is our old friend,

$$\frac{1}{r}\sin\frac{2\pi}{\lambda}\left(r - t\sqrt{\frac{n}{\rho}}\right).$$

In the first place we know that

$$\rho\frac{d^2\xi}{dx^2} = n\nabla^2\xi, \text{ etc.}$$

are satisfied, because φ and all its differential coefficients satisfy this relation.

We have then only to verify that the dilatation is zero. I will not go through the verification, but you will not make the solution your own unless you see how I obtained it. I will not say that there is anything novel in it, but it is simply the way it occurred to me. I obtained it to illustrate Stokes' explanation of the blue sky. I afterwards found that Lord Rayleigh had gone into the subject more searchingly than Stokes, and I read his work upon it.

The way I found this solution was this: $d\varphi/dx$ is clearly the displacement potential [for an elastic fluid,] corresponding to a source of the kind, a pull along the axis of x. It is like the magnetic potential of a bar magnet with its axis in the direction of Ox. The displacement function of which the displacements are the differential coefficients would take that form if this was a question, for instance, of sound and not of light. It was a question of condensational vibrations with us several days ago. I did not go into the matter in detail, but we saw that for condensational vibrations proceeding from a vibrator vibrating to and fro along the axis of x that $d\varphi/dx$ was the displacement potential; and it is obvious, if we start from the very root of the matter, that it must be so. $d\varphi/dx$ must therefore be the corresponding function that we shall have to deal with in the case of light from such a source, although that will not certainly give, by differentiation simply, the displacements we want. The displacements in the condensational wave problem are displacements which fulfil certain of the conditions, but do not fulfil all the conditions, of giving us a pure distortional wave unless we add a term or terms in order to make the dilatation zero. Just try, in the first place, for the dilatation. We have

$$\nabla^2 \varphi = \frac{\rho}{n}\frac{d^2\varphi}{dt^2} = -\frac{4\pi^2}{\lambda^2}\varphi,$$

in which we may substitute $d\varphi/dx$ for φ. Thus

$$\nabla^2 \frac{d\varphi}{dx} = -\frac{4\pi^2}{\lambda^2}\cdot\frac{d\varphi}{dx}.$$

We have verified, therefore, that the displacements given satisfy

$$\frac{d\xi}{dx} + \frac{d\eta}{dy} + \frac{d\zeta}{dz} = 0;$$

and thus we have made up a solution which satisfies the condition of being non-condensational—no condensation or rarefaction anywhere.

In the first place, taking the distant terms only we have

$$\xi \doteqdot \frac{4\pi^2}{\lambda^2}\frac{r^2 - x^2}{r^3}\sin q, \quad \eta \doteqdot \frac{4\pi^2}{\lambda^2}\frac{xy}{r^3}\sin q, \quad \zeta \doteqdot -\frac{4\pi^2}{\lambda^2}\frac{xz}{r^3}\sin q.$$

It is easy to verify that these displacements are perpendicular to the radius vector, i.e., that we have

$$x\xi + y\eta + z\zeta \doteqdot 0.$$

Just look at the case along the axis of x, and again in the plane yz. It is written down here in mathematical word painting as clearly and completely as any non-mathematical words can give it. Take $y = 0$, $z = 0$, and that makes $\xi = 0$, $\eta = 0$, $\zeta = 0$. Therefore, in the direction of the axis of x there is no motion. That is a little startling at first, but is quite obviously a necessity of the fundamental supposition. Cause a globe in an elastic solid to vibrate to and fro. At the very surface of the globe the points in which it is cut by Ox have the maximum motion; and throughout the whole circumference of the globe the medium is pulled, by hypothesis, along with the globe. But this is not a solution for that comparatively very difficult problem. I am only asking you to think of this as the solution for the motion at a great distance. It may not be a globe, but a body of any shape moved to and fro. To think of a globe will be more symmetrical. In the immediate neighborhood of the vibrator there is a motion produced in the line of vibration; the motion of the elastic solid in that neighborhood consists in a somewhat complex but very easily expressed state of things, in which we have particles in one place moving out and slipping around with motions oblique to the radius vector, as in the axis of x, and in other places moving perpendicular to the radius vector, as for points in the plane yz. All, however, except motions perpendicular to the radius vector, become insensible at distances very great in comparison with the wave length. We have taken, simply, the leading terms of the solution. These represent the motion at great distances, quite irrespective of the shape of the body, and the comparatively complicated motion in the neighborhood of the vibrating body.

Take now $x = 0$, and think of the motions in the plane yz. The vibrator is supposed to be vibrating perpendicular to this plane. We have

$$\xi \doteqdot \frac{4\pi^2}{\lambda^2}\frac{1}{r}\sin q, \quad \eta = 0, \quad \zeta = 0.$$

What does this mean? Clearly, that the vibrations are perpendicular to this plane yz. We have light spreading out uniformly in that plane, and polarized in that plane, the vibrations being perpendicular to it. That is what Stokes supposed was necessarily the theory of the blue light of the sky. Lord Rayleigh showed that it was not so clear as Stokes had supposed. He elaborately investigated the question "Is the blue light of the sky—which was supposed to be owing to particles which you may say are spherical if you like—due to the particles causing it being of density different from the surrounding luminiferous ether, or being of rigidity different from the surrounding luminiferous ether?" The real question would be, If the particles are water, what is the theory of waves of light in water; does it differ from air in being, as it were, a denser medium with the same effective rigidity, or is it a medium of the same density and less effective rigidity, or will both density and rigidity vary?

Lord Rayleigh examined that question very thoroughly, and finds, if the cause were, for instance, little spherules of water, and if in the passage of light through water the fact that the velocity of propagation is slower than in air were explained by less rigidity and the same density, we should have something quite different in the polarization of the sky from what we would have on the other supposition. On the other hand, the polarization of the sky creates the supposition (which is as much as the uncertaintitude of the experimental data allows us to judge) that the particles, whether they be particles of water, or motes of dust, or whatever they may be, act as if they were little portions of the luminiferous ether of greater density [than] and not of [different] rigidity [from] the surrounding ether.

This solution, then, is not the solution for that source of light which has such great interest as being the cause of the blue light coming from the sky. I will call attention a little more to Lord Rayleigh's explanations upon that; but it cannot be the effect of a vibrator in the source, for the reasons I have stated. We may differentiate once more with respect to x, in order to get a proper form of function that will express the motion from the vibrator vibrating to and fro like this—vibrating the hands towards and from each other. Then we shall have a [solution] which will express one single sequence of vibrations, of which the multitude constitutes the light of the source. The question is then forced upon us, what is the velocity of a group of waves in the luminiferous ether disturbed by ordinary matter? With a constant velocity of propagation each group remains unchanged. But how about the effect of a non-simple source of light in a transparent medium like glass? It is a question that is more easily put than answered. We should consider it carefully. I do not despair of seeing the answer. I think, if we have a little more patience with our dynamical problem we shall get it.

Here is a perfectly parallel problem. Commence suddenly to give a simple harmonic motion through the handle P to our system of particles $m_1, m_2, \cdots m_j$, which play the part of a molecule, of course. If you commence suddenly imparting to the handle a motion of any period whatever, avoiding only one of the fundamental periods, if there be a little viscosity it will settle into a state of things in which you have perfectly regular simple harmonic vibration. [But] if there be no viscosity whatever, what will the result be? It will be the component of simple harmonic motions in the period of our applied motion at the bell handle P, with every part in it obtained by a continued fraction. We superimpose motion upon it, and jangle it as it were, producing coexistent simple harmonic vibrations of the fundamental periods. If there is no viscosity, that state of things will go on forever. I cannot satisfy myself with viscous terms in these theories (although I believe this is the view of Lommel, Helmholtz and others), because we know that light goes on for millions and millions of vibrations. But if we have none of these viscous terms at all, whatever velocity we have must show in the vibration of something else, and that is what? In going into that sort of vibration with which we have been occupied in the other part of our course, we must account for these irregular vibrations somehow or other. The viscous terms are merely a step towards accounting for the difficulties of the theory. By viscous terms, I mean terms that assume a viscosity.

But the state of things with us is that that jangling will go on forever, if there is no loss of energy; and we want to coax our system of vibrators into a state of vibration with an arbitrarily chosen period without viscous consumption of energy. Begin thus: get it into motion with a very small range. The result will be just as I have said, only with a very small range. After waiting a little time increase the range; after waiting a little longer, increase the range again, and so go on, increasing the range by successive steps. Each of those will superimpose another state of vibration. There would be, I believe, virtually an addition of the energies of those several vibrations if you make these steps quite independent of one another.

For example, suppose you proceed thus: In the first place, start right off into vibrations of your handle P through a space, say of 30 inches. You will have a certain amount of energy in the irregular vibrations. In the second place, commence on a range of three inches. After you have kept it going on three inches any time you like, suddenly increase it to three inches more, making it six inches. Then, sometime after, suddenly increase the range to nine inches; and go on in that way for ten steps. The energy of the irregular vibrations produced by suddenly commencing through the range of three inches, which is one-tenth of 30 inches, will be one-hundredth of the energy which you would have if you commenced right away with the vibration through 30 inches. Each successive step of three inches will add the one-hundredth; and the result is that if you go by these steps to the range of 30 inches, you will have in the irregular vibrations one-tenth of the energy you would get if you began at that range right away. Thus, by very gradually increasing the range, the result will be that there will be infinitely little of the irregular vibrations.

I believe something of that kind will account for our difficulty; and I believe that that kind of thing applied to sequences of waves will without doubt show that if you commence a set of waves very gradually (through several hundred vibrations may be enough) and then make them uniform (that is, let the source go on uniformly after that), even with sea waves possibly, or with luminous waves in a transparent solid, there will be exceedingly little disturbance from the beginning [to the] end of them. It is only a vague idea I have thrown out; but I think considerations of this kind may help us to see how it is that sets of groups of waves which undoubtedly constitute the reality of light do still act as if we had a perfectly simple harmonic and continuous source of vibrations. They do act so in the propagation of light through the medium, in refraction, and reflection, and so on.

But there are cases in which we have that tremendous jangling, and that is in the fluorescence of such a thing as uranium glass, which lasts for several seconds after the exciting light is taken away, and then again in phosphorescence that lasts for hours and days. There have been exceedingly interesting beginnings, in the way of experiments already made, but I think nobody has found whether initial refraction is exactly the same as permanent refraction. For this purpose we might use Becquerel's phosphoroscope, or we might take such an appliance as Prof. Michelson has been using for light, and get something more enormously searching than Becquerel's phosphorescope, and try whether in the first hundredth of a second there is any indication of a different wave velocity from that which you would have when light passes continuously in the usual manner of refraction. If in the methods employed for ascertaining the velocity of light in a transparent body (notwithstanding the criticisms that they have received at the British Association meeting, to which I have referred several times) we apply a test for an instantaneous refraction, I have no doubt we shall not get negative results, but get properties of ultimate importance. We might take bodies in which, like uranium glass, the phosphorescence lasts only a few seconds; and then again bodies in which phosphorescence lasts for minutes and hours. With some of those we should have anomalous dispersion, gradually fading away after a time. I should think that by experimenting, and so on, we should find some very some very interesting results of this kind.

Lecture IX

We shall go on for the present with the subject of the propagation of waves from a center. Let us pass to the case of two bodies vibrating in opposite directions, in the manner which we have already had for one body which was expressed by

$$\xi = \frac{4\pi^2}{\lambda^2}\varphi + \frac{d}{dx}\frac{d\varphi}{dx}, \quad \eta = \frac{d}{dy}\frac{d\varphi}{dx}, \quad \zeta = \frac{d}{dz}\frac{d\varphi}{dx}.$$

We verified that

$$\frac{d\xi}{dx} + \frac{d\eta}{dy} + \frac{d\zeta}{dz} = 0,$$

so that this expresses rigorously a distortional wave. It is obvious that this expresses the result of a to and fro motion at the origin. Remark, for one thing, that in the neighborhood of the origin, at such moderate distance from it that the component motion in the direction Ox does not vanish, we have on the two sides of the origin simultaneously positive values. ξ is the same for a positive value of x as for x the negative of that value. At distances from the origin in the line Ox which are considerable in comparison with the wave length the motion vanishes as we have seen. This, then, expresses the result of a to and fro motion at the origin.

Pass on, now, to this case: a positive to and fro motion on the one side of the origin, and a negative to and fro motion on the other side of the origin. I will indicate these motions by arrow heads. The first case already considered

[and] the second case

The effect in the first case being expressed by the displacements ξ, η, ζ, already given, the effect in the second case will be expressed by the displacement

$$\frac{d\xi}{dx}, \frac{d\eta}{dy}, \frac{d\zeta}{dz}.$$

This displacement clearly expresses a motion which has opposite signs for equal positive and negative values of x. It will express a simultaneous outward and inward motion on the two sides of the origin and a zero motion in the plane yz. A motion for distances from the origin moderate in comparison with the wave length will be accurately expressed by these functions; but as before we shall take only the leading or distance terms, and also drop the coefficient $-8\pi^3/\lambda^3$, which we do not want. Thus

$$\xi \doteqdot \frac{(x^2 - r^2)x}{r^4}\cos q, \quad \eta \doteqdot \frac{x^2 y}{r^4}\cos q, \quad \zeta \doteqdot \frac{x^2 z}{r^4}\cos q$$

express the distant displacement of an outward and inward motion illustrated by that configuration of arrow heads (case 2), and obviously expressing a motion in which there will be zero displacement everywhere in the plane yz, with equal opposite values on the two sides of that plane. Take $y = 0$, $z = 0$, to find the motion in the axis of x, and we get as in the first case zero motion in that axis. We can easily satisfy ourselves that the radial component of the displacement is zero, i.e., that we have

$$x\xi + y\eta + z\zeta \doteqdot 0.$$

Lastly, if you think of the kind of polarization that will be produced by that motion, it is obvious that the motion will be everywhere symmetrical around the axis of x, and will be in the radial plane through Ox. $[0 \cdot \xi + z\eta - y\zeta \doteqdot 0]$ Therefore, we have light polarized in the plane through the radius of the point considered and perpendicular to the radial plane through Ox.

Look at what the magnitude of the motion will be.

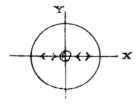

Inasmuch as the motion is symmetrical around the axis of x, we may take what goes on in the plane xy as a sample of the whole. We then have

$$\xi \doteqdot -\frac{xy^2}{r^4}\cos q, \quad \eta = \frac{x^2 y}{r^4}\cos q,$$

showing that there is zero motion in the axis of x, and zero motion in the axis of y. The expression for the amplitude of the motion is

$$\sqrt{\xi^2 + \eta^2} = \frac{xy}{r^3}\cos q.$$

Thus the displacement is distributed on the two sides of Ox and of Oy so as to be equal and opposite in adjacent quadrants. Remember that the thing is symmetrical around Ox, and you have a perfect understanding of the distribution of the motion, the distance being considerable in comparison with the wave length.

This is the simplest set of vibrations that we can consider as proceeding from any natural source of light. As I said, we might conceive of a pair of equal and oppositely torsional motions, at the two ends of a vibrating molecule. That is one of the possibilities, and it would be rash to say that any one possible kind of motion does not exist in so remarkably complex a thing as the motion of the particles from which light originates.

This motion we are considering is perhaps the most interesting, as it is obviously the simplest kind of motion that can proceed from a single vibrator. If you consider the two ends of a tuning fork, neglecting the prongs, so that everything may be symmetrical around the two moving bodies, you have a way by which the motion may be produced. Or our source might be two balls connected by a spring and pulled asunder and set to vibrating in and out; or it might be an elastic sphere which has experienced a shock. An infinite number of modes of vibration are generated when an elastic ball is struck a blow, but the gravest mode is also no doubt where the energy is greatest, and that consists of the globe vibrating from an oblate to a prolate figure of revolution.

The kind of thing that the luminiferous vibrator consists in seems to me to be a sudden initiation of a set of vibrations and a sequence of vibrations from that initiation which will naturally become of smaller and smaller amplitude. So that the graphic representation of what we should see if we could see what proceeds from one element of the source, the very simplest conceivable element of the source, would consist of polarized waves of light spreading out in all directions, according to some such law as we have here. In any one direction, what will it be? Suppose that the wave advances from left to right; you will then see what is here represented on a magnified scale.

I have tried to represent a sudden start, and a gradual falling off of intensity. Why a sudden start? Because I believe that the light of the natural flame or of the arc light, or of any other known source of light must be the result of sudden shocks from a number of vibrators. Take the light obtained by striking two quartz pebbles together. You have all seen that. There is one of the very simplest sources of light. There is some sort of a chemical or ozoniferous effect connected with it which makes a smell. As to what the cause of that may be, I suppose we are almost assured, now, that it proceeds from the generation of ozone. What sort of a thing can the light be that proceeds from striking two quartz pebbles together? Under what circumstances can we conceive a group of waves of light to begin gradually and to end gradually? You know what takes place in the excitation of a fiddle string or a tuning fork by a bow. The vibrations gradually get up from zero to a maximum and then, when you take the bow off, gradually subside. I cannot see anything like that in the source of light. On the contrary, it seems to me to be all shocks, a sudden beginning and gradual subsidence.

I say this, because I have just been reading a very interesting paper by Lommel, I think, or Sellmeier* (both touch upon this) which goes into the thing very fully. Helmholtz remarks that he gets into a little difficulty in his dynamics and does not show clearly what becomes of the energy in a certain case, but he takes hold of the thing with great power. He goes into this case very fully, and in the way with which we are all familiar. He remarks that Fizeau obtained a suite of 50,000 vibrations interfering with one another, and judges from that that ordinary light consists of polarized light, circular or elliptical, or plane polarized as I said to you myself one or two days ago, with (what I did not say) the plane of polarization, or one or both axes of the ellipse if it be elliptically polarized, gradually changing, and the amplitude gradually changes. He says *gradually* and so gradually that there is not so great a change in the course of 50,000, or 100,000, or perhaps several million vibrations in the amplitude or mode of polarization as to prevent interference. In fact, I suppose there is no perceptible difference between the perfectness of the annulments with 50,000 vibrations [and that] with 1,000; although I speak here not with confidence and I may be corrected. You have seen that, have you not Prof. Rowland?

Prof. Rowland: Yes; but it is very difficult to get the interferences.

Sir Wm. Thomson: But when you do get them, the black lines are very black, are they not?

Prof. Rowland: I do not know. They are so very faint that you can hardly see them.

Sir Wm. Thomson: What do you infer from them?

Prof. Rowland: That there is a large number. The width of the lines of

*Sellmeier; *Ann. der Phy. u. Chem.* 1872, vols. 145, 147.

the spectrum indicates how perfectly the light interferes; and with a grating of very fine lines I find exceedingly perfect interference for at least 100,000 periods, I should think.

Sir Wm. Thomson: That goes further than Fizeau. Sellmeier says that probably a great many times 50,000 waves must pass before there can be any great change. He goes at the thing very admirably for the foundation of his dynamical explanation of absorption and anomalous reflection. The only thing [on which] I do not fully agree with him in his fundamentals is the gradualness of the initiation of light at the source. I believe, in the majority of cases at all events, in sudden beginnings, and gradual endings. Prof. Rowland has just told us how gradual the endings are. Fizeau could infer that the amplitude does not fall off greatly in 50,000 vibrations. It is quite possible, from all we know, that the amplitude may fall off considerably in 100,000 vibrations, is it not?

Prof. Rowland: The lines are then very sharp.

Sir Wm. Thomson: It would not depend on the sharpness of the lines, would it?

Prof. Rowland: O, yes. It would draw them out of line.

Sir Wm. Thomson: Would it broaden them out, or throw a little light over a place that should be dark?

Prof. Rowland: It would broaden them out.

Sir Wm. Thomson: It is a very interesting subject; and from the things that have been done by Prof. Rowland and others, we may hope to see if we live a knowledge of the difficulties quite incomprehensibly superior to what we have now. I doubt, however, whether we will live to see knowledge that we can have hardly any conception of now in the way of the extinction of vibrations in reference to light. We are perfectly certain that the diminution of amplitude must be exceedingly small—practically nil—in 1,000 vibrations; we can say that it is practically nil in 50,000 vibrations; we know that it is nearly nil in 100,000 vibrations. Is it practically nil in two or three hundred thousand vibrations, or in several million vibrations? Possibly not. Dynamical considerations come into play here. We shall be able to get a little insight into these things by forming some sort of an idea of the total amount of energy there must be in one vibrator, and what sequences of waves it can supply.

In speaking of Sellmeier's work and Helmholtz's beautiful paper, which is really quite a mathematical gem, I must still say that I think Helmholtz's modification is rather a retrograde step. It is not so perhaps in the mathematical treatment; but at the same time Helmholtz is perfectly aware of the kind of thing that is meant by viscous consumption of energy. He knows perfectly well that that means conversion of energy into heat; and in introducing it he is throwing up the sponge, as it were, so far as the fight with the dynamical problem is concerned. On the other hand, Sellmeier sticks to it and I think Lommel does.

I got another quarter hundred weight of books on the subject last night. I have not read them all through. I opened one of them this forenoon, and exercised myself over a long mathematical paper. I do not think it will help us very much in the mathematics of the subject. What we want is to try and see if we cannot understand more fully what Sellmeier has done, and what Lommel has done. I see that both stick dimly to the idea that we must account for the loss of energy in the vibration of the particles themselves. That is what I am doing; and we shall never have done with it until we have explained every line in Prof. Rowland's splendid spectrum. If we are tired of it, we can rest, and go at it again.

Lommel and Sellmeier do not go into these multiple vibrations, although they take notice of them. But they do indicate that we must find some way of distributing the energy without supposing the consumption of it. That is the reason why I do not like Helmholtz's way of introducing the viscous terms. It is very dangerous, in an ideal sense, to introduce them at all. This little bit of viscosity in one part of the system might run away with all our energies long before 50,000 vibrations. If there were any viscosity connected with the moving particle it might be impossible to get a sequence of one hundred thousand or a million vibrations proceeding from one initial vibration of one vibrator.

What the dynamical problem has to do for us is to show how we can have a system capable of vibrations in itself and acted upon by the luminiferous ether, that under ordinary circumstances does not absorb the light in thousands of vibrations. That may be conceived to be the case with transparent bodies; bodies that allow waves to pass through them one hundred feet or a thousand feet, or much greater distances; transparent bodies with exceedingly little absorption. If we take vibrators, then, that will perform their functions in such a way as to give a proper velocity of propagation for light in a highly transparent body, and yet which, with a proper modification of the magnitudes of the masses or of the connecting springs [; that] will, in certain complex molecules, such as the molecules of some of those compounds that give rise to fluorescence and phosphorescence, take up a large quantity of the energy, so that perhaps the whole suite of vibrations from a single initiation may be absolutely absorbed and converted into vibrations of a much lower period [; and that] will have, lastly, the effect of heating the body, I think we shall see a perfectly clear explanation of absorption without introducing viscous terms at all; and that idea we owe to Sellmeier. I may go for a moment into this subject of arbitrary functions; but perhaps I had better leave it for the present.

I would like, in connection with the idea of explaining absorption and refraction, and lastly anomalous reflection and dispersion, to just point out, as a matter of history, the two names to which this is owing—Stokes and Sellmeier. I would be glad to be corrected with reference to either, if there is any evidence to the contrary; but so far as I am aware, the very first idea of accounting for absorption by vibrating particles taking up the energy in all modes of natural vibration of their own corresponding to the period of the light trying to pass through is from Stokes. He taught it to me at a time that I can fix in one way indisputably. I never was at Cambridge once from about June 1852 to May 1865; and it was at Cambridge walking about on the grounds of the colleges that I learned it from Stokes. Something was published about it from a letter of mine upon it which was put in a postscript by Kirchhoff to the English translation [*Phil. Mag.*, vol. 20, July, 1860] of his own paper on the subject which appeared first in Poggendorff's *Annalen* [vol. CIX, p. 275]. If you have not already read that classical paper of Kirchhoff's, I advise you to look through it at all events, whether you go all through the mathematics or not.

In the postscript you will find the following statement copied from my letter:

"Prof. Stokes mentioned to me at Cambridge some time ago, probably about ten years, that Prof. Miller had made an experiment testing to a very high degree of accuracy the agreement of the double dark line of the solar spectrum, with the double bright line constituting the spectrum of the spirit lamp burning salt. I remarked that there must be some physical connection between two agencies presenting so marked a characteristic in common. He assented, and said he believed a mechanical explanation of the cause was to be had on some such principle as the following: Vapour of sodium must possess by its molecular structure a tendency to vibrate in the periods corresponding to the degrees of refrangibility of the double line D. Hence the presence of sodium in a source of light must tend to originate light of that quality. On the other hand, vapour of sodium in an atmosphere round a source must have a great tendency to retain in itself, i.e. to absorb and to have its temperature raised by light from the source, of the precise quality in question. In the atmosphere around the sun, therefore, there must be present vapour of sodium, which, according to the mechanical explanation thus suggested, being particularly opaque for light of that quality prevents such of it as is omitted from the sun from penetrating to any considerable distance through the surrounding atmosphere. The test of this theory must be had in ascertaining whether or not vapour of sodium has the special absorbing power anticipated. I have the impression that some Frenchman did make this out by experiment, but I can find no reference on the point.

"I am not sure whether Prof. Stokes' suggestion of a mechanical theory has ever appeared in print. I have given it in my lectures regularly for many years, always pointing out along with it that solar and stellar chemistry were to be studied by investigating terrestrial substances giving bright lines in the spectra of artificial flames corresponding to the dark lines of the solar and stellar spectra." *

What I have read this far is not with reference to the origin of spectrum analysis, of which there is ample historical evidence that it was done before these dates, but the definite point of the dynamics of absorption. There is a hint there of the reaction of the vibrating particles in the luminiferous ether. Sellmeier's first title is to that effect; he takes up exactly that view for explaining absorption. He explains ordinary refraction through the inertia of these particles, and he shows how, when the light is nearly of the period corresponding to any of the fundamental periods of the vibrator, there will be anomalous dispersion. He gives a mathematical investigation of the subject, not altogether satisfactory, perhaps, but still it seems to me to form a nearer treatment of the thing. Lord Rayleigh, Helmholtz and others have quoted Sellmeier; Lommel begins afresh, I think, but he notices Sellmeier also, so the thing must have originated there, and it seems to me a very important new departure with respect to the dynamical explantion of light.

Now, let us look at this problem of vibrating particles once more; I have a little question for the ideal arithmetical laboratory. Just try the arithmetical work for this problem for 7 particles. I do not know whether it will work out well or not. I have not the time to do it myself, but perhaps some of you may find the time, and be interested enough in the thing, to do it. Take the m's in order, proceeding by ratios of 4, and the C's in order, proceeding by differences of 1:

$$m_1, m_2, m_3, m_4, m_5, m_6, m_7 = 1, 4, 16, 64, 256, 1024, 4096,$$

$$C_1, C_2, C_3, C_4, C_5, C_6, C_7, C_8 = 1, 2, 3, 4, 5, 6, 7, 8.$$

* [The following is a note appended by Prof. Stokes to his translation of a paper by Kirchhoff in *Phil. Mag.* vol. XIX, March 1860, p. 1986: "The remarkable phenomenon discovered by Foucault, and rediscovered and extended by Kirchhoff, that a body may be at the same time a source of light giving out rays of a definite refragibility, and an absorbing medium extinguishing rays of the same refrangibility which traverse it, seems readily to admit of a dynamical illustration borrowed from sound. We know that a stretched string which on being struck gives out a certain note (suppose its fundamental note) is capable of being thrown into the same state of vibration by aerial vibrations corresponding to the same note. Suppose now a portion of space to contain a great number of such stretched strings forming thus the analogue of a 'medium'. It is evident that such a medium, on being agitated, would give out the note above mentioned, while on the other hand, if that note were sounded in air at a distance, the incident vibrations would throw the strings into vibration and consequently would themselves be gradually extinguished since otherwise there would be a creation of *vis viva*. The optical application of this illustration is too obvious to need comment. G.G.S."—H.]

There will be 7 roots to find by trial. I would like to have some of you try to find some of these, if not all, also the energy ratios. You will probably find it an advantage in the calculation if you proceed thus: We have

$$a_1 = \frac{1}{T^2} - 3, \quad a_2 = \frac{4}{T^2} - 5, \quad a_3 = \frac{16}{T^2} - 7, \quad a_4 = \frac{64}{T^2} - 9,$$

$$a_5 = \frac{256}{T^2} - 11, \quad a_6 = \frac{1024}{T^2} - 13, \quad a_7 = \frac{4096}{T^2} - 15.$$

You will have to take values of T^2 by trial until you get near a root. The convergence of the continued fraction will be so rapid that you will have very little trouble in getting the largest roots in $1/T^2 = z$. Begin then with the largest root, and proceed downwards; and when several of the a's have become negative, alter the expression so as to keep positive quantities. Our standard form is

$$u_i = a_i - \frac{C_{i+1}{}^2}{u_{i+1}}.$$

If a_i is positive, well and good; you will find, at once, u_i a very large number; and to calculate for instance u_1, you may suppose u_4 to be infinite, at the same time supposing u_5 to be infinite, in calculating u_2, u_3. A very few trials will show you how many terms of the continued fraction you must take in order to get u_1 to a certain degree of accuracy. I think, to fix the ideas, and to make the demand for accuracy very moderate, we shall say that our final result shall be within 1/10th per cent, that is, 1/1000th of the absolutely true value. That would be correct to 3 decimal places. I do not want to suggest any elaborate arithmetical calculations. Work it out to four places if you like, so as to be quite sure of the third place. Take any value of z you like and calculate; then take another smaller value and you will soon find one that will make $u_1 = 0$. These are the values that we want, the values of z that make $u_1 = 0$. Take smaller values of z, and you will soon find another; take smaller values of z and you will soon find another. By this time you had better begin making the change that I now suggest, viz.: take

$$w_{i-1} = -u_{i-1}, \quad b_{i-1} = -a_{i-1}.$$

As you diminish z, you see that when z becomes less than 3 the a_1 is negative; if $x < \frac{5}{4}$, a_1 [and] a_2 are negative, etc. Instead of these negative values extending up to x_i say substitute positive values $b_1 = -a_1$, etc., at the same time altering the corresponding u's into $-w$'s. That will diminish the tendency to negative quantities among the w's; although negative values will sometimes occur. Then proceed backwards from w_i: The formula will be

$$w_{i-1} = b_{i-1} - \frac{C_i^2}{w_i},$$

or

$$w_i = \frac{C_i^2}{b_{i-1} - w_{i-1}}.$$

Put that into the form of a continued fraction if you like, but it will be easier to work step by step. Calculate w_i on the supposition that $w_1 = 0$ and u_i on the supposition that $u_{j+1} = \infty$ and equate w_i to $-u_i$—that is the process. If they are not equal, you must alter z. The value of z that makes them equal must be a root of $u_1 = 0$. In the course of the process you will have the whole formation of the u's or the w's for each root; by multiplying these in order, you have the x's for each particular root, and then you can calculate the energy ratios for each root. We shall then be able to put our formula into numbers; and I feel that I understand it much better when it is in numbers than when it is in a literal form.

I want to show you (jumping ahead a little) the explanation of ordinary refraction. Let us go back to our supposition of spherical shells, if you like—our rude mechanical model. Suppose an enormous number of spherical cavities distributed equally through the space we are concerned with. Let the quantity of ether thus displaced be so exceedingly small in proportion to the whole volume that the elastic action of the residue will not be essentially altered by that. These suppositions are perfectly natural. Now, what is unnatural mechanically, is let us suppose a massless spherical lining absolutely rigid to this spherical cavity in the luminiferous ether connected by springs—in the first place symmetrical. We shall try afterwards to see if we cannot do something in the way of aeolotropy; but as I have said before I do not see the way out of the difficulties yet. In the meantime, let us suppose this first shell m_1 to be isotropically connected by springs with the rigid shell lining of the spherical cavity in the ether. When I say isotropically connected I mean distinctly this: that if you draw this first shell aside through a certain distance in any direction, the force will be independent of the direction. Certain springs in the drawing—the smallest number would be four—placed around in proper positions will rudely represent the proper connections for us. Similarly, let there be another shell here, m_2, isotropically connected with the outer one; and so on.

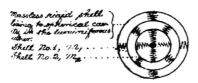

This is the simplest mechanical representation we can give of a molecule or an atom, imbedded in the luminiferous ether, unless we suppose the atom to be absolutely hard, which is out of the question. If we pass from this problem to a problem in which we shall have a continuous connection instead of a series of connections of associated particles, we shall be, of course, much nearer the reality. But the consideration of a group of particles has great advantage, for we are more familiar with common algebra than with the treatment of partial differential equations of the second order with coefficients not constant, but functions of the independent variable—which are the equations we have to deal with if we take a continuous elastic molecule instead of one made up of masses connected by springs as we have been supposing.

Let us suppose these spherical cavities to be exceedingly small in comparison with the wave length. Practically speaking, we suppose our structure to be infinitely fine grained. That will not in the least degree prevent its doing what we want. The distance also from one such cavity, a series of shells, to another in the luminiferous ether is to be exceedingly small in comparison with the wave length, so that the distribution of these molecules through the ether leaves us with a body which is homogeneous when viewed on so coarse a scale as the wave length; but it is, if you like, heterogeneous when viewed with a microscope that will show us the millionth or million-millionth of a wave length. This idea has a great advantage over Cauchy's old method in allowing an infinitely fine grainedness of the structure, instead of being forced to suppose that there are only several molecules, ten or twelve, to the wave length, as we are obliged to do in getting the explanation of refraction by Cauchy's method.

I wish to show you the effect of molecules of that kind upon the velocity of light passing through the medium. Let $m_1/4\pi^2$ denote the sum of all the masses of shell No. 1 in any volume divided by the volume; let $m_2/4\pi^2$ denote the sum of the masses of No. 2 interior shell in any volume divided by the volume; and so on. Or, if you like to say so, let $m_1/4\pi^2$ denote the amount per unit volume of No. 1 shell and so on. We will not put down the equations of motion for all directions, but simply take the equations corresponding to a set of plane waves in the which the direction of the vibration is parallel to Ox. If we denote by $\rho/4\pi^2$ the density of the vibrating medium (I am taking $\rho/4\pi^2$ instead of the usual ρ for the reason you know, viz.: to get rid of the factor $4\pi^2$ resulting from differentiation), and if n be the rigidity of the luminiferous ether, the equation of motion in the ether will be

$$\frac{\rho}{4\pi^2}\frac{d^2\xi}{dt^2} = n\frac{d^2\xi}{dx^2}.$$

Let $\ell/4\pi^2$ instead of n denote the rigidity [of the luminiferous ether], and the dynamical equation of motion will be

$$\frac{\rho}{4\pi^2}\frac{d^2\xi}{dt^2} = \frac{\ell}{4\pi^2}\frac{d^2\xi}{dx^2} + C_1(x_1 - \xi).$$

I will not go into the formal proof just now, for I am going to take up some dynamics comprehending this when we come to the subject of rotation. We shall suppose that we have gyrostatic fly wheels imbedded in these holes or cavities in the luminiferous ether, and we shall then formally go through the dynamical investigation, and see how it is that we have simply to add to the first equation and expression for the force produced by the springs connecting the lining of the cavity with m_1, which will be $C_1(x_1 - \xi)$.

For waves of period T, we have

$$\xi = C \cdot \sin 2\pi \left(\frac{x}{\lambda} - \frac{t}{T} \right).$$

The second differential coefficient of this with respect to t, x will be $-(4\pi^2/T^2)\xi$, $-(4\pi^2/\lambda^2)\xi$ respectively. Therefore our equation becomes

$$\frac{\rho}{T^2} = \frac{\ell}{\lambda^2} + C_1 \left(1 - \frac{x_1}{\xi} \right).$$

Let us find T^2/λ^2, which is the reciprocal of the velocity of propagation. You may write it $1/v^2$ if you like, or μ^2, the refracting index. We have

$$\frac{T^2}{\lambda^2} = \frac{1}{\ell} \left\{ \rho - C_1 T^2 \left(1 - \frac{x_1}{\xi} \right) \right\}.$$

Substitute our value for

$$-\frac{x_1}{\xi} = \frac{C_1 T^2}{m_1} \left(\frac{K_1^2 R_1}{K_1^2 - T^2} + \frac{K_2^2 R_2}{K_2^2 - T^2} + \cdots \right)$$

and this becomes

$$\frac{T^2}{\lambda^2} = \frac{1}{\ell} \left[\rho - C_1 T^2 \left\{ 1 + \frac{C_1 T^2}{m_1} \left(\frac{K_1^2 R_1}{K_1^2 - T^2} + \frac{K_2^2 R_2}{K_2^2 - T^2} + \cdots \right) \right\} \right].$$

This is the expression for the square of the refractive index, as it is affected by the presence of molecules arranged in that way. It is too late to go into this for interpretation just now, but, I will tell you that if you take T considerably less than K_1 and very much greater than K_2 you will get a formula with enough disposable constants to represent the index of refraction by an empirical formula, as it were, which, from what we know, and what Sellmeier and Ketteler have shown, we can accept as ample for representing the refractive index of most transparent substances. We have no means of extending its powers and introducing the effects of these other terms, so that we have a formula which is more than sufficient to give us a mathematical expression of the refrangibility in the case of any transparent body whose refrangibility is reliable.

We shall look into this a little more, and I will point out some of the applications to anomalous dispersion. We must think a good deal of what can become of vibrations in a system of that kind when the period of the vibration of the luminiferous ether is approximately equal to any one of the fundamental periods that the system could have were the shell lining in the ether held absolutely at rest.

Lecture X

We shall now think a little about the propagation of waves with a view to the question, what is the result as regards waves at a distance from the source, those at the source being discontinuous. In the first place, we will take our expression for a plane wave. The expression in our formulas showing diminution of amplitude at a distance from a source does not have an effect when we come to consider plane waves. So we just take the simple expression for plane harmonic waves propagated along the axis of y with velocity v;

$$\xi = a \cos \frac{2\pi}{\lambda} (y - vt).$$

Let us consider this question: What is the work done per period by the elastic force in any plane perpendicular to the line of propagation of the wave? We shall think of the answer to that question with the view to the consideration of the possibility of a series of waves penetrating through space previously quiescent. Suppose I draw a straight line here for the line of propagation and let this curve represent a succession of waves travelling from left to right and penetrating into space previously quiescent.

Take a plane perpendicular to the line of propagation of the waves, and think of the work done by the elastic solid upon one side of this plane upon the elastic solid on the other side, in the course of a period of the vibration. We shall take an expression for the tangential force T of the elastic solid. I am not adhering to our old notation of S, T, U, P, Q, R. We shall virtually investigate here the formula for the propagation of the wave independently of our general formula in three dimensions. Take T to denote the tangential force of the elastic medium on the one side of this plane; the direction of the arrow head which I draw being that direction in which the medium on the left pulls the medium on the right. I put infinitely near that in the medium on the left another arrow head. I cannot do that actually; it is an easy thing to understand, but not a practical thing to do. Imagine for the moment a split in the medium caused by this plane; and imagine the medium on the left taken away, and that you act upon this plane with the same force as in the continuous propagation of waves. The medium upon the left acts in this way upon the plane—that is an easy enough conception. I correctly represent that by an arrow head pointing up infinitely near to the plane on the right hand side, and an arrow head on the left pointing down. The displacement of the medium is determined by a distortion from a square figure to an oblique figure, and there is no inconsistency in putting into this little diagram an exaggeration of the obliquity, so as to show the direction of it:

The force required to do that is clearly upward on the right and downward on the left.

Let us consider now the work done by that force. Calling ξ the displacement of a particle from its mean position, $T \cdot \dot{\xi}$ is the work done by that tangential force per unit of time. $d\xi/dy$ is the shearing strain experienced in the medium, so that $n\, d\xi/dy = -T$. In this particular position which we have taken, ξ increases with y, so that the minus sign is correct according to the arrow heads.

Let there be simple harmonic waves propagated from left to right with velocity v. This is the expression for it [indicating $\xi = a \cos(2\pi/\lambda)\,(y - vt)$]. Hence,

$$\dot{\xi} = \frac{2\pi}{\lambda} va \sin q, \qquad \frac{d\xi}{dy} = -\frac{2\pi}{\lambda} a \sin q;$$

and the rate of doing work is

$$\frac{4\pi^2}{\lambda^2} a^2 v \sin^2 q.$$

That is the rate at which this plane, working on the elastic solid on the right hand side of it, does work ("per unit area of the plane" understood). Multiply this by dt and integrate through a period $T = \lambda/v$. Now

$$\int_0^T \sin^2 q\, dt = \int_0^T \tfrac{1}{2}(1 - \cos^2 q)\, dt = \int_0^T \tfrac{1}{2}\, dt = \tfrac{1}{2} T.$$

The rate of doing work, then, per period, is

$$\frac{2\pi^2}{\lambda^2} a^2 vn T = \frac{2\pi^2 a^2 n}{\lambda}.$$

If it is possible for a set of waves to advance into space previously undisturbed, then it is certain that the work done per period must be equal to the energy in the medium per wave length. Let us then work out the energy per wave length.

It is easily proved that the energy is half potential of elastic stress, and half kinetic energy; and it will shorten the matter simply to calculate the kinetic energy and double it, taking that as the energy in the medium per wave length. In our notation of yesterday, we took $\rho/4\pi^2$ as the density. Multiply this by dy, to get the mass of an infinitesimal portion (per unit of area in the plane of the wave). The kinetic energy of this mass is

$$\frac{1}{2}\frac{\rho}{4\pi^2} dy \, \dot\xi^2 = \frac{1}{2}\frac{\rho v^2 a^2}{\lambda^2} \sin^2 q \cdot dy.$$

Integrating this through a wave length ($\int_0^T \cos^2 q \, dy = 0$) and doubling it so as to get the whole energy, we have $\frac{1}{2}(\rho v^2 a^2/\lambda)$. Compare that with the work done per period, viz.: $\frac{1}{2}(a^2/\lambda)\ell$ if $\ell/4\pi^2$ be as yesterday the rigidity instead of n. That gives us correctly the velocity, $v = \sqrt{k/\rho}$. Thus the work done per period is equal to the energy per wave length.

We must not infer from this that it is possible for a discontinuous series of waves to be propagated into the elastic medium, previously quiescent. But if this did not verify, it would be impossible to have such a series of waves propagated forward without change of form into a medium previously quiescent. I wanted to verify that case, because for a moment, we shall alter to a case in which this is not verified; that is to say, when we put in our molecules. In that case, the work done per period is less than the energy in the medium per wave length, and therefore it is impossible for the waves to advance without change of form.

Before we go on to that, let us stay a little longer in an undisturbed elastic solid, and look at the well known solution by discontinuous functions. The equation of motion is

$$\rho\frac{d^2\xi}{dt^2} = \ell\frac{d^2\xi}{dy^2}.$$

Although I said I would not formally prove this now, it is in reality proved by our old equation

$$\rho\frac{d^2\xi}{dt^2} = \ell\nabla^2\xi.$$

I took the liberty of asking Prof. Ball two days ago whether he had a name for this symbol ∇^2; and he has mentioned to me *nabla*, a humorous suggestion of Maxwell's. It is the name of an Egyptian harp which was of that shape. I do not know that it is a bad name for it. Laplacian I do not like for several reasons both historical and phonetical.

I should have told you that this is the case of a plane wave propagated in the direction of OY, with the plane of the wave parallel to xz; for which case, *nabla* of ξ becomes simply $d^2\xi/dy^2$. The time honored solution of this equation is

$$\xi = f(y - vt) + F(y + vt),$$

where f and F are arbitrary functions. You can verify that by differentiation. This solution in arbitrary functions proves that a discontinuous series is possible; and knowing that a discontinuous series is possible, you could tell without working it out, that the work done per period by the medium on the one side of the plane which you take perpendicular to the line of propagation must be equal to the energy of the medium per wave length.

Before passing on to the energy solution for the case in which we have attached molecules, in which this equality of energy and work does not hold, with the result that you cannot get the discontinuous series, I want to suggest another elementary exercise for the anticipated arithmetical laboratory. It is to illustrate the propagation of waves in a medium in which the velocity is not independent of the wave length, and to contrast that with the propagation of waves when the velocity is independent of the wave length in order that you may feel for yourselves what these two or three symbols show us, but which we need to look at from a good many points of view before we can make it our own and understand it thoroughly. To realize that this equation $E = F$ gives us constant velocities for all wave lengths and that constant velocities for all wave lengths implies this equation, and to see that that goes along with the propagation of a discontinuous pulsation without change of figure, or a discontinuous succession of pulsations without change of character, I want an illustration of it, and also of the case in which the conditions of constancy of velocity for different wave lengths are not fulfilled.

I ask you first to notice the formula

$$S = \frac{\frac{1}{2}(1 - e^2)}{1 - 2e\cos q + e^2} = \frac{1}{2} + e\cos q + e^2\cos 2q + \cdots$$

which is familiar to all mathematical readers as leading up to Fourier's harmonic series of sines and cosines. Poisson and others make this series the foundation of a demonstration of Fourier's theorem. It is proved by taking $2\cos q = e^{iq} + e^{-iq}$ and resolving S into partial fractions. If $e < 1$ the series is convergent; when $e = 1$ it ceases to converge. If we take $q = (2\pi/\lambda)(y - vt)$ and draw the curve whose dependent coordinate is S, what have we? Take

$t = 0$ and measure off lengths from the origin $y = \lambda$, $2\lambda \cdots$. The curve represented will be this (heavy curve):

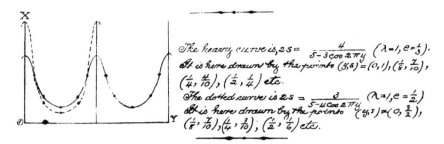

The heavy curve is, $2S = \dfrac{4}{5 - 3\cos 2\pi y}$ $\left(\lambda = 1, e = \frac{1}{3}\right)$. It is here drawn by the points $(y, s) = (0, 1), \left(\frac{1}{8}, \frac{7}{10}\right),$ $\left(\frac{1}{4}, \frac{4}{10}\right), \left(\frac{1}{2}, \frac{1}{4}\right)$ etc.
The dotted curve is $2S = \dfrac{3}{5 - 4\cos 2\pi y}$ $\left(\lambda = 1, e = \frac{1}{2}\right)$ It is here drawn by the points $(y, s) = (0, \frac{3}{2}),$ $\left(\frac{1}{8}, \frac{7}{10}\right), \left(\frac{1}{4}, \frac{3}{10}\right), \left(\frac{1}{2}, \frac{1}{6}\right)$ etc.

If you take any other value of t than zero, you merely shift the curve as it were along the axis of y. I want the arithmetical laboratory to work this out and give a graphic representation of the periodic curve for one or two different values of e if you like. Perhaps a dozen equal difference values of y will be more than enough to trace a good curve corresponding to this equation. The particular numerical case that I am going to suggest is one in which the curve will be more like this second curve which I draw (dotted curve); it is much steeper and comes down more nearly to zero. Take the extreme case of $e = 1$, and what happens? S is infinitely great for q infinitely small, and infinitely small for all other values of $q < \lambda$. I suggest to work this out for $e = (.1)^{1/10}$. The coefficient in the tenth term will be a tenth part of the coefficient in the first term. On the other hand, if you take e something smaller, say $\frac{1}{2}$, the series will converge so rapidly that long before the tenth term occurs the terms will be too small for any calculation that I would recommend to the arithmetical laboratory.

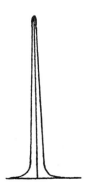

There will be no necessity to calculate the terms of this series if you have no other object than to trace this curve. Take the curve

$$2S = \frac{1 - e^2}{1 - 2e\cos q + e^2}.$$

For $y = 0$, $q = 0$ and $2S = (1 + e)/(1 - e)$; for $y = \frac{1}{2}\lambda$, $q = \pi$ [and] $2S = (1 - e)/(1 + e)$. Now, the tenth root of .1 is nearly .8, and corresponding to this value of e the maximum value of $2S$ is 9, and the minimum value is $\frac{1}{9}$; so that the case I have suggested makes the height at the origin about 81 times the minimum height here. If you want to get a still more telling expression, take a value of e still nearer unity. This problem is worth working out in itself and I would advise those also who have time to read Poisson's and Cauchy's great papers in connection with it. [Poisson, Mémoire sur la théorie des ondes. Paris, *Mem. Acad. Sci.* I, 1816, pp. 71–186; *Annal de Chemie* V, 1817, pp. 122–142. Cauchy, Mémoire sur la théorie de la propagation des ondes à la surface d'un fluide pesant d'une profondeur indéfinie [1815] Paris, *Mem. Sav. E. Étrang.* I, 1827, pp. 3–312.] Those papers are exceedingly fine pieces of paper mathematics; but they are very strong. You might have the hydrodynamical beginnings presented much more fascinatingly. If you know the theory of deep-sea waves, well and good; then take Poisson and Cauchy. Those who do not know the theory of deep-sea waves may read it up in elementary books. The best book I know is Lamb's *Hydrodynamics*. The great struggle of 1815 (that is not the same idea as la grande guerre de 1815) was, who was to rule the waves, Cauchy or Poisson. Their two memoirs seem to me of very nearly equal merit. I have no doubt the judges had some particular reason for giving the award to Cauchy, but Poisson's paper is splendid. I can see that the two writers respected each other very much and I suppose each thought the other's work as good as his own.

I want to know a little more myself whether or not we can get from this series a graphic representation of the effect of a single disturbance at sea— such a disturbance as that of throwing a stone in a deep sea. I believe there are quite valid solutions to be obtained, but there are difficulties, such as questions of convergency, and so on. That is the problem. I believe they did; it constitutes the largest part of their papers; but they go into it in the high analytical style of letting the initial condition be quite arbitrarily chosen. Every portion of an infinite area of water is started initially with a stated infinitesimal displacement from the level and a stated velocity up and down from the level, and the inquiry is, what will be the result? The solution of this constitutes the problem; but it is obvious that you have the solution of that problem from the more elementary problem: what is the result of an infinitesimal displacement at a single point, which may just as well be produced by throwing in a stone as in any other way? Let a solid, say, cause a depression in any place, the velocity of the solid performing the part of giving velocity to the particles of water, and then suddenly consider the solid annulled. The same thing in two dimensions is exceedingly simple. Take, for example, waves in an infinitely deep canal with vertical sides. Take a sudden disturbance in the canal equal all along the breadth of the canal, and inquire what will the result be. That leads toward an understanding of Cauchy's and Poisson's solution and I think it would repay any one who is inclined to go into the subject to work it out theoretically and make graphic representations. Poisson and Cauchy only give figures and do not give graphical representations.

I am going to suggest to the arithmetical laboratory to take the case of v the velocity dependent on the wave length. Let us take this as the arithmetical problem: The curve to be drawn for

$$S = \tfrac{1}{2} + e\cos q_1 + e^2\cos 2q_2 + e^3\cos 3q_3 + \cdots$$

where

$$q_i = \frac{2\pi}{\lambda}(y - v_i t).$$

For a particular case take $v_i = 1/\sqrt{i}$, and calculate the curve corresponding to any values you please of t. First give t a small value corresponding say to the time when you have a velocity 1. You might, for example, take for the first case $t = \tfrac{1}{4}$. You will find the result will be a shifting of this curve to one side about a quarter of a wave length. Then try the cases $t = 1, 2$, etc., calculating enough of the terms of this series to give you a fairly representative curve. It is not a thing that can be done quickly. It is worth putting a good deal of labor upon, and I mean myself to do it, putting the calculation into the hands of some of my assistants who will be glad to work out what I think will be a somewhat valuable representation of this interesting property.

We are going to take our molecules again and put them in the ether and look at the question a little more, what is the velocity of propagation under some suppositions which we shall make as to the masses of these attached molecules, and how much it will modify the velocity of propagation from what it would be if there were no molecules. Then we shall look at the matter, with no more work to do upon it with respect to the question of the work done upon a plane perpendicular to the line of propagation; and we shall see that the energy per wave length is much greater than the work done per period and that therefore it is impossible under these conditions for waves to spread into space previously occupied by quiescent matter.

You will find in Lord Rayleigh's book on sound the question of the work done per period and the energy per wave length gone into and the application of this principle with respect to the possibility of independent suites of waves travelling without change of form as thoroughly pointed out.

To-morrow we shall consider a piece of work that looks to the difference of velocity of propagation in different directions in an aeolotropic elastic solid for the foundation of the explanation of double refraction on the pure elastic solid idea. The thing is quite familiar to many of you, no doubt, and you also know that it is a failure in regard to the explanation of the propagation of light in biaxial crystals. It is, however, an important piece of physical dynamics, and I shall touch upon it a little, and try to show it in as simple a point of view as I can.

Now for our proper molecular question. The distance from cavity to cavity in the ether is to be exceedingly small in comparison with the wave length, and the diameter of each cavity is to be exceedingly small in comparison with the distance from cavity to cavity. Let the lining of the cavity be an ideally absolutely rigid massless shell. Let the next shell be an absolutely rigid shell of mass $m_1/4\pi^2$. I represent the thing as if we had just two of these shells and a solid nucleus.

The enormous mass of the matter of the grosser kind which exists in the luminiferous ether, or even of such a comparatively non-dense body as air, would bring us at once to very great numbers in respect to the masses which we will suppose inside this cavity in comparison with the masses of comparable bulks of the luminiferous ether. If there is time to-morrow, we shall look a little to the possible suppositions as [to] the density of the luminiferous ether —what limits of greatness or smallness are conceivable in respect to it. At present we have enough to go upon to show that even in air of ordinary density, the mass of air per cubic centimeter must be enormously great in comparison with the mass of the luminiferous ether per cubic centimeter. We must have something enormously massive in the interior of these cavities. We shall think a good deal of this yet, [and] try and find how it is we can have the large quantity of energy that is necessary to account for the heating of a body such as water by the passage of light through it; or for the phosphorescence of a body which is luminous for several days after it has been excited by light. I do not think we shall have the slightest difficulty in explaining these things. These are not the difficulties. The difficulties of the wave theory of light are difficulties which do not strike the popular imagination at all. These are the difficulties of accounting for polarization by reflection with the right amount of light reflected, and [of accounting] for double refraction. With the phenomena we have no difficulty whatever; the great difficulty in respect to the wave theory of light is to bring out the proper quantities in these effects.

People seem to think the luminiferous ether a fanciful idea. I wish to give another illustration besides shoemaker's wax; I ask you to think of glycerine. Glycerine is a substance without any coarse structure; it is molecularly fine. Glycerine takes its level if you pour it out, as accurately as water or mercury does, yet if you suddenly change its shape, it springs back. Many of you may remember Maxwell's beautiful experiments in which the effects of strain on polarized light was shown in a liquid, or rather a body which, if you give it time, takes its level absolutely, and yet if you strike it quickly, it springs back; Canada Balsam was the substance. Any of you who have any desire to do so may try the experiment. Put a stick in Canada Balsam, get the proper polarizing appliances, make a sudden turn with the stick and you will see the optical effects of double refraction produced and gradually fading away.

This is a digression from my subject, but I do not want to part from you without letting you know all I can in the way of helping you think of the luminiferous ether as a reality, and that we are speaking of real bodies and that this is not a mystification of the mind.

There is no difficulty in explaining the energy required for heating a body by radiant heat pressing through it, nor how it is that it sometimes comes out as visible light and, it may be, not so fast but that we may get light for two or three days. All these properties, remarkable as they are, seem to come out as a matter of course from the dynamical consideration. So much so that any one not knowing these phenomena would have discovered them on working out these things dynamically. He would discover anomalous dispersion, fluorescence, phosphorescence and the phosphorescence corresponding to lower periods, consisting in the heating of a body and then afterwards giving that out as heat. All these phenomena might have been discovered by dynamics; and the dynamical treatment that discovers what is afterwards verified by experiment is a very competent piece of dynamics.

I speak with confidence in this subject because it is a matter of fact. I am ashamed to say that I never heard of anomalous dispersion until after I found it lurking in the formulas. I said to myself, "these formulas would imply that, and I never have heard of it." And when I looked into the matter I found to my shame that a thing which had been known by others for eight or ten years I had not known until I found it in the dynamics.

Take our formula which we had yesterday,

$$\frac{\rho}{4\pi^2} \frac{d^2\xi}{dt^2} = \frac{\ell}{4\pi^2} \cdot \frac{d^2\xi}{dy^2} - C_1(\xi - x_1)$$

and try this with some simple harmonic motion,

$$\xi = a \sin 2\pi \left(\frac{y}{\lambda} - \frac{t}{T} \right).$$

From this we find

$$\frac{\rho}{T^2} = \frac{\ell}{\lambda^2} + C_1\left(1 - \frac{x_1}{\xi}\right),$$

which solved for the refractive index gives

$$\frac{\ell T^2}{\rho\lambda^2} = 1 - \frac{C_1}{\rho}\left(1 - \frac{x_1}{\xi}\right)T^2.$$

I want to take our formula for $-x_1/\xi$ in order to find out what position the period T may have among the fundamental periods of the vibrator on the supposition of the bounding shell held fixed, to give us a good reasonable explanation of dispersion, something in accord with the facts of observation with respect to the difference of velocity for different periods. I will not introduce the energy ratios just now, because we have not time to use them, and I will just take

$$-\frac{x_1}{\xi} = \left(\frac{q_1}{K_1{}^2 - T^2} + \frac{q_2}{K_2{}^2 - T^2} + \cdots\right)T^2.$$

In a medium which is denser than the luminiferous ether, the refractive index is always greater, the velocity smaller, If T were less than the smallest of the fundamental periods, $-x_1/\xi$ would be positive and the refractive index would be less than unity. But in all known cases the refractive index is greater than unity; therefore $-x_1/\xi$ must be negative. Take then this formula:

$$-\frac{x_1}{\xi} = \left(\frac{-q_1}{T^2 - K_1{}^2} + \frac{q_2}{K_2{}^2 - T^2} + \cdots\right).$$

In other words, we shall suppose the period T to be intermediate between the smallest and the next to the smallest of the fundamental periods, K_1, K_2. I want to see if we can get out of this a formula which will cover a range, including all light from the highest ultraviolet photographic light of about $\frac{1}{2}$ the wave length of sodium light down to the lowest we know of, which is the radiant heat from a Leslie cube with a wave length that I hear from Prof. Langley since I spoke on the subject about a week ago of about $\frac{1}{1000}$ of a centimeter or 17 times the wave length of sodium light. That will be a range of about $40:1$. The highest chemical light has a period about $\frac{1}{40}$ part of the period of the lowest invisible radiation of a radiant heat that has yet been experimented upon.

It is conceivingly possible that there are some mediums throughout every part of that range for which there are no anomalous dispersions. I think it is almost certain that for rock salt in the lower part of the range, there are no anomalous dispersions at all. In fact Langley's experiments in radiant heat are made with rock salt; and in all experiments made with rock salt, it seems as if little or no radiant heat is absorbed by it. At all events we could not be satisfied unless we can show that this kind of supposition will account for dispersion through a range of period from one to forty. It is obvious that if we are to have continuous refraction without anomalous dispersion through a wide range, T must not exceed another period. K_2 must then be 40 times as great as K_1.

If we substitute our value of $-x_1/\xi$ and work it out algebraically, we shall find

$$\mu^2 = 1 + \frac{C_1}{\rho} \left\{ q_1 K_1{}^2 - (1 - q_1) T^2 + q_1 K_1{}^2 \left(\frac{K_1{}^2}{T^2} + \frac{K_1{}^4}{T^4} + \cdots \right) \right.$$

$$\left. - (\text{terms involving } q_2, q_3 \cdots) \right\}.$$

q_1 is essentially less than unity. To agree with anything we know, q_1, $K_1{}^2$ must be large in comparison with $(1 - q_1) T^2$. This term $(1 - q_1)$ must be so small that an exceedingly large multiplication of it (for instance corresponding say to the range from the sodium D line to the lowest radiant heat $= 17^2$), must not have any very serious effect; it may be a correction upon the other terms but it must be small. We have here two disposable constants q_1, K_1. I shall look at this a little more carefully to-morrow, and think perhaps, of numerical solutions of our continued fraction and how it is we can suppose q_1 very nearly unity—I think within $\frac{1}{100,000}$ of unity.

What will that mean? That the springs between the rigid shell lining and m_1 are so strong that the static displacement of the lining (with the center of mass held at rest) makes the displacement of m_1 very nearly equal to the displacement of the lining. If you [displace] the lining to one side, m_1 will be displaced somewhat less than the lining; m_2 somewhat less than m_1; and so on. If we suppose the displacement of the lining to be exceedingly little greater than the displacement of m_1 we get an expression that will be applicable to the case.

We shall study this a little more to-morrow and think of what we can make of the graver and graver modes. Although I cannot promise you much light upon it, we must think of it in connection with this question: Suppose you give a slight shock to the lining and hold it fixed; then sometime after give another slight shock to the lining and hold it fixed, and so on; what will be the disposition of the energy? How will it creep inwards among the masses? I think that our arithmetical work will help us to see our way to the answer to some of these questions; and through them we shall be able to form perfectly definite dynamical notions of fluorescence and phosphorescence and anomalous dispersion.

Lecture XI

We shall now take up the subject of an elastic solid which is not isotropic. As I said yesterday, we do not find the consideration of the homogeneous elastic solid satisfactory or successful for explaining the properties of crystals with reference to light. It is, however, to my mind quite essential that we should understand all that is to be known about homogeneous elastic solids and waves in them, in order that we may contrast waves of light in a crystal with waves in a homogeneous elastic solid. It is one of the interesting theories in physical science to know the possibilities of aeolotropy.

Aeolotropy is in analogy with Cauchy's word isotropy which means equal properties in all directions. The formation of a word to represent that which is not isotropic was a question of some interest to those who had to speak of these subjects. I see the Germans have adopted the term anisotropy. Thus they would have us say: "An anisotropic solid is not an isotropic solid"; and this jangle between the prefix *an* and the article *an*, if nothing else, would prevent us from adopting that method of distinguishing a non-isotropic solid from one which is. I consulted Prof. [Lushington] and we had a good deal of talk over the subject. He gave me several charming Greek illustrations and we wound up on the word aeolotropy. Prof. [Lushington] pointed out that *aeolos* means variegated; and it is interesting that the Greeks used the word variegated in respect to shape, color, and time. There is no doubt of the classical propriety of the word and it has turned out very convenient in science. That which is different in different directions, or is variegated according to direction is called aeolotropy.

The consequences of aeolotropy upon the motion of waves or the equilibrium of particles in an elastic solid is an exceedingly interesting and a fundamental subject in physical science; so that there is no apology in making it a subject here except, perhaps, that it is too well known. On that account I shall be very brief and merely call attention to two or three fundamental points. I am going to take up presently, as a branch of molar dynamics, the actual propagation of a wave; and in the mathematical investigation, I am going to give you nothing but what is true of the propagation of a plane wave in an elastic solid, not limited to any particular condition of aeolotropy, but in an elastic solid which has aeolotropy of the most general kind.

Before doing that, which is strictly a problem of continuous or molar dynamics, I want to touch upon the somewhat cloud-land molecular beginning of the subject, and refer you back to the old papers of Navier and Poisson, in which the laws of equilibrium or motion of an elastic solid were worked out from the consideration of points mutually influencing one another with forces [which are] functions of the distance. There can be no doubt of the mathematical validity of investigations of that kind and of their interest in connection with molecular views of matter; but we have long passed away from the stage in which Father Boscovich is accepted as being the originator of a correct representation of the ultimate nature of matter and force. Still, there is a never ending interest in the definite mathematical problem of the equilibrium of motion of a set of points endowed with inertia and mutually acting upon one another with any given force. We cannot but be conscious of the one grand application of that problem to what used to be called physical astronomy but which is more properly called dynamical astronomy, or the motions of the heavenly bodies. We have cases in which we have these motions instead of the approximate equilibriums or infinitesimal motions which form the subject of the special molecular dynamics that I am now alluding to.

All writers who have worked upon this subject have come upon a certain definite relation or set of relations between moduluses of elasticity which seemed to them essential to the hypothesis that matter consists of particles acting upon one another with mutual forces, and that the elasticity of a solid is the manifestation of the force required to hold the particles displaced infinitesimally from the position in which the mutual forces will balance. This, which is sometimes called Navier's relation, sometimes Poisson's relation, and in connection with which we have the well known Poisson's ratio, I want to show you is not an essential of the hypothesis in question. The result for the case of an isotropic body is a most interesting one; doubtless most of you know it; it is in Thomson & Tait, and I suppose in every elementary book upon the subject. I will just repeat it.

An isotropic solid, according to Navier's or Poisson's theory, would fulfil the following condition: if a column of it were pulled lengthwise: the lateral dimensions would be shortened by one half the proportion that the length is added to; and the area of a cross section would therefore be reduced in the double ratio or would be a quarter of the elongation. Stokes called attention to the viciousness of this conclusion as a practical matter in respect to the realities of elastic solids. He pointed out that jelly and india-rubber and the like, instead of exhibiting lateral shrinkage to the extent of one-quarter of the elongation, gave only enough shrinkage to cause no reduction in value at all. That is to say, india-rubber and such bodies vary the area of the cross section in inverse proportion to the elongation so that the product of the length into the area of the cross section may remain constant.

Stokes also referred to a promise that I made, I think it was in the year 1856, to the effect that out of matter fulfilling Poisson's condition a model may be made of an elastic solid which, when the scale of parts is sufficiently reduced, will be a homogeneous elastic solid not fulfilling Poisson's condition. Stokes refers to that promise of mine, which was made very nearly 30 years ago. I propose this moment to fulfil it never having done so before. It is a very simple affair.

Let [a] box represent a rectangular parallelepipedon. The kind of elastic model I am going to suppose is this: a set of particles arranged symmetrically in rectangular order and connected by springs in a certain definite way. I am going to show you that we can connect 8 particles in the interior of an elastic solid with a sufficient number of springs to fulfil the condition of giving 18 independent moduluses; then, by transforming the coordinates from a portion of the solid made up in this partially symmetrical manner with respect to the axes to a portion of the solid taken at random, we get the celebrated 21 coefficients, or moduluses of Green. I suppose you all know that Green took a shortcut to the truth; he did not go into the physics of the thing at all, but simply took the general quadratic expression for energy with its 21 independent coefficients as the most general supposition that can be made with regard to an elastic solid.

To make a model of a solid having the 21 independent coefficients of Green's theory, think of how many disposable springs we have with which to connect 8 different particles. Let them be connected first along the 12 edges of the parallelepipedon. That clearly will not be sufficient to give any rigidity of figure whatever, so far as distortions in the principle planes are concerned. These 12 springs connecting in this way these 8 particles would give a resistance to elongation in the directions of the axes but no resistance whatever to obliquity; you could easily change it from rectangular into an oblique figure. What then must we have to give resistance to obliquity? We can connect coplanar particles diagonally. We have, in the first place, the two diagonals in each face, although the two will virtually count as but one; and then we have the four body diagonals.

Now let me see how many disposables we have got. Remark that each edge is common to four parallelepipedons. I am not going to duplicate our points. We might do it, I suppose, and build up our elastic solid in that way; but I would rather suppose these to be 8 particles of which I show the connections among themselves but not the similar connections with their neighbors in other directions. Each edge being common to four parallel-epipedons, we have only a quarter of the number of edge springs independently available.* Therefore we have virtually three disposables from the edge springs. Each force is common to two parallelepipedons; therefore from the two diagonals in each face we have only one disposable, making in the six faces six disposables. We have the four body diagonals not common to any other parallelepipedons and therefore four disposables from them. We have now 13 disposables, and I want two more. These are the two proportions of the figure, the ratios of the three principle edges. These 15 disposables are all we can absolutely get by springs arranged in this manner. We want three more, observe, in order to make up the eighteen. How I thought of the way to get the three is this:

Navier's and Poisson's theory gave an essential relation between the compressibility and the rigidity and made an incompressible elastic solid impossible. It is curious that they did not notice that jelly is practically incompressible. It is a wonder that they did not try it, and see that it did not fulfil Poisson's ratio. Their mistake was due to the vicious habit in those days of not using examples and diagrams. In the *Mécanique Céleste* you find no diagrams, nor in Langrange, nor in Poisson's splendid memoir on waves. I think I refer to it in Thomson and Tait, that if Lagrange had been in the habit of making diagrams, he never would have given out the proposition that whereas a bell is in stable equilibrium in the bottom of an elliptical dish cover, turned with the mouth up, it is in unstable equilibrium in the bottom of a cylindrical bowl. If they had been in the habit of using diagrams and thinking of their symbols more than they were, Lagrange would never have fallen into that mistake and Poisson and Navier would have found that jelly is enormously more non-compressible than their theory would make it.

What I want is to get a condition of compressibility. I must find some other disposables that will enable me to give any compressibility I please in the case of an isotropic solid. Take our 15 disposables, and reduce them down to the case of an isotropic solid, and we find that an isotropic solid made up in this way will have an absolutely definite compressibility; we can not make the compressibility what we please. We must put in something that can make it incompressible or have any compressibility we please, so that we can make our theory fit either cork or india-rubber, the extremes of natural bodies. I must confess that it is the most difficult thing in it, after I got the idea, to run a cord twice around the 12 edges of a parallelepipedon. Here you see the problem solved by these cords running around the edges of this parallelepipedon through a ring in each of the 8 corners. It cannot be done symmetrically, that is a mathematical proposition—at least I suppose it is. But just follow the cord and we will find how to do it. In fact

*[In other words, the eight connected particles forming a model of the whole medium, the bodily translation of the spring connections in the medium must give the same model—merely translated parallel springs must be equal.—H.]

I am finding out how to do it again in a certain way, myself. The following is an arrangement of the corners along the cord in succession given by their coordinates: (000) (001) (011) (010) (000) (001) (011) (010) (000) (100) (110) (010) (110) (111) (011) (111) (101) (001) (101) (111) (110) (100) (101) (100). There are plenty of other ways of doing it, but this is one way. We have got a cord thrice through each of these 8 points and the thing is done.

Suppose, for example, we wanted to make a condition of incompressibility; let this be an inextensible cord and [the] thing is done. But some one may say that we have not done it without introducing a flexible body. I will not admit any objection to this being a purely mechanical model, because we have that inextensible and perfectly flexible cord running around through hooks; but it is interesting to notice that we can do it without introducing a flexible body at all. We can do it with nothing but rigid bodies. Instead of a cord passing through rings, take wire with bell cranks everywhere where that cord bends around a corner and the thing is done. Thus by proper bell cranks fixed at the corners, and inextensible cords connecting them, you have fulfilled the condition that the sum of the 12 edges shall be constant, which, in the circumstance of being infinitely nearly a rectangular figure in all the distortions that we have, is equivalent to saying that the volume is constant.

To see that this gives us the requisite disposables, let the portions of the cords along the 12 edges be of different elasticities. That gives us 3 disposables, each edge being common to 4 others. To speak in mechanical language, let us connect the bell cranks by springs of different strengths in the directions of the three principal edges. When the body is in equilibrium, there is no pull on the springs. Each one of the 16 different independent springs that we have now got will be called into play by a perfectly general displacement of infinitely small amount. We have 18 available quantities, which will make by solution of linear equations the required 18 moduluses. Then, as I have said, with the transformation of our solid to rectangular axes in any direction, you have a solid fulfilling Green's conditions in the most general way.

Now, observe, Poisson and Navier give us the means of making a bell crank, although they do not give us means of making a jelly. They give us the means of making an elastic zig-zag spring. We can take solids fulfilling their theory and make bell cranks and springs out of them. Put these together, make the parts small enough and the number of them great enough, and you have a homogeneous elastic solid constructed out of parts satisfying Poisson's law, which, as a whole, does not satisfy it.

Although the molecular constitution of solids supposed in these remarks and mechanically illustrated in our model is not to be accepted as true in nature, still the construction of a mechanical model of this kind is undoubtedly very instructive, and we could not be satisfied unless we could see our way to make a model with the 18 independent moduluses. My object is to show how to make a mechanical model which shall fulfil the conditions required in the physical phenomena that we are considering, whatever they may be. At the time when we are considering the phenomenon of elasticity in solids, I want to show a model of that. At another time, when we have vibrations of light to consider, I want to show a model of the action exhibited in that phenomenon. We want to understand the whole about it; we only understand a part. It seems to me that the test of "Do we or not understand a particular subject in physics?" is, "Can we make a mechanical model of it?" I have an immense admiration for Maxwell's mechanical model of electro-magnetic induction. He makes a model that does all the wonderful things that electricity does in inducing currents, etc., and there can be no doubt that a mechanical model of that kind is immensely instructive and is a step towards a definite mechanical theory of electro-magnetism.

I want now to go through a piece of mathematical work which, so far as I know, is not given anywhere except in the articles on elasticity in the *Encyclopaedia Britannica*, although nearly the same was given first by Green. Green investigates the propagation of a wave in an elastic solid, but not in a perfectly general elastic solid. He gave it a certain degree of symmetry before he began this investigation; but he need not have done so. The investigation would have been almost letter for letter the same if he had made it before instead of after introducing the effects of symmetry. The investigation I refer to is that of the propagation of a plane wave—the most general possible kind of a plane wave. Green does it the same way that I am doing it, but with this difference: that I make absolutely no supposition regarding simplification by symmetrical qualities of the solid.

A plane wave in a homogeneous elastic solid is a motion in which every line of particles in a plane parallel to one fixed plane experiences simply a motion of translation—but a motion differing from the motions of particles in planes parallel to the same. Let OX be perpendicular to the plane we are going to consider. Let $x + u, y + v, z + w$ be the coordinates at the time t of a particle which, if the solid were free from strain, would be at (x, y, z). I will keep the same notation as in this article in the *Encyclopaedia Britannica*.

The strain of the solid is the resultant of a simple longitudinal strain in the direction OX, numerically equal to du/dx, and two slips parallel to OY, OZ. The motion of one plane relative to another may be thought of thus: Suppose these two books represent planes perpendicular to OX. The one part of the motion represented by u gives us a strain $= du/dx$. If, for all values of x, u is the same, the result will be only that the whole solid is pushed along. The strain, that is, the change of relative position of different parts of the solid, is expressed, so far as this part is concerned, by du/dx in the regular notation. That is the simple longitudinal strain in the direction of OX. Think now what happens parallel to OY, viz. a slipping represented by these two books slipping past each other. The two other components then are shears corresponding to dv/dx parallel to OY and dw/dx parallel to OZ. The values of these shears, according to a general principle of evaluation of strains given in this paper, are not to be reckoned by dv/dx [and] dw/dx, the simple shears. We take as unit shear the one in which the angle of distortion is $1/\sqrt{2}$, not 1, in the ordinary notation of a shear. A shear consisting in the change of shape of a square is normally represented by that angle in radians which is the diminution of one pair of right angles and the augmentation of the other pair. A simple distortion of strain, upon the principle set forth in this paper, is reckoned in terms of another unit, a unit in which $1/\sqrt{2}$ would be the unit shear without anything more than infinitesimal shears admitted. Therefore $\sqrt{2}\,dv/dx$ [and] $\sqrt{2}\,dw/dx$ are the numerical measures of the shears represented by dv/dx [and] dw/dx. I cannot go into the reasoning just now. You will find [it] distinctly set forth in Chapter X of this little article. I will just read Cor. 4 of that chapter:

"Cor. 4. A definite stress of some particular type chosen arbitrarily may be called unity; and then the numerical reckoning of all strains and stresses becomes perfectly definite." Ordinarily we choose as we please the unit. I have a reason for making all depend upon the unit which is chosen for one particular method of strain, which is fully set forth here. That is a proposition to be proved and made clear by illustrations. That being set forth, it remains for us to choose our unit. Following upon the proposition is this definition: "Def. A uniform pressure or tension in parallel lines, amounting in intensity to the unit of force per unit of area normal to it, will be called a stress of unit magnitude, and will be reckoned as positive when it is tension, and negative when pressure."

That definition being laid down, the previous proposition shows that we are no longer at liberty to represent a simple distortion by saying that it is the change of this right angle rather than some other change as for instance the elongation of the diagonal. I have two other sentences to read, so as to make my formula complete: "(4) A stress compounded of unit pressure in one direction and an equal tension in a direction at right angles to it, or, which is the same thing, a stress compounded of two balancing couples of unit tangential tensions in planes at angles of 45° to the direction of those forces, and at right angles to one another, amounts in magnitude to $\sqrt{2}$." "(5) A strain compounded of a simple longitudinal extension x, and a simple longitudinal condensation of equal absolute value, in a direction perpendicular to it, is a strain of magnitude $x\sqrt{2}$; or, which is the same thing $(\sigma = 2x)$, a simple distortion such that the relative motion of two planes at unit distances parallel to either of the planes bisecting the angles between the two planes mentioned above, in a motion σ parallel to themselves, is a strain amounting in magnitude to $\sigma/\sqrt{2}$."

Let us now consider the energy of the motion. Put

$$\frac{du}{dx} = 5, \quad \sqrt{2}\frac{dv}{dx} = \eta, \quad \sqrt{2}\frac{dw}{dx} = \zeta \quad \cdots \tag{1}$$

and let W denote the work per unit of bulk required to produce the distortion in question, irrespective of inertia. We have a quadratic function of the three components of the strain, or

$$W = \tfrac{1}{2}(A\xi^2 + B\eta^2 + C\zeta + 2D\eta\zeta + +2E\xi\zeta + 2F\xi\eta) \cdots \tag{2}$$

where A, B, C, D, E denote moduluses of elasticity of the solid. We shall consider a little more about obtaining these moduluses from the 18 moduluses of the solid. I merely say now, however, that these are the moduluses of elasticity (the definition of modulus of elasticity being "stress divided by strain") linearly obtained from any proper and sufficient data regarding the elasticity of the solid.

Let p, q, r denote the three components of the elastic traction per unit area of the wave front due to pulling these planes asunder and to their relative slipping parallel to OY and parallel to OZ. If the medium were isotropic then clearly the elastic traction resulting from these two planes would be a force opposing the traction parallel to OX and forces parallel to OY and OZ depending on the compressibility of the solid, directly opposed to the slips in those directions. But generally, each one is involved in the other in the way that is expressed so conveniently by Green by the aid of the energy function, viz.:

$$p = \frac{dw}{d\xi} = A\xi + F\eta + E\zeta,$$

$$q\sqrt{\tfrac{1}{2}} = \frac{dw}{d\eta} = F\xi + B\eta + D\zeta, \tag{3}$$

$$r\sqrt{\tfrac{1}{2}} = \frac{dw}{d\zeta} = E\xi + D\eta + C\zeta.$$

According to the notation here introduced, p, q, r being mere pulls, p, $q\sqrt{\tfrac{1}{2}}$, $r\sqrt{\tfrac{1}{2}}$ express the stress parallel to OX, OY, OZ respectively.

We want to find waves that will travel each with a given line of displacement. That is quite analogous to the problem of the fundamental modes of a vibrating body. Let us find, if we can, directions of displacements for which the return force will be in the direction of the displacement. The equations for that will be

$$\frac{dw}{d\xi} = M\xi, \quad \frac{dw}{d\eta} = M\eta, \quad \frac{dw}{d\zeta} = M\zeta \quad \cdots \tag{4}$$

where M is constant. From these equations we can eliminate ξ, η, ζ, forming the cubic equation in M:

$$(A - M)(B - M)(C - M) - (A - M)D^2 - (B - M)E^2$$
$$- (C - M)F^2 + 2DEF = 0.$$

The three real roots obtained from this cubic [equation] will solve the problem. When the solid is strained in any one of these three ways, we have

$$p = M\frac{du}{dx}, \quad q = M\frac{dv}{dx}, \quad r = M\frac{dw}{dx} \quad \cdots \tag{5}$$

or the three components p, q, r will be proportioned to the three displacements. The equations of motion, ρ being the density, are, of course,

$$\rho\frac{d^2u}{dt^2} = \frac{dp}{dx}, \quad \rho\frac{d^2v}{dt^2} = \frac{dq}{dx}, \quad \rho\frac{d^2w}{dt^2} = \frac{dr}{dx},$$

which become, on substituting the values p, q, r from (5),

$$\rho\frac{d^2u}{dt^2} = M\frac{d^2u}{dx^2}, \quad \rho\frac{d^2v}{dt^2} = M\frac{d^2v}{dx^2}, \quad \rho\frac{d^2w}{dt^2} = M\frac{d^2w}{dx^2}.$$

These, by equations (4) and (1), give the formulae

$$Au + (Fv + Ew)\sqrt{2} = Mu,$$
$$Fu + (Bv + Dw)\sqrt{2} = Mv\sqrt{2}, \tag{6}$$
$$Eu + (Dv + Cw)\sqrt{2} = Mw\sqrt{2}.$$

Let M_1, M_2, M_3 be the roots of the determinantal cubic and $b_1, C_1, b_2, C_2, b_3, C_3$ the corresponding values of the ratios v/u [and] w/u, derived from (6). Observe that

$$u = u_1, \quad v = b_1 u_1, \quad w = C_1 u_1$$

is a solution where

$$u_1 = f_1(x + t\sqrt{M_1/\rho}) + F_1(x + t\sqrt{M_1/\rho})$$

and the thing is done. That is the full investigation for one of the three waves. The velocity of preparation is $\sqrt{M_1/\rho}$. For the other two waves you can write down similar expressions corresponding to the second and third roots, M_2, M_3.

Lecture XII

We will look a little more at this wave problem. I do not know that I should have troubled you with going through a process like this, because you will find it easier to read it in the book. The conclusion is, that if you choose arbitrarily, in any position whatever relatively to the elastic solid, a set of parallel planes for wave fronts, there are three directions at right angles to one another (each oblique to the set of planes) which fulfil this important condition, that the elastic force is in the direction of the displacement; and the equations we put down express the wave motion. Each of the three waves will be a wave in which the oscillation of the matter in its front is as I am performing it now, i.e., an oscillation to and fro in a line oblique to the plane of the wave front. You will find the three waves corresponding to the three roots of the determinantal cubic are in directions at right angles to one another and in general oblique to the plane of the wave.

Green deals with the problem in a peculiar way. He expresses the conditions by means of three equations among the coefficients that in two of these three waves possible to an elastic solid the displacement is exactly in the wave front, giving two waves at right angles to each other in the wave front and the third wave in the direction perpendicular to the wave front. The result of those three relations that Green finds among the coefficients— Green does not say anything of this; but we will think of it a little—will be that in considering a disturbance from a source we have a wave of distortion proceeding outwards corresponding to, but not in all cases identical with, that of Fresnel—made identical with that of Fresnel by some other supposition which I shall not speak of now. You see, if you work out the mathematical array of figures you might put down the equations in all their generality. I do not think this has ever been done. But just take the process we have gone through with, not for a plane perpendicular to OX as we did it, but for a plane oblique to the three axes, and you get three velocities for waves perpendicular to any chosen direction. Then by taking the envelope of these with the proper mathematical conditions, which you can put down in a few moments, you get a wave surface which will differ from anything that has been thought of before, so far as I know, in the theory of elastic solids—a wave surface in which there will be three sheets corresponding to each radius vector instead of only two, as in Fresnel's wave surface; and so far as I have investigated, each part of that wave surface will involve both condensation and distortion. I have satisfied myself that this is so. It is a geometrical exercise of no contemptible character to work out this wave surface. Green's three relations cause this wave surface to split up into an ellipsoidal wave surface for a condensational wave and a wave surface like Fresnel's for a distortional wave. I say "ellipsoidal"; so far as I remember, Green does not mention it at all. That is an exceedingly interesting result, and Green's three relations that give rise to it are exceedingly interesting relations.

Look back at the formulae which we had in an isotropic solid. We have always a perfect distinction between the two waves; that is we can have a purely distortional wave spreading out with its velocity, and a purely condensational wave with its velocity. You will remember the condensational wave proceeding symmetrically from a center, for which

$$\varphi = \frac{1}{r}\sin\frac{2\pi}{\lambda}(r - vt)$$

was the displacement potential, or the condensational wave for which any differential coefficient of φ whatever,

$$\frac{d^{i+j+k}\varphi}{dx^i\,dy^j\,dz^k},$$

is the displacement potential, all of which proceed with the same velocity $v = \sqrt{(k + \frac{4}{3}n)/\rho}$. Take $n = 0$ and you have what is dealt with in Lord Rayleigh's work on sound, in which he takes not only the distance terms as I did, but terms that express the motion at distances from the center moderate in comparison with the wave length. For an isotropic solid we can have all those waves, and again simultaneously with them, we can have some of our solutions for waves of distortion. Take, if you like our solution

$$\frac{4\pi^2}{\lambda^2}\varphi + \frac{d^2\varphi}{dx^2},\quad \frac{d}{dy}\frac{d\varphi}{dx},\quad \frac{d}{dz}\frac{d\varphi}{dx}$$

as a fundamental solution, from which by differentiation you can obtain other solutions. I am going to correct something that I said as to want of interest. This solution is more interesting than I thought it would be.

In the isotropic solid the independency of the two sorts of waves comes naturally. The condensational wave goes at one speed, the distortional goes at another, and you may proceed with either as if the other wave were not there. A central disturbance in an isotropic solid will cause both sorts of waves to proceed outwards in the manner of an earthquake eruption. I do hope that before another earthquake will do as the last one did, there will be means of observing earthquake disturbances. There is a great deal to investigate in that subject. I do think it will be worth while at stations for the observation of scientific meteorology to have self recording apparatus to show the three components of apparent gravity at every instant. Beyond a doubt, if there had been records taken in this way, by instruments not too sensitive during the last earthquake, we should have had evidence of two waves through the earth, a distortional and a condensational wave.

What goes on in the isotropic solid occurs similarly in a solid which is not isotropic, but which fulfils Green's conditions. I may tell you, that these conditions are first a set of conditions of symmetry, and secondly, three equations. I have not looked into the thing to see whether, without other conditions than that of the condensational wave having its own wave surface and set of velocities, a distortional wave may be capable of propagation without condensation at all. The condition for that alone has not so far as I know been investigated. Green only gives the condition for that after having introduced certain conditions of symmetry; and I do not know whether it would come to the same result or not if he had introduced it before giving the conditions of symmetry. That is a very interesting subject and I shall attempt to work it out. I think you will find it worth while to work out the wave surface that I gave and then Green's interesting condition, what must be the condition that two out of the three waves whose fronts are parallel to a given plane shall be purely distortional. It is obvious that if this condition is fulfilled, without any other consideration of symmetry, you will have an unsymmetrical quasi-Fresnel wave surface for purely distortional waves, which, when made symmetrical with reference to OX, OY, OZ, by a series of relations, will become more like Fresnel's, but which certainly requires something more of an assumption than that to make it agree with Fresnel's.

That is a fine subject for investigation, and I am sorry I can not throw more light upon it. There is nothing more in it than would take a mathematician half a day to work out, and it would be worth doing.

But if the war is to be directed to fighting down the difficulties to the undulatory theory of light it is not the slightest use for us in solving our difficulties to have a medium which kindly permits distortional waves to be propagated through it, though it is aeolotropic. It is not enough to know that though the medium be aeolotropic it can let purely distortional waves through it, and that two out of the three waves will be distortional. What we want is a medium which, when light is refracted and reflected, will under all circumstances give rise to distortional waves alone. Green's mediums would fail in this respect when waves of light come to a surface of separation between two such mediums. All that Green secures is that there can be an outward distortional wave; he does not secure that there shall not be a condensational wave. There would be condensational waves from the source. The electric light, etc., would produce condensational waves, whether it was in an aeolotropic or [an] isotropic medium, so far as Green's conditions here spoken of go. Interesting as they are, they do not help in the slightest degree towards explaining double refraction in such a medium. What we want is a medium resisting the condensational waves; a medium with an infinite or practically infinite bulk modulus—so great that there can never be more than an amount of energy that has not been discovered by observation, developed in the shape of condensational waves.—I believe that is a correct sentence, although it is complicated.

As an essential in every reflection and refraction there may be a little loss of energy from the want of perfect polish in the surface, but as a rule, we have no loss of light in reflection and refraction. There perhaps is some and we have not discovered it. The medium that gives us the luminiferous vibrations must be such that if there is any part of the energy of the wave expended in condensational waves after refraction and reflection, the amount of it must be so small that it has not been discovered. Numerical observations have been made with great accuracy, in which, for example, Fresnel's formula for the ratio of the incident and reflected light

$$\left(\frac{\mu - 1}{\mu + 1}\right)^2$$

is verified within closer than one per cent, I think. Still a half per cent or a tenth per cent of the energy may be converted into condensational waves, for all we know; but if any per centage to speak of were converted into condensational waves, there would be a great deal of energy in condensational waves going about through space, and there would be a new force (to take an absurd mode of speaking of these things) that we know nothing of. There would be some tremendous action all through the universe produced by the energy of condensational waves if the energy of condensational waves were one-tenth, or one-hundredth, or even say one-thousandth per cent of the energy of the distortional waves. I believe that if in all instances of reflection or refraction of light at any surface or in case of violent action in the source there are condensational waves produced with anything like a thousandth or a ten-thousandth of the energy of light, we should have some prodigious effect, but which might, perhaps have to be discovered by some other senses than we have. The want of indication of any such actions is sufficient to prove that if there are any in nature, they must be exceedingly small. But that there are such waves I believe; and I believe that the velocity of propagation of electro-static force is the unknown condensational velocity that we are speaking of.

I say "believe" here in a somewhat modified manner. I do not mean that

I believe this as a matter of religious faith, but rather as a matter of strong scientific probability. If this is true of propagation of electro-static force, it is perfectly true that there is exceedingly little energy in the waves corresponding to the propagation of an electro-static force. That is going beyond our tether, however, of molecular dynamics. What I proposed in the introductory statement with reference to these lectures was to chiefly bring what principles and results of the science of molecular dynamics I could enter upon to bear upon the wave theory of light. We are sticking closely to that for the present, and we may say that we have nothing to do with condensational waves. Our medium is to be incompressible, and instead of Green's three conditions, we have one condition of incompressibility. It is obvious that one equation of incompressibility suffices to prevent the possibility of a wave of condensation at all and reduce our wave surface to a surface with two sheets, like the Fresnel surface. But before passing away from that beautiful dynamical speculation (or example of possibility I should perhaps call it) of Green's, if we think of what the condensational wave must be in an aeolotropic solid fulfilling Green's condition that it can have purely distortional waves proceeding in all directions—the condition that two of the three waves we investigated three-quarters of an hour ago shall be purely distortional—I think we shall find also condensational waves, and that the wave surfaces for them will be a set of concentric ellipsoids. It will be a single sheeted surface, that is certain, because you have only one velocity corresponding to each tangent plane at the wave surface.

I shall now leave this subject for the present. We shall come back upon it again, perhaps, and look a little more into the question of moduluses of elasticity. We shall work up from an isotropic solid to the most general solid; and we shall work down from the most general solid to an isotropic solid. We shall take first the most general value for the compressibility; we shall then come to this subject again of assuming incompressibility. We shall then begin with the most general solid possible, and see what conditions we must impose to make it as symmetrical as is necessary for the Fresnel wave surface. The molecular problem will prepare your way a good deal for this.

That puts me in mind of a correction I have to make with respect to the interest attached to this solution for distortion in an isotropic solid

$$\left\{ \frac{4\pi^2}{\lambda^2}\varphi + \frac{d^2\varphi}{dx^2}, \quad \frac{d^2\varphi}{dx\,dy}, \quad \frac{d^2\varphi}{dx\,dz} \right\}.$$

I said it was not interesting because it could not express a natural sequence of light waves. I said that to express a natural sequence of light waves we must have two bodies moving in opposite directions, so that the center of gravity may not move. I quite forgot the supposition of our shell, which does the very thing we want. Instead of passing to a higher order of differentiation, so to speak, for the most probable natural sequence of waves of light, consisting of waves of greater nodal subdivision (having a nodal circle at the equator as well as nodal points in the axis of x), I see now that this very thing is the most probable. By "probable" I mean certainly the most frequent. I look upon it as a reality that there are particles moving; and it seems to me certain that those particles are soft, and that they must have enormous mass compared with the luminiferous ether.

I had intended to prepare something about the mass of the luminiferous ether. I have not had time to take it up, but certainly shall do so before we have done with the subject. We shall go into the question of the density of the luminiferous ether, giving superior and inferior limits. We shall also consider what fraction of a gramme may be in one of these molecules and show what an enormously smaller fraction of a gramme we may suppose it to displace in the luminiferous ether. We shall try to get into the notion of this, that the molecule must be soft and that that there must be an enormous mass in its interior. Its outer part feels and touches the luminiferous ether, and the luminiferous ether feels, it may be, comparatively slight to it. It is a very curious supposition to make, of a molecular cavity lined with a massless rigid spherical shell; but that something exists in the luminiferous ether and acts upon it in the manner that is faultily illustrated by our mechanical model, I absolutely believe. I have no more doubt that something of the kind is true, than I have of my own existence.

Just think of the effect of a shock consisting say of a collision between that and another molecule. Instead of its being broken into bits, let us suppose a case around it. It will bound away, vibrating. Just imagine that the central nucleus goes in one direction while the shell is going in the other, and there will be a molecule with two parts going in opposite directions but different from what I thought of the other day in that one part is inside the other. The ether gets its motion from the outside part. Therefore I say that the most fundamental supposition we can make with reference to the origin of the sequence of waves of light is that illustrated by a globe vibrating to and fro in a straight line.

We have already investigated the solution corresponding to that. Take spherical waves; no vibrations for points in one certain diameter of the sphere; maximum vibrations in all points of the equatorial plane of that diameter and perpendicular to that plane; for all points in the quadrant of an arc of the spherical surface extending from axis to equator, vibrations in the plane of and tangent to the arc of magnitude proportional to the cosine of the latitude or angular distance from the equator and of intensity proportional to the square of the cosine of the latitude; then let the amplitude vary inversely as the distance from the center, and the intensity inversely as the square of the distance from the center, and you have a correct word-painting of the very simplest and most frequent sequence of vibrations constituting light.

———————

Let us return to the consideration of the dynamics of refraction, absorption, anomalous dispersion, and so on. We have the square of the refractive index,

$$\frac{\ell T^2}{\rho \lambda^2} = \mu^2$$

$$= 1 + \frac{C_1}{\rho} \left\{ q_1 K_1{}^2 - (1 - q_1) T^2 + q_1 K_1{}^2 \left(\frac{K_1{}^2}{T^2} + \frac{K_1{}^4}{T^4} + \cdots \right) \right.$$

$$\left. - q_2 \frac{T^4}{K_1{}^2 - T^2} - \cdots \right\}.$$

We are not going to think at present of what the values of q_1, $q_2 \cdots$ may be, except that q_1 is less than unity. Our example will illustrate that, but I think we can see it without an illustration. For explaining ordinary refraction, we probably have more than enough in the term $q_1 K_1{}^2$, and it will be convenient to suppose all the other terms small in comparison with this. When T is small in comparison with K_2, K_3, $\cdots$, the terms in q_2, q_3, $\cdots$ are negative, but I believe we can suppose q_2, q_3, $\cdots$ all exceedingly small. Remember that q_1, q_2, $\cdots$ are constants, independent of the period T, as well as K_1, K_2, $\cdots$, which are the periods in which the system would vibrate if we hold the handle P at rest, i.e., the case of $\xi = 0$ in our equations.

I want to see how we can vary T without coming to trouble. As we increase T, the negative term becomes larger and larger, and if we increase it enough, it will make $\mu^2 = 1$; increase it still more and it will make $\mu^2 = 0$; and if we increase it still more, it will make μ^2 negative. Let us put this in its other form. This form is only suitable to show its availability for modifying Cauchy's formula so as to give correct refractions. I hope we may have a little work done upon it sometime or other, in the way of seeing whether these terms will suffice for actually obtaining refractive indices. I believe Lommel has done something of this kind. I know Ketteler in 1871 had a formula quite like this, but if I remember right, his term was positive here instead of essentially negative, as I have it. It seems probable that we should be able to explain refraction through a somewhat wide range from what is here written. You are doubtless more familiar with the formula in which λ appears; but remember that λ is proportional to the period, so that this formula is simply

$$A - B\lambda^{-2} + C\lambda^{-4},$$

which falls back upon Cauchy's formula

$$A + B\lambda^{-2} + C\lambda^{-4},$$

except through a very wide range, or to meet the critical cases.

I am sorry to leave it, but we must. I have data from Langley for the refrangibility of different lights passing through rock salt, down to about three or four times the wave length of sodium light by actual observation, if I remember right, and by very probable inference from the curve obtained down to 17 times the wave length of sodium light. I received this only a day or two ago, so that I have not attempted to make a comparison with a formula of this kind.

I am going to ask you to look at the critical cases, and for this purpose replace this form by the simpler one

$$\mu^2 = 1 + \frac{C_1 T^2}{\rho}\left(-1 + \frac{q_1 T^2}{T^2 - K_1^2} - \frac{q_2 T^2}{K_2^2 - T^2} - \cdots\right)$$

in which T is greater than K_1 and less than K_2. When T is considerably larger than K_1, but small in comparison with K_2, we have ordinary refraction in a transparent body, without absorption bands, or anything of the sort. This must occur through a considerable range of values of T in the cases of glass, rock salt, etc. As T decreases we have an augmenting refractive index for ordinary normal refraction. As T approaches K_1, μ^2 approaches infinity, and you get greater and greater refraction, until you pass through K_1. When K_1 exceeds T what will the result be? μ^2 will become negative. What is the meaning of the square of the refractive index negative? Answer, waves cannot be propagated. Think of the proposition, waves cannot be propagated at all. That is clearly an absorption band.

I object to the invoking of viscous terms to get quit of the energy; for how shall we prevent them from taking away all our energy when we do not want it taken away? We can spare exceedingly little energy in the transmission of light through distilled water, if it may be propagated through 150 feet as I believe it is. Sea water is supposed to be more transparent than most bodies. It is by no means black darkness down 20 fathoms in any sea. There are about 500,000 wave lengths of sodium light in a foot of water. In 100 feet there would be 50,000,000 wave lengths. We can spare very little energy, then, in water, if we are to think of light being propagated through 50 million wave lengths before it is absorbed. If we let in viscous terms in a way that will do anything at all for us in answering the question, what becomes of the energy at the critical points, for the wave lengths that are actually absorbed, it will run away with our energy where we do not want it to. Besides that, it is throwing up the sponge in respect to the dynamical question, and confessing that we have to introduce a new force instead of dealing solely with dynamical ones. As a subordinate theory in abstract hydrodynamics, it is exceedingly interesting to introduce viscous terms; but not in molecular dynamics. We must think of what becomes of the energy. Helmholtz understands, as I said before, that the consumption of energy by the viscous terms means its conversion into heat. But I want the same vibrating molecule which gives us the ordinary laws of refraction, which gives us the anomalous dispersion at the critical points, to take up the energy also and give it out in the proper way. That is what we have been doing thus far. And I want to look at a set of vibrating particles, and see what may be obtained from them. Of course we can do far more by calculation than we can find out in that sort of way, but still, it will help us a little.

It is perfectly clear that we have a broad absorption band throughout the range of values of T smaller than K_1 which gives a negative value to μ^2. The light will first appear beyond that absorption band with very small refrangibility—exceedingly small. We have in the neighborhood of the critical point exactly that kind of inversion with anomalous dispersion, in which we have less refraction, or greater velocity of propagation, for light of periods less than a certain limit, K_1, say, and greater refractive index or less velocity of propagation for light above that period.

That is merely an indication of the fact of anomalous dispersion; it is hardly worth while to look into the details just now. That is doubtless familiar to many of you, who have read Helmholtz's paper on anomalous dispersion. The subject was worn threadbare before I knew it was discovered, in fact. Stokes originated the idea of periods of absorption corresponding to periods of molecular vibration; but there was no hint of explaining refraction in this way or anomalous dispersion. So far as I know, the first word on the reaction of these particles upon the luminiferous ether is Sellmeier's; and it is now, perhaps, a matter of general knowledge, and I should rather apologize for taking up the time in speaking of it, than to say that I am bringing anything new before you.

We shall try to see something more of the effect of light propagated through a medium of a period exactly equal to K_1. I believe each sequence of vibrations will throw in a little energy which will spread out among the different possible motions of the molecule. The continuation of the sequences, forming what we call continuous light, is not a continuous phenomenon at all. I believe that the first effect, when light begins, will be each sequence of waves of the exact period throws in some energy into the molecule. That goes on until, somewhere or other, the molecule gets uneasy. It takes in an enormous quantity of energy before it begins to get particularly uneasy. It then moves about, and begins to collide with its neighbors perhaps, and will therefore give you heat in the gas, if it be a gaseous molecule. It goes on colliding with the other molecules, and in that way imparting its energy to them. The energy will be simply carried away, by convection if you please, or a part of it perhaps. Each molecule set to vibrating in that way becomes a source of light, and so we may explain the radiation of heat from the molecule after it has been got into the molecule by sequences of waves of light. I believe we can so explain the augmented pressure of a gas, due to the absorption of heat in it.

We may consider, however, that the chiefest vibration of the molecule is that in which the nucleus goes in one direction and the shell in the opposite direction, but with a great amount of energy in the interior vibrations and very little in the shell so that the shell may go on giving out the phosphorescent energy for two or three hours or days, simply vibrating forever, except in so far as the energy is drawn off and allowed to give motion to other bodies.

I see no difficulty in answering several of the fundamental riddles of this subject by the reaction of this assumed particle in the luminiferous ether; but there is difficulty about double refraction, and I see no solution whatever of that riddle as yet.

Lecture XIII

Prof. Morley has solved the problem that I proposed for some of the fundamental periods, and you may be interested in knowing the result. He finds
roots $1/T^2 = 3.46, 1.005, .298, .087$, each root not being very different from
three times the preceeding one. A tracing of the curve, you will understand,
involves a set of asymptotes. The curve in general for any such case must be
something like this:

the curve giving in this direction (arrow) from positive to negative. At the
end $1/T^2 = \infty$. I will not go into that any further just now. I just wanted
to call your attention to what Prof. Morley has done upon the example that
I gave to the arithmetical laboratory. I think it would be worth while, also,
to work out the energy ratios. In selecting this example, I chose the case for
which the work would of necessity be highly convergent. But I chose it
primarily, however, because it is something like the kind of thing that
presents itself in the true molecule: a soft elastic body consisting of a
finite number of discontinuous masses elastically connected (with enormous
masses in the central parts, that seems certain) imbedded [in the] ether and
acted on by the ether in virtue of an elastic connection which, if this molecule
were rigid and imbedded in the ether simply like a rigid mass imbedded in
jelly, must consist of elastic bonds analogous to springs.

I think you will be interested in looking at this model which, by the
kindness of Prof. Rowland, I am now able to show you. It is made on a plan
according to which I made a wave machine which has been used for many
years in my classes, and finally modified in preparations for a lecture given
to the Royal Institution about two years ago on The Size of Atoms.

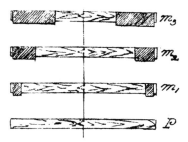

I think those who are interested in the illustration of dynamical problems will find this a very nice and convenient method. If you will look at it, you will see how the thing is done. Pianoforte wire, bent around three pins in the way you see here,

supports each bar. Those pins are slanted in such a way as to cause the wire to press in close to the bar so as to hold it quite firm. The wood is slighty cut away to prevent the wire from touching it so that there may be no impairment of elasticity due to slip of steel on wood. The wire used is fine steel pianoforte wire; that is the most elastic substance available, and it seems to me, indeed, by far the most elastic of all the materials known to us.

Prof. Rowland is going to have another machine made, which I think you will be pleased with—a continuous wave machine. This is not a wave machine, but a machine for illustrating the vibrations of several elastically connected particles. The connecting springs are represented by the torsional spring in the three portions of connecting wire and the fourth portion by which the upper mass is hung. In this case gravity contributes nothing to the effect except to stretch the wire. You will understand that these upper masses correspond to m_1, m_2, m_3. In all we have four masses here. I will just apply a moving force to this lower mass, P. To realize the circumstances of our case more fully, we should have a spring connected with a vibrator to pull P with, and perhaps we may get that up before the next lecture. I shall attempt no more at present than to cause this first particle to move to and fro in a period which is perceptibly shorter than the shortest of the three periods. The result is scarcely sensible motion of the others. I do not know that there would be any sensible motion at all if I had observed to keep the greatest range of this lowest particle to its original position on the two sides of its mean position.

The first part of our lecture this evening I propose to be a continuation of our conference regarding aeolotropy. The second part will be molecular dynamics. I propose to look at this question a little, but I want to look very particularly to some of the points connected with the conceivable circumstances by which we can account for not merely regular refraction but anomalous dispersion, and both the absorption that we have in liquids and very opaque bodies and such absorption as is demonstrated by the existing fine lines of the solar spectrum which are now shown more splendidly than ever by Prof. Rowland's gratings.

I shall speak now of aeolotropy. The equations by which Green realized the condition that two of the three waves having fronts parallel to one plane shall be distortional [are] equivalent to a very easily understood condition that I may illustrate first respecting bodies more nearly isotropic than those that we are considering in the more general problem.

I am reminded by a lady that I have said two or three times dilatational instead of distortional,* and I have just said it again. There seems to be a law by which I say dilatational when I mean distortional. Another little point with respect to yesterday's work: If you have taken the trouble to make notes, you had better cancel the $\sqrt{2}$ wherever it occurs, and let the unit tangential stress be the ordinary unit as set forth in Thomson and Tait for example, and the unit of distortion a simple shear. There is good reason for the $\sqrt{2}$, but it is a part of the theory that we are not concerned with at all, and for a special problem like that, it is better to introduce special notation. This special notation is in point of fact the more general notation.

That problem is similar to another of the very greatest simplicity, which is the well known problem of the displacement of a particle subject to forces acting upon it in different directions from fixed centres. An infinitesimal displacement in any direction being considered, the question is, when is the return force in the direction of the displacement. As we know, there are three directions at right angles to one another in which the return force is in the direction of the displacement. The sole difference between that very trite problem and that which I went through yesterday is that in the latter case the question is put with reference to a whole infinite plane in an infinite homogeneous solid which is displaced in any direction. Considering force per unit of area, we have the same question, when is the return force in the direction of the displacement, and the answer is, there are three directions at right angles to one another in which the return force is in the direction of the displacement. Those three directions are generally oblique to the plane; but Green found the conditions under which one will be perpendicular to the plane, and the other two in the plane.

I shall now enter upon the subject more practically in respect to the application to the wave theory of light, and that is, to introduce right away

---

* [This has been corrected whenever I have noticed it.—H.]

at the beginning the condition of incompressibility. Take first the well known equations of motions for an isotropic solid and express in them the condition that the body is incompressible. The equations are:

$$\rho \frac{d^2 \xi}{dt^2} = (k + \tfrac{1}{3}n) \frac{d\delta}{dx} + n \nabla^2 \xi, \text{ etc.}$$

I have another name from Prof. Ball for ∇, which is *atled*, or *delta* spelt backwards. Shall it be *nabla*, *atled*, or *Laplacian*? Laplacian, if you like.

Suppose now the resistance to compression is infinite, which means, make $k = 0$ at the same time that we have $\delta = 0$. What then is to become of the first term of the second members of these equations? We simply take $(k + \tfrac{1}{3}n)\delta = p$ and write the second member

$$\frac{dp}{dx} + n \nabla^2 \xi.$$

This requires no hypothesis whatever. We may now take $k = \infty$, $\delta = 0$, without interfering with the form of our equations. These equations, without any condition whatever as to ξ, η, ζ, with the condition $p = (k + \tfrac{1}{3}n)\delta$, are the equations necessary and sufficient for the problem. On the other hand, if $k = \infty$, the condition that that involves is

$$\frac{d\xi}{dx} + \frac{d\eta}{dy} + \frac{d\zeta}{dz} = 0,$$

which gives four equations in all for the four unknown quantities ξ, η, ζ, p.

Precisely the same thing may be done for a solid with 21 independent coefficients. We will have this equation again for an aeolotropic body, $\delta = 0$, and a corresponding equality to infinity. I am not going to introduce any of these formulae at present. In the meantime, I tell you a principle that is obvious. In order to introduce the condition

$$\frac{d\xi}{dx} + \frac{d\eta}{dy} + \frac{d\zeta}{dz} = 0$$

into our general equation of energy with its 21 coefficients, which involves a quadratic expression in terms of the six quantities that we have denoted by e, f, g, a, b, c, we must modify the quadratic into a form in which we have $(e + f + g)^2$ into a coefficient. That coefficient equated to infinity, and $e + f + g = 0$, leave us the general equations of equilibrium of an elastic solid with one fewer out of the 21 independent coefficients in virtue of this relation of incompressibility.

I want to call your attention to the kind of deviation from isotropy which is annulled by Green's equations among the coefficients which express that two out of the three waves shall be purely distortional. The next thing to an isotropic body is one possessing what Rankine calls cyboid symmetry.

Rankine marks an era in philology and scientific nomenclature. In England, and I believe in America also, there has been a classical reaction or reformation according to which, instead of taking all our Greek words through the French, changing κ into c, v into y, and several other variations that I do not remember, we spell in English, and pronounce Greek words, and even some Latin words, more nearly according to what we may imagine to be the actual usage of the ancients. We cannot however get over Kuros instead of Cyrus, Kikero instead of Cicero, in the present generation; we have not swallowed the thing altogether, yet, Rankine is a curious specimen of the very last of the French classical style. Rankine was the last writer to speak of *cinematics* instead of *kinematics*. *Cyboid* is a very good word, but I do not know that there is any need of introducing it instead of *Cubic*. *Cubic* is an exception to the regular analogy in that v is not changed into y; it should be *cybe* (I suppose $\kappa v\beta\alpha\sigma$ to be the Greek word), because *cyboid* obviously means *cubic*, and it is taken from the Greek in Rankine's manner.

Rankine gives the equations that will leave cubic asymmetry. He afterwards makes the very opposite remark that Sir David Brewster discovered that kind of variation from isotropy in analcime. I only came to this in Rankine two or three days ago. But I remember going through the same thing myself not long ago and I said to Stokes—I always consulted my great authority Stokes whenever I got a chance—"Surely there may be such a thing found to exemplify this kind of asymmetry; would it not be likely to be found in crystals of the cubic class?" Stokes—he knew almost everything—instantly said "O, Sir David Brewster thought he had found it in cubic crystals, but there was an explanation that it seemed to be owing to the effect of the cleavage planes, or the separation of the crystal into several crystalline lamina"—I do not remember what it was, but he distinctly denied that Brewster's experiment showed a true instance of cubic asymmetry. He pointed out that an exceedingly slight deviation from cubic isotropy would show very markedly on elementary phenomena of light, which might be very readily tested by means of ordinary optical instruments, and that the fact that nothing of the kind has been discovered is absolute evidence that the deviation, if there is any, from isotropy in a crystal of the cubic class, is exceedingly small in comparison with the deviation from isotropy presented by ordinary double refracting crystals. Deviation from cubic isotropy is the same thing as the conceivable cubic deviation from isotropy.

As a matter of fact, deviation from square isotropy is found in a pocket handkerchief or piece of square cloth, supposing the warp and woof to be accurately similar, a supposition that does not hold of ordinary cloth. Take wire cloth carefully made in squares and that will be symmetrical and equal in its moduluses with reference to two axes at right angles to one another. There will be a vast difference according as you pull out one side and compress the other, or pull out one diagonal and compress the other. Take the extreme case of a cloth woven up with inextensible frictionless threads, and there is a kind of absolute resistance to distortion in two directions at right angles to one another, and no resistance at all to distortion of a certain kind that is presented in changing its square shape. That is to say, a frame work of this kind

has no resistance to shearing distortion; but it has infinite resistance to the distortion produced by lengthening one diagonal and shortening the other. Just imagine a square cut out of this pattern with sides parallel to the diagonals, making a pattern of this sort.

There is a body that has infinite resistance to shearing, and zero resistance to pulling out in this direction (along the diagonal). That is not altogether a trivial illustration. Surgeons make use of it in their bandages. A person not familiar with the theory of elastic solids might cut a strip lengthwise with the thread; but cut it obliquely and you have that conveniently pliable character that allows it to serve the purpose of a bandage.

Imagine an elastic solid made up in that kind of way, with that kind of deviation from isotropy, and we have clearly two different rigidities for different distortions in the same plane. I remember that Rankine, in one of his early papers, proved that to be impossible. He proved a proposition to the effect that the rigidity was the same for all distortions in the same plane. That perhaps was founded on some special supposition as to arrangement of molecules and may be true for the particular arrangement. Rankine made too short work of the elastic solid in his first paper. He afterwards took it up very much on the same foundation that Green did, with 21 coefficients, but he uses the old proposition that rigidity is the same for all distortions in the same plane.

I will go no further into that just now than to say that if without introducing the condition of incompressibility at all, you introduce the condition that there is equality of rigidities for the two principal modes of distortion in each plane—Perhaps we shall be able to face the problem in the next lecture of introducing the relations among the 21 moduluses which are sufficient to do away with all obliquities with reference to the rectangular axes. I shall put down the figures before you somehow or other, before we have ended. But you can do this in a moment—equate to zero enough of the 21 coefficients to fulfil these conditions, that if you compress the body by uniform forces parallel to OX or OY it will remain rectangular, and that if you produce a shear in one coordinate plane it does not produce obliquity in any other, and so on, doing away with all that is necessary in order to annul obliquity. There will remain a certain number of coefficients—nine I think. Now put in your condition that in this plane the rigidity due to a shear parallel to the sides is equal to the rigidity due to a shear in a portion cut out with its sides at angles of 45° to the sides of this. There will be three equations. These equations are identical with the three equations that Green gives to express his condition as to the waves. That is really very interesting and instructive, although it does not do much for light.

I must read to you some of Rankine's fine words that he has introduced into science in his work on the elasticity of solids. That is really the first place I know of except in Green in which this thing has been gone into in a satisfactory way. It is not really satisfactory in Rankine except in the way in which he carries out the whole subject, the algebra of it and the determinants and matrices that he goes into so very nicely and, what I want to call attention to, his names. I do not know whether Prof. Sylvester ever looked at these names. I think he would be rather pleased with them. "Thlipsinomic transformations," "umbral surfaces" and so on. Any one who will learn the meaning of all these words will obtain a large mass of knowledge with respect to an elastic solid. The words of "strain and stress" are due [to] Rankine; "potential" energy also. Hear the grand words: "Thlipsinomic, Tasinomic, Platythliptic, Euthytatic, Metatatic, Heterotatic, Plagiotatic, Orthotatic, Pantatic, Cybotatic, Geniothliptic, Euthythliptic, &c."

You may now understand what cyboid asymmetry is, or as I prefer to call it, cyboid aeolotropy. Rankine had not the word aeolotropy; that came in later. Cyboid or cubic aeolotropy is the kind of aeolotropy exhibited by a cube grating, a basket woven solid with uniform cubic baskets. There is a thing that would be isotropic, except for that difference of rigidity for the two principle distortions in each one of the planes of symmetry. What I am going to do further is to point out that if we take, first of all, the condition of infinite resistance to compression, secondly, introduce the conditions necessary for symmetry, then after that annul the difference of rigidities for the principle distortions in each of the three principal planes, we shall find ourselves landed in an elastic solid with three principle moduluses which will give us a wave surface identical with Fresnel's, except that the order of procedure is different. The direction in the surface which corresponds to the direction of vibration in Fresnel's surface is a line perpendicular to the plane through the line of displacement and the perpendicular to the wave front;

I believe; but it is possibly the plane through line of displacements and the center of the wave surface. I will read it out of Green; but Green really never introduced the condition of incompressibility at all. Here it is at the bottom of page 304 of Green's collected papers: "We thus see that if we conceive a section made in the ellipsoid to which the equation (10) belongs, by a plane passing through its center and parallel to the wave's front, this section, when turned 90 degrees in its own plane, will coincide with a similar section of the ellipsoid to which the equation (8) belongs, and which gives the directions of the disturbance that will cause a plane wave to propagate itself without subdivision, and the velocity of propagation parallel to its own front. The change of position here made in the elliptical section is evidently equivalent to supposing the actual disturbances of the ethereal particles to be parallel to the plane usually denominated as the *plane of polarization.*"

Thus, in the wave surface corresponding to Green's elastic solid, draw a plane perpendicular to the wave front through the direction of displacement. The line perpendicular to that plane is the direction of displacement in Fresnel's case.

I gave you one solution of the problem of passing a cord around the eight vertices and twelve edges of a parallelepiped. It is obvious that it cannot be done by passing the cord only once along each side. To make the figure incompressible, we may suppose the cord to be perfectly flexible and inextensible. Instead of supposing the cord inextensible, we can have an elastic portion in the middle part of the cord along each side. You can thus introduce what is equivalent to three disposables in the longitudinal rigidities of the portions of the cord in question. We may dispense with the idea of a flexible body if wherever the cord changes direction we put in a bell handle, which is a mechanical principle, instead of passing the cord through a ring.

I am afraid this problem of the molecules in the elastic solid presents enormous difficulties to us. I feel that we have the utmost confidence that we can make a model that will fulfil any stated condition whatever, as to absorption, and so on. The mathematical working of it out is difficult. I am not going to solve all these problems in five minutes, but what I can do in five minutes is to show that we are quite out of our depth after all, in the thing we have been invoking. Consider the uniform isotropic elastic solid in which this molecule is imbedded. We must consider the distance from one of these imbedded shells to another to be great in comparison with the diameter of the shell and small in comparison with the wave length. If the effect is anything near sufficient to give us change of velocity through the range of 1 to 1.5, we cannot suppose the whole medium to move with the molecules. In the equation I have put down, I want to guard against the supposition that it is a rigorously correct equation. In that equation, we supposed the molecules to be so evenly distributed that, relatively to the dimensions of a thousandth part of a wave length if you like, it is practically a homogeneous solid—in other words, an exceedingly fine grained solid, so finely grained that it is practically homogeneous for portions exceedingly small in linear dimensions in comparison with the wave length. But no degree of smallness will dispense with the to and fro motion of the elastic solid relatively to the imbedded molecules.

I want to invoke Lord Rayleigh, and if we can get him to take it up, we shall have a chance of learning something about it. I suppose the medium to move together with the imbedded molecules, as will be approximately the case if the effect of the molecules is such as to produce but a small difference in the circumstances from what they would be if there were no molecules imbedded at all. In other words, if the amount of this molecular action in the medium is such as to produce but a very small change in the velocity of light in proportion to the whole velocity, then I think we are quite clear in the assumption that the whole medium moves with the molecules. If in our formulas we put $C_1 = \infty$, $C_2 = 0$, etc., you will see that it is tantamount to adding the mass of m_1 to ρ, the density of the medium, which would not be the case unless the whole of the mass added increased the average density but little in proportion to the whole density. I think I am right in saying that if the medium becomes infinitely fine grained, and if the density is but little increased, then the effect of putting in the molecules would be to add the mass per unit volume of the molecules to the density of the ether. I believe it is not so when the change of velocity is considerable in comparison with the velocity in the ether alone. And instead of our very nice, simple, mechanical arrangement that Prof. Rowland has illustrated for us here, with springs between the rigid shell and the ether, it will give us an elastic action which will be playing to and fro among these molecules, and it will be a problem extremely difficult to solve, but since Lord Rayleigh has been induced to take it up, he will give us the answer. It is not absolutely a question for any bodies whatever or even spherical bodies. But it is the question, what kind of change in the equation we have put down will be introduced by taking into account that principal as to the motion of the luminiferous ether. I wanted to warn you against thinking for a moment that we can give fundamental value to the equations that I have put before you.

We found

$$\mu^2 = 1 + \frac{C_1 T^2}{\rho}\left(-1 + \frac{q_1 T^2}{T^2 - K_1^2} - \frac{q_2 T^2}{K_2^2 - T^2} - \cdots\right).$$

When we have q_1 very nearly equal to 1, we can account for all we at present know of regular refraction, by values of T greater than K_1 and less than K_2. If q_2 be excessively small, and T not very much greater K_2, we may account for an absorption band as fine as you please. Suppose the question to be to

account for refraction by vapor of sodium, not taking into account at present
the double sodium line—that is to say, considering a substance like sodium
that gives only one line. Two terms, I believe, could be very reasonably
arranged so as to give us, by the considerations we went through yesterday,
the irregular refractions that that medium would show. The period of this
vapor would be K_2. If q_2 is very small we shall have the absorption band
appearing as a very sharp black line in the spectrum of the light coming
through this vapor. This vapor put into a prism and experimented upon for
the refractive power of the medium would give us something not distinguishable from ordinary refraction until you get near the period of the vapor,
when there would be anomalous dispersion. But I say that if you take q_2
small enough, you may make the absorptional region and the region of
anomalous dispersion as small as you please. I cannot doubt that this is the
way the thing is done in nature; there is something in nature that corresponds precisely to that course of action.

I want you to think how small q_2 must be for the sodium line, thinking
of only one sodium line. Sodium vapor shows no particular absorbing power
until you have a period differing very little from the period of sodium vapor.
How little you may judge by looking at the two sodium lines whose distance
apart is about 1.001 and whose thickness is not more than 1/50 of their
distance apart. It is apparent from that that the dispersional region corresponds to a period say

$$T = K_2 \left(1 \pm \frac{p}{50,000} \right)$$

where p is a proper fraction. This third term must be insensible for values
of T differing greatly from this. Therefore we must have $q_2 < 1/25,000$
according to these figures.

Let there be a connection of particles so as to give another mode of
vibration. It is not a hypothesis but a reality that sodium vapor has two
independent period vibrations whose periods differ by 1/50,000 of one
another. We have then the means of making something which will modify
the velocity of waves through jelly just as sodium vapor modifies waves of
light through the luminiferous ether. We have the means of making a
mechanical model of the thing. I do not say it is the explanation of it.

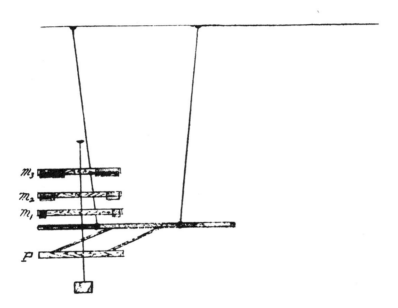

At this lecture [were] seen, immediately behind the model heretofore presented, two wires extending from the ceiling and sustaining a long heavy bar by means of closely fitting rings. By slipping these rings along the bar, the period of vibration about the bifilar suspension could be altered at will. Two pieces of wood served to transmit the motion of this vibrator to the lower bar P of the model.

This is another case from what I have been talking about. These rigid connections make the bar P go with a stated harmonic motion. I would like to have a heavy pendulum attached to the bar by a very light india rubber band. I want the vibrator to vibrate half a minute before you see any sensible motion in the model. This is another case likely, but it is quite equally interesting, and will do just as well. Let us look at this a little and see what it does. O, you can vary the period; that is very nice, that is beautiful. We are going to study these vibrations a little, just as illustrations. Prof. Rowland has kindly made this arrangement for us, and I think we will all be interested in seeing it. We have this bar P, moved by this pendulum, this pendulum being so massive that its period is but little affected, I suppose, by being connected with P. It takes sometime before the initial vibrations in the model are got quit of and the thing settles into simple harmonic motion corresponding to the period of the pendulum. If we keep this pendulum going long enough through nearly a constant range, the masses P, m_1, m_2, m_3 will settle into a definite simple harmonic motion, through the subsidence of any free vibrations which may have been superimposed upon them in the start. This now seems to be performing very nearly a simple harmonic motion. We will then superimpose another vibration on this by altering the period of the pendulum very slightly. That, you see, seems to have diminished very much the vibrations of the system. They are now increasing again. That will go on for a long time. I shall give this pendulum a slight impulse when I see it flagging to keep its range constant. When it is in its middle position, I apply a working couple. We will give no more attention to it than just to keep it vibrating, while we look at these notes which I have prepared for you, so as to shorten our work upon the board.

Lecture Notes of October 13

Homogeneous Elastic Solid of unrestricted character

$$(1)\ e = \frac{d\xi}{dx}; \quad (2)\ f = \frac{d\eta}{dy}; \quad (3)\ g = \frac{d\zeta}{dz}; \quad (4)\ a = \frac{d\eta}{dz} + \frac{d\zeta}{dy};$$

$$(5)\ b = \frac{d\zeta}{dx} + \frac{d\xi}{dz}; \quad (6)\ c = \frac{d\xi}{dy} + \frac{d\eta}{dx}.$$

$$W = \tfrac{1}{2}(e, f, g, a, b, c)^2$$
$$= \tfrac{1}{2}\{(e, f, g)^2 + 2(e, f, g)(a, b, c) + (a, b, c)^2\}$$

(for brevity).

Problem I. given $\left.\begin{array}{l} 11, 12, 13, 14, 15, 16 \\ \quad 22, 23, 24, 25, 26 \\ \qquad 33, 34, 35, 36 \\ \qquad\quad 44, 45, 46 \\ \qquad\qquad 55, 56 \\ \qquad\qquad\quad 66; \end{array}\right\}$ "tansinomic" coefficients.

required the bulk modulus (k). (NOTE, the "thlipsinomic" coefficients are more convenient for the case of incompressibility. They are more closely allied to practical moduluses.)

In (e, f, g) put

$$e = e' + \tfrac{1}{3}\delta, \quad f = f' + \tfrac{1}{3}\delta, \quad g = g' + \tfrac{1}{3}\delta$$

$$\tag{7}$$

when $\delta = e + f + g$ and therefore $e' + f' + g' = 0$.

We find

$$(e, f, g)^2 = \tfrac{1}{9}[11 + 22 + 33 + 2(23 + 13 + 12)]\delta^2$$
$$+ \tfrac{1}{3}[(11 + 13 + 12)e' + (22 + 23 + 12)f'$$
$$+ (33 + 23 + 13)g']\delta + (e', f', g')^2. \tag{8}$$

Hence

$$k = \tfrac{1}{9}[11 + 22 + 33 + 2(23 + 13 + 12)]. \tag{9}$$

Problem II. To find W for the case of incompressibility we have $\kappa = \infty$, $\delta = 0$, $\kappa\delta^2 = 0$. ($\kappa\delta$ may [be] and generally is finite. Denote it by p. We don't need it now, but shall want it, for equations of motion.) Hence

$$W = \tfrac{1}{2}\{(e', f', g')^2 + 2(e', f', g')(a, b, c) + (a, b, c)^2\}. \tag{10}$$

Problem III. (without restriction to incompressibility). To annul skewnesses relatively to OX, OY, OZ. This requires that, and is done when,

$$(e, f, g)(a, b, c) = 0$$

and

$$(a, b, c)^2 = 44a^2 + 55b^2 + 66c^2.$$

Problem IV. (without restriction to incompressibility). To annul web-oidal aeolotropy, in the case of the annulled skewnesses. In W take

$$e = \tfrac{1}{2}(r - s), \quad f = \tfrac{1}{2}(s - q), \quad g = \tfrac{1}{2}(q - r).$$

We find

$$W = \tfrac{1}{2}\{\tfrac{1}{4}\{(22 + 33 - 2 \cdot 23)q^2 + (33 + 11 - 2 \cdot 13)r^2$$
$$+ (11 + 22 - 2 \cdot 12)s^2 - 2[(11 + 23 - 13 - 12)rs$$
$$+ (22 + 13 - 12 - 23)sq + (33 + 12 - 23 - 13)qr]\}$$
$$+ 44a^2 + 55b^2 + 66c^2\}$$

case $r = 0$, $s = 0$, shows that $\tfrac{1}{4}(22 + 33 - 2 \cdot 23) = 44$

$\quad\quad s = 0, q = 0,\quad$ ''$\quad\quad$'' $\quad \tfrac{1}{4}(33 + 11 - 2 \cdot 13) = 55 \tag{11}$

$\quad\quad q = 0, r = 0,\quad$ ''$\quad\quad$'' $\quad \tfrac{1}{4}(11 + 22 - 2 \cdot 12) = 66$

the necessary and sufficient conditions. They yield

$$23 = \tfrac{1}{2}(22 + 33) - 2 \cdot 44, \quad 13 = \tfrac{1}{2}(33 + 11) - 2 \cdot 55,$$

$$12 = \tfrac{1}{2}(11 + 22) - 2 \cdot 66. \tag{12}$$

These, used in the coefficients of rs, sq, qr, give

$$\tfrac{1}{4}(11 + 23 - 13 - 12) = 55 + 66 - 44, \&c.$$

Thus finally

$$\begin{aligned} W = \tfrac{1}{2}\{ & 44(q^2 + a^2) + 55(r^2 + b^2) + 66(s^2 + c^2) \\ & - 2[(55 + 66 - 44)\,rs + (66 + 44 - 55)\,sq \\ & + (44 + 55 - 66)\,qr]\}. \end{aligned} \tag{13}$$

N.B. This is for case of no dilatation. To find W without restriction, add to (13) the terms of (8) which involve δ.

I think it would be well to go through a rather full treatment of the problem of waves in an aeolotropic elastic solid. In preparation for it, we have to-day the dynamics of a homogeneous elastic solid of unrestricted character. I think perhaps it would have been better if instead of representing these tasinomic coefficients by 11, 12, etc., we had taken the notation of Thomson & Tait, (ee), (ef), etc. I would almost advise you to use (ee) instead of 11, etc.

There is a little note here to the effect that the thlipsinomic coefficients are sometimes more convenient than tasinomic coefficients. Tasinomic coefficients, in Rankine's nomenclature, are coefficients of strain in formulas expressing stress. On the other hand, thlipsinomic coefficients are coefficients of stress in formulas expressing strain. These coefficients can be got from one another by linear equations, of course. We have for example a stress

$$P = (ee)e + (ef)f + (eg)g + (ea)a + (eb)b + (ec)c.$$

You have six equations like that for P, Q, R, S, T, U. The next will be

$$Q = (fe)e + (ff)f + (fg)g + (fa)a + (fb)b + (fc)c,$$

etc. (ee), (ef) etc. are the tasinomic coefficients. Solve these equations for e, f, g, a, b, c. The thlipsinomic coefficients will be the coefficients (PP), (PQ), etc., in the formulas

$$e = (PP)P + (PQ)Q + (PR)R + (PS)S + (PT)T + (PU)U,$$

etc. These are more convenient for working with incompressibility and are also more closely connected with the practical moduluses that we are familiar with. Young's modulus is the stress divided by the strain when the stress is a simple longitudinal force and the strain is an elongation connected with a contraction—a lengthening of the wire and lateral contraction of it. In the elementary experiment for Young's modulus, you apply a given weight and by observation find the elongation that that produces. The formula for Young's modulus is $e = (PP)P +$ terms which are zero, so that the reciprocal of a thlipsinomic coefficient $1/(PP)$ is Young's modulus.

Let us stop and look at this vibrating affair. It has been going a considerable time with the exciter going through a constant range, and you see but small motion transmitted to the system. That is an illustration of the most general solution. Our handle P is in firm connection with the large pendulum and is forced to agree with it; and is to be viewed as the virtual exciter for a system of three particles. Let us bring these at rest. Now keep the pendulum going, and in the time when the viscosity will annul the system of vibrations, representing the difference between zero and the permanent state of vibration of these three particles, they will have acquired their permanent vibration. If there were no loss of energy whatever, the result would be that this jangled state would last forever, consisting of a simple harmonic motion in the vibrator and a compound of the three fundamental modes of these three particles viewed as a vibrating system with this bar P held fixed. Let this system with the lowest bar P held fixed [be set] to vibrating in any way whatever, and its motion will be a compound of those three fundamental modes. Besides that, set this exciter going and the state of the case is this; we may have this exciter and the whole set in simple harmonic motion of the same period, or superimposed upon that, any composition of the three sets of vibrations that the system might have with the exciter fixed. We cannot improve on the mathematical treatment by observation; and really a thing of this kind is more as a help or corrective to brain sluggishness than as a means of observation or discovery. In point of fact, we can discover a great deal better by algebra. But brains are very poor after all, and this model is of some slight use in the way of making plain the meaning of the mathematics we have been working out.

The system seems to have come once more into its permanent state. Let us stop this vibrator and see how long the system will hold its vibration. The reaction of the exciter is very slight, it is very nearly the same as if that bar were absolutely fixed. But the motion communicated to it since it is not absolutely fixed will correspond to a considerable loss of energy. A very slight motion of that bar with its great length and weight has considerable energy compared even with the energy of our particle of greatest mass; so that this system will come to rest far sooner than if this bar were absolutely fixed. These particles are at present illustrating phosphorescence. You see they have gone on vibrating for a whole minute, and the lowest of these three bars must have performed a couple of dozen vibrations at least. A phosphorescence of a hundred seconds' duration is quite analogous to the giving back of vibrations by that system for two or three dozen vibrations, only instead of two or three dozen vibrations we have 40,000 million million vibrations during the hundred seconds. Now, we cannot get 1,000 vibrations out of this system, because of the loss of energy in the wire, resulting from the generation of heat in it (which in our mind's eye we can see very clearly is connected with this system and is running away with its energy). That generation of heat by viscosity is simply the conversion of energy from one state of motion into another. In our molecular dynamics, we have no underground way of getting quit of energy or carrying it off. We must know exactly what is done with it when the vibrations end after a thousand million million. We must suppose the elasticity of our matter and molecules and so on to be perfect; and we cannot in any part of our molecular dynamics admit unaccounted for loss of energy; that is to say we cannot admit viscous terms unless as an integral result of vibrations connected with a part of the system that is not convenient for us to look at.

In three minutes our system has come very nearly to rest. We infer, therefore, that in three or five minutes from the commencement of a vibration we shall have nearly the permanent state of things.

Now we vary the period of the exciter, making it as unlike any fundamental period as possible. We will keep this going in an approximately constant range for a while and look at the vibrations that produces in the system.

I have explained how the thlipsinomic coefficients are more closely allied to practical moduluses. I may say, however, that in point of fact, one of our tasinomic coefficients is the pure modulus of rigidity in an isotropic body; but it may be regarded as the reciprocal of the corresponding thlipsinomic coefficient. Take the quadratic function in a, b, c which I express shortly [as]

$$(a, b, c)^2 = \tfrac{1}{2}\{44a^2 + 55b^2 + 66c^2 + 2(56bc + 46ac + 45ab)\}.$$

In the case of an isotropic solid, $44 = 55 = 66 = n$, the rigidity; and the tasinomic equation is $S = na$. Therefore, the thlipsinomic equation is $a = (1/n)S$, and the reciprocal of the rigidity is a thlipsinomic coefficient. The tasinomic and thlipsinomic coefficients for an isotropic body are not reciprocals of each other, however, in the case of longitudinal strains. You may readily see that the two are not reciprocals in any case in which there is more than one term in the linear function by which the stress or the strain is expressed.

Now you see very markedly the difference between the vibrations of our system after it has been going for several minutes with the exciter in a somewhat shorter period of vibration than that with which we commenced.

Here is another still shorter. In the course of two or three minutes the superimposed vibrations will die out. See now the tremendous difference of this case in which the period of the exciter is approximately equal to one of the fundamental periods of the system, or the periods to the case in which the lowest bar is held absolutely fixed. The angle through which that bar turns corresponds to ξ in our formula.

Returning to our tasinomic expression required the bulk modulus. [Problem I.] Taking for the moment average pressure per unit of area all around—for instance on the three pairs of faces of a cube—as the stress, the bulk modulus is

$$\frac{\frac{1}{3}(P + Q + R)}{e + f + g}.$$

We may obtain the solution in this way: Let the actual elongations be represented in terms of elongations e', f', g' which produce no change of bulk, and δ, as in the notes before you. The work required to produce the state of things represented by $e = e' + \frac{1}{3}\delta$, etc. will be the term in δ^2 in $\frac{1}{2}(e,f,g)^2$. Let k be the bulk modulus, and consider the work done in distortion. The working pressure varies from nothing to p, the final pressure, which, according to our definition of the bulk modulus, equals $k\delta$. The average working pressure therefore equals $\frac{1}{2}k\delta$, and the work done equals $\frac{1}{2}k\delta^2$. Therefore k is equal to the coefficient of δ^2 in $(e,f,g)^2$.

For the particular case of an isotropic body, we have $11 = 22 = 33 = \mathfrak{A}$, $12 = 23 = 31 = \mathfrak{B}$; therefore, $k = \frac{1}{3}(\mathfrak{A} + 2\mathfrak{B})$. That then, coming down to the particular case of an isotropic body, is the relation between the tasinomic direct modulus and the tasinomic lateral modulus. To interpret these, let everything equal zero except f. Therefore $P = \mathfrak{B}f$, which means that $\mathfrak{B}$ is the force per unit of elongation in the direction perpendicular to the force. Again, $Q = \mathfrak{A}f$, which means that $\mathfrak{A}$ is the force per unit of elongation in the direction of the force. These are moduluses that we are not so familiar with in practice.

So much then for our first problem. Next to find W for the case of incompressibility. This is a somewhat difficult conception to deal with, since every one of our coefficients [is] infinite. For the case of incompressibility, we put $k = \infty$, $\delta = 0$, $k\delta^2 = 0$; $k\delta$ generally remains of finite magnitude and will take the place of pressure. In the case of an isotropic body, $k\delta$ is the average pressure. Putting this compressibility modulus equal to ∞, in the form of an equation, we have

$$11 + 22 + 33 + 2(23 + 13 + 12) = \infty.$$

Except in some special and exceptional cases, each one of these 6 quantities 11, 12, etc. will be infinite. But the ratios between them that are effective in the expression for the energy in the case of a pure distortion of the solid in question are finite. It is upon this account that the thlipsinomic coefficients are more convenient in the case of incompressibility. We can scarcely treat an algebraic equation of 21 quantities, each infinite, with the finite ratios between them not explicitly stated; so that we are left in a doubtful position.

Now let us look at problem IV. Without restriction as to incompressibility—with none of these infinities—to annul web-oidal aeolotropy. "Web-like" I should say. I have not a Greek dictionary with me, and have not the grand command of classical knowledge which Rankine had. Every one of Rankine's words [is] well chosen and it is a most instructive lesson on the theory of elastic solids just to read them over. I want something for "web." Can any one tell us the Greek word for *web*? Well, weblike then. That is the kind of aeolotropy we have in a piece of woven cloth. I introduce a temporary notation in the quadratic expression for the work $e = \frac{1}{2}(r - s)$, etc. This assumes e, f, g to be such as to give no change of bulk. I am not assuming that the solid is incompressible, but I am assuming the case of distortion without change of bulk. The most convenient way of expressing that is to take three quantities q, r, s and put e, f, g equal to their differences so that we have $e + f + g = 0$. You might express e, f, g in terms of two quantities by means of this relation but that is unsymmetrical. The symmetrical system is a great brain saving system in all cases in which it is useful.

I would be much obliged if mathematicians would verify this work. To understand it, take the particular case $r = 0$, $s = 0$, so that there remains only q. What is the meaning of q in this case? We have $e = 0$, $f = -\frac{1}{2}q$, $g = \frac{1}{2}q$. It was the "half business" coming in here that was my reason and justification for the notation in my paper on elasticity that I referred to, which I am not insisting upon at present. But I will give you a reference to-day to Thomson & Tait, Art. 681, which will show you the importance of the question that I answered in a very special way which unfortunately becomes too artificial in this case. The marginal statement is, "Discrepant reckonings of shear and shearing stress, from the simple longitudinal strains or stresses respectively involved." The question is to pass from positive and negative normal pressure perpendicular to two diagonal planes to the reckoning of simple stress. The reckoning of simple stress is simply the amount of tangential traction in either set of planes. On the other hand, the numerical measure of the shear or the simple distortion comes out *double* the amount of elongation or contraction in the diagonal analogue. To make them both the same, I put in a $\sqrt{2}$. For this particular application it is not worth while to do that; but in the system set forth in that paper on elasticity we have a convenient symmetrical method of reckoning all stresses and strains so that the resultant of two orthogonal shears shall be the square root of the sum of their squares.

Take then the particular case $r = 0$, $s = 0$. What is the amount of shear corresponding to f elongation in one direction, and contraction in the other?

Answer, $2f$, $2g$; that is to say, q measures the shear corresponding to $\frac{1}{2}q$ elongation in one direction and $\frac{1}{2}q$ contraction in the other. The two directions are OY, OZ.

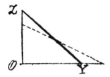

We have an extension in the direction OY and a shortening in the direction OZ and the question is, "What is the simple shear corresponding to that" and the answer is "It is numerically equal to twice the elongation, or to q." Thus q is the [shear] in the plane perpendicular to OX; but a is the strain in the plane perpendicular to OX. Therefore, the coefficient of q^2 in the equation of energy for this particular case must be equal to the coefficient of a^2 or $\frac{1}{4}(22 + 33 - 2 \cdot 23) = 44$ which is the formula stated. That condition is to express that there is such a deviation from aeolotropy as would be produced if we were to annul the differences of rigidity relatively to a shear produced by pulling out one diagonal and shortening the other compared with the shear of sliding one face past the other.

Suppose now you want to get quit of the sidelong coefficients 12, 13, 23. This equation, $\frac{1}{4}(22 + 33 - 2 \cdot 23) = 44$, you see, expresses 23 in terms of 22, 33, 44. These equations, used in the coefficients of rs, sq, qr, give

$$\frac{1}{4}(11 + 23 - 13 - 12) = 55 + 66 - 44, \text{ etc.};$$

and there remains finally, for the energy, the expression marked (13). In that expression for the energy, we have everything expressed in terms of 44, 55, 66, the three principal rigidities, and altogether independent of the moduluses that express the effects of direct pressures. We have here the most general kind of distortion, and we have the work of that distortion expressed in terms of the three rigidities; and we are ready, therefore, to go on and investigate distortional waves without further question as to whether the elastic solid is compressible or not. That question will only come up when we get to the reflection or refraction of light at the bounding surface of two mediums; or when we put in our molecules, or introduce equations which would produce condensation or rarefaction in the medium. But for the present we do not want to consider whether it is compressible or not; and that, in point of fact, is Green's position.

I had almost hoped that I would see some way of explaining double refraction by this system of molecules, but it seems more and more difficult.

I will take you into conference to-morrow, if you like, and show you the difficulties that weigh so much upon me. I am not altogether disheartened by this, because of the fact that such grand and complicated and highly interesting subjects as I have named so often, absorption, dispersion, and anomalous refraction, are all not merely explained by their means but are the inevitable results of this idea of attached molecules.

There is one thing I want to say before we separate, and that is, when I was speaking last of the subject, I saw what seemed to me to be a difficulty, but on further consideration, I find that there is no difficulty at all. Not very many hours after I told you it was a difficulty, I saw that I was wrong in making it appear to be a difficulty at all. I do not want to paint the thing any blacker than it really is and I want to tell you that that question I put as to the ether keeping straight with the molecules is easily answered when there is a large number. Our assumption was a large number of spherical cavities, lined with rigid spherical shells and masses inside joined by springs or what not, with the distance from cavity to cavity small in comparison with the wave length. It then happens that the motion of the medium relatively to the rigid shells will be exceedingly small and a portion of the medium that will contain a large number of these shells will all move together. If the distance from molecule to molecule is very small in comparison with the wave length then you may look upon the thing as if the structure were infinitely fine, and you may take it that the ether moves quite straight with them all, and not in and out among them as I said. It is evident that when the wave length of [the light traversing] the medium is moderate in comparison with the distance between the particles, it can move out and in among them. But if the stiffness of the medium is such as to make the wave length large in comparison with the distance from molecule to molecule, the stiffness is sufficient to keep them all together, and you may regard these rigid shells as bonds of attachment by which the molecule is pulled this way and that way, and that the millions and millions of these present the same effect as if the medium were made denser, so that we may suppose our reactionary forces, of which $C_1(\xi - x_1)$ is a sample, to be absolutely the same in their effect upon the medium as if they were uniformly distributed through it.

That takes away one part of the dissatisfaction of the thing. The only difficulty that I see just now is that of explaining double refraction. The subject grows upon us terribly, and so does the time, I think. If it is not too much for you I must have one of our double lectures for to-morrow.

We shall have in a short time a state of things in this model not very different from simple harmonic motion, if we get up the motion very gradually. We have now an exciting vibration of shorter period than the shortest of the natural periods. We must keep the vibrator going through a uniform range. We are not to augment it; and it will be a good thing to place something here to mark its range. Keep it going long enough and we shall see a state of vibration in which each bar will be going in the opposite direction to its neighbor. If we keep it going long enough we certainly will have the simple harmonic motion; and if this period is smaller than the smallest of the three periods, we shall, as we know, have these bars going in opposite directions. There is a longer period vibration of the largest mass superimposed on the simple harmonic motion we are waiting for. I will try and help to that condition of affairs by resisting [the] vibration of the top particle. In fact, that particle will have exceedingly little motion in the proper state of things (that is to say, when the motion is simply harmonic throughout), and it will be moving, so far as it has motion at all, in an opposite direction to the particle immediately below it. It is nearly quit of that superimposed motion now. We cannot give a great deal of time to this, but I think we may find it a little interesting as illustrating dynamical principles. Prof. Mendenhall is here acting the part of an escapement in keeping the vibrator to its constant range. We cannot get quit of the slow vibration of the particle. A touch upon it in the right place may do it. A very slight touch is more than enough. I have set it the wrong way.

Prof. Morley has been so kind as to work out a large part of the solution of this problem for the seven particles that I gave you, so that we shall be able to see the distribution of energy among the masses in the different modes of vibration, and so get a very instructive lesson, as I believe, in respect to fluorescence, phosphorescence, and the radiation from a body which has become heated by the transmission of radiant heat through it. Now we have got quit of that vibration and you see no sensible motion of the upper particle at all; these two are going in opposite directions, the lower one going in an opposite direction to the exciter. Therefore this is a shorter vibration than the shortest natural period. Now I set it to agree with the shortest of the periods, the first critical position. If we get time in the second lecture to-day, I am going to work upon this a little to try to get a definite example illustrating a particle of sodium. Before we enter upon any hard mathematics, let us look at this a little, and help ourselves to think of the thing. What I am doing now is very gradually getting up the oscillation. I am doing to that system exactly what is done to the sodium molecule, for example, when sodium light is transmitted through the vapor. We may feel quite certain, however, that the energy of vibration of the sodium molecule goes on increasing during the passage through the medium of at least two-hundred thousand waves, instead of two dozen at the most perhaps that I am taking to get up this oscillation. But just note the enormous vibration we have here, and contrast it with the state of things that we had just before. The upper particle is in motion now and is performing a vibration in the same period and phase as the lower particle, only through comparatively a very small range. The second particle, I am afraid, will overstrain the wire. By hanging up a watch, bifilarly, so that the period of bifilar suspension approximately agrees with the balance wheel, you get likewise a state of wild vibration. But if you perform such experiments with a watch, you are apt to damage it. This is a most magnificent contrast to the previous state of things when the period of the exciter was very far from agreeing with any of the fundamental periods.

We will now go to the treatment of the elastic solid. You will see a note in the paper of yesterday to which I have referred, stating that the thlipsinomic method is more convenient for dealing with incompressibility, and in point of fact it is so. I feel certain that if the k given by formula (9) is equal to ∞, the body must be incompressible, but that is the sum of six quantities each of which is generally—I believe essentially—positive for a true elastic solid. May some of these be finite, or is each one infinitely great? In all ordinary cases each one of the six quantities is infinitely great, and we are left in an unsatisfactory state as to coefficients. It will be necessary to go through a good piece of analytical work to make this clear and satisfactory. This is well worth doing, but we have not time to do it. Any of you who may wish to go into it, may proceed thus: Express the 21 coefficients in terms of 19 coefficients and k, which you can do by algebraical processes. Suppose k very great and see how things get on; then suppose k infinitely great and I think you will get some reasonable expression for incompressibility in terms of tasinomic coefficients.

I explained to you yesterday Rankine's nomenclature of thlipsinomic and tasinomic coefficients. In a certain sense, these may be all called moduluses of elasticity. I have defined a modulus as a stress divided by a strain, following the analogy of Young's modulus. If we adhere to that, then the tasinomic coefficients are moduluses, and the thlipsinomic coefficients are reciprocals of moduluses. The relations between the tasinomic and thlipsino-mic coefficients are well worked out by Rankine, but you can all do it for yourselves by going into the algebra concerned. There is not time for us to go into these matters in much detail. What we want is the essence of the dynamics. As far as symbols help us to that, we shall use symbols; and when symbols do not help us to that, we shall let them alone. We will now look at our paper:

Lecture Notes, Oct. 14

Thlipsinomic discussion of compressibility and incompressibility

$$e = (PP)P + (PQ)Q + (PR)R + (PS)S + (PT)T + (PU)U,$$
$$f = (QP)P + (QQ)Q + (QR)R + (QS)S + (QT)T + (QU)U, \qquad (15)$$
$$g = (RP)P + (RQ)Q + (RR)R + (RS)S + (RT)T + (RU)U.$$

Hence

$(e + f + g)$, or

$$\delta = [(PP) + (QP) + (RP)]P + \cdots + [(PT) + (QT) + (RT)]T + \cdots. \qquad (16)$$

Thus we see that

$$[(PP) + (QP) + (RP)], \cdots [(PT) + (QT) + (RT)], \cdots$$

of this formula are six compressibilities.

And for incompressibility each must be equal to zero, giving six equations,

$$[(PP) + (QP) + (RP)] = 0$$
$$- - - - - - - - - - - - - -$$
$$\qquad\qquad\qquad\qquad\qquad\qquad\qquad\qquad (17)$$
$$[(PT) + (QT) + (RT)] = 0$$
$$- - - - - - - - - - - - - -$$

Case of annulled skewnesses (Prob. III, of Oct. 13). The necessary and sufficient conditions are

$$(PS) = 0, \quad (QS) = 0, \quad (RS) = 0$$
$$(PT) = 0, \quad (QT) = 0, \quad (RT) = 0$$
$$(PU) = 0, \quad (QU) = 0, \quad (RU) = 0 \qquad (18)$$
$$(TU) = 0, \quad (US) = 0, \quad (ST) = 0$$

(twelve annulments leaving nine coefficients). In this case three of the compressibilities are annulled. The others are:

$$(PP) + (QP) + (RP), \quad (QQ) + (RQ) + (PQ), \quad (RR) + (PR) + (QR) \cdots.$$
$$(19)$$

————————◄▬►————————

It is startling to think of six equations to express incompressibility. I have not really noticed it before, but it is quite right, and you see the reason for it in this way: Consider an absolutely aeolotropic solid without any limitation whatever. Take [the] model of an elastic solid, if you like, that I showed you the other day, with its 18 coefficients. We will apply opposite shears to it. I shall apply a couple in this direction, and Mr. Forbes will balance that with a couple in that direction. Every one of you can understand the sort of thing that that does to the box. Suppose the axis of x is vertical. What we are doing is to shear this in the plane YZ by shears parallel to the axes. If the body be absolutely aeolotropic, doing this will alter its bulk; and again, to alter its bulk will produce that shearing effect.

Rankine did a great deal to cure the mathematical disease of aphasia from which we suffered so long; Faraday did most. The old mathematicians used neither diagrams to help people to understand their work, nor words to express their ideas. It was formulas and formulas alone. Faraday was a great reformer in that respect with his language of "lines of force," etc. Rankine was splendid in his vigor, and the grandeur of his Greek derivatives. Perhaps he over did it, but I do not like to call it an error. We cannot use all his words, but we learn from them in reading his papers. Instead of his platytatic and platythliptic coefficients, I use the much less grand and more colloquial expressions, sidelong normal and sidelong tangential coefficients. I do not know that Rankine has a word for the interaction between shears and shearing forces parallel to the faces, and direct strains. A direct strain in this case is an elongation parallel to any of the three axes. I assume you know what that means. These cross connections between shears and distorting stresses, on the one hand, and normal forces and a simple dilatation on the other, I can talk of as *sidelong*.

Look now at (15). What does (PS) mean? It expresses a relation between a distorting stress S, such as that which Mr. Forbes and I apply, and a shrinkage P. Annul everything in (15) except S, and the result is

$$e = (PS)S, \quad f = (QS)S;$$

so that (PS) is the dilatation we are causing in the direction OX. $(PS) + (QS) + (RS) = 0$ means that there is no dilatation from what we are doing.

It is clear, therefore, by this, that we have six equations to express that there is no dilatation under any kind of stress. You see also, how readily one is led to the treatment of incompressibility by thlipsinomic coefficients, while, on the other hand, it is very troublesome in terms of the tasinomic coefficients.

We next take up the case of annulled skewnesses—using a gross word as you see. I forget what Rankine's word for that would be. Skewnesses is a common word, but it is sanctioned by great mathematician so that we need not be ashamed of it. The annulment of skewnesses is set forth in problem III (Oct. 13) in short language, $(e, f, g) (a, b, c) = 0$, which means, of course, that the cross coefficients $(ea) = 0$, $(eb) = 0$, $(ec) = 0$, $(fa) = 0$, etc. That means 9 equations then, written short. Those 9 equations are obviously essential for annulment of skewnesses. Three more equations are necessary, viz.: the sidelong coefficients $(bc) = 0$, $(ca) = 0$, $(ab) = 0$, so that the quadratic $(a, b, c)^2$ reduces to a sum of squares.

To explain the tasinomic conditions, the question put is, what stress is required to produce a stated strain. Let, for example, the stated strain be a shear in the plane YZ denoted by a. If the body be aeolotropic, a stress compounded of P, Q, R, S, T, U will be required to produce it, none of the coefficients vanishing. But if the body be free from skewnesses, then it is clear that a shear of this kind requires no stress to produce it except the one corresponding to this shear. That is to say, shear a is produced by stress S, shear b is produced by stress T, [and] shear c is produced by stress U. Therefore we have 12 equations in order to annul skewnesses, bringing us down from 21 coefficients to 9. Why do we not, in avoiding skewnesses, annul the sidelong coefficients (ef), (fg), (eg)? We do not because obviously, without any skewnesses, a strain in this direction requires a stress in directions at right angles to it to prevent the body from swelling or contracting in those directions. Therefore (ef), (fg), (eg) belong clearly to the non-skew system, so that we have essentially 9 coefficients in that system.

To annul web-like aeolotropy requires the three equations (11). What may be taken as the most convenient yield of the problem are equations (12), because they allow us to get quit of the sidelong coefficients (ef), (fg), (eg) [$= 12, 23, 13$], leaving the direct coefficients 11, 22, 33 and the three principal rigidities 44, 55, 66. These equations (12) are of some importance. They are three out of Green's five equations by which he expresses that, of the three possible waves having wave front in one plane, two consist of vibrations in that plane, and one of vibrations perpendicular to it. His other two are 11, 22, 33. That shows you exactly the relation between Green's equations and the results that we have arrived at by the practical and static consideration of an elastic solid. I suppose most of you have Green's collected papers. I ask you the question because we shall use it a little in what follows. You will find these 5 equations at the foot of page 302.

We are about ready for the wave surface, but this is not an elementary class and we will not go into the geometry of the wave surface but will think of the results. As I said in the first lecture, one of the difficulties is quite refractory indeed. In the wave theory of light the velocity of the wave ought to depend on the plane of distortion. If you compare the results of the wave surface worked out for an incompressible aeolotropic elastic solid (we shall look at that a little more presently) you will see that it agrees exactly with Fresnel's wave surface if instead of the direction of the line of vibration of the particles in Fresnel's construction we have the normal to the plane of distortion.

I see no way of getting over the difficulty that the return forces in an elastic solid—the forces on which the vibration depend—are dependent on the strain experienced by the solid and on that alone. There has always seemed to me something indigestible in the way Green gets over it. I see that Stokes quotes in his report on double refraction, page 265 (British Association 1862). "In his paper on Reflexion, Green had adopted the supposition of Fresnel that the vibrations are perpendicular to the plane of polarization. He was naturally led to examine whether the laws of double refraction could be explained on this hypothesis. When the medium in its undisturbed state is exposed to pressure differing in different directions, six additional constants are introduced into the function φ, or three in the case of the existence of planes of symmetry to which the medium is referred. For waves perpendicular to the principal axes, the directions of vibration and squared velocities of propagation are as follows:

Wave normal		x	y	z
Direction of vibration $\Big\{$	x	$+A$	$N+B$	$M+C$
	y	$N+A$	$H+B$	$L+C$
	z	$M+A$	$L+B$	$I+C$

"Green assumes, in accordance with Fresnel's theory, and with observation if the vibrations in polarized light are supposed perpendicular to the plane of polarization, that for waves perpendicular to any two of the principal axes, and propagated by vibrations in the direction of the third axis, the velocity of propagation is the same."

We will try and keep this last sentence in our heads and study it. I have had an exceedingly exciting time since I saw you yesterday. I could not swallow this. It seemed to me to be absolutely wrong.* I feel this to be a

---

* [I am in receipt of a letter from Sir Wm. Thomson stating that he has thought of "extraneous forces" which can give rise to return forces dependent on rotational displacements; so that Green is here correct. The letter will be incorporated in the conclusion of this discussion of Green's second theory in Lecture [XVI].—H.]

very serious statement to make when Stokes quotes it and says that Cauchy does the same thing.

Let us see what this statement means before considering whether it may be verified, as Green supposes, by the introduction of "extraneous pressure." We are to have waves (for example N and W) perpendicular to any two of the principal axes, each propagated by vibrations in the direction of the third axis (up and down).

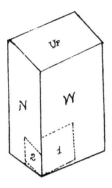

Take first the wave that is propagated South as I hold the box. There is the plane of the wave (N). The vibration up and down will consist of a distortion in this West plane (W). An upward vibration will give a shear like that (1) in which a rectangular figure becomes a rhombic figure. That represents the strain in the solid corresponding to this first state of motion. Similarly the wave propagated in this eastward direction will give rise to a shear of this kind (2), the vibration being upward. The assertion is that one set of waves is propagated at the same speed as the other. That is to say, the waves which have their shear in this West plane have the same velocity as the waves which have their shear in this North plane. The essence of our elastic solid is three different rigidities, one for shearing in this plane W, one for shearing in this plane N, and one for shearing in the other principal plane. The assumption is then that the velocities of propagation are the same in planes having different shears, i.e., do not depend on the shearing strain.

The introduction by Green (in order to accomplish this) of what he calls "extraneous force" which gives him three other coefficients has always seemed to me of doubtful ingenuity. These coefficients A, B, C occur in the little table given above, and I, M, N are the three principal rigidities. The table gives the squared velocities of propagation and waves of different wave normals and directions of vibration along the axes. The principal diagonal refers only to condensational waves, or waves in which the direction of vibration coincides with the wave normal. Taking vibrations in the direction

x, the assertion is $\mathcal{N} + B = M + C$, which, with the two corresponding equations for vibrations in the directions y, z, leads to

$$A - L = B - M = C - \mathcal{N}.$$

A, B, C are the effects of extraneous pressure. So far as I can see, they must be null. Begin with a body quite isotropic, so that we may not have our minds confused with the complicated question of aeolotropy—an elastic jelly, say, in a rectangular box. Let the box be altered in shape, still retaining its rectangular form. Will there be any difference of elasticity produced? Certainly not. The superposition of displacement will go on just as though the displacement and the external forces forming a system in equilibrium did not exist. Write down the equations if you like, expressing a stress in any portion of the solid. Superimpose a previous strain and you simply add to the formula the expression for the previous strain. The expression for stress in terms of strain is not modified by the fact that you superimpose a stress and strain upon a stress and strain previously existing. I say, therefore, it is a mistake to introduce the coefficients A, B, C if they correspond to nothing in nature. Make these annulments of A, B, C, and there is a table and a very convenient one you may find it, for the squares of the velocities in the directions stated for wave normals and vibrations. These quantities in our notation are $G = (ee)$, $H = (ff)$, $I = (gg)$, $L = (aa)$, $M = (bb)$, $\mathcal{N} = (cc)$.

I have said to myself, is it possible, after all, that this refractory difficulty can really be got over by that supposition of an extraneous force? I would not be lamenting that we could not explain double refraction if that were so; for this has long been a form of the elastic solid theory by Green in which he gets reconciliation to Fresnel's construction.

I will read two or three passages from Green. We may go back to where this is first mentioned in Green's paper "On the Reflexion and Refraction of Light." On page 248, he says, "Let us conceive a mass composed of an immense number of molecules acting on each other by any kind of molecular forces, but which are sensible only at insensible distances and let moreover the whole system be quite free from all extraneous action of every kind." That is what Green supposes first, and again he says (page 250), "The formula just found is true for any number of media comprised in this volume, provided the whole system be perfectly free from all extraneous forces, and subject only to its own molecular actions."

All this first paper is absolutely right, except the logic of those two passages that I have quoted, viz.: provided this whole system be perfectly free from all extraneous forces. If I am right in saying that the effects of his extraneous forces are null; that is logically wrong. If it is logically right, the error is mine. He uses that logic in his paper "On the Propagation of Light in Crystalized Media" read May, 1839, and in the very last paper in the book, "On the Vibration of Pendulums in Fluid Media," read at a considerably earlier date, Dec. 1833. The way he introduces it (and I have always turned from it when I saw it) is (p. 298), "If there were no extraneous pressures, the supposition that the primitive state was one of equilibrium would require $\varphi_1 = 0$, as was observed in the former paper; but this is not the case if we introduce the consideration of extraneous pressures."

Green meant a proposition of which this is a sample—take an elastic jelly; elongate it in one direction and shorten it in directions at right angles to that; and that will produce aeolotropy, introducing differences of propagation of waves in different directions in the manner his formula would show, with the A, B, C coming in along with the L, M, N. Here is how you might introduce aeolotropy into a jelly: viz., by compressing it beyond the limits of its elasticity. That is quite another affair. Even then, you just make it a crystalline solid, and you will come back upon a case in which the velocity of propagation will depend on the direction of the strain in this previously isotropic solid which has been rendered aeolotropic by stress. This mode of aeolotropy fulfils other than the conditions Green wants.

What the result of the introduction of extraneous force may be is of very great importance. If it be what it seems to me it is, it cuts away the last ground for explanation of the propagation of waves in a strained or unstrained elastic solid so as to fulfil Fresnel's law that the velocity of propagation depends on the direction of vibration rather than upon the plane of distortion.

As for the future of this work, I almost think it will pay us better not to trouble ourselves much more about wave surfaces. It is very pretty geometry; and if we had another week or fortnight, we might do more upon it. I may give you another leaf like this, putting down the wave surface on the elastic solid theory. But what we want to do is to think of the wave surfaces that we may get by other conceivable suppositions, suppositions that make the velocity of propagation depend on something else than the distortion of an elastic medium; and to think of whether by any of these methods we can get a wave surface agreeing with Fresnel's absolutely, or as nearly as the limits of accuracy of observation on which the belief in Fresnel's wave surface is founded require.

Before leaving this, I want you to notice that our equations of yesterday bring us virtually down to the assertion that when the wave is one of distortion alone and when the solid is symmetrically related to the axes, i.e., when we have got rid of skewness and web-like asymmetry, the problem is reduced to dependence on the three rigidities, so that all we want to know is 44, 55, 66. The 11, 22, 33 disappear either in the compressibility affair or in the condensational wave which may be propagated independently of the distortional wave in a true elastic solid. We do not care to get quit of the condensational wave so far as the theory of waves in crystals is concerned. It is only when we come to the subject of reflection and refraction that we require the condition of incompressibility.

Lecture XVI

I want to call your attention to this, that Green's formula for the energy on page 299 expresses the energy virtually as a function of strain components and rotation components. It is not explicitly put in terms of rotation components; it is put in terms of true strain components and certain differential coefficients which are neither pure strains nor pure rotations. I hope to get something written out on that by to-morrow, to put in your hands, to show precisely what Green's formula means, and you will see that it express energy in terms of rotation, as well as strain. If you think of the thing physically, you will see, I think, that it is quite impossible that a portion of the solid has been turned around by the extraneous forces. There is no relation to any non-rotating body by which we can possibly get terms in the potential energy depending on rotations. If there are terms in the potential energy depending on rotations, there would have to be terms in the expression for the forces required to hold the body displaced, depending on the angle through which a portion is turned, and that is obviously not the case.

Green does not discuss his energy formula as we are accustomed to do. He had not risen to the ideas of potential energy and the systematic interpretation of the coefficients that are now so familiar to us. He is one of those who led the way, but who died before going so far on it as has been done by his successors.

Now I want to think a little more about the possibility of explaining the phenomena of light by our system of detached molecules. As we have been touching so near upon double refraction, I shall continue upon it, and show you my difficulty as I promised. If time permits, in the next few days that remain, we shall put down a little more definitely, perhaps, the wave surface, and so on, that we are led to by such aeolotropy as we can get. I want, somehow or other, to extort an aeolotropy which shall be available for double refraction out of our supposition of molecules imbedded in an isotropic medium.

Take for our detached molecule the very simplest case of one particle, m_1. This is equivalent to making the remote attachment of spring C_2 a fixed point, or to making $m_2 = \infty$ in our equations. We thus find directly

$$\frac{x_1}{\xi} = \frac{C_1}{C_1 + C_2 - m_1/T^2},$$

which, substituted in the formula for the square of the refrangibility, gives

$$\mu^2 = 1 + \frac{C_1}{\rho}\left(\frac{x_1}{\xi} - 1\right)T^2 = 1 + \frac{C_1 T^2}{\rho}\left(\frac{m_1/T^2 - C_2}{C_1 + C_2 - m_1/T^2}\right).$$

The period of the molecule is given by $K_1{}^2 = m_1/(C_1 + C_2)$. If T^2 lies between $m_1/(C_1 + C_2)$ and m_1/C_2, all that I have said in favor of the more general expression, with reference to its availability for representing in a reasonable manner the facts of the ordinary refraction, applies as well to this; but in the former we can help ourselves, if necessary in any case to explain the facts of refraction, by a critical period considerably greater than the longest period with which we have to deal. It is probable that if we go into the thing very fully, examining such results as Langley's with rock salt, etc., we shall have need of something of that kind. There is not the same wealth of coefficients in this to explain the observed vibrations of refraction that we have in the general solution; but I do not know that we can get much out of the general solution that we cannot get out of this, so far as ordinary refraction is concerned.

Our supposition is that a smaller velocity of propagation than in the luminiferous ether is due to molecules being attached by a something to the ether. If it is explained by imbedded molecules, difference of velocity for waves in different directions (or in other words double refraction) must be explained in the same way. Let us try to do so. First $\mu^2 - 1$ must be something that is nearly constant for variations of T. The greatest dispersion is from 1.5 to 1.6; 4% of [the] difference in velocity from extreme red to extreme violet is very high dispersion; for ordinary refraction the difference in most cases is not more than a few per cent. On the other hand, there is a little difference between the double refractions (in iceland spar we have, for the ordinary and [the] extraordinary ray, 1.4 and 1.6, which shows at once a difference of 1/7 between the two refractive indices)—but double refraction is not a phenomenon of prismatic colors, and the difference between the two refractive indices for the extreme cases in iceland spar, although it does differ for the different wave lengths, does not differ enormously. If it did, double refraction would be obviously a colored phenomenon, as is helical change of the plane of polarization and as is rotational magneto-optic change of the plane of polarization. These two last mentioned phenomena are entirely dispersive, and the amount of dispersion is more than four times as great for violet light as for red light. We shall come to that hereafter.

In double refraction therefore there is very little dispersion to consider, and $\mu^2 - 1$ is very nearly constant. Writing this in the form

$$\frac{m_1 - C_2 T^2}{1 - (K_1/T)^2}$$

with a constant coefficient which I need not put down just now, we must have T considerably greater than K_1, so much greater that, writing this in the form

$$(m_1 - C_2 T^2)\left(1 + \left(\frac{K_1}{T}\right)^2 + \left(\frac{K_1}{T}\right)^4 + \cdots\right),$$

the first term $(K_1/T)^2$ will be sufficient to explain the dispersion. This gives a formula

$$(m_1 - C_2 K_1^2) - C_2 T^2 + (m_1 K_1^2 - C_2 K_1^4)\frac{1}{T^2}$$

$$+ \text{ another constant into } \frac{1}{T^4} \&c.,$$

which agrees with Cauchy's formula because T is proportional to the wave length. It is quite certain that $C_2 T^2$ must be very small in comparison with m_1 in order that $\mu^2 - 1$ may be very nearly constant.

If we were to depend at all upon this term, $C_2 T^2$, for explaining the difference of refractive index in different directions, we should have that difference directly proportional to the square of the wave length, or four times as much for red as for violet light, which is not verified by observation. Not being able to help ourselves by that term, can we help ourselves in virtue of the appearance of C_2 in K_1? No, because C_2 is small in comparison with K_1^2/T. The only thing that might help us is difference of values of C_1 in different directions. That will give for difference of refractive indices

$$\mu^2 - \mu'^2 = \frac{C_1 - C_1'}{\rho} \cdot \frac{T^2(C_2 - m_1/T^2)^2}{(C_1 + C_2 - m_1/T^2)(C_1' + C_2 - m_1/T^2)}.$$

Now can we in any way get anything constant out of that? Remark first that the factors of the denominator do not differ very much from our main denominator. Our main denominator expanded gives the comparatively exceedingly small change of value that corresponds to ordinary refraction, so that the denominator is approximately constant. Secondly, since m_1/T^2 is large in comparison with C_2 this difference is roughly

$$-\frac{C_1 - C_1'}{\rho DD'} \cdot \left(\frac{m_1}{T}\right)^2.$$

Thus the difference of our two refractive indices will be inversely proportional to the square of the wave length, and double refraction would be as colored a phenomenon as the effect of quartz upon polarized light, producing the brilliant effects you know so well. This is absolutely out of the question for explaining double refraction.

I have been working in silence for a considerable time on this molecular theory. I became more and more interested in it and it has been a very great incentive to keep me at work upon it to have had the prospect of speaking upon the subject to you. I cannot but feel that there is a great reality in the theory of detached molecules. I cannot believe that the theory that does what it does in the way of explaining two or three of the phenomena that I have named, which have been the most enigmatical of all the phenomena of light according to the ordinary considerations, can be passed over; I cannot but believe that it is really true. But the explanation of double refraction remains ungiven by it.

I am able to explain the very finest lines that Rowland can show us, as well as the broad bands. Others have explained that, so that I am only ambitious to point out what others have done in this direction; but what I wish to make noticeable is what others, I think, have not noticed so much, viz.: that we can do it without making away with energy. What seems to me important is to see how we can explain everything connected with observations of light by a definite communication of vibration to a system whose motions we can explain. As I have said two or three times before, the test of completeness and satisfactoriness in this kind of theory is, can we make a mechanical model of it. Take a perfectly elastic jelly—that is a known thing, that we can have and look at and experiment upon. Fill it up with myriads and myriads of things like these molecular shells, and you can produce a solid which will transmit vibrations at a slower velocity than if the jelly were not modified by their presence; and if the rate of diminution of velocity thus produced follows somewhat nearly the law of the velocity of light in an ordinary medium, and if besides we can account for the energy that is not transmitted as waves in a particular case, with periods approximately so and so, like the case of sodium vapor, by showing that it exists in the molecules and that it reappears afterwards, and if we can account in that way for all variety of dispersions, and so on, then I say we can make something like a mechanical model illustrative of waves of light, so far as our theory is concerned.

I want to go somewhat into detail as to periods and magnitudes of masses and energies, so that there may be nothing indefinite in our ideas upon this part of the subject. I want in the first place to call attention to two or three points connected with the possible density of the luminiferous ether. If any person present has seen a paper of mine, "Note on the Possible Density of the Luminiferous Ether and on the Mechanical Value of a Cubic Mile of

Sunlight," * I would be much obliged by him or her holding up a hand. I see Prof. Forbes. No one else?

The very title of it is peculiar. In a reprint of it in a lithographed volume that was about ready to come out when I left England, I find a note of date Dec. 22, '82 to the following effect: "The brain wasting perversity of the Insular system which still condemns British Engineers to reckonings of miles and yards, and feet and inches and grains and pounds and ounces and acres is curiously illustrated by the title and numerical results of this article. The sacrifice of this Insular system that you heard discussed yesterday at the Congress would be made not only by us but Americans would make very much the same sacrifice. I believe engineers would save such an immense amount of labor in their calculations that in whole departments of drawing offices and designing offices in engineering establishments their occupation would be gone. The distinguishing feature of an engineer is the quickness with which he can reduce from square feet to acres, and so on. If his brain were free from that, he might do more elsewhere, and have more time to find out about the properties of matter. In illustration of this I have been here wasting brain on cubic miles and cubic feet instead of walking about and getting rested for this lecture. I am not going to go through that, however, but I am going to try and make some estimate that you can understand, assuming that there must be a medium etc. I then thought that medium must be a continuation of our atmosphere. I could not say anything like that now.

"The first question that would naturally occur is, What is the density of the luminiferous ether in any part of space. I am not aware of any attempt having hitherto been made to answer this question, and the present state of science does not in fact afford sufficient data. It has, however, occurred to me that we may assign an inferior limit to the density of the luminiferous medium in interplanetary space by considering the mechanical value of sunlight as deduced in preceding communications to the Royal Society of Edinburg from Pouillet's data on solar radiation and Joule's mechanical equivalent of the thermal unit."

I want to ask in what proportion we may increase the numbers that depend on Pouillet's estimate. I think it is $1\frac{1}{2}$ or $1\frac{1}{3}$. For instance 83 foot-pounds per second per square foot becomes not far from 100 foot-pounds per second per square foot, so that if the whole light and heat from the sun on a square foot is all absorbed, we have a heating effect corresponding to about 100 foot-pounds per second. That is a very definite experimental question. There are many doubts as to the accuracy of Pouillet's results, but not sufficient to shake them as being in the main a rough approximation to the truth. Many observers have repeated them, and the tendency of observa-

* *Transactions of the Royal Society of Edinburgh* 1854.

tions since his time is to get larger and larger results. My impression is that Langley is inclined to reduce the figures. However, I am going to keep the figures as I have them here.

The mechanical value of a cubic mile of sunlight is 12,000 foot-pounds, equivalent to the work of a one-horsepower engine for one-third of a minute. There is something curious and interesting in that. The greatest volume of space lighted by the electric light is enormously short of the illuminating power of the sun over a cubic mile. It would be rather interesting to think of how many arc lights you must get into a certain space to have an initial illumination of, let us say, $\frac{1}{100}$ of a horsepower of work.

This result may give some idea of the actual amount of mechanical energy of the luminiferous motions and forces within our own atmosphere. Merely to commence the illumination of three cubic miles requires an amount of work equal to that of a horsepower for a minute; the same amount of energy exists in that space as long as light continues to [traverse] it; and if the source of light be suddenly stopped, must be emitted from it before the illumination ceases. Similarly we find (the law of this being the inverse square of the distance) 15,000 horsepower for a minute as the amount of work required to generate the energy existing in a cubic mile of light near the sun—45,000 times as much as for a cubic mile of the sunlight at the earth's distance. The matter which possesses this energy is the luminiferous ether. If, then, we knew the velocities of the vibratory motions, we might ascertain the density of the luminiferous medium; or conversely, if we knew the density of the medium we might determine the average velocity of the moving particles. Without any such definite knowledge, we may assign a superior limit to the velocities, and deduce an inferior limit to the quantity of matter, by considering the nature of the motions which constitute waves of light. For it appears certain that the amplitudes of the vibrations constituting radiant heat and light must be but small fractions of the wave lengths, and that the greatest velocities of the vibrating particles must be very small in comparison with the velocity of propagation of the waves. Let us consider, for instance, [homogeneous] plane polarized light, and let the greatest velocity of vibration be denoted by v, the distance to which a particle vibrates on each side of its position of equilibrium by A, and the wave length by λ. Then, if V denote the velocity of propagation of light of radiant heat, we have

$$\frac{v}{V} = 2\pi \frac{A}{\lambda}$$

and therefore if A be a small fraction of λ, v must also be a small fraction (2π times as great) of V. The same relation holds for circularly polarized light, since in the time during which a particle revolves once round in a circle of radius A the wave has been propagated over a space equal to λ. Now the whole mechanical value of homogeneous plane polarized light in an infinitely small space containing only particles sensibly in the same phase of vibration, which consists entirely of potential energy at the instants when the particles are at rest at the extremities of their excursions, partly of potential, and partly of actual energy when they are moving to or from their positions of equilibrium, and wholly of actual energy when they are passing through these positions, is of constant amount, and must therefore be at every instant equal to half the mass multiplied by the square of the velocity the particles have in the last mentioned case. But the velocity of any particle passing through its position of equilibrium is the greatest velocity of vibration, which has been denoted by v; and therefore, if ρ denote the quantity of vibrating matter contained in a certain space, a space of unit volume for instance, the whole mechanical value of all the energy, both actual and potential, of the disturbance within that space at any time is $\frac{1}{2}\rho v^2$. The mechanical energy of circularly polarized light at every instant is (as has been pointed out to me by Prof. Stokes) half actual energy of the revolving particles and half potential energy of the distortion kept up in the luminiferous medium; and therefore v being now taken to denote the constant velocity of motion of each particle, double the preceding expression gives the mechanical value of the whole disturbance in a unit of volume in the present case. *Actual energy* was Rankine's word. The expression, kinetic energy, I am answerable for. I called that mechanical energy then. I had not begun to talk of kinematics as the science of motions and dynamics as the science of force, and I then used "mechanics" as it was generally used in books and universities and as it is sometimes used still.

"Hence it is clear (here is the point) that for any elliptically polarized light the mechanical value of the disturbance in a unit of volume will be between $\frac{1}{2}\rho v^2$ and ρv^2, if v still denotes the greatest velocity of the vibrating particles. The mechanical value of the disturbance kept up by a number of coexisting series of waves of different periods, polarized in the same plane, is the sum of the mechanical values due to each homogeneous series separately, and the greatest velocity that can possibly be acquired by any vibrating particle is the sum of the separate velocities due to the different series. Exactly the same remark applies to coexistent series of circularly polarized waves of different periods. Hence the mechanical value is certainly less than *half* the mass multiplied into the square of the greatest velocity acquired by a particle, when the disturbance consists in the superposition of different series of plane polarized waves; and we may conclude, for every kind of radiation of light or heat except a series of homogeneous circularly polarized waves, that *the mechanical value of the disturbance kept up in any space is less than the product of the mass into the square of the greatest velocity acquired by a vibrating particle in the varying phases of its motion.* How much less in such a complex radiation as that of sunlight and heat we cannot tell, because we do not know how much the velocity of a particle may mount up perhaps even to a considerable value in comparison with the velocity of propagation, at some instant by the superposition of different motions chancing to agree; but we may be sure that the product of the mass into the square of an ordinary maximum velocity, or of the mean of a great many successive maximum velocities of a vibrating particle, cannot exceed in any great ratio the true mechanical value of the disturbance. Recurring, however, to the definite expression for the mechanical value of the disturbance in the case of homogeneous circularly polarized light, the only case in which the velocities of all particles are constant and the same, we may define the mean velocity of vibration in any case as such a velocity that the product of its square into the mass of the vibrating particles is equal to that whole mechanical value, in actual and potential energy, of the disturbance in a certain space traversed by it; and from all we know of the mechanical theory of undulations, it seems certain that this velocity must be a very small fraction of the velocity of propagation in the most intense light or radiant heat which is propagated according to known laws. Denoting this velocity for the case of sunlight at the earth's distance from the sun by v, and calling W the mass in pounds of any volume of the luminiferous ether, we have for the mechanical value of the disturbance in the same space $(W/g)v^2$, where g is the number 32.2 measuring in absolute units of force the force of gravity on a pound. Now we found above $83/V$ for the mechanical value in foot-pounds of a cubic foot of sunlight; and therefore the mass in pounds of a cubic foot of the ether must be given by the equation

$$W = \frac{32.2 \times 83}{v^2 V}.$$

If we assume $v = (1/n) V$, this becomes

$$W = \frac{32.2 \times 83}{V^3} \times n^2 = \frac{32.2 \times 83}{(192{,}000 \times 5{,}280)^3} \times n^2 = \frac{n^2}{3{,}899 \times 10^{20}}$$

and for the mass in pounds of a cubic mile we have

$$\frac{32.2 \times 83}{(192{,}000)^3} n^2 = \frac{n^2}{2{,}649 \times 10^9}.$$

It is quite impossible to fix a definite limit to the ratio $1/n = v/V$; but it appears improbable that it could be more, for instance, than $\frac{1}{50}$ for any kind of light following the observed laws. We may conclude that probably a cubic foot of the luminiferous ether in the space traversed by the earth contains not less than $1/(1{,}560 \times 10^{17})$ of a pound of matter, and a cubic mile not less than $1/(1{,}060 \times 10^6)$."

The statement is not that these are the number of pounds of luminiferous ether in the cubic foot and mile, but that the number of pounds cannot be less than these figures, or else that the velocity of the vibrations will be more than $\frac{1}{50}$ of the velocity of light. Let us see what this ratio is. The corresponding statement as to amplitude and wave length would be

$$2\pi \times \text{amplitude} = \tfrac{1}{50} \times \text{wave length},$$

or

$$\text{amplitude} = \tfrac{1}{300} \times \text{wave length}.$$

I think we can scarcely conceive of light coming away from the sun with vibrations through much greater amplitude than $\frac{1}{300}$ of the wave length. If it is not greater than that at the sun, then that mass of the luminiferous ether at the sun is 45,000 times the number of pounds here given per cubic foot, or $3/10^{16}$ pounds, so that we may say that the luminiferous ether cannot contain less than this amount of matter in the neighborhood of the sun, and probably through the solar system. There are strong reasons for supposing that the density of the luminiferous ether is pretty nearly the same all through the solar system. In fact, all we know about the propagation of light seems to show that the refraction depends on the difference of effective density of the luminiferous ether, and in so far as there is no sensible refraction, in all probability the luminiferous ether is very nearly of the same density.

I wish to make a little calculation, to show how much the luminiferous ether is condensed by the sun's attraction. We are accustomed to call [the ether] imponderable. How do we know it is imponderable? If we had never dealt with air except by our senses, air would be imponderable to us. But we can show that the weight of a column of air is sufficient to cause a difference of pressure on the two sides of a glass receiver. We have not the slightest reason to believe the luminiferous ether to be imponderable; it is just as likely to be attracted to the sun as air is. I do not like to make too many statements of that kind. At all events, the onus of proof rests with those who assert that it is imponderable. I think we shall have to modify our ideas of what gravitation is if we have a mass spreading through space with mutual gravitations between its parts without being attracted by other bodies. In the meantime, it is an interesting and definite question to think of what the weight of a column of luminiferous ether of infinite height resting on the sun will be, supposing the sun cold and quiet.

That is the same problem as that of the weight of the terrestrial atmosphere, supposing it of equal density throughout. You all know the theorem for mean gravity in calling the energy at different distances inversely as the square of the distance. That applied to the case of one of the distances infinite gives the ordinary potential law. Take a column of height h and one square foot section resting on the surface of a body of the size of the sun (radius $= r$). The mean gravity will be

$$\frac{h}{r(h + r)} \times 28.6 \times r^2,$$

the density of the sun compared with terrestrial density being 28.6. Make $h = \infty$ and this becomes $28.6 \times r$—I beg your pardon for going through all that; I ought to have known this result without finding it out. Unfortunately, I only remember the sun's radius in miles from the woeful defect of notation that is common to England and America. Call it 441,000 miles or $.44 \times 10^6$. To reduce to feet multiply by 5,280, and that by 28.6; and then that into the number of pounds in a cubic foot of the luminiferous ether. Will some of you kindly work that out. I make it 2×10^{-5} * * * I am very glad to find that I am right, but I thought the possibilities were $100:1$ that I was not.

I think we may say pretty safely that if the luminiferous ether is subject to gravity according to the same laws as are other bodies, the pressure per square foot on the sun's surface (setting aside the heat and motion of the sun) will be $2/10^5$ of the terrestrial weight of a pound. Compare that with the atmospheric pressure, which is 2,000 pounds. We find $2,000 \div (2/10^5) = 10^8$, so that the atmospheric pressure is one-hundred million times the ether pressure on the sun on the suppositions we have made.

Now, we have been supposing the luminiferous ether practically incompressible for light; but it does not follow at all that such a comparatively enormous pressure as $\frac{1}{50,000}$ of a pound per square foot might not condense it. Of course this is very far beyond our knowledge. But if the luminiferous ether has the density indicated, the pressure certainly at the surface of a body like the sun would be one hundred millionth of an atmosphere.

Lecture XVII

I have written out a statement regarding Green's expression for the effect of extraneous pressures. The formula for energy that Green gives on page 299—not that Green called it that; he had not that name and merely called it a quadratic function—commences with the three terms which are written at the top of this paper, involving A, B, C. I have called this $2W$ for convenience. The other terms are those that we are familiar with for the case of symmetry, but not farther reduced. I have not thought it necessary to write down more than the special terms I wanted to comment upon.

If you look at those terms, you see something quite unlike what appears in the equation of energy for an elastic solid as we know it. If we examine the meaning of those terms by taking our previous notation, a, b, c, for strains, and ω, ρ, σ for rotations, we have the second set of formulas in this paper. What is meant here by *rotations* is not angular velocities as in the vortex motion theory, but angular turnings. For instance, the half of $d\eta/dx - d\zeta/dy$ expresses the angle through which the corresponding portion of the medium must be turned to bring it back to such a position that what it has experienced is merely an irrotational strain. In other words, if ξ, η, ζ be the actual displacements of any particle in the medium, viewed as functions of x, y, z, the dislocation of the material consisting in displacements of every particle to the positions designated by ξ, η, ζ, whatever strain it involves, involves a rotation through an angle equal to ω, ρ, σ.

Find $d\eta/dz$, etc., in terms of strain and rotation and we have the third set of formulas. Substitute these in the expression for $2W$ and there results the last formula. This Green's formula, if it is true, implies that a certain amount of work would need to be obtained from the mere turning of each element, irrespective of the elastic forces between it and its neighbors. There is nothing that I can see in Green's assumption to correspond to that; there is no indication of any force that would produce it. The only way I see for producing anything of the kind would be by having two mediums mutually penetrating the space occupied and possessing some properties, of course not understood by us, according to which one of those mediums might resist the

Fac-simile of Lecture Notes, Oct. 15th.

Green's expression for
Effect of "Extraneous Pressures"

(p299)

$$2w = A\left\{\left(\frac{d\xi}{dx}\right)^2 + \left(\frac{d\xi}{dy}\right)^2 + \left(\frac{d\xi}{dz}\right)^2\right\}$$

$$+ B\left\{\left(\frac{d\eta}{dx}\right)^2 + \left(\frac{d\eta}{dy}\right)^2 + \left(\frac{d\eta}{dz}\right)^2\right\}$$

$$+ C\left\{\left(\frac{d\zeta}{dx}\right)^2 + \left(\frac{d\zeta}{dy}\right)^2 + \left(\frac{d\zeta}{dz}\right)^2\right\}$$

Put $\quad a = \dfrac{d\eta}{dz} + \dfrac{d\zeta}{dy} \quad ; \quad 2\varpi = \dfrac{d\eta}{dz} - \dfrac{d\zeta}{dy}$

$\qquad b = \dfrac{d\zeta}{dx} + \dfrac{d\xi}{dz} \quad ; \quad 2\rho = \dfrac{d\xi}{dx} - \dfrac{d\xi}{dx}$

$\qquad c = \dfrac{d\xi}{dy} + \dfrac{d\eta}{dx} \quad ; \quad 2\sigma = \dfrac{d\xi}{dy} - \dfrac{d\eta}{dx}$

We deduce

$$\frac{d\eta}{dz} = \tfrac{1}{2}a + \varpi, \qquad \frac{d\zeta}{dy} = \tfrac{1}{2}a - \varpi$$

$$\frac{d\xi}{dx} = \tfrac{1}{2}b + \rho, \qquad \frac{d\xi}{dz} = \tfrac{1}{2}b - \rho$$

$$\frac{d\xi}{dy} = \tfrac{1}{2}c + \sigma, \qquad \frac{d\eta}{dx} = \tfrac{1}{2}c - \sigma$$

Hence

$$2w = A\left(\frac{d\xi}{dx}\right)^2 + B\left(\frac{d\eta}{dy}\right)^2 + C\left(\frac{d\zeta}{dz}\right)^2$$

$$+ A\left\{\tfrac{1}{4}(c^2 + b^2) + (c\sigma - b\rho) + \sigma^2 + \rho^2\right\}$$
$$+ B\left\{\tfrac{1}{4}(a^2 + c^2) + (a\varpi - c\sigma) + \varpi^2 + \sigma^2\right\}$$
$$+ C\left\{\tfrac{1}{4}(b^2 + a^2) + (b\rho - a\varpi) + \rho^2 + \varpi^2\right\}$$

$$\text{I} \qquad\qquad \text{II} \qquad \text{III}$$

turning of the other relatively to it. But from the passage that I read to you yesterday from Green it is perfectly clear that he did not think of any thing of that kind. In the first place, as I said yesterday, the application of extraneous forces to a homogeneous isotropic solid cannot cause any difference in respect to the forces that would be produced by any dislocation superimposed upon that produced by the supposed extraneous forces— always provided the amount of the displacement is so small that the return force is simply proportional to the displacements of stresses represented by linear functions of the strains. If, however, this condition be not fulfilled, if stresses were applied so as to go beyond the proper limits of elasticity, or take first the case if there were a body that had proper elasticity through so wide a range that stresses might cease to augment in simple proportion to the strain and augment through more or less than a simple proportion, and if we were to apply extraneous forces to it sufficiently large to allow the deviation from simple proportionality to have any sensible effect, then A, B, C, terms such as those of Green would come into play. But under no circumstances that I can see could the rotational parts of Green's expression be true; and the only part of Green's expression that would have reality would be the first line, and the column marked III, in the last formula of this paper. But III, observe, would correspond merely to a modification of the principal rigidities. In other words, that column may be written in the form

$$(B + C)\tfrac{1}{2}a^2 + (C + A)\tfrac{1}{2}b^2 + (A + B)\tfrac{1}{2}c^2;$$

so that it would be merely equivalent to adding $\tfrac{1}{2}(B + C) \cdots$ to our rigidities $(aa) \cdots$. Also the first line is merely equivalent to adding A to our (ee), etc. I do not say, however, that we can adhere to Green's formula to this extent, that when A, B, C are the additions made to the direct tasinomic moduluses, then the additions to the rigidities would be $\tfrac{1}{2}(B + C), \tfrac{1}{2}(C + A), \tfrac{1}{2}(A + B)$.*

Take the other case of the weakening effects of stress applied to a body beyond its limits of elasticity. By hammering, you develop in all probability aeolotropy in a body previously isotropic. You will see mentioned in my article on elasticity an experimental proof of aeolotropy developed by such an action, showing the development of sidelong rigidities by torsion. A long straight steel pianoforte wire was twisted round through a great many turns—far beyond its limit of elasticity—and left to itself. Then it was found that when a weight was hung on it, it turned slowly in one direction and when the weight was taken off it turned back again. That was proof of a development of aeolotropy in rigidity that made itself manifest in an obvious enough way by sidelong coefficients of rigidity. I do not feel that this expression of Green's goes towards expressing the physical theory of the

* Change above III into I and $\tfrac{1}{2}$ to $\tfrac{1}{4}$.

introduction of aeolotropy in the properties of elastic solids such as is produced by hammering, with which we are all familiar, etc. It would be interesting for physics if it were.

[The terms II and III do not, as I first thought, express an impossibility. There is, in the case of an elastic medium subject to Green's "extraneous force," a dynamical relation to directions fixed with reference to the boundary of the portion of the medium concerned, which gives rise to return forces dependent on rotational displacements, analogous to the return force developed in a stretched cord by pulling it with equal force in opposite transverse directions, at two points very near one another, so as to produce an infinitesimal rotation of the intervening portion. Then what is called Green's second theory (pp. 305, 306 of his collected papers) does open a door for explaining the dependence of propagational velocity on direction of vibration instead of on the plane of distortion of the ether in a crystal. Stokes' explanation of this affair at the top of page 265 of his Report on Double Refraction (British Association, Cambridge, 1862), referred to, also on page 129 of his Burnett Lectures on Light (London, Macmillan, 1884), should be carefully read.]*

I want to pursue a little further the dynamics of an elastic solid, especially with reference to the wave theory of light. Before going on to that, I turn to questions of aeolotropy. Weblike aeolotropy is, I believe, a very interesting and important subject in practical mechanics. The theory of it for a continuous elastic solid helps us in working out ideas that are important in respect to structures. In a structure as a whole, properties corresponding to aeolotropy are produced in virtue of the manner of the structure. In fact, all structures of iron work, ties, and bracings, etc., are such that if we imagine a myriad of them put together—built up, as it were, like bricks—we should have an aeolotropic elastic solid. Our somewhat abstract questions of aeolotropy are closely connected with very important practical questions as to the mode of yielding of a body under the influence of certain definite forces. For example, take that of a tower made of diagonal bracings, etc., like that of the electric light tower to light the passage of Hell Gate in the harbor of New York. If any great weight is put upon the top of it, it will illustrate to us the same kind of sidelong aeolotropy in rigidity that the permanent twisting of a wire beyond its limits of elasticity develops in it. Generally the independent bracings of a tower are all placed symmetrically, so that nothing of the kind would happen, but take a tower braced unsymmetrically with diagonals all slanting one way, and there will be that kind of aeolotropy. I merely mention this as a somewhat crude illustration, just to show you that the theory of the continuous elastic solid is closely connected with subjects of great importance in engineering.

As to the physical properties of matter, which are more properly subjects

* Added Nov. 24th 1884.

of interest and the subjects that we occupy ourselves with, I say it is an investigation of very considerable importance to find whether or not there is any of this weblike aeolotropy in crystals. Take crystals of the cubic class—crystals which have perfect equality and symmetry—with respect to a cube. This is not a question of aeolotropy such as we have in the optical properties of biaxial crystals. The optical properties, as we have seen, are symmetrical with respect to the three axes. Our mechanical properties may or may not be so symmetrical. Take such crystals, which to appearance are absolutely similar in all their properties with reference to the six sides of a cube, and in reality seem to be absolutely similar in all physical properties as well; are they isotropic or not? There remains possible for them weblike asymmetry; and it seems not improbable that there will be weblike asymmetry of elasticity in cubic crystals. It may be very easily tested—or rather it is very easy to imagine a test. Think of what weblike asymmetry is with respect to a cube. I means more or less easier yielding to the distortion corresponding to a shear parallel to the faces than to the distortion corresponding to a shear parallel to the diagonals. Cut bars out in proper directions from such a crystal and test their flexuval rigidity—that would be one way. This is not so easily done, however, because it is exceedingly difficult to get crystalline specimens, and to cut bars out of them. Other ways may be thought of. I merely speak of this thing to point out an interesting subject of research: Are there or are there not aeolotropic properties in respect to elasticity in crystals of the cubic class? We can make models, as we have seen, of every kind of aeolotropy expressible by our 21 coefficients, and there is nothing easier than to make a model with weblike asymmetry. In fact, build up any structure with cubes—build up a structure of packing boxes, and there is preeminently a structure with weblike asymmetry. Take a structure built of cubic bricks and the fact of there not being absolute continuity through the mortar gives to that structure most distinctly a weblike asymmetry. The elastic properties of solids are nearly related to the perfect elasticity developed in idea, at least in connection with infinitely small displacements.

I do not need to put the question, is there *any* deviation from isotropy in crystals of the cubic class. The very first question of crystallography shows that there is. I remember a fine specimen of crystalline spar which Dr. Wm. Cooper showed us quite 50 years ago, and knocked off a corner with a hammer. The fracture proved aeolotropy of strength. That well known elementary experiment shows us that the crystal is stronger in one direction than in another. That being so, does it not seem improbable that its moduluses of elasticity are all equal? It is a question of interest, and I had hoped to find ways of experimenting—I have not time to think of it now—and to experiment and find whether there were three moduluses of elasticity in crystals of the cubic class, and to get approximations to their magnitudes.

We have passed over preliminary considerations regarding double refraction. It is not necessary to spend any of the time that remains in going into the well known geometrical treatment of Fresnel's wave surface, whether we do it as Fresnel did it or get it from the elastic solid. That is sufficiently entered into in the various elementary works upon the subject. But I want you distinctly to consider this question: What reasons have we for judging as to whether the direction of vibration is perpendicular to or is in the plane of polarization? To understand the meaning of the question, we must know what we mean by the plane of polarization. That is a mere technicality. The plane of incidence and reflection when light is polarized by reflection is called the plane of polarization. With that definition the question has a meaning. Of course, otherwise it might be confounded with the question how are you going to define the plane of polarization. I wish the question had come to us otherwise. I wish the plane of polarization had been defined in the beginning with respect to the vibrations, not [that] the question had been put more distinctly [as] a physical question, in respect to light polarized by reflection, viz. Is it when vibrations are in the plane of reflection, not refraction, that at a certain angle no light or but very little light is reflected, or is it when the vibrations are pendicular to the plane of reflection and refraction that at a certain angle but little or no light is reflected? That is the physical question.

Mathematical literature has been loaded with a great deal of bad writing on this subject. A great number of investigations and statements called theories have been made, in which a piece of dynamical work is gone through; and then a condition is arbitrarily introduced; and that is called Cauchy's theory, and something else is called Neumann's theory and something else is called MacCullagh's theory. I have perhaps done injustice in this statement to the possessors of these three names who have done such great things. I support myself, however, in this statement by reading a few lines from Lord Rayleigh's paper on the Reflection of Light from Transparent Matter. It is rather a celebrated thing. "Quite different from the foregoing is the theory of MacCullagh and Neumann, which is given in accessible form in Lloyd's 'Wave Theory of Light.' The following principles are laid down as the basis of investigation: I. The vibrations of polarized light are *parallel* to the plane of polarization. II. The density of the ether is the same in all bodies as *in vacuo*. III. The *vis viva* is preserved; from which it follows that the masses of the ether put in motion, multiplied by the squares of the amplitudes of vibration, are the same before and after reflection. IV. The resultant of the vibrations is the same in the two media; and therefore in singly refracting media the refracted vibration is the resultant of the incident and reflected vibrations." One of these principles is simply an arbitrary assumption absolutely inconsistent with the dynamical conditions of the problem. If you want not to put too fine a point on it, you may call it MacCullagh's mistake or Neumann's mistake. Here is Lord Rayleigh's remark upon it: "When the vibrations are normal to the plane of incidence, and therefore parallel in all three waves, the application of these principles gives rigorously Fresnel's tangent expression. If the vibrations are in the plane of incidence, the fourth principle alone leads to Fresnel's sine-formula. This only shows that the fourth principle is inconsistent with the others; for, as we shall see, unexceptionable reasoning founded on I and II leads to an altogether different result. The very particular case of IV required when the vibrations are normal to the plane of incidence happens to be correct."

Lord Rayleigh, I see, has the thing wrong so that I cannot show all the niceties of the wrongnesses of it. Everything about reflection and refraction of waves of light at the bounding surface separating two elastic solids is absolutely definite, and not hypothetical at all. Nobody can introduce a principle; it is a thing in which we have absolutely definite conditions to fulfil. I hope to put before you in a short form by to-morrow the conditions to be fulfilled and perhaps part of the work. You find the thing done absolutely correctly by Green, and you find Green's theory reproduced, with some very important analytical improvement in the treatment of it, by Lord Rayleigh, and in Lord Rayleigh's paper you find the thing worked out in a direction that Green left it unworked. Green, I may say, in a somewhat lax and not very well considered statement, assumes that the rigidities are equal in the two mediums and that the difference of wave length is due to difference of density. The only knot in this paper of Green's is manifest in what I read to you here: "The formulae which we have obtained are quite general and will apply to the ordinary elastic fluids by making $B = 0$ [that is, rigidity $= 0$]; but for all the known gases A is independent of the nature of the gas, and consequently $A = A_1$. If, therefore, we suppose $B = B_1$, at least when we consider those phenomena only which depend merely on different states of the same medium, as is the case with light, our conditions become, etc." There is a note here: "Though for all known gases A is independent of the nature of the gas, perhaps it is extending the analogy rather too far to assume that in the luminiferous ether the constants A and B must always be independent of the state of the ether, as found in different refracting substances. However, since this hypothesis greatly simplifies the equations due to the surface of junction of the two media, and is itself the most simple that could be selected, it seemed natural first to deduce the consequences which follow from it before trying a more complicated one, and, as far as I have yet found, these consequences are in accordance with observed facts."

Now the analogy with gases is quite nonsense. I am rather surprised that Green put that in his paper as a reason for making $A = A_1$, because in his paper on the Reflection and Refraction of Sound he takes the reflection of sound at a surface of separation between air and water, in which the relation corresponding to this does not hold, and he points out how enormously far from holding is any such relation as this. I spoke of the disease of aphasia. This is a manifestation of it. What does one know of the meaning of A and B who can only speak of the properties of matter by "A" and "B". If Green had thought of the thing itself and not of the letters he would have saved himself that reference to gases at all. He would simply have said this, "Let us try the case of equal rigidities, and unequal densities," and he might have added, "This simplifies the formulae, and so far as I know, the results of the formulae with this simplification agree with observation." That is the state of the case. Everything else in Green is perfect. Lord Rayleigh improves the mathematical treatment by adopting that most valuable piece of shorthand, the imaginary symbol. Without the imaginary symbol, you have 8 equations in 8 unknown quantities. A skillful pilot will pilot himself among these 8 unknown quantities and pretty quickly find that they reduce to 4. But that is rather artificial work even for a skillful pilot among mathematical symbols. Lord Rayleigh's way can be followed by anybody acquainted with the mathematical forms and theorems he uses who is no pilot at all. The enormous value of this mathematical shorthand—I owe that expression to Lord Rayleigh himself—is illustrated by no case better than by this. I do not care to use it when it does not help us; I prefer the sines and cosines; but when it saves ink and paper and brain let us by all means use shorthand.

Lord Rayleigh considers the question, Can you account for the known phenomena of the reflection of light, polarized and unpolarized, other than by supposing the rigidities equal in the two mediums and the densities unequal? He discusses the question penetratingly, and by a particular test case he finds that it is impossible to get anything approximately of the same character as the real phenomena by the other extreme supposition which is admissible, that the difference of velocity in the two mediums depends on one of them being more rigid than the other, while their densities are equal. One of these suppositions, as Green found, gives results which somewhat approximately agree with the phenomena; the other, Lord Rayleigh proves, gives results exceedingly far from the phenomena.

Here is the state of the case: With the vibrations perpendicular to the plane of incidence in a wave of incident light, the supposition of equal rigidities and unequal densities in the mediums gives exactly Fresnel's law for light polarized in the plane of reflection. Alter this now by supposing the densities equal and the rigidities unequal and you get exactly Fresnel's formula for light polarized perpendicular to the plane of reflection. In other words, the polarization of light by reflection could be accounted for by supposing the densities equal and [the] rigidities unequal and the vibrations of polarized light in the plane of reflection, because in this case the light which is wholly transmitted and none of which is reflected consists in vibrations perpendicular in the plane of incidence. So far, therefore, we cannot judge between the two suppositions. But take the formula for vibrations in the plane of incidence. If the densities are unequal and the rigidities equal that gives us Fresnel's formulas. Those formulas are—one of them rigorously, the other approximately—the results of the full dynamical investigation corresponding to this supposition. But if we now take the other supposition, we get only one of Fresnel's formulas fulfilled, and the other excessively far from being fulfilled. It is absolutely impossible to get anything near to Fresnel's formula by supposing the vibrations of polarized light to be in the plane of incidence and reflection.

It remains to be considered whether by an intermediate supposition we can get any improvement in the result. For instance, suppose the density to be greater in one of the mediums and the rigidity to be greater but much less greater in proportion than the density. We might in that way get an improvement on the imperfect agreement for one of Fresnel's formulas without losing the perfect agreement for the other. But a full examination of that case leads to no satisfaction whatever.

We have an approximate agreement with Fresnel's formula on the supposition that the vibrations are perpendicular to the plane of incidence and that the action of the two media upon one another is that of homogeneous elastic solids; but the agreement is only approximate. Take Green's expression for the square of the ratio of the reflected and incident light

$$\frac{\beta^2}{\alpha^2} = \frac{K + (\mu^2 - 1)^4 \cdot b^2/a^2}{H + (\mu^2 - 1)^4 \cdot b^2/a^2}.$$

K vanishes at the polarizing angle; but what remains corresponds to a considerable deviation from zero in the amount of the reflected light. You find this gone into in an appendix to Green's paper on reflection and refraction of light. For the case of air and glass we find as much as $\frac{1}{49}$ for β^2/α^2 at the polarizing angle. The amount of light reflected at the polarizing angle is very much less than that.

We cannot spend much more time upon this. Between Green and Lord Rayleigh we have the thing quite complete, so if I have explained it very badly to-day, you may make amendments to my explanations by reading Green and Lord Rayleigh. There are enough reasons here to make it very difficult to avoid the conclusion that the vibrations are perpendicular to the plane of polarization. But there are still stronger reasons than we have here. The strongest reason is of the kind first suggested by Prof. Stokes. It is closely related to his celebrated experiment on diffraction. I cannot say that it cannot be answered, but it seems to me that it is unanswerable. Good reasons for considering it unsatisfactory have certainly been given, but I think it probable that when the thing is fully examined it will be found that the conclusion may be still considered as rendered very probable, if not absolutely certain, by Stokes' diffraction experiment. But the experiment that seems most decisive is that on the polarization phenomena analogous to the blue of the sky. Stokes first suggested this, I believe, as a reason for supposing that the direction of vibration is perpendicular to the plane of polarization, but, as Lord Rayleigh has shown, it was not so clear as Stokes supposed it to be. The view is this: Imagine a color to be produced by an enormous number of particles of diameters small in comparison with the wave length. The colors of the blue sky are only seen when the particles are known to be small in comparison with the wave length, which is not the case with colored dusts and halos, etc. Stokes' view is that if the luminiferous ether is moving to and fro in the neighborhood of a particle the effect will be the same as if the ether were at rest and the particle moving, the relative motion of the two being all that we have to consider. That being the case, it is obvious that the effect of a single spherule like that in the air, or of a vast number, would be to produce the kind of waves that we first considered. That is to say, waves with a zero motion in this direction and this ↔ and oscillations to and fro perpendicular to the equatorial plane. You remember our formula with polarization in the equatorial plane. That is the kind of vibration we should have if the effect of the particles were as assumed in that view of Stokes. Therefore the light from a particle must consist of vibrations per-pendicular to the plane which is perpendicular to the line through the center of the particle in the direction of the vibrations of the ether at the particle—the effect of the relative motion is that and cannot be anything else but that. Therefore all we have to do to find the direction of vibration in plane polarization is to test the polarization of light in the equatorial plane. The blue sky is complicated by the reflection from the surface of the earth, white clouds, etc., but in the main the light of the blue sky presents an almost complete polarization and a polarization in the plane through the sun— There is an almost complete polarization when we look in a direction at right angles to the direction of the sun. Experiments made on blue precipi-tates of various kinds all agree in this respect; Lord Rayleigh, however, points out that there is another way of viewing the thing. We might in the first place assume that we have a dense mass whose inertia prevents it from moving, but Lord Rayleigh looks more particularly into the nature of the thing and considers this body as in many cases transparent. He considers the initiation of light upon it and passes continuously from the case of large drops of rain to the smaller drops of cloud white and the little particles of sodium or salt, or spherules of dust or whatever they may be, which cause the blue of the sky. He investigates fully the case when the particles are exceedingly small in comparison with the wave length. You must think of the light as *reflected* and *refracted* from the particle when it is large, and we are just brought back to the question I have put before you of the reflection of light at a transparent body. But when the particle is small in comparison with the wave length the theory of reflection and refraction at bounding surfaces does not at all follow. Lord Rayleigh works out the problem for equal rigidity and differing density and again for equal density and differing rigidity. The one is shown to come out exactly as Stokes pointed out but did not go into so fully, which is represented here by the to and fro vibration. The other case is curious and is worth special consideration. I will put it down here and contrast it with the other—Suppose the spherule to differ from the rest of the medium in not having the same rigidity. What sort of vibrations will be produced? At the place of maximum displacement there is zero strain, but at the time when there is zero displacement there is a maximum of strain. Now when the difference is a difference of density this spherule will tell by its presence at the time when the acceleration of the medium to right or left is greatest, and the only effect on the medium is continuity of strain. On the other hand, if the densities are equal the motion of the ether will have no effect at the times of maximum acceleration and zero distortion; it will have maximum effect at the time of maximum distortion. Let us put down an indication of the distortion of the luminiferous ether. We will have a slipping of one of these parallel lines with reference to the other. Suppose this spherule has not the same rigidity as the luminiferous ether, it will be slewed from side to side in the manner I am indicating. It will be drawn out there (a) and in there. It is a bad drawing but it shows the principle.

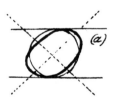

A circle is made oblique by sliding all the chords parallel to one diameter in one directions. A particle will then alternately be made oblique in this direction and made oblique in the other direction. I will put in dotted lines for the obliquities on either side. The vibration consists in an alternate elongation in a direction of 45° from the vertical on one side, and an elongation in the direction of 45° on the other side of the vertical, with zero of change in the direction perpendicular to the board. Think of spherules yielding to the distortion of the ether, but having more or less rigidity. That will cause them to act upon the medium in the same way as a vibrating body alternately getting longer and shorter in this direction and shorter and longer in this direction (dotted lines). That was one of our fundamental oscillations, our second case of motion, you will remember. There will be zero of effect in two lines at right angles to one another and maximum effect in directions perpendicular to those. Lord Rayleigh has pointed out that there will be no phenomenon corresponding to the zeros in this solution. We may consider this test of Lord Rayleigh as settling the thing that Stokes overlooked. While Stokes says so and so, Lord Rayleigh says it is not so clear, but on looking into the thing, finds it must be so.

Thus we have absolutely proved that the direction of vibration is perpendicular to the plane of polarization, because we find that the plane of polarization, defined in the usual way and tested by Nicol's prism or what not, is the plane through the sun in the case of light reflected at right angles to the direction of illumination by a body consisting of minute spherules separate from the luminiferous ether. I shall try to put a little more clearly on paper the state of the case in reflection and refraction to-morrow. I had intended to say something upon molecular dynamics to-day, but alas, the time has all gone. I have used my opportunities very imperfectly in bringing this subject before you, but we must make the best of it, notwithstanding.

Lecture XVIII

I have tried to put down something regarding the reflection and refraction of waves at the surface separating two homogeneous mediums, vibrations to be in the plane of the three rays. I took that case at once because the other cases are so exceedingly easy that it does not matter much whether we take them or not. You will find them thoroughly and simply worked out in Green and also Lord Rayleigh.

Lecture Notes of Oct. 16

Refraction and Reflection at Interface between Two Homogeneous Elastic Mediums

I—Vibrations in the plane of the three rays—This plane zy

$$\xi = \frac{d\varphi}{dx} + \frac{d\psi}{dy}$$

$$\eta = \frac{d\varphi}{dy} - \frac{d\psi}{dx} \qquad \delta = \nabla^2 \varphi \tag{1}$$

$$(k - \tfrac{2}{3}n)\delta = p \tag{2}$$

($-p$ corresponds to fluid pressure.)

$$P = p + 2n\frac{d\xi}{dx}, \quad Q = p + 2n\frac{d\eta}{dy}, \quad R = 0$$

$$S = 0, \quad T = 0, \quad U = n\left(\frac{d\xi}{dy} + \frac{d\eta}{dx}\right) = n\left(2\frac{d^2\psi}{dx\,dy} + \frac{d^2\psi}{dy^2} - \frac{d^2\psi}{dx^2}\right) \tag{3}$$

$$\rho\frac{d^2\xi}{dt^2} = \frac{dP}{dx} + \frac{dU}{dy}, \quad \rho\frac{d^2\eta}{dt^2} = \frac{dQ}{dy} + \frac{dU}{dx} \tag{5}$$

These without accents refer to upper medium
with   ”   ”   ” lower   ”

$$\psi = A\varepsilon^{\iota(az+by+\omega t)} + A_1\varepsilon^{\iota(-ax+by+\omega t)}$$

$$\varphi = B\varepsilon^{-b'x+\iota(by+\omega t)}$$

$$\psi' = \varepsilon^{\iota(a'x+by+\omega t)}$$

$$\varphi' = B'\varepsilon^{b'x+\iota(by+\omega t)} \quad \text{where } \omega = 2\pi/T$$

(6)

By (1), (2), (3) and (5) we have

$$\rho\frac{d^2\psi}{dt^2} = n\nabla^2\psi$$

$$\rho\frac{d^2\varphi}{dt^2} = (k + \tfrac{4}{3}n)\nabla^2\varphi$$

(7)

From (6) and (7) we find

$$b^2 - b'^2 = \rho\omega^2/(h + \tfrac{4}{3}n)$$

$$\rho\omega^2 = n(a^2 + b^2) = n(a'^2 + b^2)$$

(9)

At interface $(x = 0)$ we have

$$\xi = \xi', \quad \eta = \eta', \quad \text{by continuity of matter}$$

and

$$P = P', \quad U = U', \quad \text{by balance of forces}$$

(10)

We have, by (8), and (2), and (8) and (9), the part of P dependent on φ,

$$= \{(k - \tfrac{2}{3}n)(b'^2 - b^2) + 2nb'^2\}B$$

$$= \{-\rho\omega^2 - n(b'^2 - b^2) + 2nb'^2\}B$$

$$= n\{-(a^2 + b^2) + (b'^2 + b^2)\}B$$

$$= n(b'^2 - a^2)B$$

(11)

Hence, and from (6) and (1),

$$\xi = \xi' \text{ gives} \quad -b'B + \iota b(A + A_1) = b'B' + \iota b$$

$$\eta = \eta' \quad " \quad \iota bB - \iota a(A - A_1) = \iota bB - \iota a'$$

(12)

$$P = P' \quad " \quad n\{(b'^2 - a^2)B - 2ab(A - A')\} = n'\{(b'^2 - a'^2)B' - 2a'b\}$$

$$U = U' \quad " \quad n\{-2\iota b^2 B + (a^2 - b^2)(A + A_1)\} = n'(2b^2 B' + a'^2 - b^2)$$

(13)

For case of sound don't use (11); but *early* take $n = 0$. Results green, quite simple and perfect.

For case of incompressibility take

$b' = b.$ \hfill (13)

For general supposition of finite compressibility, b' must, by (8), be either wholly real, or wholly imaginary.

Case I. ω too small to allow condensational wave to be generated by the reflection and refraction.

Case II. $\rho\omega^2/(k + \frac{4}{3}n) > b^2$; a condensational wave is generated. If ℓ be its wave length

$$\frac{2\pi}{\ell} = \sqrt{(b^2 - b'^2)} = \sqrt{\frac{\rho\omega^2}{k + \frac{4}{3}n}}$$ \hfill (14)

as we know long ago.

I will say a few words in explanation of the different formulae. In the first place, we have motion in two dimensions alone, and our formulae belong therefore to the general formulae with that limitation to two dimensions and coordinates x, y—z not appearing. In our original division of the solution into a condensational part and a distortional part, φ in equation (1) corresponds to the first, and ψ to the second; for observe that ψ expresses here a solution for which $d\xi/dx + d\eta/dy = 0$, which is the condition of no dilatation. We have separated the solution, therefore, merely by a functional device, into these two parts. As we are going to apply these solutions to the case in which the medium is incompressible, so that condensation is impossible, I will introduce a new word. Instead of "condensational wave," we will talk of "pressural wave"; and we shall find that at the bounding surface of a medium we have a pressural wave, even if the medium be incompressible. I have brought in $p = (k - \frac{2}{3}n)\delta$ because it does not become infinite at all when k becomes infinite, the other factor δ becoming zero. Verify the values of P and Q which appear in equations (3) by substituting for p the value $(k - \frac{2}{3}n)\delta$ and you will find that they come to forms written out for them in one of the earlier lectures. I put down a form for P, you may remember, that I said was convenient for some purposes [lecture I]. This being a case of motion in two dimensions, the shear is wholly in the plane xy. Therefore $R = 0$, $S = 0$, $T = 0$. The value of U obtained from its fundamental form completes equations (3). In point of fact and as matter of arrangement, I need not have written down the general equations of motion (5) at all, but might simply have taken equations (7) from our old friends the differential equations [lecture I] which φ and ψ must fulfil. However, it is well to put it down from the beginning and to verify for yourselves that equations (7) are derivable from (1), (5), etc. All these formulas used without accents refer to the upper medium; all with accents refer to the lower medium. I use the words upper and lower merely for convenience and as corresponding to this diagram, without reference to the actual positions of the mediums. We might have, for instance, a case in which water was the upper medium of our diagram and the lower medium air. This is a case in which the introduction of the analytical shorthand is very valuable. Try this case without it, and you will find you have 6 equations in 6 unknown quantities. The analytical shorthand reduces the problem to 4 unknown quantities, A, A_1, B, B'. I use the symbol ι for $\sqrt{-1}$, i [and] i' being reserved for the angles of incidence and refraction. ω is the angular velocity of the relatively circular motion, or $\omega = 2\pi/T$. The object is to express a simple harmonic motion. The advantage of the mathematical shorthand consists in the fact that a similar set of formulae holds for $-\iota$ as well as for ι. You can realize by adding, obtaining sine and cosine formulae. Just remark the term $\varphi = B\varepsilon^{-b'x+\iota(by+\omega t)}$. If B comes out in our result a real quantity, change the sign of ι and add. That gives a cosine. If B comes out a pure imaginary, change ι into $-\iota$ and subtract. That gives a sine. In reality B will come out mixed real and imaginary, and there will result this form:

$$\varepsilon^{-b'x}\{C\cos(by + \omega t) + D\sin(by + \omega t)\}.$$

I have taken out this term because we want to look a little more particularly at it.

Let us think of the meaning of these different terms. There is only one plane distortional wave in the lower medium, because by hypothesis the light is incident in the upper medium upon the separating interface. We must then have in the lower medium an expression for a refracted plane wave, and, if we cannot get quit of it, we must have a pressural wave. That is then what is denoted by ψ' and φ'. For the sake of symmetry I have chosen the refracted distortional wave as being the given one of the set. That is why it appears with no second coefficient; and also not wanting to use more coefficients than necessary, I let the single coefficient of ψ' be unity. The remaining coefficients bring in the 4 unknown quantities. There is something more to be said as to what is known and unknown. What of the a, b, b', ω? We shall suppose ω to be known, and the moduluses to be known. The equations of motion then give us the a, b, b', as in equations (8), (9). In strict analytical propriety we must not know the x, y, t, in the second medium; but we do the accented a's and b's. We accent a because it is clearly different in the two mediums. We do not accent the b because it has clearly the same value in the two mediums. Put down the values of a and b in terms of the wave length and then the thing will be perfectly clear. Let λ be the wave length of the plane distortional wave in the upper medium; we have then in the upper medium

$$-(ax + b'y) = \frac{2\pi}{\lambda}r,$$

r being the perpendicular distance from the focus to the wave front. Since i is the inclination of the wave front to the axis of y, we have

$$a = \frac{2\pi}{\lambda}\cos i, \quad b = \frac{2\pi}{\lambda}\sin i.$$

For perpendicular incidence, we have $i = 0$, giving the augment

$$\frac{2\pi x}{\lambda} + \omega t$$

which is correct. For grazing incidence, $i = 90°$, giving

$$\frac{2\pi}{\lambda}y + \omega t,$$

which is correct. We may treat similarly the pressural wave, in the cases in which it extends into the second medium as a plane wave, letting ℓ be the wave length as in the paper. You might just add the above to the paper in explanation of the notation a, b in equations (6). It is so difficult to write with the jelly-graph ink that I economized as much as possible.

At the interface, that is to say, the position $x = 0$, we have continuity of matter. Hence $\xi = \xi'$, $\eta = \eta'$. Again, we have nothing to do with Q at the interface between two mediums, because Q is a force that acts on the surface perpendicular to the interface. If we consider the forces on the interface, there must be a balance between them. Therefore $P = P'$, $U = U'$. These are the conditions to be solved. That leaves a clean simple problem of dynamics, and yet people have been working at it for 50 years and have left it in a very sadly muddled condition, with the exception of the clear, accurate, and very comprehensive papers of Green and Rayleigh. The thing that has introduced the difficulty, and makes this a more complicated difficulty than the other cases, is the pressural wave. The pressural wave, in fact, has been the *bète noir* of this problem. I do not know how Cauchy treats the animal. Somehow, he introduces fallacious terms involving consumption of energy. MacCullagh and Neumann killed the animal with bad treatment. Sam Haughton yoked it to an Irish Car and it would not go. Green and Rayleigh have treated it according to its merits and it has escaped whipping at their hands.

There is a little novelty in this way of treating it expressed in (11). I have got the thing into a form in which I avoid the question of compressibility or incompressibility until we are supposed to take it up definitely. In equations (11), I want to get the part of P dependent on φ. That is the thing in which a little management is required to avoid difficulties. It works out rigorously from the preceding fundamental formula. Note the two little b's and distinguish between them for the present. In the first modification we introduce $\rho\omega^2$, by (8). In the next modification we get rid of $\rho\omega^2$ by using that of the two values of (9) which belongs to the upper medium, viz.: $\rho\omega^2 = n(a^2 + b^2)$. Thus the part of P depending on φ takes the simple form $n(b^2 - a^2)B$.

I have worked this problem out in this paper more fully than has been done in Lord Rayleigh's paper. It would take too long to go all through it for you I have done it at various times, chiefly in steamers and on railways. I came in that way quite unexpectedly upon this result—I am not going to give much time to it though, because it is not really of importance for light—I found a very curious expression which gives us a case of complete polarization. At the regular polarizing angle, the following relation involving unequal rigidities

$$na(a - 3a') = n'a'(a' - 3a)$$

brings about a vanishing of the imaginary part of A, and therefore leads to a case of complete polarization by reflection. The a, a' must have the values corresponding to the angle of polarization, which is the same as Fresnel's angle. For $n = n'$ the result is null. This relation implies a greater rigidity in the medium of slower wave velocity, and, as the medium of slower wave velocity has a greater refractive index, it implies greater density also—but not greater in the same proportion as the rigidity. Now I have looked at the light that would be reflected at direct incidence and find that it is very much in excess of what would be given by this. The ratio of the intensity of reflected to refracted ray for direct incidence on the supposition of equal rigidities and unequal densities is

$$\left(\frac{\mu - 1}{\mu + 1}\right)^2.$$

For the case of glass take $\mu = 1.5$ and that becomes $i/25$. I tried, and as nearly as my rude experiment allowed me to judge, something like a tenth part of the light was reflected from a piece of ordinary glass. The whole light reflected from two surfaces should be approximately double that from one. Thus my own rough experiments showed that Fresnel's formula was so nearly correct that I was quite unlikely to make anything out of this supposition of unequal rigidities. From that moment the algebra lost its interest for me. I shall put it in form sometime or other; whether in time to be incorporated in the report of these lectures or not is not of great consequence to you. I just tell you about it, however. It is worth knowing that a thing may be examined in this way and that way, and what sort of possibilities there are in it. I would not altogether discard the possibility of the rigidity being different in the two mediums for all cases. Our knowledge of transparent bodies is, in fact, very limited, and that knowledge is confined chiefly to visible light. When we investigate these things for invisible chemical light, and for dull radiant heat, we may find something very different from what we at present suppose to be the state of things as regards the answers to these fundamental questions. I note, indeed, that the reflection of radiant heat from rock salt seems to be much greater than according to Fresnel's formula. Green takes $n = n'$ because it simplifies the work. Lord Rayleigh supposes the rigidities to be equal and unequal densities to be the cause of difference of velocity. In this paper on the reflection of light from small particles his reasoning is very urgent and seems exceedingly binding on this subject; we can scarcely get away from the conclusion that the rigidities are equal or very nearly equal and the difference of velocity does depend on difference of density. He shows that if we make any considerable deviation from the position of equal rigidities, we induce effects not known to observation and lose known effects. Most particularly telling is the polarization of light from fine particles. By Lord Rayleigh's work it seems that if there be any sufficient difference of rigidities to be worth thinking of in the way of explaining outstanding difficulties of another kind, the polarization that we have will be annulled and we shall not have nearly a good enough approximation to the polarization to represent the state of the case.

That being the case, the question is left, what can we make of the results of these equations. The results are given in Lord Rayleigh and Green. I unfortunately, yesterday, did not come upon the right paper. I will call your attention to it once more because I want to speak of magnitudes. I want to show you that we are very far indeed from an agreement with observation in the formula derived from these processes. We ought to find from these processes that our reflected light very nearly vanishes at a certain angle of incidence. Green works it out and gives a formula. The actual minimum value of that formula is not quite that which Green gives, but in an appendix by Ferrers the true value is given. For the case of air and water, $\mu = 4/3$, and Green finds for the minimum value of the intensity of the reflected light $\frac{1}{151}$. Now compare that with the light reflected from water by direct incidence. By Fresnel's formula derived by the same mathematics, that is $(\frac{4}{3} - 1)^2/(\frac{4}{3} + 1)^2 = \frac{1}{49}$. How could Green say that his result was, as nearly as he knew, conformable with observation, when he finds that the light at the polarizing angle is a third part of that reflected by direct incidence? It is nothing like a third part. Speaking roughly, I do not believe the light reflected at the polarizing angle is a 20th part, from the nullness of the amount of light that is left at the polarizing angle when you apply the light in the usual way. Try it and you find that proportion is enormously less than the proportion Green gives. Ferrers helps a little by saying that instead of Green's minimum value of $\frac{1}{151}$ we have more accurately $\frac{1}{166}$; but that is at an angle not quite agreeing with Fresnel's polarizing angle, which does not make matters much better. It is, moreover, so small an approach to the annulment of light that we have that it cannot show anything satisfactory. Take the case of glass ($\mu = 1.5$) in which the intensity of the reflected light at the polarizing angle is $\frac{1}{49}$ and for direct incidence it is $\frac{1}{25}$. Actually in the case of glass there is not at the polarizing angle anything like half the light at direct coincidence. The formula is simply a failure. Green did not notice this; he had switched off on something else, I dare say. To be sure $\frac{1}{151}$ is a small number and it looks as if it might be right, but if he had considered how small the reflection really is, he would have seen that that is no approach to a satisfactory explanation. I will just give you the formulas, because some of you may not have access to Lord Rayleigh's papers [*Phil. Mag.* Aug. 1871].* The ratios of the amplitudes of the reflected and incident vibrations is given by

$$\frac{R'^2}{R^2} = \frac{\cot^2(i + i') + M^2}{\cot^2(i - i') + M^2}, \quad \text{when } M = \frac{\mu^2 - 1}{\mu^2 + 1}.$$

It may be interesting for you to work that out for yourselves, which you can do from our equations.

Besides the minimum ratio attained when we vary the direction of the incident light from normal to grazing incidence, there is a change of phase. If we had complete polarization the state of things would be this: [the] phase remaining perhaps constant until the intensity diminishes to zero, then the phase changing suddenly as the inclination passes through the zero position. What really happens according to the formula is: [the] phase varies gradually; at the minimum it is, roughly speaking, midway between the phase corresponding to direct incidence and the phase corresponding to grazing incidence. The want of complete fulfilment is connected with the gradual change of phase. In observations we can only take the relative phase—the difference of phase between the two component rays, i.e. the component consisting of vibrations in the plane and perpendicular to the plane. Lord Rayleigh refers to Jamin here and says, "Now what is observed in experiments is the acceleration or retardation of one polarized component with regard to the other and is therefore given simply by difference between the two angles. The ambiguity must be removed by the consideration that when the incidence is normal, there is no relative change of phase, though throughout Jamin's papers it is assumed that there is in that case a phase

---

*Other papers of Lord Rayleigh's referred to are in *Phil. Mag.*, Feb., April, and June 1871 [and] May 1872.

difference of half a period. I am at a loss to understand how Jamin could have entertained such a view, which is inconsistent with continuity, inasmuch as when $i = 0$ the distinction between polarization in the plane of reflection and polarization in the perpendicular plane disappears."

In this paper I have only given you the reflection and refraction for the case of vibrations in the plane of the three rays. The case is so exceedingly simple for vibrations perpendicular to the plane of the diagram that you will not regret my not having given it to you. It brings out [an] exceedingly simple formula, which agrees exactly with Fresnel's sine formula when we suppose the rigidities equal and the densities unequal. Very curiously, densities equal and rigidities unequal give you Fresnel's tangent formulas; and it gives you complete polarization—that is a most interesting result. What is more, it gives you the same intensity for light reflected at direct incidence as Fresnel's formula. You might think that would be a good foundation for allowing that the vibrations were in the plane of polarization. But alas for that supposition, Lord Rayleigh has shown that it is absolutely impracticable in the problem of vibrations in the plane of the three rays to get anything approaching to Fresnel's formula at all, if you take the densities equal and the rigidities unequal.

We cannot but conclude, from all we have before us, that the theory of the homogeneous elastic solid is quite unsatisfactory in respect to polarization, the approximation to explanation of the extinction of the ray consisting of vibrations in the plane of the three rays being so exceedingly, so monstrously rude as we have seen. I am surprised that it has not been denounced more by others who have touched upon the subject.

I would like to call your attention to Green's formula for refraction of sound. You have got the formulas down here passing over (13), and that beautiful result of Green becomes exceedingly simple. The ratios of the intensities, or the squares of the ratios of the displacements is

$$\left(\frac{\rho'}{\rho} - \frac{\cot i'}{\cot i}\right)^2 \bigg/ \left(\frac{\rho'}{\rho} + \frac{\cot i'}{\cot i}\right)^2.$$

For the case of all gases,

$$\frac{\rho'}{\rho} = \frac{\sin^2 i}{\sin^2 i'}$$

and the above formula reduces to

$$\frac{\tan(i - i')}{\tan(i + i')}$$

or Fresnel's tangent formula. There then is an agreement with one [of] Fresnel's most remarkable formulas for sound reflected at an interface of separation between two gases of different densities. On the other hand, if we have anything like the law of relation between bulk moduluses on the one hand and densities on the other that we have between air and water, or between two different liquids, we have no approach to this formula. It is not easy to see how that formula for sound can be verified by experiment, but still the result is in itself exceedingly interesting.

For the case of incompressibility, we must take $b' = b$. That gives us our relation $\delta = \nabla^2 \varphi = 0$. It is interesting to remark that without taking $b' = b$ we have a set of formulas that may be used. In some cases those formulas will give condensational waves; in others not. Instead of saying what is under case I you might cancel it, and say according as $\rho\omega^2/k + \frac{4}{3}n$ is less than or greater than b'^2 we have case I or case II. You will see then that inasmuch as $b = 0$ for direct incidence, that for incidences not too oblique we always have condensational waves; for very oblique waves we have no condensational waves. For direct incidence the condensational wave, as you may easily see from working out the formula, is null; it is necessarily null. But for incidence nearly direct, the condensational wave is not null and it can only be annulled by making $k = \infty$.

With respect to my very faulty expression regarding Sam Haughton having yoked his animal to an Irish car, I meant to say that he tried to make this condensational wave help the car out of the ditch in which it is lodged—that is to say, he tried to get us out of our difficulty by aid of the difference between b' and b; but it would not work.

We will put this condition for the existence or nonexistence of a condensational wave in a better form. We have

$$b = \frac{2\pi}{\lambda}\sin i,$$

where λ is the wave length of the distortional wave. The relation between λ and ω is as follows: $\underline{v}$ (the velocity of propagation of the distortional wave) $= \omega\lambda/2\pi = \sqrt{n/\rho}$; therefore

$$\rho\omega^2 = \frac{4\pi^2 n}{\lambda^2}.$$

These values substituted in

$$\frac{\rho\omega^2}{k + \frac{4}{3}n} - b^2$$

give

$$\frac{4\pi^2}{\lambda^2}\left\{\frac{\pi}{k + \frac{4}{3}n} - \sin^2 i\right\};$$

therefore the equation for finding critical angle is

$$\sin^2 i = \frac{\pi}{k + \frac{4}{3}n}.$$

We have the conclusion that if the angle of incidence is anything less than that given by this formula, there is a condensational wave, unless the angle is zero—then we have no condensational wave. And if i is greater than that critical angle, there is no condensational wave. That I think absolutely settles the whole question with regard to the condensational wave.

There are two or three things that I wish to speak about. I want to clear off at once the question of helicalness on the plane of polarization, commonly called the non-magnetic rotary effect. I have objected to the name rotary, because it is not properly applied, and have taken the name helical because the phenomenon essentially depends on a screw like form somehow or other. So far as I know, the first place where this distinction is pointed out and the essential connection of the Faraday property with rotation shown is in a paper of my own in the *Proceedings of the Royal Society of London,* May, 1856. I just read two or three sentences from that paper: "The elastic action of a homogeneous strained solid has a character essentially devoid of all helical and of all dipolar asymmetry. Hence the rotation of the plane of polarization of light passing through bodies which either intrinsically possess the helical property (syrup, oil of turpentine, quartz crystals &c.) or which have the magnetic property induced in them, must be due to elastic reactions depending on the heterogeneousness of the strain through the space of a wave or to some heterogeneousness luminous waves," etc. But here is the point to which I wish to come. I imagine for example a liquid filled homogeneously with spiral fibers or a solid with spiral passages through it of steps—I said here—of not less than [a] forty millionth of an inch—meaning steps of a screw not less than a thousandth of the wave length. This "might be certainly expected to cause either a right handed or a left handed rotation of ordinary light." There can be no doubt that this is the correct explanation. For a rough mechanical model of a medium posessing helical properties, take a jelly and bore ever so many cork screw holes in it—that will introduce a heterogeneousness of structure with a definite spiral character. Take another jelly and bore it with left handed cork screw holes and that will induce a definite spiral structure also. One of those mediums seen in a looking glass would look like the other; we have that kind of want of symmetry that there is between the right and left hand. Another example is to take a bunch of spiral springs and fill up the interstices with mortar, jelly, or something of that kind, and you will have that property. If the wave length be enormously great in comparison with the dimensions of heterogeneousness, the turning effect on the plane of polarization will be exceedingly small. It will be null if the wave length is infinite in comparison with the dimensions corresponding to the heterogeneousness. It seems almost certain that this worked out would give us, I will call it, the rotary effects (although I protest against the name) of quartz, etc., somewhat nearly according to the well known formula of inversely as the square of the wave length. You know that in reality the practically constant quantity, square of the wave length into the amount of the rotation, does gradually increase as the frequency increases for the substances that have been experimented on, so that the rotation varies more than according to the inverse square of the wave length. I see it is stated that Biot has worked this out. If he has worked it out right, it is exceedingly interesting and important. The other question of the magnetic influence on light I shall say nothing about.

I had hoped to bring forward an addition to our molecular theory, showing you definitely the rotation of the plane of polarization produced by introducing an enormous number of gyrostats into our jelly. I will show you how the thing may be done, and I will tell you why I do not give the mathematics of it. One reason is that to-morrow is our last day, otherwise I would try to give you the mathematics of it, unsatisfactory as it is.

Suppose we have here distortional waves. The arrow heads indicate the to and fro motion of a wave in the plane in question. Besides the distortion there will be rotation. Suppose we have our massless rigid shell lining or spherical cavity in the ether; and in that lining let us pivot by a proper shaft a fly wheel like the fly wheel of a gyroscope and suppose that to be rotating with enormous rigidity. The jelly may move this way or that way without inclining the axis of that flywheel. But force the axis to turn in the plane of the board, and that introduces a tort pressing upon the bearings of the ends of the fly wheel in a plane perpendicular to the board. The plane of that tort is, of course, the plane of the axis perpendicular to the plane of our diagram.

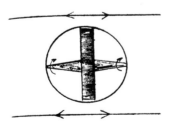

That transverse force is very easily introduced into the equations of motion and it gives us just what we want if we only want to show rotation of the plane of polarization. It gives you rotation of the plane of polarization following Faraday's law that if you send the light in one direction you get a rotation of the plane of polarization. Send it backwards or forwards in the directions of the revolution of the plane of polarization and it goes on rotating, as you all know. That is satisfactorily explained by this gyrostat—nothing could be more satisfactory or clear than it is.

In the time that I have been talking about it, I might have put down the symbols. Why do I not go into it, and try to make it a part of our molecular dynamics? I answer: because I cannot bring out the law of inverse proportionality to the square of the wave length, which observation shows to be somewhat approximately the law of the phenomenon. If you deal with it in this simple way, it comes out inversely as the wave length and not inversely as the square of the wave length. Until a week ago, I thought that by putting a fly wheel somehow or other into our molecule I could get a rotary effect according to which the magnitude would vary according to two terms $c/\lambda + c'/\lambda^3$. If that were so, I could bring the thing to vary according to observation, because there is no rigorous agreement to the inverse square of the wave length; it varies more than that and it is possible that it will be expressed by some such formula. But alas, my results give me the other law, not more effect with greater frequency, but less effect with greater frequency, according to the inverse wave length. I therefore lay it aside for the present, but with perfect faith that the principle explanation of the thing is there. I cannot pretend that the very simple matter of molecular dynamics at which I am driving has accomplished the solution of any great difficulties, but I do think it is of high importance and interest.

[Referring to rotational, or Faraday-magneto-optic, effect: When I said, "I get persistently $1/\lambda$ for the law, but it is to be approximately but varies a little more than that, etc.," I was under a misapprehension; to be now corrected as follows:

This result I have found is that circularly polarized light travels with different velocities according as the orbital motions are with or against the amperian currents: this difference of the two velocities being, for lights of different homogeneous colors, directly as the frequency of the vibrations. The resultant of two circular motions of equal periods in opposite directions in the same circle is simple harmonic vibrations along a diameter in the same period;* and therefore two circularly polarized rays in opposite directions travelling with different velocities, as stated above, are equivalent to a plane polarized ray travelling with mean velocity, and having its plane of polarization rotated at the rate $\delta/\lambda v$ per unit of distance travelled, if δ denote the difference of the two velocities, v the mean of the two, and λ the wave length in the medium.† But for lights of different homogeneous colors, I

---

*Thomson & Tait § 73.
†Remark how *very* small δ/v is in all known cases of the Faraday effect in transparent mediums; but how *not very small* it is found by Kundt (*Phil. Mag.* Oct. 1884) to be for light passing through an excessively thin film of metallic iron magnetized transversely. This case seems splendidly in accordance with the molecular dynamics of metallic reflection and the transmission of light through metals suggested in my last lecture, and developed in the addition I am going to send. [See Appendix.]

found δ to vary as $1/T$, that is, as $1/\lambda$. Call it t'/λ. Hence rotation per unit of distance travelled equals $A/v\lambda^2$. And by ordinary dispersion

$$\frac{1}{v} = \text{refractive index} = \mu_0 + \frac{C}{\lambda^2}.$$

Hence rotation per unit of distance travelled equals

$$\frac{A\mu_0}{\lambda^2} + \frac{AC}{\lambda^4},$$

which agrees with the result of observations, showing that $\lambda^2 x$ amount of rotation per &c. is, to a rough approximation, constant, and augments as we go up from red to violet.

I have a great improvement however on the mechanical model for gyrostatic effect, which I described in my lecture of Oct. 16. I had thought of it while striving to make something fairly satisfactory of the gyrostatic affair for the lectures but have only succeeded in developing it satisfactorily since their termination. I shall, if possible, write it to-morrow before I sail. I hope at all events to write it during the voyage and post it in time for incorporation in your report.* The same also for metallic reflection &c. and Kerr's magnetic reflection.—W. T.]†

To-morrow I think we shall see that the anomalous dispersions and reflections and the heatings that I have been speaking of by the absorption of light passing through a not perfectly transparent medium are all going to be explained simply and well and that this molecular theory has the merit of telling us things we did not know before. It seems not at all improbable that we shall find thin transparent bodies in which the velocity of propagation of light is greater than in the luminiferous ether. If you look at the formula when it is ready for you, you will see that when T^2 is somewhat less than the shortest of the critical periods κ^2, we have μ^2 negative. μ^2 is $-\infty$ when T^2 is just less than κ^2; decrease T^2 a little more and you get $\mu^2 = 0$; decrease it still more and you get $\mu^2 < 1$, or the velocity of propagation is greater than in the ether. I think we ought to find that phenomenon. I think Quincke found that in some metals the velocity of propagation is greater than in the ether. There has been very little prismatic examination of the bodies that show anomalous dispersion. It has been alluded to by some of those who have done most in that subject, but there is more to be learned. I think it will not be at all improbable that we shall find zero refractive index and a refractive index less than unity in the neighborhood of some of these critical points. I do not say it is a very fundamental phenomenon, but it is worth looking for. Quincke says that

---

* See Appendix.
† Added Oct. 21, 1884.

there is a very distinct acceleration, showing a greater velocity of propagation in metals than in the luminiferous ether.

What seems to me to be the true theory of absorption is a storing for a moderate time of energy in the attached molecules. Instead of putting in viscous terms in our equations with resistances depending on the velocities, I am disposed to admit no such terms as I have already said and to look for the explanation of absorption in the manner I have indicated. Looking at it in that way, and taking in connection reflection, it seems to me that we should have total reflection for those rays whose frequencies are just a little above a critical frequency—rays which are such as to make μ^2 negative. We may put down the mathematics of that for you to-morrow perhaps. That corresponds to a case in which light cannot get into the medium at all and it must be totally reflected unless there is absorption. It seems to me not very improbable that the great proportional amount of light reflected from polished silver surfaces may be explained in that way. Why is so much absorbed and lost in other metals? We cannot tell. But I think that somehow or other, if we take natural suppositions as to attached molecular systems with particles massive enough and lightly enough connected by means of springs, and suitably connected somehow or other by springy connections with the medium, that, not only in the neighborhood of critical values, but through a very wide range of frequency of vibration, we shall find a great amount of conversion of the energy into vibrations, i.e. of absorption. There is no real loss of energy, absorption distinctly going to the heating of the body by generation of vibrations in it. I do not despair of seeing an explanation of metallic reflection in this way. I am going to say a little about that to-morrow, but we shall not have a mathematical lecture at all to-morrow.

I want to show you some of this work that Mr. Morley has gone through. He has found five of the seven roots and the results are most interesting. The roots are 3.4618, 1.0048, .2986, .02556, .007256. I think the 2 roots that are not found are between the two last and the three preceding. It is interesting in connection with the continued fraction, and the form of working I pointed out to you, that the u's are, as we know they must be, all positive for the smallest root or root of the greatest frequency $1/T^2 = 3.4618$. That means that the particles are all moving in opposite directions. For the next root $1/T^2 = 1.0048$, the first u is negative, and all the rest are positive. That means that number 1 particle moves in the same direction as Number 2 particle, while particles 2, 3, 4, 5, 6, 7 are all moving in opposite directions; and so on.

As to the distributions of energy, taking the successive roots, the particles that have the greatest energy are farther and farther away from the end from which we work. The consideration of the distribution of the energy in these different modes is of vital importance in respect to the application I desire to make in this subject. I thought the working out of an example of that kind would help us greatly, and I am sure we are under obligations to Mr. Morley for having made these calculations. I hope some of you will not forget another question that I suggested to the arithmetical laboratory, because it will throw great light upon the theory of deep sea waves. What I say to-morrow will be upon that subject. I am going to show you that when we attach molecules to the ether, the work done on the medium per period is much less than the energy per wave length and that therefore a front of a succession of waves cannot penetrate into the medium with constant velocity and undiminished amplitude as it does in the familiar case of this wave machine which Prof. Rowland has had constructed for us [consisting of some 50 or 60 bars attached equi-distantly along a pianoforte wire in the manner already described in the case of the molecular model, and suspended from the ceiling]. There we have waves penetrating with constant velocity, and without change of form. Work done by the wave front per period equal to the energy per wave length is the condition that is necessary and sufficient for the propagation of waves of all lengths at the same velocity, and the same condition is sufficient for the propagation of a pulse without change of form. The question of velocity of groups, which was discussed at Montreal, is touched upon here. I do not know whether I can throw any light upon it in connection with Mr. Michelson's observations or not. The thing is of enormous importance in connection with the theory of light, besides being exceedingly interesting in itself as a problem.

Lecture XIX

We now have completed (see following [table]) the problem of the determination of the periods, displacements and energies for the seven particles that I gave you. I was under a misapprehension in supposing that there were two roots in a certain gap. Prof. Franklin noticed that the first root is $3\frac{1}{2}$ times the second, the second, rather more than 3 times the third, the third, about $3\frac{1}{2}$ times the fourth. The fifth is as we now know about $3\frac{1}{2}$ times the fourth, the sixth about $3\frac{1}{2}$ times the 5th and the seventh about $3\frac{1}{2}$ times the sixth. They are not exactly in that geometrical ratio of $3\frac{1}{2}$ but it is curious that they are not far from being so. I gave you root 3 and 5, and said there were two roots in between. Prof. Franklin saw that [this] was very improbable, and we find that there is another root less than the last root I gave you yesterday.

The maximum displacements in the first mode of vibration corresponding to the greatest value of $1/T$ (that being the frequency in Lord Rayleigh's language) are alternately positive and negative. That must be the case in any system whatever of a similar linear character to this. In the last mode they are essentially all positive. The tendency is to have one fewer change of sign in each successive mode than in the one before it. I cannot give you that as the general rule, because there may be cases in which a node coincides with one of the particles. That is a very common case. In the gravest mode it is obvious that all are swinging in one direction. I will hold this lower particle P at rest, and try for the gravest mode. I can almost hit it off—not quite—by merely disturbing the uppermost one; it brings the others with it. That is very nearly the gravest mode. There is a little wiggling to the lowest of the movables, it has not quite got it. It is not quite in order. These two come together at the end of their ranges. They should all be going out together and coming in together. I will bring it to rest. Now we more nearly hit off the gravest mode. But one has a wiggle on it—wiggle is not my word, if you please, I adopt it. "A little wiggle superimposed on the graver mode" tells better the state of the case than more dignified words could. There, then, is the gravest mode with a little wiggle superimposed.

Solution for Fundamental Periods, Displacements & Energy Ratios
of a System of Springs connected Particles.
m = 1, 2, 4, 16, 64, 256, 1024, 4096.
C = 1, 2, 3, 4, 5, 6, 7, 8.

By Edward W. Morley, Cleveland, Ohio.

Fundamental Periods Corresponding to Outer Ends of Springs 1 to 8 held fixed

$\frac{1}{P^2} =$	3.4618	1.00483	0.29849	0.080078	0.0266607	0.0075644	0.0014701
Displacement Ratios or values of $\left(\frac{x_i}{x_1}\right)$							
x_1	1.	1.	1.	1.	1.	1.	1.
x_2	−.231	1.0000	1.351	1.456	1.487	1.496	1.499
x_3	.014	−.341	1.047	1.589	1.761	1.813	1.829
x_4	−.III27	.025	.481	1.129	1.787	1.997	2.066
x_5	−.V15	−.III50	.033	.511	1.223	1.960	2.216
x_6	−.VIII26	−.V30	.III68	.040	.581	1.322	2.203
x_7	−.XI13	−.VIII51	.V89	−.III81	.045	.628	1.717
Energy Ratios or values of $\frac{m_i x_i^2}{m_1 x_1^2}$							
$m_1 x_1^2$	1.	1.	1.	1.	1.	1.	1.
$m_2 x_2^2$	.213	3.998	7.30	8.48	8.85	8.96	8.99
$m_3 x_3^2$	.II33	1.864	17.62	40.41	49.64	52.68	53.54
$m_4 x_4^2$	.147	.039	11.88	81.66	204.85	255.94	273.14
$m_5 x_5^2$	.IX61	.II65	.28	66.73	382.71	963.10	1057.82
$m_6 x_6^2$	.XIV7	.VIII 9	.III47	1.62	945.60	1788.13	4448.41
$m_7 x_7^2$	.XX7	.XII 1	.VIII63	.002	8.42	1616.99	12070.04
Sum	1.21	6.90	38.00	199.90	1000.57	4706.10	17927.87

Lecture Notes, October 17th

Comparison of Work with Energies

$$\xi = R \sin \frac{2\pi}{\lambda}(y - vt)$$

v being the velocity of propagation as modified by embedded molecules

$\dfrac{c}{4\pi^2}$ = density of the ether.

$\dfrac{c}{4\pi^2}$ = rigidity " " "

I Work done by wave surface in one (fixed) plane reckoned per unit area of the plane; and per period of the motion

$$= \int_0^\tau dt \; \frac{c}{4\pi^2}\frac{d\xi}{dy}\left(-\dot{\xi}\right)$$

$$= \frac{R^2}{2}\tau \cdot \frac{c}{\lambda} \cdot \frac{v}{\lambda} 1 = \frac{R^2}{2}\frac{c}{\lambda}$$

$$T = \frac{c}{4\pi^2}\frac{d\xi}{dy}$$

Rate of doing work $= T \times (-\dot{\xi})$

II. Potential Energy of the (distorted) ether, per wave length

$$= \int_0^\lambda \left(\frac{1}{2}T \cdot \frac{d\xi}{dy}\right) dy = \frac{1}{2}\int_0^\lambda dy \cdot c \left(\frac{d\xi}{dy}\right)^2 \frac{1}{4\pi^2} = \frac{R^2}{2} \cdot \frac{1}{2}\frac{c}{\lambda}$$

III. Kinetic Energy of the (moving) ether per wave length

$$= \int_0^\lambda dy \cdot \frac{1}{2}\frac{c}{4\pi^2}\dot{\xi}^2 = \frac{R^2}{2} \cdot \frac{1}{2}\frac{cv^2}{\lambda}$$

$$I - (II + III) = \frac{R^2}{2}\left\{\frac{c}{\lambda} - \frac{1}{2}\left(\frac{c}{\lambda} + \frac{cv^2}{\lambda}\right)\right\} = \frac{R^2}{\lambda} \cdot \frac{c}{2} \cdot \frac{1}{2}\left(1 - v^2/c\right)$$

I think some of you will be induced to reject these experiments, whether Professors or not. See how easily this model is made. Do the work at home with your own hands; then you will have a very interesting piece of miscellany. I wanted to make a fundamental part of our subject (and I only wish we had a few more days to introduce that and some other similar things) the transmission of waves along a row of particles instead of a continuous line. For example, the transmission of a set of waves like waves in a cord, along a necklace. Instead of a rope with matter uniformly distributed through it, take a necklace with beads strung along it. It is exceedingly easy; we have the equations for it. What we have to do is, in our equations, to take $m_1 = m_2 \cdots$ and $C_1 = C_2 \cdots$, and we get a pretty set of initial conditions, and a charming piece of work that I would have liked to have spent an hour upon, and I think you would have liked it too. Along an infinite row of such particles waves can be propagated in any period longer than the period of that vibrations in which every two particles are turning in opposite directions. Start them with equal amplitudes in opposite directions, and think of the time of vibration. You can calculate that from your initial data, one particle with a twisting force C towards a fixed point on the one side and a twisting force $2C$ towards the fixed point on the other side. The theory leads us to this conclusion that waves in any period less than a critical period cannot be propagated along an infinite row of particles mutually acting each on its predecessor in the series. Equal particles, and equal forces, that is Lagrange's system of linearly connected bodies; and that is a very sweet problem in mathematics—a lovely problem.

If you try to send a wave in a period shorter than the critical period, what is the result? This figures in that paper of mine on the size of atoms. I think some of you may have seen it in Nature. This model of a wave machine is not constructed for illustrating that particular thing because the period is too short. But I give you a hint, if you want a pretty thing. If you are asked to give a popular lecture on waves of light or anything of that kind, you cannot have a better illustration than this to make people understand what you are speaking of. If you make this machine see that the pins have proper obliquities to press the wire in close to the wood. Then cut away the wood where the wire touches it in coming away from the pins. I would suggest that you have the bars very close, so that you can scarcely see anything through them; the vibrations will be prettier. There are three kinds of models: one on this plan, then the wiggler, and lastly another one with particles placed at considerable distances apart—perhaps six inches—and with masses such that the critical period with any particle held at rest may be moderately large. Then you will see the result of trying to send a set of waves along it smaller than the smallest period for which waves can be transmitted. This wiggler, I want you to notice, will not send waves along at all. You get simply a vibration of one particle in the direction opposite to its neighbor. Your finite difference equation for waves has imaginary roots when waves are transmitted, and real roots when the period of the exciter is shorter than the shortest period for which waves can be transmitted. The finite difference equation gives you a simple algebraic equation. Choose the simplest possible quadratic equation—the products of the two roots equal to unity; work that out and you will see that when you have imaginary roots, you have one case, and when you have real roots, the other. A beautiful solution it is. When you have real roots, the displacement of the i'th particle takes the form ρ^i into the displacement of the first particle, ρ being the one of the two roots which is smaller than unity. The general solution is

$$C\rho^i + e'\rho'^i, \quad \rho\rho' = 1.$$

In cases of excitement at one end, and zero motion at an infinite distance, the answer takes, I think, this form, displacement of particle $i = (-1)^i C\rho^i$, ρ being the least root (and positive). Turn to the table of displacements and you will see how different that is from the problem of waves along a row of equal particles that I have been telling you about. With equal particles a wave is transmitted through if the period be anything less than the one critical period corresponding to that case. That is not quite the same problem we have here because this is a problem of a finite number of particles, with the remote particle connected by a spring to a fixed body. But we can see how things are going on just as well as if we had an infinite row, by taking the first two or three terms. There is in this first mode very little disturbance spread beyond the third particle, indeed it scarcely reaches that particle. It begins to perceptibly reach the last particle only in the fifth mode. But the mass of this last particle is so great in comparison with that of the first that its energy is 8 times the first, with only $\frac{1}{20}$ of the displacement.

I have written down some things I wish particularly to speak to you about, I will tell them to you before-hand. First, the pressural wave, sometimes called the surface wave. It is a surface wave only when it is a condensational wave. The subject I want to speak upon is the condensational wave, the annulment of it into a mere pressural wave and the nature of this pressural wave. We have spoken of it; you have seen it in the formula; I do not know that you have all got it clearly into your heads what it is. Next, I want to show you, very roughly, the formula of reflection and refraction with vibrations perpendicular to the plane of the three rays. Third, I want to speak about the aeolotropy of inertia, first suggested by Rankine, afterwards, independently, by Lord Rayleigh, to account for differences of velocity on different directions manifested by double refracting crystals. Fourth, Stokes' negative to that interesting hypothesis. Fifth, just a very brief summing up of the points of difficulty, and then we will go to our little sheet of lecture notes.

I can best explain the condensational wave by showing you something about waves in general in an elastic solid. Let the space below this line

be a portion of an elastic solid. We will think of the propagation of waves into it—or vibrations either; the mathematical solution is very indifferent to vibrations or waves. The general formula of yesterday is just as ready to be converted into a formula for vibrations as it is into a formula for waves. But I want, without the formula at all, to think of the propagation of a wave in such a medium. Suppose, in the first place, you disturb the surface and hold it disturbed. There will be a certain static problem to solve for the disturbance of the surface. I rub away the original boundary, and leave this wave boundary.

The problem of the waves would include the whole problem, and it is very easily worked out. Make [an] ever so slow or sudden static disturbance of any given shape: that problem is not very difficult to work, and is very interesting as a problem of dynamics with a view to physical applications, and it is also valuable on its own account.

Now apply a corrugated rigid form to the surface, and slip that form along at a certain speed. If we slip it along too slowly no waves will be sent into the solid at all. The velocity of propagation of waves is, you know, $\sqrt{n/\rho}$ or $\sqrt{(k + \frac{4}{3}n)/\rho}$ according as we have distortional or condensational waves, irrespective of the wave length—because we are not in molecular dynamics just now, we are in molar dynamics. If we slip our form along at a speed less than the velocity of propagation of a wave, no waves at all will be sent into the medium.

There is a certain charm about the mathematical analysis that gives us the general solution of a problem like this by consideration of mixed real and imaginary quantities. If you calculate the effects of applying a form and moving it at a speed less than the velocity of propagation of a wave in the medium, you will have the result in the form of exponentials with real indices—quite analogous to our problem of a finite number of particles, where we have real roots of the equation with nothing spreading into the interior. Here we have different cases. The most difficult case is for vibrations in the plane of the board. The second case is simpler.

I repeat again, get the effect of a sudden shock upon the medium—of course, if you twist it out of shape, you send an earthquake through it, but without twisting it out of shape, give it a shock, or whatever you may call it, just as I do when I displace this handle P of our model and cause it suddenly to begin performing a simple harmonic motion. Then get the simple harmonic motion again which every particle performs consistently with the surface being affected by a corrugated rigid form carried along at a constant rate. Our formulas of yesterday are only adapted to giving us the simple harmonic motion, and that is what we are considering—the simple problem of real periodic waves. There can be no periodic waves sent into the medium if the speed of the form is less than the velocity of propagation of the wave—no distortional wave if the speed be less than $\sqrt{n/\rho}$, and no condensational wave if the speed be less than $\sqrt{(k + \frac{4}{3}n)/\rho}$. We might work the thing out, and a very pretty problem it is.

* [Referring to the motion of a corrugated rigid form along the surface of an elastic solid, I said "greater than" when I should have said less than.[†] The true statement is as follows:

I (Applicable to vibrations either in the plane perpendicular to the bounding plane and containing the direction of motion of the form—the plane of the board—or perpendicular to that plane).

If the velocity (V) with which the form is carried along is less than the velocity (v) of a distortional wave, no wave will be propagated inwards: only a disturbance of which the magnitude diminishes from the surface inwards according to the logarithmic of exponential law.

II (Vibrations in the plane of the board)

If the velocity (V) of the form exceeds that of the distortional wave (v) but is less than that of the condensational wave (u) a distortional plane wave is propagated inwards, but no condensational wave. The inclination of the wave front of the distortional wave to the bounding surface of the medium is $\sin^{-1}(v/V)$.[‡]

III (Vibrations in the plane of the board.)

If $V > u$, two plane waves are propagated inwards—distortional and condensational—the inclinations of their wave fronts to the bounding surface of the medium being respectively $\sin^{-1}(v/V)$ and $\sin^{-1}(u/V)$.

This subject ought to be carefully and thoroughly illustrated by diagrams,

* Added Oct. 21, 1884.
† This has been corrected in the report.
‡ Omit the restriction to $< u$, and II becomes unqualifiedly applicable to vibrations perpendicular to the plane of the board.

showing the wave fronts, and the corrugated lines of particles which are in straight lines when undisturbed—all this for vibrations both in and perpendicular to the plane of the board.—W.T.]

The problem of reflection and refraction is a small part of this matter. It is more interesting as a problem of mathematical dynamics than anything else. I was saying to Stokes that I wanted this worked out more than it had been done before. He said, "Cui bono." I say there is the *cui bono*: it is interesting and instructive to work it out. We are all forced to feel that we are rather in a hole—I will not call it the slough of despond because we do not despond, really, as to the explanation of refraction and reflection; and although this will not explain refraction and reflection, let us see what it will do. Books on dynamics could well be devoted to work of this kind. If I am able to go on with the work on Natural Philosophy that I have in mind, I intend to make this investigation.

The question of applying a form and moving it along, and so on, does not exhaust the data for this problem. Our conditions are, a certain form slipped along with certain geometrical conditions to be fulfilled as to the change of shape of the surface, it being always made to fit the form. That will correspond to our first set of equations $\xi = \xi'$ developed yesterday. But with respect to the horizontal component, the form may drag the particles with it. You may vary your data thus: let there be a stated tangential force between the form and the solid at every point. Let the form be so constituted that while it is being moved along it will shove back in some places and shove forward in other places, producing a given distribution of tangential force all over its surface. The given distribution of tangential force must vary according to a simple harmonic motion in order that we may get a simple problem. It must vary as the sine of the angle corresponding to the variation, or it must be expressed as an exponential logarithm.

We need not go further with that sort of problem. You can see what it is. Without thinking of it as a corrugated form applied to the medium, think of it that you act upon every element of the surface of a medium with a normal and a tangential force after you have given it any displacement you please constituting a *given* set of waves in the medium. We took that as a reason yesterday for making a coefficient unity. Somebody might have said, "Why do you not take the incident ray as given, and the refracted and reflected rays as the unknowns?" I answer, in the lower medium there is only one plane wave, unless there be a condensational wave. It is convenient then to take that medium in the first place and the other in the second place; and furthermore, we get a perfect symmetry if we take unity say for the coefficient of the refracted wave and then leave two quantities for the reflected wave. The two ratios of the three things is all we want.

Have you ever thought (it is a curious enough explanation this) what sort of an arrangement would have to be made in order to have one incident ray giving rise to a refracted ray, and a quasi-reflected ray? Think of the case thus: reverse the motion of every particle in the problem as put here. We cannot produce such a thing, but there is not the slightest difficulty in imagining it. If the motion of every particle concerned was to be reversed, the refracted wave would travel back; our originally incident ray would travel back, and the reflected ray would travel in the corresponding reverse direction. That is a sample of what is introduced in the mathematical treatment of all such questions; but as to getting a source of light with its vibrations and relations so timed that that would be the result—there is no such thing. You will notice, also, that the work done by the wave front in any part of the incident wave per period is equal to the energy per wave length in the first medium, and according to our formula as worked out, that would hold for the second medium; so that the sum of the energies per wave length in the reflected and refracted rays is equal to the work done per period in the incident ray. In reversing, we must take that into account, so that we must supply a state of energy at the surface in order to make things come out in the way I have stated.

I will now put in a medium above our medium which we have been considering. The displacements in the interface of the two mediums are the same—not merely the normal components of the displacements, but the tangential components. That gives two particular equations. The upper medium pulls upon the lower with normal and tangential components of force. You might imagine other cases. Although it would be not at all an interesting problem, you might say, let there be a possibility of finite slip between one and the other, or rather you might imagine the two not cohering together but to be separated and to be perfectly smooth. The result would be zero tangential force in each medium, giving two equations, with a third and a fourth equation, viz., normal components of displacements and normal pressures in the two mediums equal. It is not an interesting view, because a finite slip between the two mediums is an inconceivable arrangement for our optical application at all events. Yet I do not think it is a less interesting problem merely as a problem of mathematical dynamics to suppose the two mediums to be separate and perfectly smooth. We cannot do away with the equality of normal pressures—we cannot get a mathematical problem according to that—because it would be inconsistent with harmonic motion. It is not inconsistent with reality, as anybody who has tried to ring cracked glass will see. Stroke cracked glass and notice the jarring. That comes from the cracks bending and slipping together. These are not the kind of problems I want to look at now.

You see that the problem we are solving comes out wonderfully simple when put into the form of a problem with only four unknown quantities by means of imaginaries. I put down yesterday what our condensational wave becomes when realized:

$$\varphi = \varepsilon^{-b'x}\{C\cos(by + \omega t) + D\sin(by + \omega t)\}.$$

I want to get quit of this. I do not want to go into details, but will just call your attention to the last equation in yesterday's paper, and the form I put it in afterwards that for all angles of incidence between zero and $\sin^{-1}\sqrt{n/(k + \frac{4}{3}n)}$ we have a plane condensational wave going into the interior with the distortional wave, and also a reflected condensational wave. The only way to get quit of that for all angles of incidence is to suppose $k = \infty$. If k be very large in comparison with n, a condensational wave will only be generated between zero and a very small angle, viz.: $\sin^{-1}\sqrt{n/k}$. I do not know as to our right to say that k is infinite, although there is no doubt we have a right to say that it is very large. Stokes went into that very fully in his report on double refraction and has given really the substance of every conceivable illustration of it. He shows that in every reflection and refraction, at all events with not too great obliquities, there will be a condensational wave generated from light falling on a body which consists of merely distortional waves; and he shows that according to the supposition that Cauchy made, which assumes Poisson's and Navier's ratio for elastic solids, that the condensational wave has a magnitude of very considerable energy compared with the distortional wave. Even if the ratio of n to k were enormously less than Poisson's ratio would make it, the energy of the condensational wave would still be so much as to produce immense effects. If you take an exceedingly intense light, that would produce a condensational wave of small energy in comparison to its own, perhaps a ten-thousandth of its own energy. But take sunlight falling upon a piece of glass—waves having a ten-thousandth of the energy of sunlight would have still very large energy compared with ordinary light, and that again, going at a velocity different from what we know—enormously greater—would, in falling upon a body, develop distortional light, and we should have distortional light springing up in places where there was no visible cause for it. We know of no such phenomenon. We are perfectly certain that if there is any such phenomenon it is of exceedingly small energy compared with light. I think we may safely say, whatever condensational wave there may be, its energy cannot amount to more than one-hundred-thousandth of the actual energy of the distortional light that produces it. It might or might not amount to a much larger proportion than that. But all I say, you understand, is that we have no such agency going about through the universe—enormous quantities of it coming from the sun with sunlight—from the fact that we have no trace of it in nature, and no evidence of such a force coming from the sun. There being no trace of it resulting from the combination of materials in practical experiments, we infer with certainty that if there is a condensational wave at all, it is of excessively small energy in comparison with the energy of the distortional wave accompanying it or giving rise to it.

Therefore we say k is practically infinite, and our attempt to introduce a condensational wave has been a woeful failure. Make now $b' = b$ and the result is

$$\varphi = C\varepsilon^{-bx}\cos(by + \omega t),$$

which is simply the well known expression for the displacement potential of deep sea waves corresponding to wave length $2\pi/\ell = b$. If you look at our formula of yesterday, you will see that $2\pi/b$ is the length from crest to crest. In our expressions you will see that the coefficient of y is the same throughout. In each particular expression it is $= 2\pi \sin i$ [divided] by the wave length, which corresponds to the wave length ℓ in the extreme case of grazing incidence.

Take the extreme case of grazing incidence, and if the wave length in the refracting medium is longer than in the other we have a wave travelling in one and in the other not, and it comes out a case of total internal reflection. If the wave length is shorter in the lower medium, the true state of the case will be this: We would have a vertical set of wave fronts in the upper medium and a case of light refracted into the lower medium with inclined wave fronts, the wave length being shorter: We have $\sin i = (1/n)\sin i'$, and i being 90°, we have $\sin i' = 1/n$, the well known case. As to the treatment of total internal reflection, that comes out with extraordinary ease from the analytical method, as you all know.

The condensational wave has become no longer such by the supposition $b' = b$; it is what may be called a pressural wave. Lord Rayleigh calls it a surface wave. It is a wave that spreads into one medium and the other so as to produce disturbances in condensations through a range comparable with the wave length—comparable with this quantity $1/b$, which is comparable with the wave length for any angle of incidence of considerable obliquity. I have put down the form for the upper medium. For the lower medium it is

$$\varphi = C'\varepsilon^{bx}\cos(by + \omega t).$$

x is negative in the lower medium, which justifies the change from $+b$ to $-b$. ε^{-bx}, which occurs in φ, occurs also in its differential coefficients, and is therefore the coefficient of diminution of the displacements as we recede from the interface. Take $x = \pm\ell = \pm 2\pi/b$ and we have a coefficient $\varepsilon^{-2\pi}$— what is the magnitude of that? In my own classes when I am lecturing on this subject, I ask my boys to write in the first page of their note books the values of ε, ε^2, $\cdots \varepsilon^{1/2}$, $\varepsilon^{1/4}$, $\cdots$ also ε^π, $\varepsilon^{2\pi}$, $\cdots \varepsilon^{(1/2)\pi}$, $\varepsilon^{(1/4)\pi}$, $\cdots$. Thackeray says, no person ever calculates his own logarithms. Quite wrong; every mathematician calculates his own logarithms; he must calculate them in order to have them. Thackeray did not know that. But notwithstanding it is not true, that expression is a good one for illustrating the subject. I only remember two figures of the value of ε. It is about 2.7—that raised to the power 2π is a large number. $\varepsilon^{-2\pi}$ then is a small fraction. Our displacements are then very small when $x = \ell$. Take $x = 2\ell$ and the coefficient of diminution is excessively small.

This is precisely the case of a deep sea wave, and you see that the motion of the water at a depth of the wave length is very small. Even at half the wave length the coefficient is $\varepsilon^{-\pi}$; or at a depth of half a wave length the disturbance is only about $\frac{1}{27}$ of what it is at the surface. The diminution is enormously rapid. That is exactly the case with this pressural wave. It produces a disturbance in each medium which is sensible at distances comparable with the wave length; insensible at distances a considerable multiple of the wave length. There is no difficulty in thinking of pressural waves in an incompressible solid, an elastic jelly for instance. We cannot have a pressural wave at all in the interior of an infinite incompressible solid. It must get away somewhere. If it is free on one side, there is no difficulty about it—I must withdraw that remark that a pressural wave cannot originate in the interior of an incompressible solid. Move a body about in the interior of such a solid and you have a solution—φ for the case of k infinite is a solution with definite displacements corresponding to a pressural wave. But none of that kind of effect appears at distances from the source considerable in comparison with the wave length. We can not prevent the introduction of this pressural wave, or quasi-water wave as others call it, in order to allow of the two components of displacement and two components of force on the two sides of the interface being equal. The extinctional formula by which Cauchy gets rid of the condensational wave, and those hypotheses of Neumann and MacCullagh that are still (as if there was any importance or weight to be attached to them) spoken of as if they were theories, are merely mistakes. I am a little aroused because I read, not two hours ago, an article in the *Compte Rendue*, by a new name, taking up with all gravity Neumann's theory and MacCullagh's theory and giving great weight and importance to them, finding that they come within an exceedingly small fraction of so and so, and so on. Go into it as analyzed by Lord Rayleigh, and you will see that their theories consist in introducing conditions that are inconsistent with two portions of matter pressing against one another with equal force, one pressing against the other with the same force that the other presses against it. We ask nothing more than that action and reaction are equal and opposite at the separating surface together with continuity of matter. Those are the only particles, notwithstanding the four principles of MacCullagh that I read to you the other day from Lord Rayleigh. These are the only principles, that action and reaction are equal and opposite, giving two equations, and that matter is continuous and does not slip, giving two more. These are comprised in one, viz.: mutual impenetrability of the two homogeneous mediums. I think we have spoken of that *bête noir* sufficiently. Leave it alone and you see it is a good enough animal after all.

I wanted to give you the reflected and refracted rays; but I need not do so because most of you have access to Lord Rayleigh's paper on the Reflection of Light from Transparent Matter. You see how charmingly short it is: there is the whole of it. Read that and then look at the conclusion. Green makes his simplification $n = n'$ entirely too soon; otherwise he might have got this result and said "This is what I got in the case of reflection of sound." Nothing could be simpler than this. $n = n'$ is a very slight simplification for the comparatively not very difficult case of only four unknown quantities. In case you do not have Lord Rayleigh's book at hand, note this if you think it worth while.

Ratio of amplitudes of reflected to incident vibrations

$$= \left(\frac{\tan i'}{\tan i} - \frac{n'}{n} \right) \Big/ \left(\frac{\tan i'}{\tan i} + \frac{n'}{n} \right),$$

which becomes Green's sine formula for $n = n'$. That lovely formula, as I call it, is given first so far as I know by Lord Rayleigh. I am pretty certain that he is the first who has given it correctly, because I know of no other writer except Green who worked at this problem without introducing impossibilities that vitiate the whole affair; and Green did not do it. Vibrations, then, perpendicular to the plane of incidence for two elastic solid media—no matter whether compressible or incompressible—give the same law as to intensity of reflection as two fluids destitute of rigidity (and therefore giving us a case in which the vibrations are essentially in the plane of the three rays). Vibrations purely compressional in a medium without rigidity are essentially in the plane of the three rays and give identically the same expression for the ratio between incident and reflected vibrations as does an elastic solid with vibrations perpendicular to the plane of incidence.

Having obtained this formula, Lord Rayleigh takes up the cases. About four days ago, I got hold of that thing wrong side up and it was only a few hours ago that I took up the cases right, and I find everything is true, interesting, intelligible and instructive.

Case I is Green's $n = n'$, which gives his sine formula

$$\left(\frac{\sin (i' - i)}{\sin (i' + i)} \right)$$

for the ratio of reflected to incident ray. Case II is MacCullagh's $\rho = \rho'$. MacCullagh is a very clever and able man, but he ignored dynamics vitally in the most peculiar parts of his work. We have

$$\frac{n'}{\rho'} \Big/ \frac{n}{\rho} = \mu^2;$$

n/ρ, n'/ρ' being the square of the velocities in the two mediums and μ^2 the refractive index. Take then $\rho = \rho'$ and we have

$$\frac{n'}{n} = \frac{1}{\mu^2} = \frac{\sin^2 i'}{\sin^2 i}.$$

Substitute that in the tangent formula, and it reduces to

$$\frac{\tan(i' - i)}{\tan(i' + i)}.$$

We have therefore this case of equal densities and unequal rigidities giving complete extinction at the angle of polarization. I have told you about that. It is satisfactory as a mathematical problem, but it is a failure for what we wish to account for in the theory of light. But we must stop here I am afraid.

Lecture XX

I have down next in my notes Rankine's very beautiful suggestion of aeo-lotropy of inertia. We want to explain aeolotropy in a crystal. We know that the velocity of propagation depends on the direction of vibration and not on the plane of distortion. Rankine's idea was this: let there be connected with the ether, or imbedded in it, gross molecules. I do not say ponderable or imponderable, but I use the word *gross* not meaning to throw any obloquy on them but simply to say that they are large. I do not say that I am giving Rankine's way of doing it. He mixes it up with molecular vortices and so on, and it is the molecular vortices that we can not very well get an idea of. I do not think I would like to suggest that Rankine's molecular hypothesis is of very great importance. The title is of more importance than anything else in the work. Rankine was that kind of genius that his names were of enormous suggestiveness; but we can not say that always of the substance. We cannot find a foundation for a great deal of his mathematical writings, and there is no explanation of his kind of matter. I never satisfy myself until I can make a mechanical model of a thing. If I can make a mechanical model I can understand it. As long as I cannot make a mechanical model all the way through I cannot understand; and that is why I cannot get the electro-magnetic theory. I firmly believe in an electro-magnetic theory of light, and that when we understand electricity and magnetism and light, we shall see them all together as part of a whole. But I want to understand light as well as I can without introducing things that we understand even less of. That is why I take plain dynamics. I can get a model in plain dynamics, I cannot in electro-magnetics. But as soon as we have rotators to take the part of magnets, and something imponderable to take the part of magnetism and realize by experiment Maxwell's beautiful ideas of electro-displacements and so on, then we shall see electricity, magnetism and light closely united and grounded in the same system.

Suppose here a massless rigid lining of our ideal cavity in the luminiferous ether. Let there be a massive heavy molecule inside, with fluid around it. The main thing is that this molecule, which only affects the effective inertia of the ether by adding its own mass to the moving mass of the ether, has aeolotropy of inertia. Imagine this spherule moving first in a horizontal direction. The effective inertia of this sheath will be altered if it moves to and fro in a vertical direction, there being by hypothesis liquid between it and the ether. The density of this mass must be greater than the density of the liquid, that is all. If there is danger of its coming to the sides of the cavity let there be springs to keep it in place if you like but let its connection with the lining of the cavity be in the main through fluid pressure. Then its effective inertia is different in different directions. This fluid lining seemed to hit off the very thing we wanted. Now comes Rankine's want of strength. He cut around the edges of it, and I think rather jumped at it, and put down a wave surface the same as Fresnel's and said that it came to that. But alas, Stokes (long before Lord Rayleigh suggested it) showed that it would give a different surface from Fresnel's—Lord Rayleigh, in repeating Rankine's suggestion, showed his strength where Rankine was not so strong, in mathematical powers of grappling with a different dynamical problem. Lord Rayleigh is a man who grapples with a difficulty and sees how much he can do with it. He puts it aside if he cannot solve it; but he never shirks it. Rankine was not a mathematician in that sense at all. Lord Rayleigh finds, not Fresnel's wave surface, but a wave surface differing from Fresnel's by certain terms appearing in reciprocals instead of directly. Lord Rayleigh could not pick up a thing of that kind without seeing the end of it, and he says in conclusion: "Between the theory here advanced and that of Fresnel observation ought to decide; but it does not appear that any experiments hitherto made are competent to do so. As Prof. Stokes points out, all the measurements which are to be combined in one calculation should refer to the same specimen of the crystal; otherwise an element of uncertainty is introduced sufficient to render the application of the test ambiguous. Should the verdict go against the view of the present paper, it is hard to see how any consistent theory is possible, which shall embrace at once the laws of scattering regular reflection, and double refraction." *

In the course of that paper Lord Rayleigh finds that Stokes had written that up and he is greatly surprised. The way he refers to Stokes is rather interesting: "I had got about as far as this in my original work when, on reference to Prof. Stokes' report, I was greatly surprised to find allusions to a theory of double refraction mathematically, if not physically, identical with that here advanced." After insisting on the importance of precise measurements, he says: "I will not read all that." Here is something: "Were the law of wave velocity expressed for example by the construction already mentioned having reference to ellipsoid (12), the wave surface (in this case a surface of the 16th degree) would still have plane curves of contact with

* *Philosophical Magazine*, June (Supplement) 1881, "On Double Refraction."

the tangent plane, which in this case also, as in the wave surface of Fresnel, are, as I find, circles, though that they should be circles could not have been foreseen." That is in respect to conical refraction, which Stokes says is thus no test of Fresnel's construction. Stokes told me of all this. It was he who first called my attention to the fact that Rankine was doubtful. He had not made his experiments then; but sometime after he told me of them. It seemed to me that they were experiments of very great accuracy, and I implored him to publish them. It was very hard to get him to do it. Every time I went to Cambridge I asked him to publish his results. Finally he did, and here is the whole of it, just 12 lines in the *Proceedings of the Royal Society*, June, 1872, under the title "Law of Extraordinary Refraction in Iceland Spar," and he has never published a word more about it. "It is now some years since I carried out in the case of iceland spar the method of examination of the law of refraction which I described in my report on Double Refraction, published in the Report of the British Association for the year 1862, page 272. A prism approximately right angled isosceles was cut in such a direction as to admit of scrutiny across the two acute angles in directions of the wave normal within the crystal comprising respectively inclinations of 90° and 45° to the axis. The directions of the cut faces were referred by reflection to the cleavage planes and thereby to the axis. The light observed was the bright D of a soda flame.

"The result obtained was that Huygens' construction gives the true law of double refraction within the limits of errors of observation. The error, if any, could hardly exceed a unit in the fourth place of decimals of the index, or reciprocal of the wave velocity, the velocity in air being taken as unity. This result is sufficient *absolutely to disprove* the law resulting from the theory which makes double refraction depend on difference of inertia in different directions.

"I intend to present to the Royal Society a full account of the observations; but in the meantime, the publication of this preliminary notice and the result obtained may be useful to those engaged in the theory of double refraction."

That was in 1872. 12 years have passed and nothing more has been published. You should be grateful to me for getting so much; you owe it to me.

I have next to consider some of the difficulties. What are they? Without the question of double refraction at all, consider simply the problem of reflection and refraction at the separating surface of transparent mediums. Take the theory that you know, work out every detail on the supposition $n = n'$ and that gives us at best only a rough approximation to Fresnel's results. They do not come near expressing the extinction at the polarizing angle.

Of one thing we are sure, the only way of coming at all within one-hundred miles of explaining the known facts of polarization is by supposing the vibrations to be perpendicular to the plane of the three rays. We are certain that if light is to be explained by the problem of an elastic solid, the vibrations must be perpendicular to the plane of the three rays unless we are to alter our facts altogether. I tried it with my molecules, and it makes no difference. My molecules give exactly the same result as the theory before you—no modification whatever. We cannot help ourselves at all by the molecules.

Then comes the difficulty (if you call it that) of making the line of vibration perpendicular to the plane of the three rays in the case in which we have no approach to extinction of the reflected ray. The difficulty is to get so near an approach to extinction as we have at the polarizing angle for light vibrating in the plane of incidence, and to explain the results of observation or the supposed results of observation that we have on the subject. These, according to Jamin's experiments, are very curious and noteworthy. According to his experiments there is a certain critical case for refraction, in which the refractive index is 1.4. If they are all right there would be perfect polarization for refractive index $\mu = 1.4$ and the phase going opposite ways from that—I am speaking very badly, but you will understand, the order of things as regards change of phase would be opposite for refractive index exceeding 1.4 to what it is for refractive index less than 1.4. Something like that results from Jamin's work; but his work was done a long time ago, and some people think it not altogether trustworthy. I do not know as Jamin himself would be fully satisfied with it now.

More work is wanted in the subject. Do not let us break our wings in battling against, and in trying to explain, facts which may not turn out to be facts. We can work on the theory, and try to get all we can out of it, with its 21 coefficients; but let us also work together and get some of the facts. I hope you will all make observations on the polarization of light. That expression *elliptic polarization* should always be coupled with elliptic polarization in reflected light when the incident light has been plane polarized with its plane neither in nor perpendicular to the three rays. Elliptical polarization is a confusing expression—find what is understood by that. Somebody must do it; I hope some of you will do it. Make also photometric experiments as to the quantity of light. Prof. Rood has made some splendid experiments of that kind. I meant to speak of those yesterday instead of my own rude experiments. He found for reflection of light from one or two substances at direct incidence a fulfilment of Fresnel's formula

$$\left(\frac{\mu - 1}{\mu + 1}\right)^2$$

to within a fraction of a per cent. He made experiments on several bodies but has not published them except for ground glass. Do make him publish them for iceland spar and plane glass. Anything from Rood is certain not to be rude. Like Stokes, he was not satisfied and did not publish his experiments although he made them ten or twelve years ago. After what I have obtained from Stokes, I hope all of you will try and extract the results from anybody who has good things in the shape of results.

I made many years ago a measurement of the celebrated v, the number of electro-static units in an electro-magnetic unit. I have just heard that the measurement has been made here with the whole system of apparatus and with the accuracy applied to electro-static measurement, which seems inconceivably superior to any measurements that have been made anywhere else so far as I know. I intend to get it for the Royal Society of London, which will not preclude its being published in any American publication.

The explanation of polarization at all by reflection, that is a difficulty. After that comes the other difficulty to explain double refraction; to find out how we can get it reasonably without introducing a fallacy of any kind, without introducing some other feature that is contrary to observation: to account for difference of velocity in different directions in a crystal by such a dynamical theory that the velocity of propagation shall be a function of the direction of vibration, and not of the direction of the strain. To read Rankine's splendid failure in this is most instructive and valuable.

If you were to ask me what other difficulties there were in the undulatory theory of light, I could say, I do not know that there is any other difficulty. The only other one is the old difficulty of the ether—how the planets can go through it; or, how the molecules of the kinetic theory of gases, going at velocities of from one-hundred to five-hundred meters per second (say half a kilometer per second), can go through it without any resistance, so far as we know, and that yet the maximum velocity of the molecular vibrations which produce light must be a small fraction of 300,000 kilometers per second, the velocity of light. The velocity of the vibrating molecules might amount to 1/50th of the velocity of light; more probably it is not a thousandth of it; probably in faint light it is not a three-hundred thousandth of it, or not more than a kilometer per second. You see I am taking you into my confidence; I am concealing nothing from you that I see. Here we have the particles going with a velocity of half or a quarter of a kilometer per second in the kinetic theory of gases, and yet we have the molecules creating waves of light by vibrations of a velocity which may not be more than one kilometer per second and probably cannot be as much as a thousand kilometers per second.

I only put the thing before you. Many of you have been thinking of this, no doubt, as a difficulty. I do not want to gloss over anything. But putting this aside, let us come down to ordinary matter. If you make a vibration in glycerine quick enough it will act like a perfectly elastic solid. I do not speak of the velocity of the vibration, I mean the period of the vibration. If the period of the vibration is short enough, I suppose glycerine would act like a perfectly elastic solid. Again, Maxwell's kinetic theory of gases leads us almost to say that for quick enough motions of a molecule in a crowd of molecules—motions by which the theory is explained—we may have a quasi-elasticity as of a solid coexisting with the gas.

But I fall back on glycerine. I tried last winter a new kind of galvanometer, and I made a woeful failure of it, I am sorry to say. I made very many uses of glycerine in checking the vibrations of the needle. The needle would, however, attain its full velocity, make two or three oscillations about a false pole and gradually come back. I could not look at that without being taught by it that the difficulty of a luminiferous ether would turn out not to be a difficulty at all. It is the shortness of the period of the to and fro motion in the luminiferous ether that allows it to act as a perfectly elastic solid for the luminiferous vibrations. For motions of particles of corresponding space not much greater, or perhaps of equal or less space, there is a perfect line with respect to absolute velocity when the force applied to a molecule acts for a long enough time to get it into motion. Why does a collision between molecules in the kinetic theory of gases give rise to velocities of one or two kilometers per second, or change the velocity one or two kilometers per second? Answer, because the whole time of collision is enormously greater than the four hundred million millionth of a second or than the slowest of the vibrations that Langley has found. In a paper that I have from Langley—I want to speak of it, it is so interesting—he has stated that as 17 times the period of sodium light. Make it 20 times: that gives the rate of 20 million million vibrations per second as the most sluggish vibration we know of in light and radiant heat.

The medium's being perfectly elastic for the to and fro recovery of motions in the 20 million millionth of a second is perfectly consistent, it seems to me, with its being like a perfect fluid in respect to forces acting perhaps for one millionth of a second.

Imagine what is the force of the collision between molecules. Take two billiard balls, and not allowing for the heat of collision, we can calculate the force of it roughly from our knowledge of the elasticity of the materials. Now, imagine the molecules of oxygen and nitrogen to be about as hard as billiard balls. I think if we only were to see the thing as it is, the collisions between molecules on the kinetic theory of gases would appear very gentle influences. Two molecules would come slowly together and be gradually stopped; and if you were to think of the viscosity in relation to all this, and calculate it out, you would see relations we cannot stop to take up just now. But compare that with a to and fro motion twenty million times as rapid. A million is not inconceivable; but it is a tremendous number. Think of once per second as compared with 30 times per second, and you need not think it incredible that the medium act as if it were perfectly elastic relatively to one vibration and perfectly yielding with reference to the other.

Our molecular theory will fit this. Go back to our spherical molecule with its central spherical shells—that is the rude mechanical illustration, remember. I think it is very far from the actual mechanism of the thing, but it will give us a mechanical model. By working at it and helping ourselves by such work as this of Prof. Morley's we shall see how every sequence of waves leaves a little more and a little more of energy in the gravest modes of the compound molecule until the energy is absorbed in modes of which the period is perhaps the millionth of a second instead of the 20 million millionth, or the 400 million millionth of a second. Think of the molecules, while they are doing work for light, as also moving about with a velocity of as much as a kilometer per second, say. Well, two of them come into collisional distance and one gives the other a gentle shove in the course of a millionth of a second and causes it to change its speed. Part of the energy that these molecules had from light vibrating at the rate of 20 or 400 million million times per second has been got into the form of long vibrations—so long that when the two come into collision they give to one another the gentle kind of shove required for the kinetic theory of gases.

Thus we can see perfectly how absorption will lead us down through fluorescence, phosphorescence, the heating up of the molecules so that they will give it out again by radiation all around through the ether, and then again still lower degredations, down to the sluggish vibration according to which, two molecules, swinging something like this $\longleftarrow\hspace{3cm}\longrightarrow$, their centres going one way and their shells the other, come together in the period of a millionth of a second, gently shove one another, and go off in other directions, adding their inertias to the velocity or taking it from the velocity or turning the course around at right angles. Thus I can see how our compound molecules act not only to increase the temperature when you increase the pressure according to the kinetic theory, but how the same molecules act to give us fluorescence and phosphorescence and then again the radiant heat from a body which is heated by rays passing through it.

I intended (but the time is too short to carry out that intention) to have worked out a mechanical model for sodium light. I will tell you how to do it so as to show quite an exceedingly sharp effect—as sharp as the two D lines are shown in Prof. Rowland's spectrum. If we had a day or two longer, we would hang on our particle m_1 a little pendulum—we would have to invoke gravity to help us here. If we are too proud to use gravity we can hang on a little springy molecule whose vibration is a certain period. Stick on beside it another springy molecule whose period varies by 1/800th or a thousandth from the first and another whose period is ever so little compared with either—say one whose period is 1/899th of a second and another whose period is one second exactly. Let these be so small that they produce no sensible effect until the period of the vibrator is within a hundred thousandth of either. Then it will begin to be enlivened up, and begin to make vibrations that will tell. While it is within a hundred-thousandth of the period of one, the period of vibration differs a hundred times as much from the period of the other and the energy of the vibrations produced in the other will be enormously smaller. Think then of adding to our first particle two molecules, with the period of one within a thousandth part of the period of the other, and another whose period is ever so little, and in saying good-bye to this illustration we will have a perfect model of a molecule that will produce sodium light, and produce the effect that is produced by sodium vapor upon light.

I have brought a book which I intended to make a subject of our lecture. I'm afraid it will be passed over. The book is Stokes' paper "On the Metallic Reflexion Exhibited by Certain Non-metallic Substances." * I only wanted to tell you that this molecular theory explains the colors of aniline and this wonderful thing that Stokes experimented on—this safflower-red. I wanted to read about the bright lines in the light reflected from safflower-red discovered by Stokes. I was thinking about this three days ago, and said to myself, there must be bright lines of reflection from bodies in which we have these molecules that can produce intense absorption. Speaking about it to Lord Rayleigh at breakfast, he informed me of this paper of Stokes and I looked and saw that what I had thought of was there. It was known perfectly well, but the molecule first discovered it to me. I am exceedingly interested about these things, since I am only beginning to find out what everybody else knew, such as anomalous dispersion and those quasi-colors and so on. There is no difficulty about explaining these things; we can predict them from the consideration of the molecule without experimental knowledge. And here again is a thing that suggests itself to me, that most probably there are bodies in which light is propagated faster than in the luminiferous ether.

I wish we could go into the dynamics of that, but we cannot. Take our old formula that we had about a week ago, $\mu^2 =$ so and so—if I write it out I would get it wrong, certainly. We found that μ^2 was a negative infinite for values a little above the frequency of the highest critical period, or any other critical period. What does $\mu^2 = -\infty$ mean? It corresponds to a total reflection. Put "μ^2 is negative" into your analytical formula and you find the case in which vibrations cannot be propagated. We want a mechanical illustration of that. Do it by taking two heavy stretched cords connected by slight elastic bands—or rather take one stretched cord to show transverse vibrations, connected by very fine elastic bands with fixed points, and you will find that you cannot get a wave to go along it at all above a certain frequency, just as we cannot get a wave to go along this wave machine above a certain frequency, but for a different reason, and in a different way. But just work that out—it will take about three-quarters of an hour to do it

* *Phil. Mag.*, Dec. 1853.

nicely—and think of the interpretation of μ^2 negative: it will correspond to the case in which waves cannot get into the medium at all, and we have total reflection. We find an imaginary symbol introduced in the kind of solution we are familiar with. The corresponding kind of real symbols would express the thing. The use of the imaginary symbol for explaining the ordinary total internal refraction is perfectly straight-forward. It used to be made a very difficult thing; but now everybody knows what mathematicians were puzzled over 40 or 50 years ago, and that is the interpretation of a true dynamical formula whenever an imaginary symbol comes into it. You know that perfectly well. Green took that up and made it clear. Green was the first, I think, to give the total internal reflection of glass and so on. Precisely the same kind of analysis that gives you total internal reflection at very oblique incidences gives you total reflection even at direct incidences for certain frequencies a little above any of the critical periods. That agrees, I believe, with observations. That ought to be the case with metals, although there are observations that go against the totality of the reflection; but if you look at appearances, it seems as if there ought to be total reflection. Silver is a shining instance; silver is total reflection all over. The molecular explanation of that property of silver would be simply that the highest mode, the shrillest mode, of vibration of the molecules with which silver loads the luminiferous ether is graver than the mode of the gravest light or radiant heat that we have ever had reflected from silver. That is all. Is it improbable that the shortest period of the molecule in silver may not be greater than the twenty million-millionths of a second—is it not very probable that the quickest mode of vibration of the molecules in such a heavy body, a body of such high specific gravity as silver, may be at least 20 times as long as in the molecule of sodium? That is all that is assumed; surely that is probably enough.

But now, what if you get a little light through—take a piece of silver whose thickness is less than the wave length and some light will get through. I have not worked this theory out, but I hope to do so in going home so that you may have it in the report.* We shall find no doubt that the light will get through that faster than in the luminiferous ether. Take gold leaf, say, of the thickness of half a wave length, or a quarter of a wave length—I have a specimen of such a leaf here, given to me by Prof. Trowbridge. I have an interest for some of you to see this specimen. Quincke has experimented upon very thin pieces of metal and has found that light passes through them with an acceleration. These are rather interesting experiments with gold of thickness engraved upon them of about the tenth or twentieth of a wave length. I am sorry we have not time to study them. I would have liked to have brought them before you all.

* See appendix.

Suppose we have not μ^2 negative with total reflection, but μ^2 less than unity; first we have $\mu^2 = 0$, and then going on up to unity. In the positions for vibrations corresponding to μ^2 between 0 and 1, which is for periods a little shorter than a critical period, we should have acceleration in the substance, a velocity of propagation greater than in the luminiferous ether. If we had an hour to more carefully study the quantities concerned in the absorption of light by, for instance, sodium vapor, we should arrive at some very curious and interesting conclusions and thoughts. I am afraid we must leave it, but think of a sodium flame in a hollow space in the interior of a glass globe, provided properly with air, with sodium vapor filling the globe so as to literally extinguish the flame in all directions. All the light that comes from that flame is absorbed into the sodium vapor. Think of the energy thus laid up, and you will get some very instructive lessons.

I will leave this sheet of to-day in your hands; it speaks for itself. You all understand that there must be continually work done in sending a wave in one direction. Take any portion of the wave front, and work must be done by the medium on one side of the wave front upon the medium on the other side to the extent exactly equal to the energy transmitted into the space beyond. If, then, the thing that is transmitted into the space is a succession of waves beginning abruptly and then perfectly regular and continuous—a succession of waves representing an arbitrary function, if you like—then the work done by the plane of the wave front per period must be equal to the sum of the kinetic and potential energies of the medium per wave length. That is the case in our ordinary formulas when we have $v^2 = e/\rho$, as we verified the other day. Now I call attention to this, that when the medium is loaded with molecules, the work done by the wave front exceeds the work done in the ether itself by the amount written down in this last formula. That is the amount of work then that goes to give energy to the attached molecules. It depends upon the spring arrangements, periods and so on, whether the energy taken by the molecules is not much greater than, or somewhat less than, or enormously greater than the energy in the elastic medium itself. When you come to the question of absorption bands, etc., the molecules will take thousands or perhaps millions of times as much energy as the energy of elastic action and motion in the ether itself. It is to prepare the way for this sort of thought that this paper is put in your hands. I think it sufficiently prepares the way. Suppose, for example, the energy of the molecules is two or three times the energy of the medium. Then it is perfectly clear that a succession of waves would go on advancing into the medium uniformly. The motion must be got up gradually. The result will be that if you commence a source of light and continue it quite constant for a length of time, there will be a gradual change in the first thousand, or the first hundred thousand, or the first million waves, but after a certain time it will be simply periodic. That will be the difference of circumstances from the circumstances we have to consider in the plane theory without attached molecules, or with a homogeneous medium in which the work done by the wave front per period is equal to the energy per wave length and in which we advance without change of form of a single wave or group of waves. I am afraid the thing is very imperfect, but it is a most practical and important subject that we have to think of, such as it is.

I want you to look at this drawing of Langley's. This is a thing that is most important. Here are relations of wave lengths and refrangibilities, but this is the thing I want you to see. We are all familiar with that drawing. There is the thing we know so well that Herschel worked out showing where we have the maximum heat in the solar spectrum. Here again is the energy of a Leslie cube—a cube of hot water. There is the maximum, way down in 37 of the scale. It is most important to see the wave length corresponding to the maximum energy in the spectrum of a Leslie cube and to compare it with that of the solar spectrum.

I am exceedingly sorry that our 21 coefficients are to be scattered; but though scattered far and wide, I hope we will still be coefficients working together for the great cause we are all so much interested in. I would be most happy to look forward to another conference, and the one damper to that happiness is that this is now to end and we shall be compelled to look forward for a time. I hope only for a time and that we shall all meet again in some such way. I would say to those whose homes are on this side of the Atlantic, come on the other side and I will welcome you heartily and we may have more conferences. Whether we have such a conference on this side or on the other side of the Atlantic again, it will be a thing to look forward to as this is looked back upon, as one of the most precious incidents I can possibly have. I suppose we must say farewell.

[Sir Wm. Thomson's allusion to the 21 Coefficients will be explained by the following humorous poem read at a dinner party of the previous day, which was given to Sir William Thomson and the physicists in attendance upon his lectures by President Gilman of the Johns Hopkins University. The author is Prof. G. Forbes, of London, England.

The Lament of the 21 Coefficients in Parting from each other and from their esteemed Molecule.

An aeolotropic molecule was looking at the view
Surrounded by his coefficients, twenty-one or two,

And wondering whether he could make a sky of azure blue
With platitatic *a b c* and thlipsinomic 2.

They looked like sand upon the shore with waves upon the sea
But the waves were all too wilfull and determined to be free,
And in spite of *n*'s rigidity they never could agree
In becoming quite subservient to the thlipsinomic *P*.

Then *web*-like coefficients and a loaded molecule
With a noble wiggler at their head worked hard as Haughton's mule,
But the waves all laughed and said a wiggler thinking he could rule
A wave was nothing better than a sidelong normal fool.

So the coefficients sighed and gave a last tangential skew
And *a* shook hands with *b* & *c* and *S* and *T* and *U*,
And with a tear they parted, but they said they would be true
To their much beloved wiggler and to thlipsinomic 2.

Signed (g.f.), a cross-coefficient now annulled.

President Gilman passed favorable verdict upon the versification, Sir
Wm. Thomson said the mathematics seemed all right, and the coefficients
unanimously concurred in the sentiments expressed. I therefore consider its
insertion justifiable even in a more solemn and heavy scientific work than
this purports to be.

—H.]

Appendix

Improved Gyrostatic Molecule*

The efficiency of the gyrostatic molecule described in my lecture of 16th Oct. is obviously in simple proportion to the amount of moment of inertia per unit volume of the medium: this is clear on the supposition that the axes of all the molecules are parallel, and their rotations in the same direction. When the axes are turned in all directions, the sum of components of moments of momentum round three axes at right angles to one another may be first taken, and their resultant in the usual manner of dealing with problems of moment of momentum. It is to the amount of this resultant moment of momentum per unit volume that the required efficiency is proportional, whatever be the distribution of axes of the molecules through the medium. Further it is easily proved that the rate of rotation of the plane of a distortional wave advancing through the medium per unit of distance travelled is, for different directions of the wave-normal, proportional to the cosine of its inclination to the direction of the resultant axes, determined in the manner just described. With these understandings, it will be convenient for the sake of simplicity to deal particularly with the extreme case of the axes of all the molecules parallel, and their rotations in the same direction; also to suppose them all equal and similar. Let *a* be the distance between the pivotted ends of the flywheels of each molecule (or the diameter of the spherical sheath imagined in the little diagram of Oct. 16). Let *k* be the radius of gyration of the flywheel, and let *m* be the sum of the masses of all the flywheels, distributed through a volume of the ether. To admit of definite calculation, we must (as before in respect to our compound spring molecules) suppose the sum of the volumes of the spaces occupied by the sheaths of the molecules to be infinitely small in comparison with the volume of the space filled with the homogeneous ether around them. It is easy to prove that the equations for wave motion, with wave front perpendicular to the

---

* Preliminary regarding molecule of Oct. 16, added Nov. 1st, 1884.

axes of the rotations, are

$$\frac{\rho}{4\pi^2}\frac{d^2\eta}{dt^2} = \frac{e}{4\pi^2}\cdot\frac{d^2\eta}{dx^2} + \frac{mk^2\gamma}{8\pi^3}\cdot\frac{d^3\zeta}{dt\,dx^2},$$

$$\frac{\rho}{4\pi^2}\frac{d^2\zeta}{dt^2} = \frac{e}{4\pi^2}\cdot\frac{d^2\zeta}{dx^2} - \frac{mk^2\gamma}{8\pi^3}\cdot\frac{d^3\eta}{dt\,dx^2},$$

(1)

where x denotes distances from a fixed plane parallel to the wave front, and η, ζ the components of displacement parallel to two fixed lines at right angles to one another in that plane. As previously, $e/4\pi^2$ denotes the rigidity of the ether, and $\rho/4\pi^2$ its density; including now however the masses of the sheaths and gyrostatic molecules; so that $\rho/4\pi^2$ is the average density of the whole material medium and imbedded molecules.

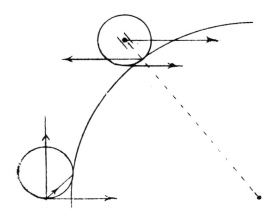

The most convenient way of dealing with these equations is to apply them at once to investigate circularly polarized light. For this purpose let

$$\eta = \sin 2\pi\left(\frac{x}{\lambda} - \frac{t}{T}\right),$$

$$\zeta = -\cos 2\pi\left(\frac{x}{\lambda} - \frac{t}{T}\right).$$

(2)

With this assumption, either of equations (1) gives

$$\frac{\rho}{T^2} = \frac{e}{\lambda} + \frac{mk^2\gamma}{T\lambda^2};$$

(3)

hence

$$\frac{\lambda^2}{T^2} = \frac{e}{\rho} + \frac{mk^2\gamma}{\rho T} = \left(1 + \frac{mk^2\gamma}{\ell T}\right)\frac{e}{\rho}$$

(4)

and therefore very approximately

$$\frac{\lambda}{T} = \left(1 + \frac{1}{2}\frac{mk^2\gamma}{\ell T}\right)\sqrt{\frac{e}{\rho}}. \tag{5}$$

Similarly, if λ' denote the wave length for circularly polarized light, with orbital motions in the opposite direction to that expressed by equations (2), we find

$$\frac{\lambda'}{T} = \left(1 - \frac{1}{2}\frac{mk^2\gamma}{\ell T}\right)\sqrt{\frac{e}{\rho}} \tag{6}$$

and instead of (2) we may take for this case

$$\xi' = \sin 2\pi\left(\frac{x}{\lambda'} - \frac{t}{T}\right),$$

$$\eta' = \cos 2\pi\left(\frac{x}{\lambda'} - \frac{t}{T}\right). \tag{7}$$

The resultant (ξ'', η'') of the motions (2) and (7) superimposed is expressed by

$$\xi'' = \xi + \xi' = \sin 2\pi\left(\frac{x}{\lambda} - \frac{t}{T}\right) + \sin 2\pi\left(\frac{x}{\lambda'} - \frac{t}{T}\right),$$

$$\eta'' = \eta + \eta' = \cos 2\pi\left(\frac{x}{\lambda} - \frac{t}{T}\right) - \cos 2\pi\left(\frac{x}{\lambda'} - \frac{t}{T}\right). \tag{8}$$

But now,

$$\frac{1}{\lambda} = \frac{1}{\ell} - \frac{1}{a}, \quad \frac{1}{\lambda'} = \frac{1}{\ell} + \frac{1}{a}. \tag{9}$$

We find from (8)

$$\xi'' = 2\cos\frac{2\pi x}{a}\sin 2\pi\left(\frac{x}{\ell} - \frac{t}{T}\right),$$

$$\eta'' = 2\sin\frac{2\pi x}{a}\sin 2\pi\left(\frac{x}{\ell} - \frac{t}{T}\right). \tag{10}$$

These express the motion in a wave of transverse rectilinear vibrations of which the velocity of propagation is

$$\frac{\ell}{T} = \sqrt{\frac{e}{\rho}} \tag{11}$$

(call this v) and in which the direction of the vibrations is constant in every part of the medium, but turns round the direction of propagation, at the rate of one round per distance equal to a of which the value, found from (9), (5) and (6), is

$$a = \left\{ \frac{1}{2}\left(\frac{1}{\lambda} - \frac{1}{\lambda'}\right) \right\}^{-1} = \frac{\lambda \lambda'}{\frac{1}{2}(\lambda - \lambda')}$$

$$= \frac{\ell \sqrt{e/\rho}}{mk^2 \gamma} T^2 = \frac{\ell}{mk^2 \gamma} v T^2.$$

(12)

Thus we see that the efficiency in rotative effect on the plane of polarization is equal to $mk^2 \gamma / \ell$. Suppose now the molecules to be made smaller and smaller

(Continued on page [247])

[In the lectures, $\ell/4\pi^2$ denotes the rigidity of the ether, I having mistaken the e in my notes for an ℓ. I took it to be the same in the manuscript of the above, and in handing it to the copyist instructed him to make it more plainly an ℓ. In reading the proof, however, it seems to me that formula (9) introduces a new letter. The three letters ℓ, e, c are very confusing in manuscript. Witness the following *tracings* from (1) and (9):

$$\frac{e}{4\pi^2}, \quad \frac{\ell}{c}.$$

The traced diagram above [p. 219] was found upon the back of the manuscript page opposite without reference marks.

I have received the following correction from Sir Wm. Thomson. "Please omit 'I do not find it quite &c.' and 'For instance Lord Rayleigh &c' (at top of p. 16). I think I found that I had misunderstood or misremembered one sentence of Lord Rayleigh's, and that what he said on this particular point was quite unobjectionable."

I infer (from a marginal note by W. T.) the following from Lord Rayleigh on Double Refraction is the sentence referred to:

"Fresnel and Green were inconsistent. The latter has given two rigorous theories of double refraction which differ from one another in important points, but agree in this, that neither of them can be reconciled with his explanation of reflection; for both assume that the forces which resist displacement within a crystal vary according to the direction of displacement. Precisely the same remark applies to investigations of Cauchy."

—H.]

To test the molecular hypothesis for the reflection of light at the surfaces of metals, and the transmission of light through thin metal foils.

I. Metallic Reflection
  (1) Vibrations perpendicular to the plane of incidence
  (2) Vibrations in the plane of incidence
Notation and explanations as in the leaf of notes for lecture of Oct. 16th.

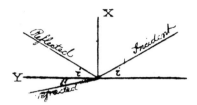

Adopting now the suppositions of incompressibility we have $b' = b$; and the equations become

$$\left.\begin{aligned}\psi &= A\varepsilon^{\iota(ax+by+\omega t)} + A_1\varepsilon^{\iota(-ax+by+\omega t)}\\ \varphi &= B\varepsilon^{-b'x+\iota(by+\omega t)}\end{aligned}\right\} x \text{ positive}$$

for upper medium; and for lower medium

$$\psi = \varepsilon^{\iota(a'x+by+\omega t)}; \quad \varphi = B'\varepsilon^{b'x+\iota(by+\omega t)} \quad (x \text{ negative})$$

Then,

$$\xi\left(=\frac{d\varphi}{dx} + \frac{d\psi}{dy}\right)$$

equated for upper and lower mediums gives

$$-Bb + \iota b(A + A_1) = B'b + \iota b \cdots .$$

Then

$$\eta\left(=\frac{d\varphi}{dx} + \frac{d\psi}{dy}\right)$$

equated for upper and lower mediums gives

$$\iota Bb - \iota a(A - A_1) = \iota B'b - \iota a' \cdots .$$

These yield

$$A + A_1 = 1 - \iota(B + B'); \quad A - A_1 = \frac{a'}{a} + \frac{b}{a}(B - B') \tag{1}$$

and so, conveniently, the problem is reduced to the determination of the interfacial wave (B, B').

The other two equations are found as follows:

$$P\left(=p^* + 2n\frac{d\xi}{dx}\right)$$

equated for upper and lower mediums gives

$$n\{(a^2 - b^2)B + 2ab(A - A_1)\} = n'\{(a'^2 - b'^2)B' + 2a'b\}. \qquad (P)$$

$$U\left[=n\left(\frac{d\xi}{dy} + \frac{d\eta}{dx}\right)\right]$$

equated for upper and lower mediums gives

$$n\{-2\iota b^2 B + (a^2 - b^2)(A + A_1)\} = n'\{2\iota b^2 B' + a'^2 - b^2\}. \qquad (U)$$

Eliminating $A - A_1$ and $A + A_1$ from these by (1) we find

$$n(a^2 + b^2)B - \{n'(a'^2 + b^2) - 2(n' - n)b^2\}B$$

$$= 2(n' - n)a'b$$

and   (2)

$$n(a^2 + b^2)B + \{n(a^2 + b^2) + 2(n' - n)b'\}B'$$

$$= \iota[n'a'^2 - na^2 - (n' - n)b^2].$$

These two equations determine B and B', and the results in (1) give A and A_1, so completing the solution of the problem. It is interesting and important, not only for the wave theory of light, but for the dynamics of elastic solids, to work out explicitly and to thoroughly interpret this solution without any restriction as to the rigidities (n, n') or the densities (ρ, ρ'). Meantime for reasons already considered we shall suppose $n = n'$, by which at this stage a great simplification is produced, reducing (2) to

$$(a^2 + b^2)B - (a'^2 + b^2)B' = 0 \qquad (3)$$
$$(a^2 + b^2)(B + B') = \iota(a'^2 - a^2) \qquad (4)$$
$$\left.\right\} \text{(case } n = n'\text{).}$$

Now we have

$$a^2 + b^2 = \frac{4\pi^2}{\lambda^2}, \quad a'^2 + b'^2 = \frac{4\pi^2}{\lambda'^2} \qquad (5)$$

---

*By work on the leaf of notes of Oct. 16th, we saw that $p = \rho\omega^2\varphi = -n(a^2 + b^2)\varphi$; in upper medium and the same with accents for the lower medium.

If λ, λ' denote the wave length in the upper and lower mediums. Hence (3) gives

$$\frac{B}{\lambda^2} = \frac{B'}{\lambda'^2} = \frac{B + B'}{\lambda^2 + \lambda'^2} = \frac{B - B'}{\lambda^2 - \lambda'^2} \tag{6}$$

and (4) gives

$$B + B' = \iota\frac{\lambda^2 - \lambda'^2}{\lambda'^2}. \tag{7}$$

From this and (6) we find

$$B - B' = \iota\frac{(\lambda^2 - \lambda'^2)^2}{\lambda'^2(\lambda^2 + \lambda'^2)} \tag{8}$$

and this again with (6) gives

$$B = \iota\frac{\lambda^2(\lambda^2 - \lambda'^2)}{\lambda'^2(\lambda^2 + \lambda'^2)}; \quad B' = \iota\frac{\lambda^2 - \lambda'^2}{\lambda^2 + \lambda'^2}. \tag{9}$$

Denote now by μ the index x of refraction from the upper to the lower medium. We have

$$\frac{\lambda}{\lambda'} = \mu \tag{10}$$

Hence (7), (8) and (9) become

$$B + B' = \iota(\mu^2 - 1), \quad B - B' = \iota\frac{(\mu^2 - 1)^2}{\mu^2 + 1} \tag{11}$$

and

$$B = \iota\frac{\mu^2(\mu^2 - 1)}{\mu^2 + 1}, \quad B' = \iota\frac{\mu^2 - 1}{\mu^2 + 1}. \tag{12}$$

Remembering that

$$a = \frac{2\pi}{\lambda}\cos i, \quad b = \frac{2\pi}{\lambda}\sin i, \quad a' = \frac{2\pi}{\lambda'}\cos i', \quad b' = \frac{2\pi}{\lambda'}\sin i',$$

$$\frac{b}{a} = \tan i, \quad \text{and } \frac{a'}{a} = \mu\frac{\cos i'}{\cos i} = \frac{\tan i}{\tan i'} \tag{13}$$

and using (11) in (1) we find

$$A + A_1 = \mu^2, \quad A - A_1 = \frac{\tan i}{\tan i'} + \iota\tan i\frac{(\mu^2 - 1)^2}{\mu^2 + 1}. \tag{14}$$

Hence, finally, for our solution, we have in upper medium

$$\psi = \tfrac{1}{2}\left\{\mu^2 + \frac{\tan i}{\tan i'} + \iota \tan i \frac{(\mu^2 - 1)^2}{\mu^2 + 1}\right\} \varepsilon^{\iota(ax+by+\omega t)},$$

$$\psi_1 = \tfrac{1}{2}\left\{\mu^2 - \frac{\tan i}{\tan i'} - \iota \tan i \frac{(\mu^2 - 1)^2}{\mu^2 + 1}\right\} \varepsilon^{\iota(-ax+by+\omega t)}, \qquad (15)$$

$$\varphi = \iota \frac{\mu^2(\mu^2 - 1)}{\mu^2 + 1} \varepsilon^{-bx+\iota(by+\omega t)},$$

[and] in lower medium

$$\varphi = \iota \frac{\mu^2 - 1}{\mu^2 + 1} \varepsilon^{bx+\iota(by+\omega t)},$$

$$\psi = \varepsilon^{\iota(a'x+by+\omega t)}, \qquad (16)$$

$$\psi_1 = 0.$$

To realize for the case of μ real, change ι into $-\iota$ and add the results to the preceding. Altering the notation correspondingly, to let ψ, φ, &c., denote real functions, we thus find in upper mediums

$$\psi = \left(\mu^2 + \frac{\tan i}{\tan i'}\right)\cos(ax + by + \omega t) - \frac{(\mu^2 - 1)^2}{\mu^2 + 1}\sin(ax + by + \omega t),$$

$$\psi_1 = \left(\mu^2 - \frac{\tan i}{\tan i'}\right)\cos(-ax + by + \omega t) + \frac{(\mu^2 - 1)^2}{\mu^2 + 1}\sin(-ax + by + \omega t),$$

$$\varphi = -\frac{\mu^2(\mu^2 - 1)}{\mu^2 + 1}\varepsilon^{-bx}\sin(by + \omega t) \qquad (17)$$

and in lower mediums

$$\varphi = \frac{\mu^2 - 1}{\mu^2 + 1}\varepsilon^{bx}\sin(by + \omega t), \qquad (18)$$

$$\psi = 2\cos(a'x + by + \omega t); \quad \psi_1 = 0.$$

Let now W be the resultant displacement of any part of either medium at a distance from the interface large in comparison with the wave length. The interfacial wave, φ, contributes nothing sensible towards this resultant and we have, as is easily seen from $(\cdots)$, $(\cdots)$ above,

$$W = \eta \sec i = -\frac{d\psi}{dx}\sec i. \qquad (19)$$

Before using this, reduce ψ and ψ_1 for the upper medium to the normal simple-harmonic form $R \cos(q + e)$ and $R_1 \cos(q_1 - e)$, by the notation

$$\tan e = \frac{(\mu^2 - 1)^2}{\mu^2 + 1}\bigg/\left(\mu^2 + \frac{\tan i}{\tan i''}\right); \quad R = \left(\mu^2 + \frac{\tan i}{\tan i''}\right)\sec e$$

$$\tan e_1 = \frac{(\mu^2 - 1)^2}{\mu^2 + 1}\bigg/\left(\mu^2 - \frac{\tan i}{\tan i''}\right); \quad R_1 = \left(\mu^2 - \frac{\tan i}{\tan i''}\right)\sec e_1.$$

(20)

Then, by (19) and (13), we find in upper medium

$$W = \frac{2\pi}{\lambda} R \sin(ax + by + \omega t + e) \qquad \text{incidental wave}$$

$$W_1 = -\frac{2\pi}{\lambda} R_1 \sin(-ax + by + \omega t - e_1) \quad \text{reflected wave}$$

(21)

and in lower medium

$$W = 2\frac{2\pi}{\lambda'}\sin(a'x + by + \omega t) \qquad \text{refracted wave}$$

which agrees with Green's original solution. The formula which I quoted (Oct. 16) from Lord Rayleigh comes immediately from it; as also his formula for retardation of phase which I had not time to quote. But our present affair is the case of $-\mu^2$ real positive numeric, for which we must now realize the symbolic formulas (15), (16).

Because $\sin i'$ is now imaginary, we may conveniently replace in (15) $\tan i / \tan i'$ by a'/a, its value according to (13); and for a' as follows;

$$a'^2 + b^2 = -v^2(a^2 + b^2), \quad \text{whence } \iota a' = h$$

(22)

where

$$v^2 = -\mu^2, \quad h = \{(v^2 + 1)b^2 + v^2 a^2\}^{1/2}$$

$$= \frac{2\pi}{\lambda}(v^2 + \sin^2 i)^{1/2}$$

$$= a(v^2 \sec^2 i + \tan^2 i)^{1/2}.$$

Thus (15) and (16) become in the upper medium

$$\psi = \frac{1}{2}\left\{-v^2 - \iota\frac{h}{a} - \iota \tan i\frac{(v^2 + 1)^2}{v^2 - 1}\right\}\varepsilon^{\iota(ax + by + \omega t)},$$

$$\psi_1 = \frac{1}{2}\left\{-v^2 + \iota\frac{h}{a} + \iota \tan i\frac{(v^2 + 1)^2}{v^2 - 1}\right\}\varepsilon^{\iota(-ax + by + \omega t)},$$

(23)

$$\varphi = -\iota\frac{v^2(v^2 + 1)}{v^2 - 1}\varepsilon^{-bx + \iota(by + \omega t)}$$

and in the lower medium

$$\varphi = \iota \frac{v^2 + 1}{v^2 - 1} \varepsilon^{bx + \iota(by + \omega t)},$$

$$\psi = \varepsilon^{hx + \iota(by + \omega t)}, \tag{24}$$

$$\psi_1 = 0.$$

Adding solutions for $\pm \iota$ and altering the notation of ψ and φ to let them denote real functions, we find in upper medium

$$\psi = -v^2 \cos(ax + by + \omega t) + \tan \iota \frac{(v^2 + 1)^2}{v^2 - 1} \sin(ax + by + \omega t),$$

$$\psi_1 = -v^2 \cos(-ax + by + \omega t) - \tan \iota \frac{(v^2 + 1)^2}{v^2 - 1} \sin(-ax + by + \omega t), \tag{25}$$

$$\varphi = \frac{v^2(v^2 + 1)}{v^2 - 1} \varepsilon^{-bx} \sin(by + \omega t)$$

and in lower medium

$$\varphi = -\frac{v^2 + 1}{v^2 - 1} \varepsilon^{-bx} \sin(by + \omega t),$$

$$\psi = 2\varepsilon^{hx} \cos(by + \omega t), \quad \psi_1 = 0. \tag{26}$$

To reduce to normal simple harmonic form, put

$$\left[\frac{h}{a} + \tan \iota \frac{(v^2 + 1)^2}{v^2 - 1} \right] v^2 \tan f, \quad \mathscr{S} = v^2 \sec f \tag{27}$$

and we find in upper medium

$$\psi = -\mathscr{S} \cos(ax + by + \omega t + f),$$

$$\psi_1 = -\mathscr{S} \cos(-ax + by + \omega t - f). \tag{28}$$

Hence, as above, we find for the resultant displacements of parts of the upper medium at distances from the interface great in comparison with the wave length

$$W = -\frac{2\pi}{\lambda} \mathscr{S} \sin(ax + by + \omega t + f) \quad \text{incident wave}$$

$$\tag{29}$$

$$W_1 = \frac{2\pi}{\lambda} \mathscr{S} \sin(-ax + by + \omega t - f) \quad \text{reflected wave}$$

****

Having thus completed the work with the simplification of which, following Green, we introduced into equations (3), and have kept in all up to equations (29), it is worth while now to take the general solution without this simplification, which I have worked out, in the first place for the sake of endeavoring to judge whether or not there is advantage to be gained for the wave theory of light, by supposing the effective rigidities different in different mediums, and in the second place because the general solution is in itself interesting in the theory of elastic solids. Going back to equations (2) put for brevity

$$a^2 + b^2 = 1; \frac{n'}{n} = r; \text{ and } \mu = \frac{n\rho'}{n'\rho}; \text{ which makes } \sqrt{(a'^2 + b^2)} = \mu;$$

and therefore

$$(30)$$

$$a' = \sqrt{(\mu^2 - b^2)}; a = \cos i; b = \sin i.$$

Put also

$$2(r - 1)b^2 = u. \tag{31}$$

We may now write equations (2) as follows:

$$B - (r\mu^2 - u)B' = \frac{a'}{b}u$$

and

$$(32)$$

$$B + (1 + u)B' = \imath(r\mu^2 - 1 - u).$$

From these we find

$$(1 + r\mu^2)B' = \imath(r\mu^2 - u - 1) - \frac{a'}{b}u,$$

$$(33)$$

$$(1 + r\mu^2)B = \imath(r\mu^2 - u)(r\mu^2 - u - 1) + \frac{a'}{b}u(1 + u),$$

and using these (33) in (1) above we find

$$(1 + r\mu^2)(A + A_1) = r\mu^2 + (r\mu^2 - u)^2 - \imath\frac{a'}{b}u^2,$$

$$(34)$$

$$(1 + r\mu^2)(A - A_1) = \frac{a'}{a}\left\{r\mu^2 + (1 + u)^2 + \imath\frac{b}{a}(r\mu^2 - u - 1)^2\right\}$$

Abbreviated by putting $r\mu^2 = D$, by which D shall denote the ratio of density of the lower medium to density of the upper medium. Thus finally

$$2(1 + D)A = D + (D - u)^2 + \frac{a'}{a}\{D + (1 + u)^2\}$$

$$+ \imath\frac{b}{a}\left\{(D - u - 1)^2 - \frac{aa'}{b^2}u^2\right\},$$

$$2(1 + D)A_1 = D + (D - u)^2 - \frac{a'}{a}\{D + (1 + u)^2\}$$

$$- \imath\frac{b}{a}\left\{(D - u - 1)^2 + \frac{aa'}{b^2}u^2\right\},$$

(36)

which is the final solution, in form convenient for being realized in either [the case] μ^2 real positive or [the case] μ^2 real negative. The realized forms for the case of μ real (D real and positive) are obvious from the equations, and need not be written down here. In the case of $r = 1$, u vanishes and we fall back on equations (14) with their consequences (20), (21) alone, as a particular case of these (36).

In the particular case of $r = 1/\mu^2$, which makes the densities equal in the two mediums, we ought, as we shall see below, to find a result not differing greatly from Fresnel's "sine-formula," if MacCullagh's admirable but seductive explanation of Fresnel's tangent-formula, by vibrations perpendicular to the plane of the three rays, were correct. How wildly wide of agreement with Fresnel's sine-formula or with anything in nature respecting the reflection of light is the supposition of equal densities and unequal rigidities in the two mediums was discovered by Lorenz and Rayleigh, partly from examination of the particular case of $\mu - 1$ infinitely small and vibrations *in* the plane of the three rays, in the problem of reflection and refraction at a plane surface, and partly by Rayleigh's dynamics of the blue sky. Our general solution (36) agrees of course for the particular case referred to with Lorenz's and Rayleigh's; and it serves to accentuate their important conclusion by showing equally wild results for all values of μ. I have worked it out for several angles of incidence, for the cases of $\mu = 1.225$ and $\mu = 1.5$, and have found curiously interesting results, which I need not give here as they have no importance for the wave theory of light unless as confirming what scarcely needed confirmation.

Our general solution (36) is also useful in dispelling the idea that, if Rood's experimental verification of Fresnel's formula

$$\left(\frac{\mu - 1}{\mu + 1}\right)^2$$

for the intensity of light reflected nearly at right angles from transparent bodies did not bar the way, we might, by giving r some value differing largely from either 1 or $1/\mu^2$, get something available, whether for light polarized

in the plane of the three rays or perpendicularly to it, out of the case of vibrations *in* the plane of the three rays. We see [that] in fact $r = 1$, or $r \fallingdotseq 1$, is
the only supposition that gives any approach to agreement to anything in
nature respecting the reflection or refraction of light in transparent mediums.
But, alas, we see also that the approach which the supposition $r = 1$ (Green's
theory) gives to explanation of the known phenomena of polarization is
sadly distant, and that no either small or large change from the exact value
1, for r, can better it.

For our immediate purpose of trying to see something of dynamical explanation for metallic reflection, let us realize (36) for μ^2 real and negative.
As in (22), (29) above, with our present abbreviations $a^2 + b^2 = 1$, we now
have

$$\mu^2 = -v^2; \quad h = (v^2 + b^2)^{1/2}; \quad a' = -2h; \quad u = 2(r-1)b^2; \quad r = \frac{n'}{n};$$

$$a = \cos i; \quad b = \sin i; \quad \text{put also } -D = \chi \tag{37}$$

so that, as the *effective density* of the lower medium is now negative, χ is the
corresponding positive ratio to the density in the upper medium.

Thus equations (36) and (33) become

$$2(1 - \chi)A = -\chi + (\chi + u)^2 - \frac{h}{b}u^2$$

$$+ \imath \left\{ \frac{h}{a}[\chi - (1 + u)^2] + \frac{b}{a}(\chi + u + 1)^2 \right\}$$

$$2(1 - \chi)A_1 = -\chi + (\chi + u)^2 - \frac{h}{b}u^2$$

$$- \imath \left\{ \frac{h}{a}[\chi - (1 + u)^2] + \frac{b}{a}(\chi + u + 1)^2 \right\} \tag{38}$$

$$(1 - \chi)B = \imath \left\{ (\chi + u)(\chi + u + 1) - \frac{h}{b}u(1 + u) \right\}$$

$$(1 - \chi)B' = -\imath \left(\chi + u + 1 - \frac{h}{b}u \right).$$

To realize as usual, put

$$\frac{h}{a}[\chi - (1 + u)^2] + \frac{b}{a}(\chi + u + 1)^2 = H,$$

$$-\chi + (\chi + u)^2 - \frac{h}{b}u^2 = K \tag{39}$$

and (modifying the ψ, φ notation suitably for realization,) we have [in the upper medium]

$$2(1 - \chi)\psi = H\sin(ax + by + \omega t) + K\cos(ax + by + \omega t),$$

$$2(1 - \chi)\psi_1 = H\sin(-ax + by + \omega t) - K\cos(-ax + by + \omega t),$$

$$(1 - \chi)\varphi = \left[(\chi + u)(\chi + u + 1) - \frac{h}{b}u(1 + u)\right]\varepsilon^{-bx}\cos(by + \omega t)$$

[and in the lower medium] (40)

$$(1 - \chi)\varphi = -\left(\chi + u + 1 - \frac{h}{b}u\right)\varepsilon^{bx}\cos(by + \omega t),$$

$$\psi = \varepsilon^{hx}\sin(by + \omega t).$$

Put now

$$K = R\sin f, \quad H = R\cos f,$$

whence

$$\tan f = \frac{K}{H}, \quad R = v(H^2 + K^2),$$

which gives (41)

$$2(1 - \chi)\psi = R\sin(ax + by + \omega t + f),$$

$$2(1 - \chi)\psi_1 = R\sin(-ax + by + \omega t - f)$$

and, by the fundamental equations, preceding (1) above, we have for the components of displacement:

incident wave

$$2(1 - \chi)\eta = -aR\cos(ax + by + \omega t + f),$$

$$\xi = -\frac{b}{a}\eta;$$

reflected wave

$$2(1 - \chi)\eta = aR\cos(-ax + by + \omega t - f),$$

$$\xi = \frac{b}{a}\eta;$$

interfacial wave in either medium with proper values of φ from (40) (42)

$$\xi = \frac{d\varphi}{dx},$$

$$\eta = \frac{d\varphi}{dy};$$

motion of lower medium in lieu of refracted wave

$$\eta = -h\varepsilon^{hx}\sin(by + \omega t),$$

$$\xi = b\varepsilon^{hx}\cos(by + \omega t).$$

Before interpreting this result let us find the corresponding result [(56) below] for the much easier case of vibrations perpendicular to the plane of polarization. The differential equations for the upper and lower mediums respectively are:

for upper medium (x positive)

$$\rho\frac{d^2\zeta}{dt^2} = n\left(\frac{d^2\zeta}{dx^2} + \frac{d^2\zeta}{dy^2}\right)$$

for lower medium (x negative)

$$\rho'\frac{d^2\zeta}{dt^2} = n'\left(\frac{d^2\zeta}{dx^2} + \frac{d^2\zeta}{dy^2}\right).$$

(43)

The stresses for this case of motion clearly involve solely tangential force in the plane of the wave front, and perpendicular to (YX) (the plane of the diagram).

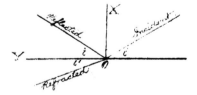

The component of this stress in any plane parallel to the interface between the two mediums, being the T of our general notation, is as follows for the two mediums:

upper

$$T = n\frac{d\zeta}{dx}$$

(44)

lower

$$T = n' \frac{d\zeta}{dx}.$$

Our solution in the case of vibration in the plane of the three rays might have been worked out from the beginning for a wave represented by any arbitrary periodic function, but it was more convenient for the ordinary analytic method of imaginaries which we used to work it out in the first place exponentials and simple harmonic functions. But the fact that the resultant laws of refraction and reflection do not involve the wave length, suffice to prove them true for waves or pulses represented by arbitrary periodic or non-periodic functions. In the present case there is no advantage, in point of simplicity or convenience, in expressing our work in terms of exponentials or simple harmonic formulas. Let us follow Green therefore in taking the arbitrary solution as follows:

upper medium

$$\zeta = A f(ax + by + \omega t) + B f(-ax + by + \omega t)$$
$$\underset{\text{incident wave}}{} \qquad \underset{\text{reflected wave}}{}$$

(45)

lower medium

$$\zeta = f(a'x + by + \omega t).$$
$$\underset{\text{refracted wave}}{}$$

[This] satisfies equation (43) provided

$$\rho\omega^2 = n(a^2 + b^2),$$
$$\rho'\omega^2 = n'(a'^2 + b^2).$$

(46)

The conditions to be satisfied at the interface, being equality of ζ on the two sides of it and equality of T on the two sides of it, are expressed by the equations

$$A + B = 1,$$
$$na(A - B) = n'a',$$

(47)

by which we find

$$A = \frac{1}{2}\left(\frac{n'a'}{na} + 1\right),$$
$$B = -\frac{1}{2}\left(\frac{n'a'}{na} - 1\right).$$

(48)

Still denoting as before by i and i' the angles of refraction and incidence, and putting now

$$\sqrt{a^2 + b^2} = c, \quad \sqrt{a'^2 + b^2} = c'$$

we have

$$c' = c\sqrt{\frac{n\rho'}{n'\rho}} = c\mu, \quad \sin i' = \sin i/\mu,$$

$$b = c\sin i = c'\sin i', \tag{49}$$

$$a = c\cos i, \quad a' = c'\cos i', \quad \frac{a'}{a} = \frac{\tan i}{\tan i'} = c(\mu^2 - \sin^2 i)^{1/2}.$$

Using this in equation (48), we find

$$\frac{B}{A} = -\frac{n'\tan i - n\tan i'}{n'\tan i + n\tan i'}, \tag{50}$$

which expresses the ratios of the amplitude of the refracted [ray] to the amplitude of the incident ray. For the particular case of $n = n'$, this gives

case I

$$\left. \begin{array}{l} n = n' \\ \dfrac{\rho'}{\rho} = \dfrac{\sin^2 i}{\sin^2 i'} = \mu^2 \end{array} \right\} \quad \frac{B}{A} = -\frac{\sin(i - i')}{\sin(i + i')}, \tag{51}$$

which is Fresnel's celebrated sine-formula. By squaring each member we have his expression for the ratio of the intensity of the reflected to the intensity of the incident light. The negative sign shews change of plane by half a period on the reflected ray relatively to the transmitted ray; of course there is no distinction in the circumstances between retardation and acceleration of half a period and therefore we cannot say which it is.

case II

$$\left. \begin{array}{l} \rho = \rho' \\ \dfrac{n}{n'} = \dfrac{\sin^2 i}{\sin^2 i'} = \mu^2 \end{array} \right\} \quad \frac{B}{A} = \frac{\sin^2 i\tan i' - \sin^2 i'\tan i}{\sin^2 i\tan i' + \sin^2 i'\tan i}$$

$$= \frac{\sin i\cos i - \sin i'\cos i'}{\sin i\cos i + \sin i'\cos i'}$$

$$= \frac{\tan(i - i')}{\tan(i + i')}. \tag{52}$$

The last member is Fresnel's celebrated tangent-formula, which he gives for vibrations *in* the plane of the three rays. The very curious result that this formula expresses rigorously the law of reflection for vibrations *perpen-*

dicular to the plane of the three rays, in the case of equal densities and unequal rigidities in the two mediums, seems to have been first discovered by MacCullagh. It is most tempting in respect to the explanation of polarization by reflection. It tempts us to suppose with MacCullagh the line of vibration to be *in* the plane of polarization, because at angle of incidence = $\tan^{-1}\mu$ a wave of vibrations perpendicular to the plane of the three rays gives rise to no reflected light and is transmitted without loss of energy into the lower medium of our diagram. But if this were the case the law of reflection of a wave of vibrations in the plane of the three rays should agree with Fresnel's sine-formula, or at all events should not differ from it more than observation allows us to suppose that light polarized in the plane of the three rays can in reality differ from that formula. But, alas, Lorenz and Rayleigh* have shewn that instead of fulfilling Fresnel's sine-law, the reflected ray in a wave of vibrations *in* the plane of the three rays would vanish at angles of incidence equal respectively to one-quarter and three-quarters of a right angle, when the index of refraction from one medium to the other differs little from unity ($\mu \doteqdot 1$). This they find by working out formulas equivalent to our equations (1) and (2) above for the case $\rho = \rho'$ and $\mu = \sqrt{n/n'}$. They therefore, with a cogency of which the force is clearly irresistable, concluded that the difference of the velocity of light in different mediums cannot be due to the difference of effective rigidities with equal effective densities or with approximately equal effective densities of the vibrating substance in the two mediums, and that in polarized light the vibrations are perpendicular to the plane of polarization. Lorenz went farther and concluded not merely that the difference of velocity is not due to difference of effective rigidity, but that it is wholly due to difference of density *"in all transparent uncrystalline substances."* Rayleigh, accepting this conclusion, refused to limit it to uncrystalline substances; his words are, "Lorenz draws the conclusion that the elastic force of the ether is the same in all transparent uncrystalline substances as *in vacuo* and that the vibrations of light are performed normally to the plane of polarization. He might, I think have omitted the word *uncrystalline.*"

I cannot myself quite admit this as a conclusion from the premises. I do not see that there is sufficient ground in any of the phenomena referred to by either Lorenz or Rayleigh for inferring that the effective rigidity is exactly, or is even very approximately, equal in the two mediums. It might be for instance that the rigidity is greater in the denser medium, but not greater in the same proportion as the density. This would make the velocity of propagation less in the denser medium, and it would give another available constant besides the index of refraction (and another is imperatively needed) to account for the enormously great difference between the results of observation, and of Green's theory, as expressed in equations (20) and (21) above,

---

*See the Hon. J. W. Strutt (now Lord Rayleigh), *Phil. Mag.* Aug. 1871.

in respect to the law of reflection of light polarized perpendicularly to the plane of the three rays, that is to say light of which the vibrations are in the plane of the three rays. But any such difference of rigidity, to be sufficient to go any considerable way towards accounting for the prodigious discrepancy between observation and Green's theory, would cause the reflected light at approximately perpendicular incidence to be vastly greater than

$$\left\{\frac{\mu - 1}{\mu + 1}\right\}^2$$

of the incident light, which Green's theory, on the supposition of equal rigidities, makes it. I know of no observations bearing upon this point except those of Prof. Rood of Columbia College, New York.* They alas, make the agreement with the

$$\left\{\frac{\mu - 1}{\mu + 1}\right\}^2$$

law exceedingly close for crown glass, and as Prof. Rood himself informed me, when I had the good fortune to see him in New York, immediately after the conclusion of my lectures in Baltimore, that the unpublished observations on flint glass and quartz to which he referred in his paper confirm the same law for them also to a somewhat close degree of accuracy, notwithstanding the imperfection of adjustment to which he alludes, as the reason which caused him to withhold them from publication. It seems therefore that after all we must accept the conclusion of Lorenz and Rayleigh, that the rigidity of the luminiferous ether is equal, or is at all events very approximately equal, in ordinary transparent solids. It remains, however, for the experimental examination to find whether or not the rigidity is also equal in transparent liquids and in extreme cases of transparent solids such as diamond ($\mu = 2.47$ to 2.75) and sulphuret of arsenic ($\mu = 2.454$), and I see no way of deciding the question except by photometric experiments such as Rood's.

In the meantime Jamin's beautiful discovery of what he calls positive and negative reflection[†] remains without dynamical explanation. It violates Cauchy's formulas, but they are empirical and not dynamical. They have great merit as empirical formulas; and, no dynamical law being fulfilled by Cauchy's "theory," none is broken by the modifications which Jamin, and Quincke[‡] in pursuing similar investigations, have given to Cauchy's formula to cause them to agree with observation.

* "On the amount of Light transmitted by plates of polished Crown Glass at a perpendicular incidence," *American Journal of Science and Arts*, Vol. L, July 1870.
[†] *Annales de Chimie et de Physique*, Vol. 29, 1850, page 263.
[‡] *Annalen der Physik und Chemie*, Vol. 119, 1863, p. 368; Vol. 127, 1866, pp. 1 and 199; Vol. 128, 1866, pages 355 & 541; Vol. 129, 1866, pages 44 and 177.

But metallic reflection is our present subject and therefore let us realize the solution (48) for the case of $\mu^2 = -v^2$, v real. Take now instead of (45) with its arbitrary function, f, the ordinary exponential imaginary formulas, thus:

upper medium

$$\zeta = A\varepsilon^{\iota(ax+by+\omega t)} + B\varepsilon^{\iota(-ax+by+\omega t)}$$

$$(53)$$

lower medium

$$\zeta = \varepsilon^{\iota(a'x+by+\omega t)}.$$

Looking to (49) we see that i' is now imaginary, but i remains real, and by this the expression for a' becomes

$$a' = -\iota c(v^2 + \sin^2 i)^{1/2}. \tag{54}$$

Eliminating a' by this and a by its expression in (49), (48) becomes

$$A = \tfrac{1}{2}\{1 - \iota r(v^2\sec^2 i + \tan^2 i)^{1/2}\}$$
$$B = \tfrac{1}{2}\{1 + \iota r(v^2\sec^2 i + \tan^2 i)^{1/2}\}$$

$$(55)$$

where

$$r = n'/n.$$

To realize as usual put

$$\tan e = r(v^2\sec^2 i + \tan^2 i)^{1/2}$$

and

$$R = \tfrac{1}{2}\{1 + r^2(v^2\sec^2 i + \tan^2 i)\};$$

we find

incident ray

$$\zeta = R\cos(ax + by + \omega t - e),$$

reflected ray

$$\zeta = R\cos(-ax + by + \omega t - e),$$

and (in lieu of refracted ray)

motion in lower medium

$$\zeta = \varepsilon^{xc(v^2+\sin^2 i)^{1/2}}\cos(by + \omega t).$$

$$\left.\begin{array}{l} \\ \\ \\ \\ \\ \\ \\ \\ \end{array}\right\} \quad (56)$$

When we return a little later to the molecular theory developed in the lectures, we shall see that for periods slightly less than any one of the critical periods (κ_1, κ_2 &c of our former notation) the value of μ^2 is negative, and that for a wide proportionate range, say from $T = \kappa_1$ to $T = \kappa_1/\mathcal{N}$, where $\mathcal{N}$ denotes some large numeric, we may have μ^2 negative, diminishing from $-\infty$ at $T = \kappa_1$ to zero at $T = \kappa_1/\mathcal{N}$. We shall also see that from $T = \kappa_1/\mathcal{N}$ to $T = $ zero, μ^2 augments from zero to one. All this was I believe developed by Sellmeier ten or twelve years ago. Our molecular theory gives no dynamical foundation for the assumption of a mixed real and imaginary numeric, which Cauchy has used for explaining metallic reflection; but it by no means follows that some modified molecular theory may not give some dynamical foundation for this assumption, which acquires great importance, and is at all events rendered exceedingly interesting, by the remarkable success of Cauchy's formulas for metallic reflection even if viewed only as merely empirical.

For the present, however, we confine ourselves to assumptions for which we see a definite dynamical foundation, and of which we can, as it were, construct a mechanical model, according to the molecular hypothesis we have been considering. We shall therefore restrict ourselves to μ^2 a real positive or negative integer, and try what we can do towards explaining the translucency of thin metallic films, the known phenomena of metallic reflection, and Kerr's discovery regarding the reflection of light from a polished magnetic pole, by supposing that for metals $-\mu^2$ is a real positive integer, μ^2 according to our notation of equations (37) and (49) above. We do not now assume $r = 1$, as it is only for transparent substances that any reason for this supposition has been discovered from observation or theory; and we may imagine that the effective rigidity of the ether acting in the interstices between the molecules should be largely different from the true rigidity of the homogeneous matter constituting the ether: In fact it is clear that if the round massless sheath of our molecule is infinitely rigid the effective rigidity of the ether in the interstices would be much greater than the true rigidity of continuous ether; but on the other hand if the sheath of each molecule be not rigid, but more or less yielding and quite perfectly elastic, the effective rigidity of the ether in the interstices might be either greater than, or equal to, or less than the true rigidity of continuous ether.

Now looking to our formulas (42) and (56) above we see that when $-\mu^2$ is positive, the intensity of the reflected ray is equal to that of the incident ray, both for vibrations, (42), in the plane of the two rays, and for vibrations, (56), perpendicular to this plane. Thus reflection at the surface of a medium for which $-\mu^2$ is positive is total. This totality is for all angles of incidence, and therefore the case is far from being analogous to that of total internal reflection in a transparent medium; the totality in this case being essentially confined to incidence exceeding the critical angle $\sin^{-1}(1/\mu)$. The reflection of light when polarized in the plane of incidence or perpendicular to it at a well polished silver surface involves, as has long been well known, very little loss of light; about 8 or 10 per cent has been generally supposed to be the amount of the loss.

Sir John Conroy has shewn that the loss is really much less than this, when the metal is very pure and the polish of the surface very perfect. Thus he succeeded in getting so good a polish on a double silver film deposited on glass (*Proc. Roy. Soc. of London*, May 15, 1884) that with light polarized in the plane of incidence, the loss by reflection was only 2.7 per cent, when the angle of incidence was 30°; and was not discoverable by very delicate observations, and seems to have been proved to have been less than a half per cent at angles of incidence of from 50° to 75°. With the same reflector and light polarized perpendicularly to the plane of incidence, he found no loss of light at incidences of 30°, and losses of from 2.5 to 6 per cent at incidences of from 40° to 75°. Whether a somewhat thicker film, or still more perfect polish, would annul these losses, or nearly annul them, is a very interesting subject for inquiry, and it is much to be hoped Sir John Conroy will continue his observations. Meantime we may take silver as a body which is certainly not far from fulfilling the totality of reflection given by our supposition of $-\mu^2$ positive, with no assumption of conditions causing the extinction of light. At the same time it is obvious that for any other metal than silver, extinction of a large percentage of the incident light is an essential and most serious condition of the problem. It is easy to imagine that our molecular hypothesis can be adapted, without any unnatural straining to directly take into account this condition. For the present, however, I must confine myself to the case of no extinction, and to silver as our one illustration.

Looking now to formulas (42) and (56), we see that for vibrations in the plane of the two rays the reflected ray is retarded in phase relatively to the incident ray, by an amount which, reckoned in radianal measure, is equal to $2f - \pi$, while for vibrations perpendicular to the plane of the two rays the phase of the reflected ray is accelerated relatively to that of the incident ray by an amount $2e$. Hence if the incident light be polarized in any plane oblique to the plane of incidence the reflected ray consists of two plane polarized components, of which the one consisting of vibrations in the plane of incidence is in phase behind the ether by an amount equal to

$$2f - \pi + 2e, \tag{57}$$

and by the formulas (44) and (56) for f and e we see that

$$2f - \pi + 2e = 2\left\{\tan^{-1}\frac{K}{H} - \tan^{-1}\frac{\cos i}{r(v^2 + \sin^2 i)^{1/2}}\right\}. \tag{58}$$

The retardation of phase of the component consisting of vibrations in the plane of incidence, relatively to that of the component consisting of vibrations perpendicular to the plane of incidence expressed by this formula, vanishes, as it must do, when $i = 0$; because for normal incidence there is no distinction between the two polarized components. If we increase i from 0 to 90°, the retardation increases from zero to π, which agrees with observation. If we suppose both v and rv, being the χ of (37) &c., to be very large numerics, we have $h \doteqdot v$ and therefore by (39)

$$\frac{H}{K} \doteqdot \frac{\sec i}{rv} + \tan i. \tag{59}$$

Hence, with the same approximation in the second term of its second member, (58) becomes

$$2f - \pi + 2e = 2\left\{\tan^{-1}\left(\frac{\sec i}{rv} + \tan i\right) - \tan^{-1}\left(\frac{\cos i}{rv}\right)\right\}. \tag{60}$$

Remark now that unless r be very small rv is very large, and therefore the second member of (60) increases very suddenly, from zero when $i = 0$ to being very little short of π when i is still quite small, and then completes the small difference of growth up to π as i increases to 90°. This is not consistent with observation, and therefore we must suppose r very small, small enough to make rv be of moderate dimensions. For example, if we take $(rv)^{-1} = 3.65$ we find that the second member of (60) increases probably from 0 to $\pi/2$ as i is increased from 0 to 75°48′, and completes the growth up to π as i is increased further up to 90°. This is precisely the case for silver according to Sir John Conroy's observations (*Proc. Roy. Soc. of London*, May 15th, 1884), the value which he finds for the "principal incidence" * in the case of his double silver film being 75°47′. It is probable that the law according to which the relative retardation increases up to $\pi/2$, and again from $\pi/2$ to π as the incidence increases to 75°47′ and again from 75°47′ to 90°, may be found as accurately expressed by our formula (60) with the value 2.935 for $(rv)^{-1}$, as it can be determined by observation; but observation is needed to test this supposition. Should the result shew insufficient agreement with the approximate formula (60), it would [do] to adapt (58) to give the requisite agreement with observation by supposing v not so large as to allow $\sin i$ to be neglected in comparison with it in the expression $v^2 - \sin^2 i$ for h^2, a supposition which would also give in (58) a perceptible effect to the terms $-\chi$, u, $u + 1$, &c., of (39), which are neglected in (60).

---

* "Principal incidence" is the name technically given by Jamin, Quincke, and others to express the angle of incidence at which the difference of phase of the two reflected components is a quarter of the period. Thus if light be polarized at such an azimuth that the two polarized components [of] the reflected ray are of equal intensity, and if the angle incidence be the principal incidence, the reflected light is circularly polarized. The azimuth thus defined for light at the principal incidence is called the "principal azimuth." The reflection being so nearly total as we have seen it to be, the principal azimuth for the silver surface ought to be very nearly 45°. Sir John Conroy's measurement of principal azimuth gives 44° for his double silver film.

For other metals than silver with the different values of the principal incidence found for them by observation, it would be also easy by the approximate formula (60) to find values of rv which would give the observed value for the principal incidence and if necessary to introduce the necessary modification by the more complete formula (58) to obtain agreement with observation. It is interesting to observe that the general law of metallic reflection, which has been found by observation, according to which the component of the reflected light whose plane of polarization is perpendicular to the plane of incidence is retarded relatively to the other component by an amount which augments from zero to π as the angle of incidence increases from zero to 90°, is brought out without any strained supposition by our formula (58) provided the direction of vibration be perpendicular to the plane of polarization, as we have been compelled by other reasons to believe it to be.

It seems then as if we might be very happy in our molecular explanation of metallic reflection: but alas, one most serious and seemingly essential characteristic of metallic reflection remains unexplained, and that is the fact that there is in it very little of what we might call chromatic dispersion, which in this case would shew itself in differences of the principal incidence for light of different periods. Our dynamical theory makes $v^2 + 1$ vary for different colors approximately in proportion to T^2, when T is very small compared with κ_1. We have no dynamical theory advanced enough to give the law of relation between r (the effective rigidity of the ether in the interstices between the molecules) and the period of the vibrations. It is difficult to conceive how any natural or acceptable theory could bear the strain of being forced to make the [product] rv as nearly the [same] through the wide range of periods presented by the different colours of visible light as is necessary to account for the known facts of metallic reflection. We are thus forced to admit that our dynamical theory of metallic reflection is a failure for the present; but it is not unsuggestive and it may possibly help to the true dynamical explanation which is so much desired. That it does indeed contain part of the essence of the true dynamical theory can scarcely be doubted after we have considered the next two subjects on which we are going to try it: the translucency of thin metallic films, and the effect of magnetism on polarized light incident on polished magnetic poles, or transversing thin films of magnetized iron, nickel or cobalt. The three remarkable discoveries of Quincke, Kerr and Kundt in this subject are as we shall see all brought out directly and without strain from our molecular theory.

Translucency of Thin Metallic Films *

To avoid circumlocution we shall continue to use the words "upper" and "lower," and suppose the light to be incident in our upper medium, with a horizontal interface between it and a denser medium below the interface. We shall now suppose the denser medium to be in the form of a plate between two parallel faces, and the medium below the plate to be the same as the medium above it. There is no difficulty in working out, by the general method expressed in the equations (1), (53) and (54) above, the problem for the reflection of light from the plate into the upper medium, and the transmission of light through the plate into the lower medium, for the two cases of vibrations perpendicular to the plane of the three rays, and vibrations in this plane. If we work this out for either case and for μ real, we find with great ease the ordinary formula expressing the wave theory of Newton's colors of thin plates. The only difference between the two cases is that the intensity of the reflected light for a simple reflection at one surface varies differently with the angle of incidence in the two cases. The complication of different acceleration or retardation of phase at different incidences, presented by the case of vibration in the plane of the three rays, does not involve any additional complication when we pass from reflection and refraction at a single interface to the problem of the plate.

Working out the problem for $-\mu^2$ real and positive, and equal to v^2 as above, and taking δ to denote the thickness of the plate, $\chi = 0$ to correspond to its upper side, and $\chi = -\delta$ to correspond to its lower side, we find as follows for the whole motion of the mediums due to a plane wave incident in the upper medium; all our other notations being the same as before; and now for brevity put $q = by + \omega t$.

upper medium

$$\zeta = \tfrac{1}{2}\sec e \big\{ \underbrace{\cos(ax + q - e) - \varepsilon^{-2h\delta}\cos(ax + q + 3e)}_{\text{Incident wave}}$$

$$+ \underbrace{(1 - \varepsilon^{-2h\delta})\cos(-ax + q + e)}_{\text{Reflected wave}} \big\} \qquad (62)$$

motion in plate

$$\zeta = \varepsilon^{hx}\cos q - \varepsilon^{-h(2\delta+x)}\cos(q + 2e)$$

wave transmitted into lower medium

$$\zeta = 2\sin e \cdot \varepsilon^{-h\delta}\cdot \cos[a(x + \delta) + q + e - \pi/2]$$

* Added Dec. 4–11, 1884.

whereas in (37) and (56) above

$$h = \sqrt{(v^2 + \sin^2 i)} \tag{63}$$

and

$$\tan e = rh/a. \tag{64}$$

Taking only this last equation into account, it is easy to verify that equations (62) fulfil, at each interface, the proper interfacial conditions, which are that on the two sides of each interface the values of ζ are equal, and the value of $r\,d\zeta/dx$ in the plate equals the value of $d\zeta/dx$ in the contiguous medium on the other side of the interface.

Reflection from and transmission through a plate for the case of vibrations in the plane of the three rays.

The result so far as the waves in the upper and lower mediums must clearly be identical with that expressed in (62) with $\pi/2 - f$ substituted for $e; f$, as in equations (41) and (39) above, being found by the following formula:

$$\tan f = \frac{-\chi + (\chi + u)^2 - (h/b)u^2}{(h/a)\,[\chi - (1 + u)^2] + (b/a)\,(\chi + u + 1)^2}. \tag{65}$$

The corresponding value of the four coefficients corresponding to B and B' of (38), which are now required to express the double interfacial wave, are easily written down by aid of (38), but they are not required for our present purpose.

Looking back now to (62), whether with e as in the formula for vibrations perpendicular to the plane of the three rays or with $\pi/2 - f$ in the place of e to suit the case of vibrations in the plane of the three rays, we see that when δ is infinitely small the reflected wave vanishes, and the wave transmitted into the lower medium agrees with the incident wave in amplitude and phase: that is to say the film has no effect, which is of course the correct result for this case.

Next suppose δ to be large enough to [make] $\varepsilon^{-h\delta}$ exceedingly small. The wave transmitted into the lower medium becomes infinitely small and the reflected wave in the upper medium agrees infinitely nearly with what we found above in (42) and (56) for the case of reflection at a single metallic surface.

When $\varepsilon^{-h\delta}$ is a small fraction of unity, not zero, the amplitude of the transmitted wave is approximately $4 \sin e \cos e \varepsilon^{-h\delta}$ of the amplitude of the incident wave for the case of vibrations perpendicular to the plane of the three rays; and the same with f for e for the case of vibrations in this plane. The phase of the transmitted wave is accelerated by an amount approximately equal to $a\delta + 2e - \pi/2$ in the former case, and equal to $\pi/2 - f$ in the latter case. The amount of the acceleration thus calculated for each case is that by which the transmitted wave is in advance of an ideal continuation of the incident wave with the plate removed. The unit of reckoning is the radian. To reduce [the] space travelled in the medium on either side of the plate we must divide by $a\sec i$. Hence, remarking that

$$a = \frac{2\pi}{\lambda}\cos i \tag{66}$$

where λ is the wave length in the medium on either side of the plate, we find [the following] for the amounts of the advance of phase in the two cases.

vibrations perpendicular to the plane of the three rays

$$\cos i \cdot \delta + \left(\frac{e}{\pi} - \frac{1}{4}\right)\lambda \tag{67}$$

vibrations in the plane of the three rays

$$\cos i \cdot \delta + \left(\frac{1}{4} - \frac{f}{\pi}\right)\lambda. \tag{68}$$

We have seen that when $i = 0$, e and f are each positive acute angles and complements of one another; and each augments to the value $\pi/2$ when i is increased from $0°$ to $90°$. Hence the second members of (67) and (68) vanish respectively for two particular values of i. In these cases the advance of the corresponding polarized component is equal to $\cos i \cdot \delta$. To explain this let ab be a wave front in the upper medium and $a'b'$ [be] the position it will reach in the lower medium after any particular time t. Now imagine the plate to be annulled and the lower medium to be moved perpendicularly to the line of the plate so as to fill up the gap. The phase of the transmitted wave $a'b'$, in its actual position, is the same as would be the phase at the same time t, in the altered position of $a'b'$, with the plate annulled.

When the second term of (67) or (68) is positive, there is an advance of the transmitted ray even more than that corresponding to the annulment of the plate. There is positive advance, though of less amount than corresponding to annulment of the plate, when the second term is negative, but of less absolute value than the first. The general result of advance of phase produced by a metallic film upon light transmitted through it was discovered experimentally by Quincke 21 years ago; but alas for our dynamics, the details of his results seem very far from agreeing with anything I can make out of our formulas. We must not, however, be discouraged by this. At all events the nearest approach to the explanation of Quincke's result, on the supposition of a real refractive index, makes the refractive index vary with the angle of incidence—a brilliant *reductio ad absurdum*—and gives it values ranging from 3 to 8 or 9 for different metals or even for different specimens of the same metal!

Our dynamical theory perfectly explains Kerr's result, for normal reflection from a metallic pole, crossed, whether normally or obliquely, by lines of magnetic force; which is that plane polarized light, incident normally or nearly normally, produces a plane polarized reflected ray, with plane of polarization turned slightly in the direction opposite to that of the "Amperian currents" of the magnetisation. The effect of magnetisation of the iron must be to give different values to v for circularly polarized light, according as the direction of the orbital motion is with or against the "Amperian currents." Thus, while according to our formulas there is for every ray total reflection, the effect of the magnetisation is to change, in the act of reflection, not the intensity but the phase of [a] circularly polarized ray. Hence plane polarized light incident normally, ideally resolved into two opposite circularly polarized rays, gives rise to two opposite circularly polarized reflected rays, differing slightly in phase and therefore equivalent to a plane polarized ray in a plane of polarization turned through a small angle. On the other hand, if we imagine the iron to act as a transparent medium, with real refractive index, the only possible effect in the case of normal incidence is to give different intensities to the two circularly polarized components of the reflected ray, and so to give a slight degree of ellipticity to the reflected ray, with [the] major axis of the ellipse precisely coincident with the line of vibration of the incident light. This is FitzGerald's result, which, as remarked by FitzGerald himself, and by Kundt, is absolutely at variance with Kerr's experimental discovery. It is therefore quite certain that iron does not act as a transparent medium with real refractive index. It is, however, quite conceivable that the extinctivity which the iron must have (to give it its practical opacity), if it has a real refractive index, may, under the influence of magnetisation, give the difference of phase required to explain Kerr's result. That extinctivity must indeed be invoked (as Cauchy long ago invoked it) seems in this new case probable; because, though our dynamical formulas, without extinction, perfectly explain Kerr's result, they are utterly at variance with Kundt's,[*] according to which the plane of polarization of light passing normally through a thin iron plate is turned through a not very small angle (amounting in some of his experiments to as much as $3\frac{3}{4}°$ for iron, $2°$ for cobalt and $\frac{3}{4}°$ for nickel), in the *opposite* direction to that found by Kerr in reflection. But it does not seem possible to abandon our pure imaginary for refractive index in metals ($-\mu^2$ real), and we may hope that extinctivity on a true dynamical foundation in connection with our molecular theory, which it must be remembered is due originally to Sellmeier, may serve to solve the numerous difficulties in

[*] *Berlin Sitzungsberichte,* July 10, 1884; or *Philosophical Magazine,* October, 1884.

connection with metallic reflection and transmission, which give us so much anxiety. Extinctivity however cannot help to solve the great difficulty as to reflection at the interface between two transparent mediums, in the case of vibrations in the plane of the three rays. Green's attempt to explain this difficulty by gradualness in the transition of physical quality from one medium to another seems to me most unpromising if not utterly hopeless. There remains Green's other suggestion of "extraneous force," by which as we have seen he opened a door for explaining how the velocity of light in a crystal can depend on the direction of the line of vibration, irrespectively of the line of propagation. If this suggestion becomes realized it must modify the circumstances at the interface which determine the reflection. Is it possible that it can lead to the true law for reflection of waves consisting of vibrations in the plane of the three rays?

In continuation of MSS "Improved Gyrostatic Molecule" despatched 1st Nov. 1884*

[See page 221 above.]

and the number greater and greater, with the same total mass, the angular velocity γ must be augmented in inverse proportion to k^2.

Now for our improved gyrostatic molecule imagine two kinetically equal and similar rotators mounted by means of ball and socket joints in the interior of a rigid spherical sheath; and, by axles projecting from them towards the centre of the sheath, let them be jointed together in the manner indicated; that is to say by a ball projecting from the one fitting in a cylindrical projection from the other. To make them kinetically equal and similar as supposed, notwithstanding this slight difference of form, their masses and moments of inertia round corresponding axes are to be exactly equal. To avoid all complexity, we shall suppose the outsides of the sheath to be perfectly smooth and of truly spherical figure, so that when embedded in ether it may not be affected by the rotational part of the·motion of the ether, and that it may experience merely translational forces in lines through its centre in virtue of the translational motion of the ether. We shall, however, in investigating the kinetic properties of our new compound molecule, not restrict ourselves to the supposition of perfect smoothness in the sheath, and shall consider the result of the giving of any motion, whether translational or rotational, to the sheath.

First suppose the interior rotators to be given at rest with their axes in one line as indicated by the strong lines in the diagram.

The diameter of the molecule through the centre of the ball and socket joints will, for brevity, be called the axis of the molecule. Suppose now a torque to be applied to the sheath round an axis perpendicular to the line of axis of the interior rotators. This torque will cause the sheath to commence turning round the axis of the torque; and the two rotators, each resisting by its inertia, will each carry the other round by the mutual action of the ball-and-cylinder joint between them. Thus the whole system will turn as a rigid body, and receive acceleration from the supposed torque according to the law of acceleration of a rigid body. Suppose now that a force be applied to the sheath in a direction perpendicular to the axes of the rotators. They will clearly lag in the motion thus produced, and their axes turning in opposite directions will make an increasing obtuse angle with one another till the rotators strike the sheath. It is curious to see how this mode of jointing gives perfect quasi-rigidity relatively to rotatory motion of the sheath and absolute limpness to all translational motion of the sheath except along

* Added Dec. 11 to 13, 1884, Continued from page [221].

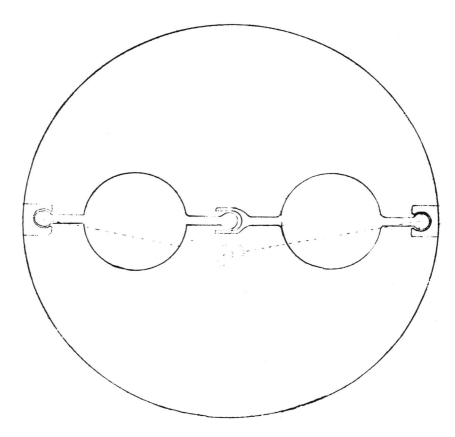

the line of axes of the rotators (which, be it remembered, we had initially in one line). Suppose now the two rotators, given with their axes in one line, to be set into rapid rotation round this line. The quasi-rigidity relatively to rotation of the sheath still remains perfect; and therefore, for all rotational motions of the sheath with its centre unmoved, the rotators will act precisely as if they were rigidly connected, so that the compound molecule will act merely as a simple gyrostat.

Relapsing now into the supposition of the sheath perfectly smooth on its outside, let any forces be applied to it. It is clear that its motion will be purely translational when we consider the symmetry of the reactions in the two ball-and-socket joints. A result of any acceleration of the centre of the sheath not exactly along the axis of the molecule must be to disalign the axes of the rotators; but if the regular velocity of the rotators be very great, their gyrostatic action will give rise to an exceedingly great quasi-rigidity against the disalignment. It is easy to write down the equations of the translational motion of the sheath and of the whole motion—rotational and translational—of the rotators, under the influence of any given forces, applied normally as supposed to the sheath. For our present purpose it will be sufficient to write down these equations of motion for the case of infinitesimal disalignment of the axes of the rotators, but it will help us to understand all the circumstances if we first take the rigorous solution for the case of steady precessional motion of the rotators with their axes inclined at any finite angle, θ, to the axis of the molecule. This steady motion involves uniform circular motion of the sheath; or in one particular case zero motion of the sheath

OL is perpendicular to OB;
$BOA = \theta; BOI = u;$
$\gamma =$ component angular velocity round OB;
$\zeta =$ component angular velocity round OL;
$\omega =$ angular velocity of the plane BOL round OA.

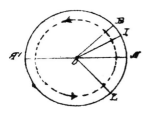

Let O be the centre of one of the vibrators, and $A'OA$ a line through it parallel to the axes of the molecule. This, on account of the symmetry, is the line joining the centres of the two rotators. Let OB and OI be respectively the axis of figure of the rotation and its instantaneous axis of revolution at any time. The supposed motion will be the same as that of a cone having OB for its axis and BOI for its semi-vertical angle rolling on a fixed cone having OA for its axis and IOA for its semi-vertical angle (compare Thomson & Tait's *Natural Philosophy* § 105). Suppose now the component angular velocity round OB to be of any given magnitude γ. This remains absolutely constant because the ball-and-socket and ball-and-cylinder joints are perfectly frictionless. Suppose now in the case of motion investigated the angle BOA to be given equal to θ, and the precessional angular velocity of given magnitude ω. Draw OL perpendicular to OB in the plane BOA, and let it be required to find the component angular velocity of the rotation round OL, which we shall denote by ζ.

Make CL, OA, OI, OB each equal to unity. The linear velocities of the matter at L and at B are respectively equal to the angular velocities γ and ζ. Now the same matter is also at the point B, and therefore the required angular velocity ζ is simply equal to the linear velocity of the point B in the diagram. Supposing the rotational and precessional motions viewed from A to lie in the direction opposite to the hands of a watch, B and I move perpendicularly to the paper outwards, and L perpendicularly to the paper inwards as follows (the second expression for the velocity of B being found by considering that the velocity of B is also the velocity of the matter of the rotator at B, and that the velocity of the matter of the rotator at I is zero):

$$\zeta = \text{linear velocity of } B$$

$$= \omega \sin\theta = \sqrt{(\gamma^2 + \zeta^2)} \sin u$$

$$= \gamma \tan u, \tag{13}$$

$$\text{linear velocity of } I = \omega \sin(\theta - u), \tag{14}$$

$$\text{linear velocity of } L = \omega \cos\theta. \tag{15}$$

Of these expressions the only ones we require are the first, for ζ, and the third, for the velocity of L. The others are put down merely to illustrate the circumstances.

Let m be the mass of the rotator and mk^2 and $m\ell^2$ its moments of inertia respectively round OB and OL. The component moments of momentum round the axes OB and OL are respectively $mk^2\gamma$ and $m\ell^2\zeta$. Hence, as the points B and L have absolute velocities perpendicular to the plane of the paper outwards and inwards equal respectively to ζ and $\omega \cos\theta$, the moments of the couples required to produce the corresponding changes of direction of the two components of momentum are respectively $mk^2\gamma\zeta$ and $m\ell^2\zeta\omega \cos\theta$. These couples are both in the plane of the diagram, the first in the direction of the arrow head and the second in the opposite direction, and therefore the whole couple required to cause the rotation to move as it does is

$$m(k^2\gamma - \ell^2\omega \cos\theta)\zeta = m(k^2\gamma - \ell^2\omega \cos\theta)\omega \sin\theta. \tag{16}$$

Now let us suppose the sheath of the compound molecule to be kept so moving that the centres of the ball-and-socket joints revolve with uniform angular velocity ω, in circles each of radius r, perpendicular to the axis of the molecule; and let it be required to find what must be the value of θ, in order that the rotator may move with steady precessional motion in the manner supposed. Let F denote the force towards the centre of each of these circles, with which the socket acts upon the ball turning within it. Imagine, after Poinset, pairs of equal balancing forces F to be applied in parallel lines through the centres of inertia of the two rotators. We thus have a force F at the centre of inertia of each rotator, and a couple whose moment is $Fa \cos\theta$, if a denote the distance from the centre of the ball-and-socket joints to the centre of inertia of the rotator. The centre of inertia of each rotator in these circumstances moves with angular velocity ω in a circle whose radius is

$r + a\sin\theta$; and the centreward force required to cause it to so move (or the force balancing its centrifugal force) is F. Hence

$$F = m\omega^2 (r + a\sin\theta). \tag{17}$$

The function performed by the couple is to change directions of moments of momentum in the manner explained above, and therefore it must be equal to the formula (16):

$$Fa\cos\theta = m(k^2\gamma - \ell^2\omega\cos\theta)\omega\sin\theta. \tag{18}$$

These two equations serve to determine F and θ.

For our present purpose it is sufficient to work out the result for θ infinitely small. Thus, by taking θ for $\sin\theta$ and 1 for $\cos\theta$ in (17) and (18), we find

$$\theta = \frac{ra\omega}{k^2\gamma - (a^2 + \ell^2)\omega} \tag{19}$$

and

$$F = m\omega^2 r\left[1 + \frac{a^2\omega}{k^2\gamma - (a^2 + \ell^2)\omega}\right]. \tag{20}$$

We conclude that for circularly polarized light the effect of rotation within the gyrostatic molecule is to cause it to have the same influence on the motion of the ether as if its mass, instead of $2m$, were

$$2m_1 = 2m\left[1 + \frac{a^2\omega}{k^2\gamma - (a^2 + \ell^2)\omega}\right]. \tag{21}$$

Hence, supposing $\rho + m_1$ and $\rho + m_1'$ to be the effective density of the ether with its embedded molecules, for two circularly polarized rays with opposite orbital motions, and v [and] v' the velocities of propagation of these rays, we have

$$\frac{v}{v'} = \sqrt{\frac{\rho + m_1}{\rho + m_1'}} \tag{22}$$

where m_1' is the same as m_1 as given by (21), with the sign of ω changed. Hence, the ratio being exceedingly nearly equal to unity, we have approximately

$$\frac{v}{v'} = 1 + \frac{1}{2}\left[\frac{a^2\omega}{k^2\gamma - (a^2 + \ell^2)\omega} + \frac{a^2\omega}{k^2\gamma + (a^2 + \ell^2)\omega}\right]. \tag{23}$$

If ω could be large enough to make $(a^2 + \ell^2)\omega$ equal to or greater than $k^2\gamma$, we should have something analogous to "anomalous dispersion" in the magneto-optic effect. It does not however appear probable that any such critical condition can be at all approximated to by the highest ultra-violet light known to exist; and for the present it is convenient to suppose

$(a^2 + \ell^2)\omega$ infinitely small in comparison with $k^2\gamma$, which reduces (23) to

$$\frac{v}{v'} = 1 + \frac{a^2\omega}{k^2\gamma}. \tag{24}$$

This, as worked out in (8) to (12) above, leads to the true law of relation between the rate of turning of the plane of polarization and the period of vibration of the light. It is very curious to remark that the gyrostatic efficiency of our improved double-rotator molecule, depending as it does on translational, not on rotational, motion of the sheath, is inversely proportional to the angular velocity of the rotators, provided this angular velocity be great enough for gyrostatic domination (Thomson & Tait's *Natural Philosophy*, second edition, § 345), while the gyrostatic efficiency of our crude original gyrostatic molecule (depending as it does on the rotational motion of the sheath) was directly proportional to the angular velocity of the rotator.

Going more into detail, we see that with the crude original gyrostatic molecule the proportional alteration of the velocity of light due to circular polarization depends on $\omega\gamma k^2/v^2$, whereas with the improved gyrostatic molecule it depends simply on ω/γ (really upon $\omega a^2/\gamma k^2$, but we may suppose a^2/k^2 to be some constant numeric of moderate value a little more or less than unity). If now the improved gyrostatic molecule, instead of being perfectly smooth on the outer boundary of its sheath, as for simplicity we took it in the investigation, be now supposed to be adhesively embedded in the ether so that it shall be carried round with the ether in the infinitesimal rotations which the ether experiences in the course of luminiferous vibrations, it will act in respect to these rotations as if it were a simple vibrator like the unimproved molecule, and at the same time it will have efficiency in virtue of its translatory motion, according to the result of the preceding investigation—the same efficiency in respect to translational movements as if its outer surface were smooth as supposed in the investigation. And now, what is most important, we see that if the linear dimensions of the molecule be made small enough, without changing the angular velocity of its rotators, the influence of the rotational motion on the sheath becomes smaller and smaller, and quite insensible in comparison with the gyrostatic effect due to translational motion of the sheath; this last remaining unchanged with the diminution of linear dimensions, provided that not only the angular velocity but the ratio of the mass of the rotators to the whole mass of the molecule is kept unchanged.

The kinetic properties of the improved gyrostatic molecule are exceedingly interesting, but we have had all of them that are essential to our present purpose and time presses so much that I close this final despatch without even writing down the cartesian equations of its motion.

Since my return, Prof. E. W. Morley has kindly sent me a diagram of curves, giving a complete graphical representation of $-\xi/x_1$ [for the problem proposed on page [89] of the lectures] both as a function of T and $1/T = z$. This I hope to make good use of in attempting to explain extinction, anomalous dispersion, and fluorescence and phosphorescence. The results of Prof. Morley's calculations are here appended.—W. T.

[The table of roots of $-\xi/x_1 = 0$ and their corresponding displacement and energy ratios is given on page [193]. Another table is here added for branches of the curve in z. The branches are numbered so as to bring the corresponding fundamental periods, κ_1, κ_2, $\cdots \kappa_7$, in ascending order of magnitude. These branches are given by full lines in the diagram. The dotted lines are for the reciprocal branches in T, which are drawn upon a longitudinal scale $\frac{1}{10}$ of that in the former set of curves, so as to bring the two sets upon the same diagram.

—H.]

$x^2 = \frac{5}{y^2}$	$-\frac{5}{x} = y$	$z = \frac{1}{y}$
SEVENTH BRANCH.		
.0	−1.7678	.0
.001	−1.3944	.03162
.00125	−1.1781	.03536
.0014	−0.7273	.03741
.0014701	+0.000023	
.0015	+0.8512	.03873
SIXTH BRANCH.		
.0016	−5.41	.04000
.0018	−2.235	.04242
.002	−1.9564	.04472
.0025	−1.7665	.05000
.005	−1.5128	.07081
.007	−0.7104	.08366
.0072	−0.2802	.08485
.0072564	−0.000070	
.0075	+3.4476	.08660
FIFTH BRANCH		
.008	−3.82	.08944
.009	−2.13	.09487
.01	−1.922	.1
.015	−1.6676	.12247
.016	−1.5956	.1264
.02	−1.4601	.1414
.022	−1.2172	.1483
.025	−0.4366	.1581
.0255607	+0.000062	
.0265	+1.9560	.1628
FOURTH BRANCH		
	−12.268	.1700
	−2.731	.1750
.03	−2.278	.1826
.04	−1.9235	.2
.05	−1.7130	.2236
.07	−1.315	.2640
	−0.983	.28
.08801	+0.00042	
.09	+0.4497	.30
	+2.086	.305

$x^2 = \frac{5}{y^2}$	$-\frac{5}{x} = y$	$z = \frac{1}{y}$
THIRD BRANCH.		
.1	−9.3911	3162
	−5.66	.32
.1122	−3.0162	.335
.1225	−2.454	.35
.16	−1.877	.40
.25	−1.0385	.50
.28	−0.540	.5292
.29	−0.278	.5385
.29849	+0.00006	
.3136	+0.7714	.56
.3	+4.90	.5773
SECOND BRANCH.		
.36	−20.38	.60
.4225	−3.450	.65
.49	−2.460	.7
.5625	−2.001	.75
.7225	−1.345	.85
1.00483	−0.000004	
1.21	+2.643	1.1
FIRST BRANCH.		
1.44	−22.13	1.2
1.69	−4.38	1.3
1.96	−2.66	1.4
2.56	−1.2444	1.6
2.89	−0.726	1.7
3.4618	−0.000045	1.8605
3.5	+0.0463	1.8708
4.	+0.631	2.
5.	+1.791	
14	+6.89	
2a	+16.95	

For rest of branch the equation becomes $y = z^2 - \frac{1}{z^4} - 2$, and the curve is merely a parabola

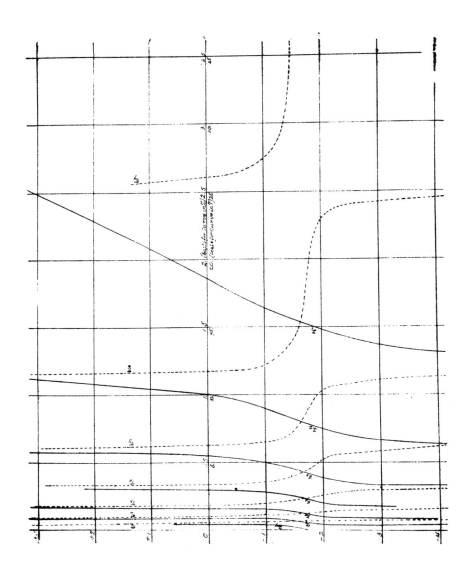

Index of Lectures
(With some references in connection with the subject matter)

Lorenz, "Lehre vom Licht." *Rankine*, Mis. Sci. Papers. *Rayleigh*, Ency. Brit. "Optics." *Stokes*, Math. & Phys Papers Vols I, II, III. *Tait*, "Light" Ency. Brit and Text-Book. *Verdet* Lecons d'Optique Physique.

Difficulties of the Wave Theory.

Cauchy Mim. Acad. Sci. 1850, C.R. 1836, '42, '50. *Helmholtz* "Wiss Abh" Pogg Ann. Vol. 154. *Ketteler*, Ann. Phys Chem 1870, 1874. *Kirchhoff*, Phil. Mag. 20, 1860. Pogg. Ann CIX *Lommel*, Ann. Phys. Chem. III, 1878. *Meyer*, Pott. Ann Vol. 415. *Rayleigh* Phil. Mag. 1871–2. *Sellmeier* Pogg. Ann. Vols. 145, 147 *Stokes*, Math Papers, *Verdet*, Lecons d'Optique Phys. *Wüllner*, "Helmholtz's Dispersion Th." Ann. Phys. Chem. 1884.

Abney, Phil. Mag. 1876–'83. *Maxwell*, Ency. Brit. 9 Edit. "Ether" Elec. & Mag. Art. 806. *Newcomb*, Am. Astro Ephemeris Vol. II. *Rayleigh*, Nature, Aug. 25, Nov. 17, 1871. *Thomson*, Sir Wm, Trans. R.S. Edinb. XXI.

Cauchy, C.R. 1849; *Fresnel, Green, Rayleigh, Verdet, Rood* "On the amount of Light transmitted by plates of Crown glass at Perpendicular Incidence" Am. J. Sci L, 1870

Fresnel, Green, Rankine, Rayleigh (Phil. Mag. 1871) *Stokes*, Brit Assoc. Rep. 1862, *Hollefreund* (nach Lommel'-schen "Riebungstheorie") Nova Acta XLVI. No. 1.

Clebsch "Elasticität fester Körper." *Gibbs*, Trans. Conn. Acad. Arts & Sci. 1874–78. *Green*, Math Papers. *Lame* "L'Elasticite des corps Solides." *Rankine* Mis. S. Pap. *Poisson*, 20 Cahier Journ. de l'Ecole Polytech. *Stokes*; Equilibrium and Motion of Elas. Solids—Math. Pap. *Thomson*, Sir Wm Ency. Brit. 9 Ed. "Elasticity." *Thomson & Tait*, Nat. Phil. *Saint Venant* Memoires des Savants Etrangers 1855

Becquerel, Ann. de Chim. XII, 1877. *Biot* Mem. de la prem. Classe
de l'Inst. XIII. *Briot* C.R.L. *Kundt*, Phil. Mag. 1884. *Lommel*, Ann.
Phys. Chem. XIV, 1881. *Thomson, Sir Wm.* Proc. R.S. London,
1856, Phil. Mag. 4th Series XIII. *Verdet* Lecons d'Op. Phy.

Becquerel (Infra-red emission spectra of metallic vapors) C. A.
Aug. 1884 Phil. Mag. Oct., 1884, Am. J. Sci. 1884. *Cauchy* C.R.
1839, '48. Pogg. Ann. 1849 Liois J. Math. 1842. *Jamin* Ann de
Chemie Vol. 29, 1850, *Rayleigh*, Phil. Mag. May 1872, Stokes,
Phil. Mag. 1853. *Quincke*, Ann. Phys. Chem. Vols. 119, 127, 128,
129.

Essays

P. M. Harman   **Mathematics and Reality**
                    in Maxwell's Dynamical
                    Physics

Now I want you kindly to advise me as to what subject or subjects should be chosen. I presume it is to be something in mathematical physics? Would waves and vibrations of fluids, equilibrium and motion (vibrations & waves) of elastic solid suit? Or something of vortex motion?... I suppose I am to reckon on mathematics enough for more than anything I would give on any of these subjects.

William Thomson to J. J. Sylvester, September 5, 1882[1]

In writing to J. J. Sylvester, then Professor of Mathematics at Johns Hopkins University, about the invitation to deliver his "Baltimore lectures," William Thomson (the future Lord Kelvin) outlined a framework of topics that brought together interests that had episodically but intensively drawn his attention since the beginning of his career as a physicist. The theory of the dynamics of the elastic solid ether, hydrodynamical analogies for electromagnetism, and the link between hydrodynamical and elastic solid theories of electromagnetism and light by appeal to the theory of vortex motion within a continuous ether—these topics formed the subject of the "Baltimore lectures." The central theme in these lectures is Kelvin's treatment of the relationship between ether and matter, investigating the mutual actions between a continuous elastic solid ether and the molecules of matter. Kelvin sought a "comprehensive dynamics" embracing the ether, electromagnetism, and the wave theory of light.[2]

The richness and the generality of this theme clearly place the Baltimore Lectures in the mainstream of nineteenth-century physics. The program of mechanical explanation, often termed the dynamical world view, which supposed an ontology of particles of matter in motion as the substratum of physical reality and which aimed to explain physical phenomena by the structure and laws of motion of a mechanical system, dominated physical theorizing in the nineteenth century.[3] Kelvin's early physics had indeed been instrumental in shaping the dynamical physics of the nineteenth century, and the Baltimore Lectures set the capstone on his contribution to ether physics. Of all nineteenth-century physicists, none developed the theory of dynamical physics more cogently than Kelvin's younger contemporary James Clerk Maxwell, who had died in 1879. Maxwell discussed the objectives and limitations of the program of mechanical explanation in a sophisticated and influential manner, with a keen awareness of the tensions between physical and mathematical models of mechanical systems and the gap between the structure of physical reality and the encompassing net of mechanical representations.[4] This method received its clearest expression in Maxwell's work on the theory of the electromagnetic field and his development of the electromagnetic theory of light. Kelvin, however, clearly aimed to set the Baltimore Lectures,concerned with "molecular dynamics and the wave theory of light," in opposition to the style of mathematical physics developed in Maxwell's *Treatise on Electricity and Magnetism* (1873).

At the root of Kelvin's critique of Maxwell lies his concern "to make a mechanical model which shall fulfil the conditions required in the physical phenomena." For this reason he expresses "immense admiration" for Maxwell's theory of molecular vortices in "On Physical Lines of Force" (1861−62) as "a step towards a definite mechanical theory of electromagnetism." He considers Maxwell's electromagnetic theory of light, however, to be "rather a backward step" for failing to provide dynamical foundations adequate to the canons of mechanical explanation that he deemed necessary: "as long as I cannot make a mechanical model all the way through I cannot understand; and that is why I cannot get the electromagnetic theory [of light]." He contrasts Maxwell's electromagnetic theory of light, which he believes to be insufficiently grounded on mechanical principles, with the "absolutely definite mechanical notion that is put before us by Fresnel and his followers," who had developed the elastic solid theory of light.[5] While he declares "I firmly believe in an electromagnetic theory of light," he considers that the hoped-for unification of electricity, magnetism, and light will be based on the "true elastic solid" and "plain matter-of-fact dynamics" rather than on Maxwell's electromagnetic theory.[6]

There is of course a paradox in Kelvin's critique of Maxwell's electromagnetic theory of light and his opposition to the tradition of mathematical physics flowing from Maxwell's *Treatise on Electricity and Magnetism*. Maxwell's mathematical and physical ideas on electromagnetism, from his early paper "On Faraday's Lines of Force" (1856) to the *Treatise*, were profoundly shaped by Kelvin's mathematical physics. In his use of a geometrical and physical analogy for lines of force, in his adoption of a program of the mechanical representation of electricity and magnetism, in his development of the theory of molecular vortices as a model of the electromagnetic field, and in his application of analytical dynamics to the theory of electromagnetism, Maxwell's debt to Kelvin was deep.[7] Nevertheless, Kelvin recognized that Maxwell had fundamentally restructured and transformed the scope and style of the physics which he had been bequeathed. The subtle twists that Maxwell gave to the program of dynamical explanation; the development of vectorial and geometrical representations of electromagnetic quantities; and the creation of the electromagnetic theory of light, and its expression in a form which Kelvin considered to be insufficiently grounded on mechanical principles—these fundamental elements of the mathematical style and physical content of the *Treatise* presented a framework of mathematical physics that Kelvin came to reject.

The *Treatise on Electricity and Magnetism* offers more than a synthesis of the theories of electricity and magnetism that Maxwell had elaborated in his series of great papers, most notably in "On Faraday's Lines of Force" (1856), "On Physical Lines of Force" (1861–62), and "A Dynamical Theory of the Electromagnetic Field" (1865). The core of the theoretical argument of the *Treatise* is the mathematical expression of physical quantities freed from their direct representation by a mechanical or a geometrically visualizable model. Maxwell's use of quaternions, integral theorems, and topological arguments and his interpretation of the Lagrange-Hamilton method of analytical dynamics build upon his own earlier physical and mathematical methods. But the blend of geometrical, vectorial, and generalized dynamical arguments in the *Treatise* offers a systematic framework of mathematical physics in which individual elements having their origin in his disparate early papers are transformed into a new mode of representation in mathematical physics. This transformation had as its consequence the reappraisal of methods and models which he had derived from Kelvin, and which had significantly shaped the strategy of his early papers on the theory of the electromagnetic field. The enriched mathematical vocabulary of the *Treatise* embraces but also transforms his earlier modes of representation, opening the way to a new development of the theme that plays throughout Maxwell's physics: the relation between mathematics and physical reality.

Maxwell's Physical Geometry

Maxwell and Kelvin had a common background in Scottish physics and Cambridge mathematics. The use of physical analogies as a heuristic method is a common feature of their work, and may reflect the analogical approach to physics which was a notable feature of the lecture courses they attended at Scottish universities.[8] Their debt to Cambridge mathematics was profound. Cambridge "mixed mathematics" embraced the problems of mathematical physics, construed as the mechanics deriving from Newton's *Principia* and including geometrical and physical optics and astronomy. Topics of current physical research, such as electricity and heat, were excluded from the Tripos by the reformed syllabus of the late 1840s, and hence Maxwell—who entered Peterhouse (before migrating to Trinity) in 1850—would not have been exposed to these subjects during formal teaching. His notebooks on mechanics, hydrodynamics, astronomy, and optics—these latter topics being given special emphasis by his tutor William Hopkins (who had tutored Kelvin some ten years previously in the early 1840s)—give evidence of Maxwell's study of the mathematical aspects of these physical subjects.[9] Geometry and "mixed mathematics" formed a substantial part of the syllabus of the Mathematical Tripos, attempts by the reformers of the Analytical Society to give Cambridge mathematics a more thoroughgoing analytical focus having been thwarted.[10]

Maxwell arrived in Cambridge with a considerable, if disordered, knowledge of mathematics,[11] and a remarkable expertise in mathematical physics. His masterly paper "On the Equilibrium of Elastic Solids" (1850), which was written when he was only 18 years old and an undergraduate at Edinburgh and which owed its origin to his interest in the color fringes produced when strained glass is viewed with polarized light, deployed a mathematical theory of elasticity which demonstrated his familiarity with the formidable contemporary work on the subject by Cauchy, Poisson, Lamé and Clapeyron, and Stokes.[12] Like the Cambridge mathematician George Gabriel Stokes, who had based his theory of the elasticity of solids on the assumption of the independence of the elasticity of shape and volume of solids, Maxwell rejects any theory of an elastic solid (such as Poisson's) based on a "physical mechanics" (the term is Poisson's[13]) of molecules acting at a distance. Maxwell espouses a theory based on axioms derived from experiments, stating experiential relations between the pressure and compression of elastic media, the concepts of pressure and compression being defined in experiential terms.[14] Stokes justly described the paper as "involving a profound acquaintance with some of the most abstruse branches of physical investigation."[15]

The combination of physical insight and mathematical sophistication apparent in this early paper, and the emphasis on physical and geometrical topics fostered by his mathematical education at Cambridge, make Maxwell's burgeoning interest[16] in the geometry of surfaces and the science of electricity after his graduation in 1854 quite unsurprising. Electricity was open to incorporation within "mixed mathematics," and Maxwell's mathematical treatment of Faraday's concept of "lines of force" was a "geometrical model" of lines of forced based on propositions about lines and surfaces. Maxwell's geometrical representation of lines of force in "On Faraday's Lines of Force" parallels his use of the concept of "lines of bending" as a mode of representation of the change of form of a surface in "On the Transformation of Surfaces by Bending" (1856). Congruent with the methodology of his paper on elastic solids, he did not expound a physical hypothesis of the field based on a physical mechanics of interacting particles. In a draft of "On Faraday's Lines of Force" (1856) he observed that "Faraday treats the distribution of forces in space as the primary phenomenon, and does not insist on any theory as to the nature of the centres of force round which these forces are generally but not always grouped."[17] Maxwell's "geometrical model" of lines of force avoided the extremes of the "rashness in assumption" of a physical hypothesis and the tendency to "lose sight of the phenomena to be explained" in the pursuit of mathematical subtleties.[18]

Maxwell's approach, and the program of extending the scope of Cambridge "mixed mathematics" to a wider range of physical phenomena, may have been shaped by his early reading of Joseph Fourier's *Théorie analytique de la chaleur* (1822)[19] and, more potently, by Kelvin's "allegorical representation of the case of electrified bodies by means of conductors of heat," as Maxwell expressed it.[20] In Fourier's theory of heat, the conduction of heat in a solid is characterized by the flux of heat between contiguous infinitesimal elements of the solid, and is represented by differential equations of heat flow. Kelvin, in a series of papers written during the 1840s, had established that the continuity equation for fluid flow could be applied to the theories of heat and electricity: the continuity or conservation of the flux of heat implied the conservation of electric force.[21] Maxwell thus acquired an analogy for the flux of electric force, and, more generally, a mathematical formalism that linked field theory to the mechanics of continuous media, to hydrodynamics and elasticity, and to the continuity equation established by Euler for the flow of a fluid. In his *Treatise on Electricity and Magnetism* (1873) he explained that the mathematical language of partial differential equations provides an expression of the concept of the physical field. The equations of the field involve quantities that are continuous functions of their variables, and these equations express the continuous propagation of force or energy between neighboring infinitesimal elements in the field. Thus "the differential equation is the appropriate expression for a theory of action exerted between contiguous parts of [the field]."[22] In his paper "On Faraday's Lines of Force" he expresses the analogy between heat and electricity as a mathematical analogy with the flow of a fluid, not an analogy between the phenomena of thermal conduction and electrostatic attraction. Maxwell expresses the analogy as "a physical analogy ... that partial similarity between the laws of one science and those of another which makes each of them illustrate the other." There was therefore "a resemblance in mathematical form" between the flux of heat and the continuous flow of electric force across isothermal or equipotential surfaces.[23]

Maxwell expounds his mathematical analogy in terms of "a geometrical model of the physical phenomena," appealing to the analogy with the mathematical theory of fluids. He grounds his theory of lines of force on the concept of an incompressible fluid moving in tubes formed by lines of force, the analogy with fluid flow being "purely geometrical" and the fluid itself being "merely a collection of imaginary properties"—a wholly imaginary substance which would enable theorems in fluid flow to be grasped more readily, and their application to the phenomena of electricity and magnetism to be grasped.[24] Maxwell's drafts show that the paper was first formulated as a purely hydrodynamical study, a treatment of the equations of motion of an incompressible fluid.[25] Little of the mathematical argument in these drafts (concerning the conditions of stability of an incompressible fluid) remained in the published paper, but the hydrodynamical argument provides the basis for Maxwell's concept of "lines of fluid motion,"[26] which he uses to express a quantified theory of Faraday's lines of force. Writing to Kelvin in September 1855, Maxwell remarked that his theory was "in itself a collection of purely geometrical truths embodied in geometrical conceptions of lines, surfaces &c," a physical geometry of "lines of force &c which may be *afterwards* applied to Electricity, Heat or Magnetism."[27] This mathematical theory of the motion and stability of fluids, shaped by his Cambridge study of the pressure, motion, and equilibrium of fluids and of the continuity equation as applied to hydrodynamics, established that the mathematical theory of electricity could be incorporated within the explanatory framework of "mixed mathematics."[28]

In the second part of the paper Maxwell turns to a discussion of "Faraday's 'Electro-tonic State,'" a concept Faraday had used to explain "the peculiar state of tension or polarity"[29] into which bodies are thrown by the presence of magnets and currents. Maxwell develops this concept in an important new way, to represent the number of lines of magnetic force that pass through a surface enclosed by an electrical circuit. The electro-tonic state provided "the means of avoiding the consideration of the quantity of magnetic induction [the number of lines of force] which *passes through* the circuit" by "considering the current with reference to quantities existing in the same space with the current itself." Instead of an "artificial conception" of lines of force passing through the surface enclosed by the circuit, Maxwell sought the "natural [method]" based on the current in the circuit: "an expression for the quantity of magnetic induction passing through a closed circuit in terms of quantities depending on the circuit itself, and not on the enclosed space." Maxwell thus sought the connection between quantities which had relation to surfaces and quantities which had relation to the boundaries of those surfaces, to "have an integration *round the curve itself* instead of one *over the enclosed surface*."[30] Using terminology which he was subsequently to adopt in the *Treatise*, he was seeking connections between the surface integral of a quantity taken over a finite surface (to represent the number of lines of magnetic force) and a line integral of another quantity taken round its boundary.

The origin of Maxwell's argument can be found in a letter he wrote to Kelvin in November 1854.[31] In the published paper he employs the concept of the electro-tonic state to represent the quantity manifested in the circuit, in the same space as the current, and expresses the relation between the number of lines of magnetic force and the electro-tonic state as an electromagnetic law: "The entire electro-tonic intensity round the boundary of an element of surface measures the quantity of magnetic induction which passes through that surface, or, in other words, the number of lines of magnetic force which pass through that surface."[32] This law expresses the connection between the surface integral of the magnetic intensity (the number of lines of force) taken over a surface enclosed by a circuit and the line integral of the electro-tonic state round the boundary of the circuit. Maxwell did not formulate this law in an analytic form in "On Faraday's Lines of Force" (though he was later to do so, significantly, in the *Treatise*), but he no doubt had in mind an integral theorem he had encountered in the Smith's Prize Examination set by Stokes at Cambridge in February 1854:

If X, Y, Z be functions of the rectangular co-ordinates x, y, z, dS an element of any limited surface, l, m, n, the cosines of the inclinations of the normal at dS to the axes, ds an element of the bounding line, shew that

$$\iint \left\{ l\left(\frac{dZ}{dy} - \frac{dY}{dz}\right) + m\left(\frac{dX}{dz} - \frac{dZ}{dx}\right) + n\left(\frac{dY}{dx} - \frac{dX}{dy}\right) \right\} dS$$

$$= \int \left(X\frac{dx}{ds} + Y\frac{dy}{ds} + Z\frac{dz}{ds} \right) ds, \tag{1}$$

the differential coefficients of X, Y, Z being partial, and the single integral being taken all round the perimeter of the surface.

This theorem is now known as Stokes' Theorem, though it was first stated by Kelvin in a letter to Stokes in 1850.[33] Within the mathematical framework of the *Treatise*, this integral theorem came to have an important meaning for the mathematical expression of electromagnetic quantities. The transformation of line integrals into surface integrals became a key feature of Maxwell's geometrical representation of electromagnetic field theory, expressing the relation between quantities having relation to circuits and quantities having relation to the surfaces enclosed by circuits.

Maxwell's Physical Mechanics

The hydrodynamical analogy of "On Faraday's Lines of Force" illustrates the spatial distribution of lines of force by reference to a mechanical analogy of the flow of an incompressible fluid. In remarking that he ultimately aimed to formulate a "mechanical conception" of the electro-tonic state as the substratum of the field, Maxwell indicated, however, that he wished to develop a physical representation of the lines of force that would explain their physical constitution and hence provide not merely a geometrical (hydrodynamical) analogy but a physical model of the structure of the field. Nevertheless he emphasized that the representation of the electro-tonic state by "any mechanical system" would be "only a kind of artificial notation" and would involve "no physical theory."[34] In "On Physical Lines of Force" (1861–62) he advances from his treatment of the physical geometry of lines of force to a physical mechanics based on the motion of particles, the "theory of molecular vortices." This shift in outlook is significant. In 1856, in "On Faraday's Lines of Force," he had indicated that the kind of mechanical explanatory theory he had in mind would be similar to Kelvin's mathematical theory of the transmission of electrical and magnetic forces in terms of the linear and rotational strain of an elastic solid.[35] In "On a Mechanical Representation of Electric, Magnetic, and Galvanic Forces" (1847)[36] Kelvin had employed the mathematical methods that Stokes had developed for the treatment of rotations and strains in continuous media. This was, however, "merely a sketch of the mathematical analogy," Kelvin had told Faraday,[37] rather than a physical theory of electricity and magnetism based on a molecular model of the ether. But in seeking a physical mechanics of the ether in "On Physical Lines of Force," rather than a mere mathematical analogy, Maxwell was again following Kelvin's lead.

In November 1857 Maxwell had suggested to Faraday that the extension of Faraday's theory of the field involved "questions relating to the connexion between magneto-electricity and certain mechanical effects which seem to me [to be] opening up quite a new road to the establishment of principles in electricity and a possible confirmation of the physical nature of magnetic lines of force." He continued: "Professor W. Thomson seems to have some new lights on this subject." [38] He was here alluding to Kelvin's recent explanation of the Faraday effect, the rotation of the plane of polarization of linearly polarized light by a magnetic field, by means of a hypothesis of a vortical motion in the ether. Kelvin had claimed that "the explanation of all phenomena of electromagnetic attraction or repulsion, and of electromagnetic induction, is to be looked for simply in the inertia and pressure of the matter of which the motions constitute heat." The dynamical theory of heat, that heat is caused by the motions of the particles of matter, provided the model for a "dynamical illustration" of the electromagnetic field.[39] Kelvin's theory of the Faraday effect, which suggested "that the cause of the magnetic action on light must be a real rotation going on in the magnetic field" (as Maxwell put it), lay at the root of the theory of "molecular vortices" in "On Physical Lines of Force," which provided a physical mechanics of the ambient ether as the substratum of the field.[40] Maxwell informed Kelvin in December 1861: "I have been trying to develope the dynamical theory of magnetism as an affection of the whole magnetic field according to the views stated by you." [41]

In Maxwell's physical model of vortex tubes and "idle wheel" particles, the mechanical analogy of the electromagnetic field provides mechanical correlates for electromagnetic quantities. The angular velocity of the vortices corresponds to the magnetic field intensity, and the translational motion of the idle-wheel particles corresponds to the flow of an electric current. The electrotonic state has a central role in the field equations and in Maxwell's mechanical imagery. He states the mathematical relations between the electro-tonic state and the lines of magnetic force (represented by the magnetic intensity α, β, γ) as follows:

$$\frac{dG}{dz} - \frac{dH}{dy} = \mu\alpha,$$

$$\frac{dH}{dx} - \frac{dF}{dz} = \mu\beta, \tag{2}$$

$$\frac{dF}{dy} - \frac{dG}{dx} = \mu\gamma,$$

where "μ is proportional to the density of the vortices and represents the 'capacity for magnetic induction' in the medium" and where the quantities F, G, and H are "the resolved parts of that [quantity] which Faraday has conjectured to exist, and has called the *electrotonic state*." The electro-tonic state "is what the electromotive force would be if the currents, &c to which the lines of force are due, instead of arriving at their actual state by degrees, had started instantaneously from rest with their actual values." The electromotive force could be deduced by differentiating the electro-tonic state with respect to time. Electromotive force is defined in terms of the action between the vortices and the idle-wheel particles between them, and Maxwell suggests that the interpretation of the electro-tonic state "from a mechanical point of view" would be "the *impulse* which would act on the axle of a wheel in a machine if the actual velocity were suddenly given to the driving wheel, the machine being previously at rest." He terms this impulse the "*reduced momentum*"; just as the force on any part of a mechanism arising from the variation of motion is calculated by differentiating the reduced momentum with respect to time, the electromotive force bears the same relation to the electro-tonic state.[42]

Maxwell emphasizes that his mechanical model of idle-wheel particles which by their translatory motion explain the interaction and rotation of the vortices (the rotatory motion of the vortices representing the magnetic field intensity) is only a "provisional and temporary" hypothesis. He observes that the model might appear "awkward," but he states that he did "not bring it forward as a mode of connexion existing in nature." The supposition of a "mechanically conceivable" model of the ether demonstrated the possibility of a mechanical explanation of the electromagnetic field.[43] It was for this reason that Kelvin expressed admiration for Maxwell's mechanical model of the electromagnetic field in the Baltimore Lectures.

The key feature of the model is that it enabled Maxwell to explain the relationship between "magnetism as a phenomenon of rotation" (as implied by the Faraday magneto-optic rotation) and "electric currents as consisting of the actual translation of [the idle-wheel] particles" between the molecular vortices. The model therefore provides a mechanical embodiment or pictorial illustration of a fundamental mathematical law between the electric current and the magnetic field intensity (Ampère's law):

$$p = \frac{1}{4\pi}\left(\frac{d\gamma}{dy} - \frac{d\beta}{dz}\right),$$

$$q = \frac{1}{4\pi}\left(\frac{d\alpha}{dz} - \frac{d\gamma}{dx}\right), \tag{3}$$

$$r = \frac{1}{4\pi}\left(\frac{d\beta}{dx} - \frac{d\alpha}{dy}\right),$$

where α, β, γ are the components of magnetic intensity and p, q, r are the components of electric currents. "If α, β, γ represent the rotatory velocities of vortices whose centres are fixed, then p, q, r represent the velocities with which loose particles placed between them would be carried along." The key mathematical feature of the relationship between these quantities is that if p, q, r have a "linear" character then α, β, γ have a "rotatory character." This connection between linear and rotational quantities corresponds to the connection between magnetism and electricity, this relation between magnetism as a rotational phenomenon and electricity as a linear phenomenon being embodied in the theory of molecular vortices and idle-wheel particles. Maxwell argues that electric currents must be conceived as linear in character, giving electrolysis as an example, while he maintains that the Faraday effect establishes that magnetism is a rotational phenomenon, as is shown by "the rotation of the plane of polarized light when transmitted along the lines of magnetic force." Assuming that "all the direct effects of any cause which is itself of a longitudinal character, must be themselves longitudinal" (electric currents being caused by the linear translation of particles in the model) and that "the direct effects of a rotatory cause must be themselves rotatory" (magnetism being explained by the rotation of molecular vortices), Maxwell's theory of molecular vortices provided a mechanical embodiment of the mathematical law connecting the distribution of lines of magnetic force and that of electric current, a model for the connection of the linear phenomenon of electricity with "magnetism as a phenomenon of rotation" as implied by the Faraday effect.[44]

Maxwell's Analytical Mechanics

While Maxwell unwaveringly held to the belief that there was "a real rotation going on in the magnetic field," which implied "some kind of mechanism" performed by "a great number of very small portions of matter, each rotating on its own axis," he was conscious of the speculative cast of his "theory of molecular vortices."[45] In a letter to Peter Guthrie Tait in December 1867 he remarked that his ether model was "built up to show that the phenomena can be explained by mechanism" and that "the nature of the mechanism is to the true mechanism what an orrery is to the Solar System."[46] He later observed, in the *Treatise*, that his attempt to imagine a working model of this mechanism "must be taken for no more than it really is, a demonstration that [a] mechanism may be imagined capable of producing a connexion mechanically equivalent to the actual connexion of the parts of the electromagnetic field." [47] The mechanical model of ether particles is therefore purely illustrative, demonstrating the possibility of a mechanical explanation of electromagnetism and providing a physical embodiment of the connection between linear and rotatory motions in the ether. Writing to Kelvin in October 1864, Maxwell observed that "the tendency in my rotatory theory of magnetism was towards the to me inconceivable and ∴ no doubt to the misty." [48] By this time he had moved to a new method of demonstrating the application of mechanical principles to electromagnetism, in "A Dynamical Theory of the Electromagnetic Field" (1865). His caution over his physical model is understandable: the specificity of the theory of molecular vortices provided an embodiment of the relation between lines of magnetic force and electric currents, but the experimental phenomenon of the Faraday effect and the mathematical relationship between linear and rotational quantities in electromagnetism remained logically independent of the physical model of molecular vortices and idle-wheel particles.

Maxwell therefore developed the theory of the electromagnetic field from analytical equations of mechanical systems, using the Lagrangian formalism of analytical dynamics, without employing a specific mechanical model to represent the structure and motion of the ether. This method of analytical dynamics assumes that "motion is communicated from one part of the [ethereal] medium to another by forces arising from the connexions of those parts," and he now disclaims any mechanical model of "a particular kind of motion and a particular kind of strain" in the ether. The theory is, however, a "dynamical" theory in that Maxwell assumes that electromagnetic phenomena are produced by particles of "matter in motion" constituting "a complicated mechanism capable of a vast variety of motion." [49]

The discussion of the concept of the electro-tonic state follows from the introduction of Lagrangian dynamics. The electro-tonic state is now termed the "electromagnetic momentum," for Maxwell argues again that "the effect of the connexion between the current and the electromagnetic field surrounding it, is to endow the current with a kind of momentum." Although he draws the analogy with the operation of a machine, he now illustrates the electromagnetic momentum dynamically, drawing on Lagrange's "general equation of dynamics," which relates the forces, displacements, and velocities of a system of particles. He concludes that "this dynamical illustration is to be considered merely as assisting the reader to understand what is meant in mechanics by Reduced Momentum." [50] He had deleted from the manuscript of the paper a passage in which the interaction between two currents in the field is compared to the action of two horses pulling on each arm of a lever attached by its fulcrum to a carriage; "then if one horse increases its speed the immediate effect will be to produce a tension in the traces of the other horse tending to pull him back."[51] The argument is thus couched only in the most general dynamical terms, and expressions such as "electromagnetic momentum" are employed "merely to direct the mind of the reader to mechanical phenomena which will assist him in understanding the electrical ones." Thus, "all such phrases in [the paper] are to be considered as illustrative, not as explanatory."[52]

In disclaiming any appeal to a specific mechanical model, Maxwell was aware, as he explains in the *Treatise*, that there was in principle no limit to the number of possible mechanical models: "The problem of determining the mechanism required to establish a given species of connexion between the motions of the parts of a system always admits of an infinite number of solutions."[53] The dynamical theory of the field is therefore now developed in an abstract rather than a concrete representation. The philosophical stance of the paper is also entirely consistent with the concepts of "dynamical theory"— implying independence from hypotheses—of mid-nineteenth-century British physics,[54] as well as with Maxwell's own early work, in which he had adopted an appropriately cautious and critical attitude to the incorporation of speculative physical hypotheses in mathematical physics. Maxwell's work, with the exception of "Illustrations of the Dynamical Theory of Gases" (1860) and "On Physical Lines of Force" (1861–62), shows a preference for "analytical mechanics" rather than "physical mechanics" (the terms are Poisson's[55]), for Lagrangian analytical mechanics rather than Laplacian models of molecular motions and interparticulate forces. Reviewing "A Dynamical Theory of the Electromagnetic Field" for the Royal Society in March 1865, Kelvin found it "most decidedly suitable for publication in the Transactions."[56] Nevertheless, he came to see Maxwell's approach to dynamical theory, elaborated and refined in the *Treatise on Electricity and Magnetism* (1873), as insufficiently realistic. Not only was Kelvin's attitude in regard to the relation between physical and mathematical modes of representation in the theory of electromagnetism changing (his demonstration that magneto-optic rotation was produced by the rotatory motion of vortices in the ether increasingly inclined him toward a realistic approach to the status of physical models[57]); his attitude in regard to the conceptual structure of Maxwell's mathematical physics in the *Treatise* was changing as well.

The Mathematical Physics of the *Treatise on Electricity and Magnetism*

Thus far I have discussed some of the central mathematical and physical ideas which guided Maxwell's work on electromagnetic field theory. There have been a number of themes: geometrical representation, the relation between mathematical and physical modes of representation, the status of mechanical models as physical illustrations, the status of dynamical theory as a mechanical explanation of the field, the representation of electromagnetic quantities by mathematical formalisms, and the development of equations of the electromagnetic field. In his *Treatise on Electricity and Magnetism* (1873) Maxwell aimed to present a systematic treatment of the subject. The mathematical methods appropriate to the construction of a theory of electricity and magnetism and "the relations between the mathematical form of this theory and that of the fundamental science of dynamics" had pride of place in this endeavor.[58] To this end he developed a framework of mathematical physics that enlarged upon his earlier methods and permitted a deeper explication of the mathematical expression of physical quantities (into "that still more hidden and dimmer region where Thought weds Fact, where the mental operation of the mathematician and the physical action of the molecules are seen in their true relation"[59]). He enlarged upon his use of the Lagrangian formalism of dynamics for the representation of the dynamical theory of the electromagnetic field, clarifying and amplifying his account of the relation between the mathematical formalism of analytical dynamics and the physical concepts thereby employed.

The main methods and their results and applications are expounded in the mathematical "Preliminary" to the *Treatise*, in the chapters on Lagrangian dynamical theory as applied to electromagnetism, and in the chapters on the equations of the electromagnetic field. Taken as a whole, Maxwell's argument develops a mode of representation for mathematical physics which, while embracing and developing his own earlier methods and building on the work of others (especially Kelvin), elaborates a mathematical style that emphasizes the mathematical expression of physical quantities freed from their direct representation in terms of a mechanical model. This dynamical and geometrical style embraces four fundamental mathematical arguments: the use of quaternions, integral theorems, and topological ideas, and the Lagrange-Hamilton method of analytical dynamics. To characterize the mathematical style of Maxwell's dynamical physics, I shall discuss these mathematical ideas in turn.

Quaternions and Vectors

William Rowan Hamilton had developed his calculus of quaternions from his work on algebra in the 1830s. Concerned with the concept of number and the definition of complex numbers, Hamilton aimed to give an algebraic definition of numbers and to base algebra on the ordinal character of numbers. From his study of complex numbers Hamilton sought to extend the complex-number system to three dimensions. In 1843 he invented "quaternions"—hypercomplex numbers with one real and three (imaginary) complex parts. He interpreted the three imaginary numbers as "vectors" directed along three mutually perpendicular lines in space. The real part of the quaternion corresponded to a line in space of only one dimension; this was the "scalar" part of the quaternion. A quaternion is therefore considered as the sum of its own vector and scalar parts, and the quaternion calculus *"selects no one direction in space as eminent* above another, but treats them all as equally related to that *extra-spatial* or simply SCALAR direction."[60] Quaternions provide a form of analytical geometry that requires no prior selection of coordinates.

In a letter to Tait in November 1870, Maxwell signaled his keen interest in quaternions. Tait, Maxwell's closest scientific friend since their school days in Edinburgh, was Hamilton's disciple in the popularization of quaternions. Maxwell remarked that he wished to "make statements in electromagnetism," and a week later he wrote: "I want to leaven my book [the *Treatise*] with Hamiltonian ideas." "The value of Hamilton's idea of a Vector," he went on to say, "is unspeakable."[61] Maxwell was interested in the conceptual role of quaternions for the geometrical representation of physical quantities. He later emphasized the value of quaternions as a "mathematical method ... of thinking" that "calls upon us at every step to form a mental image of the geometrical features represented by the symbols." In the *Treatise* he points out that the physical quantities of electrodynamics—the electric and magnetic forces, the electro-tonic state (now renamed the vector potential), and the displacement current—are vector quantities, and urges the use of quaternions as a means of fixing the mind "on a point of space instead of its three coordinates, and on the magnitude and direction of a force instead of its three components." This vectorial "mode of contemplating geometrical and physical quantities" is "more primitive and natural" than the method of Cartesian coordinates, and this geometrical method provides a direct representation of physical quantities congruent with "physical reasoning" and involving "the continued construction of mental representations."[62]

Maxwell was especially interested in the quaternion expression

$$\nabla = i\frac{d}{dx} + j\frac{d}{dy} + k\frac{d}{dz}, \tag{4}$$

where *i, j, k* are unit vectors at right angles to each other—the three imaginary parts of the quaternion. In the *Treatise*, as a preliminary to considering the application of quaternions to electrical and magnetic quantities, he discusses the effect of the operator ∇ on a vector function σ, finding that $\nabla\sigma$ consists of two parts, one scalar and the other vector. Maxwell calls the scalar part $S\nabla\sigma$ the "convergence" of the vector function, for, as he explained in "The Mathematical Classification of Physical Quantities" (1871), "the convergence of a vector function is a very good name for the effect of that vector function in carrying its subject inwards towards a point." He terms the vector part $V\nabla\sigma$ the "curl" of the vector function; this vector represents "the direction and magnitude of the rotation of the subject matter carried by the vector σ." He chose the term "curl" rather than "rotation" so as not to "connote motion";[63]

earlier he had remarked to Tait that the term "curl," unlike "twirl," would not be "too dynamical" for "pure mathematicians" such as Arthur Cayley.[64]

Maxwell applies this vectorial method throughout the chapters on the electromagnetic field equations. The method can be illustrated by considering his treatment of the relation between the electric current **C** and the magnetic force **H**. Maxwell states this equation, derived in "On Physical Lines of Force" and interpreted as representing the relation between the linear character of electric currents and the rotatory character of magnetism, by the quaternion equation

$$4\pi\mathbf{C} = V\nabla\mathbf{H},\tag{5}$$

which is the quaternion equivalent of equation 3. In "On Physical Lines of Force" the relation between electric currents and magnetism was explained in terms of the rotation of molecular vortices. In the *Treatise*, Maxwell employs quaternion notation to express the relation between linear electric currents and the rotatory character of magnetism.[65] The rotational character of magnetism is explicitly represented by the term $V\nabla\mathbf{H}$, which may be written curl **H**. This relation between electricity as a linear phenomenon and magnetism as a rotational phenomenon is a central feature of Maxwell's theory of electromagnetism. In discussing the application of vectorial methods he places special emphasis on "the distinction between longitudinal and rotational properties," which, he observes, is "very important in a physical point of view." He points out that the electric current that causes or accompanies electrolysis is a longitudinal effect, whereas "the effect of magnetism in rotating the plane of polarized light distinctly shews that magnetism is a rotational phenomenon."[66] The method of quaternions highlighted this geometrical relationship.

In the second edition of the *Treatise* Maxwell replaced the term "curl" by "rotation,"[67] a term he had previously rejected as inappropriate because it implied motion. But, as he had written to Kelvin, "it is very remarkable that in spite of the *curl* in the electromagnetic equations of all kinds Faraday's twist of polarized light will not come out without what the schoolmen called local motion."[68] In the *Treatise* he continued to affirm that the Faraday effect implied that "we have good evidence for the opinion that some phenomenon of rotation is going on in the magnetic field, that this rotation is performed by a great number of very small portions of matter, each rotating on its own axis, this axis being parallel to the direction of the magnetic force, and that the rotations of these different vortices are made to depend on one another by means of some kind of mechanism among them."[69] Quaternions thus expressed a fundamental feature of electromagnetism. On reviewing George Francis FitzGerald's paper "On the Electromagnetic Theory of the Reflection and Refraction of Light" (1880) for the Royal Society in February 1879, Maxwell remarked to Stokes that he had found that, if " no bodily motion of a medium is introduced," "we find that the terms which seem to lead up to a rotation of the plane of polarization, though they enter into the preliminary equations, disappear from the final ones, so that Faraday's phenomenon remains un-explained."[70] The mathematical formalism of quaternions thus expresses the fundamental geometrical and dynamical features of Maxwell's theory of the electromagnetic field.

Vectors and Integral Theorems

In the mathematical "Preliminary" to the *Treatise*, Maxwell pays further atten-tion to the properties of vectors, and introduces a classification of vector quantities so as to demonstrate the application of Stokes' Theorem to electro-magnetism. The origin of his argument can be traced to Maxwell's paper "On Faraday's Lines of Force," in which he had distinguished between electrical and magnetic quantities having relation to lines and surfaces. He had employed the terms "quantity" and "intensity" to distinguish between the tubular surface formed by a system of lines of force and the lines of force themselves.[71] In the *Treatise* he develops this distinction more formally: "Physical vector quantities may be divided into two classes, in one of which the quantity is defined with reference to a line, while in the other the quantity is defined with reference to an area." The displacement of a particle is measured with reference to the line drawn between its original and final positions, while the displacement of a quantity of a substance would be measured by determining the quantity of the substance that passed across a given area during the displacement. He there-fore distinguishes between quantities (such as magnetic and electric force) that are defined with reference to lines, which he termed "forces,"[72] and quanti-ties (such as electric and magnetic induction and electric currents) that are defined with reference to areas, which he termed "fluxes."[73]

Maxwell uses this distinction, now given clear physical meaning as a classi-fication of vector quantities, to apply Stokes' Theorem to electromagnetism. He discusses two important mathematical operations involving these two classes of vectors: the operation of integration of the resolved part of a vector quantity along a line (the line integral of a force), and the integral, over a surface, of a flux through every element of the surface (the surface integral of a flux). The discussion of line and surface integrals opened the way to a further enlargement of the role of vector quantities in the mathematical theory of electromagnetism: the explication of Stokes' Theorem, which establishes a connection between the surface integral of a flux over a surface and the line integral of a force taken round the boundary of the surface.

This theorem had been stated by Thomson (Kelvin) and Tait in their *Treatise on Natural Philosophy* (1867), though Maxwell remembered Stokes' Smith's Prize Examination of 1854. Writing to Stokes in January 1871, he asked whether Stokes had not "set the theorem about the surface integral ... over a surface bounded by ... [a] curve ... being equal to [its corresponding line integral]." He remarked: "I have had some difficulty in tracing the history of this theorem. Can you tell me anything about it?"[74] Stokes' memory was perhaps rusty after twenty years. In April 1871 Maxwell mentioned the matter to Tait, who had given the theorem in quaternion form in his paper "On Green's and Other Theorems" (1870)[75]: "But the history of

$$\int\int\left\{l\left(\frac{dZ}{dy} - \frac{dY}{dz}\right) + m\left(\frac{dX}{dz} - \frac{dZ}{dx}\right) + n\left(\frac{dY}{dx} - \frac{dX}{dy}\right)\right\} dS$$

$$= \int\left(X\frac{dx}{ds} + Y\frac{dy}{ds} + Z\frac{dz}{ds}\right) ds$$

[equation 1] ascends (at least) to Stokes Smiths Prize paper 1854 and it was then not altogether new to yours truly. Do you know its previous history? Poisson? on light ???."[76] Maxwell is here alluding to the complex early history of integral theorems: Poisson had indeed employed integral transformations in his papers on magnetism, electrostatics, and elasticity (the work on elasticity having been known to Maxwell in 1849) when he had written "The Equilibrium of Elastic Solids" (1850).[77] In reply Tait declared that he "really thought [the theorem] due to [Thomson], and first published by [Thomson and Tait]."[78]

In proving Stokes' Theorem, Maxwell clearly has an eye for the physical application of the theorem to the theory of electromagnetism. He first proceeds formally in the mathematical "Preliminary" to the *Treatise*, taking X, Y, Z as the components of a vector quantity **A** whose line integral is taken round a closed curve s; S is a continuous finite surface bounded entirely by s, and ξ, η, ζ are the components of a vector quantity **B** related to X, Y, Z by the following equations:

$$\xi = \frac{dZ}{dy} - \frac{dY}{dz},$$

$$\eta = \frac{dX}{dz} - \frac{dZ}{dx}, \tag{6}$$

$$\zeta = \frac{dY}{dx} - \frac{dX}{dy}.$$

(These equations correspond to equations 2 above.) He then states that "the surface-integral of **B** taken over the surface S is equal to the line-integral of **A** taken round the curve s," and proves this theorem in the form

$$\int\int (l\xi + m\eta + n\zeta)\, dS = \int\left(X\frac{dx}{ds} + Y\frac{dy}{ds} + Z\frac{dz}{ds}\right) ds, \tag{7}$$

where l, m, n are the direction cosines of the normal to an element of the surface dS and where "the first integral is extended over the surface S, and the second round the bounding curve s."[79] Comparing equations 2, 6, and 7 makes it clear that Maxwell has prepared the ground for the expression of a relation between the flux of magnetic induction (lines of force) through a surface and the electro-tonic state round its boundary, and this is what he proceeds to do in explicating his equations of the electromagnetic field.

In developing the physical application of Stokes' Theorem to electromagnetism, Maxwell writes the components of the vector **A** as F, G, H, the symbols he had earlier used for the components of the electro-tonic state (electromagnetic momentum). He defines the vector **A** as representing the "time-integral" of the electromotive force a particle placed at a point would experience if the current passing were stopped, and he terms **A** the "vector potential of magnetic induction." He then introduces three components a, b, c of a vector **B** defined by the equations

$$a = \frac{dH}{dy} - \frac{dG}{dz},$$

$$b = \frac{dF}{dz} - \frac{dH}{dx}, \tag{8}$$

$$c = \frac{dG}{dx} - \frac{dF}{dy},$$

which correspond formally to equations 2 and 6. Thus, by comparison with equation 7, "we may express the line-integral of **A** round any circuit in the form of the surface-integral of **B** over a surface bounded by the circuit, thus . . .

$$\int \left(F\frac{dx}{ds} + G\frac{dy}{ds} + H\frac{dz}{ds} \right) ds = \iint (la + mb + nc)\, dS." \tag{9}$$

He goes on to interpret this equation in electromagnetic terms. Electromagnetic phenomena depend on the variation of the number of lines of magnetic force passing through the circuit; because "the number of these lines is expressed mathematically by the surface-integral of the magnetic induction through any surface bounded by the circuit," "we must regard the vector **B** and its components a, b, c as representing what we are already acquainted with as the magnetic induction and its components." While the vector **B** belongs in the category of "fluxes" and occurs in a surface integral, the vector **A** belongs in the category of forces since it appears in a line integral. The relation between the flux of magnetic induction through a surface and the electro-tonic state round its boundary, first stated in "On Faraday's Lines of Force," is now expressed formally in terms of vectors ("forces" and "fluxes") and by Stokes' Theorem: The line integral of the electro-tonic state taken round a closed circuit is equal to the surface integral of the flux of magnetic induction over the surface bounded by the circuit.[80] The concept of the electro-tonic state or vector potential, which became an auxiliary quantity in post-Maxwellian electrodynamics (being eliminated from the field equations by Heaviside and Hertz), is, however, central to Maxwell's exposition of his mathematical theory of the electromagnetic field.

The Topology of the *Treatise*

Maxwell's treatment of the theory of electromagnetism in terms of "forces" and "fluxes," lines and surfaces, and vectors emphasized that electromagnetism could be conceived in terms of "purely geometrical relations" between lines in space representing circuits and lines of force. The mathematical treatment of the properties of lines in space could best be presented by the use of a method "specially adapted to the expression of such geometrical relations— the *Quaternions* of Hamilton." [81] The representation of electric and magnetic quantities as vectors had, therefore, a special meaning: The physics of electricity and magnetism could be first presented in an essentially geometrical form, depending on the purely geometrical relations of lines and surfaces in space— a physical geometry that could then be applied to electricity and magnetism. The *Treatise* therefore carried through the program of research that Maxwell had envisaged in the mid 1850s, but in a deeper and richer form. Not only did the *Treatise* incorporate the program of providing a dynamical theory of electricity and magnetism, again in a form that considerably enlarges upon Maxwell's earlier arguments, but the transformed physical geometry of the *Treatise* embraces ideas first developed in "On Physical Lines of Force" and presented there in terms of the physical mechanics of the ether.

An important example of this process of conceptual enlargement is found in the discussion of electricity as a linear phenomenon and of magnetism as having a rotational character. In "On Physical Lines of Force" this is discussed in terms of the linear translation of idle-wheel particles and the rotation of molecular vortices. In the *Treatise*, Maxwell continues to affirm his commitment to the idea that the Faraday effect is caused by the rotation of vortices in the ether (while disclaiming the precise theory of molecular vortices and idle-wheel particles proposed as a hypothetical model in "On Physical Lines of Force"); however, the relation between electricity as a linear phenomenon and magnetism as a rotational phenomenon is now expressed geometrically, in vectorial (quaternion) terms, and is directly linked to the geometrical foundations of the mathematics of the *Treatise*, based on vectors, quaternions, and integral theorems embracing geometrical considerations of lines and surfaces in space. Following from this geometrical treatment of linear and rotational phenomena in electromagnetism, Maxwell was concerned to define the directions of linear and rotational motions in space, enlarging his physical geometry to include a topological treatment of lines and surfaces.

Maxwell discussed the problem of defining the directionality of linear and

rotational motions in postcards to Tait in May 1871, at first expressing confusion over the different directional conventions adopted by Hamilton and by
Thomson and Tait: "I am desolated! I am like the Ninevites! Which is my right
hand? Am I perverted? a mere man in a mirror, walking in a vain show? What
saith the Master of Quaternions [Hamilton]?" He soon clarified the issue, and
provided Tait with a draft of section 23 of the *Treatise*, where the question of
directionality is formally discussed: "In this treatise, the motions of translation
along one axis and of rotation round that axis will be assumed to be of the
same sign when they are related to each other in the same way as the motions
of translation and rotation of a right handed screw.... This is the right handed
definition of directions and is adopted in this treatise...."[82] Maxwell observes
in the *Treatise* that this system of relations in space could be called the system
of the vine, as the tendrils of the vine are right-handed screws, and that the
opposite left-handed system would be the system of the hop.[83]

This issue had been discussed by the German mathematician J. B. Listing,
who, as Maxwell notes in the *Treatise*, had called the operation of passing from
the one system to the other "perversion." Maxwell had remarked to Tait that
"if we confound the one [system] with the other, every figure will become
perverted (a phrase of L. denoting an effect similar to that of reflexion in a
mirror)."[84] This discussion of the systems of direction in space is especially
significant in Maxwell's treatment of the Faraday effect. He points out that a
plane-polarized ray of light could be represented by two circularly polarized
rays, one right-handed the other left-handed (as regards the observer), and
notes that two such rays having the same wavelength are "geometrically alike
in all respects, except that one is the *perversion* of the other, like its image in
a looking glass." To summarize the argument, he maintains that the Faraday
effect is an essentially directional phenomenon that cannot be explained on
the supposition that one of the rays has a shorter period of rotation than the
other, and concludes that the Faraday magneto-optic effect does "not depend
solely on the configuration of the ray, but also on the direction of the motion
of its individual parts." The rotation of light in the Faraday effect is therefore
"affected by the relation of the direction of rotation of the light to the direction
of the magnetic force." This leads to the conclusion that "in a medium under
the action of magnetic force something belonging to the same mathematical
class as an angular velocity ... forms a part of the phenomenon." Thus, "some
rotatory motion is going on," and the angular velocity must be conceived as
the rotation of very small portions of the medium: "This is the hypothesis of
molecular vortices."[85] The geometrical and dynamical arguments therefore
coalesce.

Maxwell referred to Listing's paper "Der Census räumlicher Complexe" (1861)
in the *Treatise* in an attempt to classify the relations between curves and
surfaces (which were of central importance in Maxwell's physics, where the
relation between "forces" acting along lines and "fluxes" acting across surfaces

was fundamental). His interest in what Listing had termed the "geometry of position" and "topology"[86] had been stimulated by well-known papers by Helmholtz and Kelvin on vortex motion in a fluid. In his paper "On Vortex Motion" (1869) Kelvin had drawn upon Helmholtz's treatment of the laws of vortex motion, in which it had been demonstrated that vortex filaments in a perfect fluid would be immune to destruction or dissipation.[87] Kelvin's discussion of the mathematical theory of a finite mass of an incompressible frictionless fluid enclosed in a rigid fixed boundary had involved a treatment of curves, surfaces, and boundaries.[88] Maxwell developed these ideas in a more systematic and fundamental way.

The origins of these topological investigations lay in Leibniz's interest in the use of geometric algorithms to express geometrical location; in Euler's work on the classification of polyhedra in terms of a relation between the number of edges, faces, and vertices of a closed convex polygon; and in Riemann's discussion of continuity in geometry, in which an attempt had been made to classify surfaces and their boundaries by their topological properties.[89] Maxwell indicated the direction of his ideas in letters to Kelvin in September and October 1868.[90] His drafts show him grappling with the problems of curves and surfaces and the definition of polygons, and adopting Listing's complex terminology for the classification of the geometrical relations between boundaries and surfaces.[91] He also discussed surfaces and the relation between the edges, faces, and vertices of a polyhedron in his paper "On Hills and Dales" (1870), his interest undoubtedly having been aroused by the application of these topological ideas to maps and contour lines. These topological ideas are presented in some detail in the *Treatise*, where Maxwell deploys definitions of lines, boundaries, and surfaces, emphasizing their application to the construal of line and surface integrals and providing further geometrical foundations for the physics. Discussions of the number of surfaces bounding finite regions of space, of the continuity of surfaces and their connectivity by closed curves, and of the closure of finite surfaces by external surfaces, and definitions of the continuity of lines and surfaces, provide the geometrical basis for the use of integral theorems.[92] Quaternions, integral theorems, vectors, and topological concepts form the mathematical foundations of the *Treatise*, a framework of geometrical concepts expressing the vectorial character of electrical and magnetic quantities.

Dynamics and Physical Reality

These mathematical modes for the representation of physical quantities present a deepening of the physical geometry of lines of force first developed in "On Faraday's Lines of Force." The argument of the *Treatise* also encompasses a transformation of Maxwell's physical mechanics of vortices in the ether, providing a development of the generalized Lagrangian theory of the field as presented in "A Dynamical Theory of the Electromagnetic Field." Maxwell's appeal in the *Treatise* to the Lagrangian form of analytical dynamics is constrained by his continued commitment to the aim of providing a physical expression of the symbols of analytical dynamics. For Maxwell, any symbolic representation had to provide a physical interpretation of nature. A purely symbolic theory employing "the machinery with which mathematicians have been accustomed to work problems about pure quantities" was inadequate, and the mathematical symbolism had to be "clothed with the imagery ... of the phenomena of the science itself."[93]

In the *Treatise* Maxwell stresses his aim of formulating a dynamical theory of the field that will emphasize the primacy of the concepts of momentum, velocity, and energy. He therefore declares his goal of translating the results of the Lagrangian method "from the language of the calculus into the language of dynamics" so that "our words may call up the mental image, not of some algebraical process, but of some property of moving bodies."[94]

To achieve this aim, Maxwell develops the method of dynamical explanation that had been established by Kelvin and Tait in their *Treatise on Natural Philosophy* (1867). He contrasts Lagrange's method, which he considers to be a formalism of generalized equations of motion conceived as "pure algebraical quantities" in a manner "free from the intrusion of dynamical ideas," with the method by which Kelvin and Tait sought to "cultivate" dynamical ideas.[95] Whereas the Lagrangian method provided a purely mathematical formalism that avoided reference to the concepts of momentum, velocity, and energy after they had been replaced by symbols in the generalized equations of motion, Kelvin and Tait placed the emphasis on dynamical concepts. The basic mathematical axiom in their treatment of dynamics was a minimum theorem discovered by Kelvin that related the variation of a system by impulsive force to the kinetic energy of the system. Because an impulsive force acts in an infinitesimal time increment, the configuration and the potential energy of the system would not be altered by the action of an impulsive force; only the kinetic energy would change. Generalized equations of motion are derived from the supposition of impulsive forces.[96] In Maxwell's view, although this method "kept out of view the mechanism by which the parts of the system are connected" and hence followed his aim of avoiding the formulation of a mechanical ether model, the method did keep "constantly in mind the ideas appropriate to the fundamental science of dynamics," and hence it satisfied the criterion for dynamical explanation.[97] Concepts of energy, momentum, and velocity were fundamental to Maxwell's aim of formulating a "consistent representation" of the field, an expression he had derived from a remark by Gauss on the need for a "constructible representation [*construirbare Vorstellung*]" of the manner in which the propagation of electric action takes place.[98] This would require the expression of "dynamical ideas from a physical point of view," and an emphasis on the link between the mathematical formalism of abstract dynamics and the physical reality depicted; "we must," wrote Maxwell, "have our minds imbued with these dynamical truths as well as with mathematical methods."[99]

In seeking to provide an exposition of dynamics from a physical point of view, Maxwell, in the chapter "On the Equations of Motion of a Connected System" in the *Treatise*, sought to develop a new framework for dynamical theory. His argument here should be considered in relation to the draft paper "On the Interpretation of Lagrange's and Hamilton's Equations of Motion" (probably dating from 1873), which provides some amplification of his published argument in the *Treatise*.[100] In the draft he observes: "Our popular dynamical ideas are far too exclusively drawn from the dynamics of a particle.... it is unfortunate that in expressing the relations between these ideas we have sometimes adopted a form of expression, which, though true for a particle, is not easily applicable to a connected system [of particles]." As Maxwell explains in the *Treatise*, in attempting to achieve this objective he favors the mode of expressing kinetic energy in terms of the velocity and momenta of the particles constituting a material system, obtaining $\dot{q} = dT_p/dp$, where "the velocity corresponding to the variable q is the differential coefficient of T_p [the kinetic energy expressed in terms of the coordinate q and momentum p] with respect to the corresponding momentum p." Thus "the velocities and momenta, depend on the actual state of motion of the system at the given instant, and not on its previous history."[101] This mode of expressing kinetic energy was due to William Rowan Hamilton, and Maxwell obtains an equation derived by Hamilton for the impressed forces acting on the material system:

$$F_r = \frac{dp_r}{dt} + \frac{dT_p}{dq_r}. \tag{10}$$

In the draft Maxwell notes that the first term on the right "expresses the fact that part of the force is expended in increasing the momentum," and that "the second term indicates that if the increase of the variable q has a direct effect in increasing the kinetic energy, a force will arise from this circumstance." He prefers this Hamiltonian form of expressing the equations of motion of a connected system to the alternative form due to Lagrange. In the Lagrangian form, momenta are expressed in terms of the velocities, yielding an expression for kinetic energy in terms of the velocities and the variables q, the kinetic energy being denoted $T_{\dot{q}}$; this yields the equation of motion

$$F_r = \frac{dp_r}{dt} - \frac{dT_{\dot{q}}}{dq_r}. \tag{11}$$

Maxwell argues: "According to Lagrange's expression it would appear as if the kinetic energy had a tendency to increase, and to do work as it increases. This arises from the fact that the kinetic energy is expressed in terms of the velocities." He prefers the form of the equation of motion given by Hamilton because of the physical meaning of the equation, which is based on the momentum rather than (as with Lagrange) on velocity:

Now it is not the velocities which obey Newton's law of persevering in their actual state, but the momenta or "quantities of motion." Hence if we wish to apply Newton's law we must express the kinetic energy in terms of the momenta and use Hamilton's form of the equations of motion. We then see at once that the second term [in equation 10] indicates that if a given displacement has a direct tendency to increase the kinetic energy, the momenta remaining the same, a quantity of work, equal to this increase of kinetic energy is performed by the external force F during the displacement.

The dynamical theory of the electromagnetic field is therefore grounded on concepts of momentum, velocity, and energy, and the equations are translated into "language which may be intelligible without the use of symbols"; by contrast, the original Lagrangian method had sought "even to get rid of the ideas of velocity, momentum and energy, after they have been once for all supplanted by symbols in the original equations." Thus, the "language of dynamics" would express "some property of moving bodies." [102] The physical meaning of the equation of motion is justified by its expression in terms of momentum; hence, Newton's first and second laws of motion, rather than Lagrangian symbols, express dynamical theory.

Conclusion: Maxwell and Kelvin

In a famous statement in his original lectures at Johns Hopkins University, Kelvin commented unfavorably on Maxwell's electromagnetic theory of light, and by implication on the *Treatise on Electricity and Magnetism*, in the following terms: "I never satisfy myself until I can make a mechanical model of a thing. If I can make a mechanical model I can understand it. As long as I cannot make a mechanical model all the way through I cannot understand; and that is why I cannot get the electromagnetic theory.... I want to understand light as well as I can, without introducing things that we understand even less of. That is why I take plain dynamics. I can get a model in plain dynamics; I cannot in electromagnetics." [103] For this reason he espoused the elastic-solid theory of light as providing the appropriate model for theory construction in electromagnetism; only in this way would it be possible to develop a unified theory of electricity, magnetism, and light that would be based on "plain dynamics." These remarks, and others with a similar tone, express Kelvin's dissatisfaction with the theory of Maxwell's *Treatise*: Maxwell's theory of electricity and magnetism was not sufficiently grounded on "plain dynamics." Indeed, Maxwell himself had admitted as much. He had pointed out that, according to the theory of the *Treatise*, electrical action was "a phenomenon due to an un-known cause, subject only to the general laws of dynamics"; but that a "complete dynamical theory" of the electromagnetic field would represent the hidden structure of the material system constituting the field, so that "the whole intermediate mechanism and details of the motion [would be] taken as the objects of study."[104]

Maxwell's fundamental disagreement with Kelvin can be gauged by con-trasting Kelvin's declaration about "plain dynamics" as necessarily founda-tional with Maxwell's response to FitzGerald's electromagnetic theory of the reflection and refraction of light. Writing to Stokes in February 1879, Maxwell called his own theory of the Faraday magneto-optic effect "a hybrid theory, in which bodily motion of the medium is made to cooperate with electric cur-rent." By contrast, "in Mr FitzGerald's work I cannot find that he ever intro-duces the bodily motion of the medium." Maxwell declared that such an explanation of the Faraday effect "by a purely electromagnetic hypothesis" would be "a very important step in science." While Maxwell was prepared to envisage a further weakening of the link between electromagnetism and mechanical representation, nevertheless he remarked that the value of Fitz-Gerald's mathematical theory would be greatly increased by an interpretation in terms of a "dynamical hypothesis."[105] Kelvin, by contrast, was wholly hos-tile to attempts to "take up the so-called electromagnetic theory of light";[106] he expressed his exasperation with such developments—probably with Hertz and, especially, Heaviside in mind—"uncompromisingly" in a letter to Fitz-Gerald in April 1896: "It is mere nihilism, having no part of lot in Natural Philosophy, to be contented with two formulas for energy, electromagnetic and electrostatic, and to be happy with a vector and delighted with a page of symmetrical formulas."[107] Kelvin required mathematical relations to be firmly and unambiguously based on mechanical visualisations.

It would be wrong to exaggerate the differences between the Maxwell of the *Treatise* and the Kelvin of the Baltimore Lectures, and to see Maxwell's com-ments on FitzGerald as prescient anticipations of a demechanized mathemati-cal physics. As those comments show, Maxwell remained committed to a mechanistic ontology of particles of matter in motion as providing the substra-tum for physical explanation; his molecular theory of gases and his dynamical theory of the electromagnetic field make this clear. But the *Treatise on Electri-city and Magnetism*, compounded of geometrical, vectorial, and generalized dynamical arguments, presents a framework of mathematical physics that contrasts markedly with the approach advocated by Kelvin in the Baltimore Lectures, where mechanical illustration is taken as the paradigm of explanation in theory construction. It is, perhaps, an irony of history that Maxwell's mathe-matical way—which has provided a model for subsequent mathematical phys-ics, was in large measure built on the shoulders of Kelvin.

Acknowledgments

I am grateful to the Syndics of the University Library, Cambridge, to the Master and Fellows of St. John's College, Cambridge, to the Librarian of the Royal Society, and to the Librarian of Glasgow University Library, for kind permission to quote from documents in their keeping. I am grateful to the Council of the Royal Society for generously supporting my work on the Maxwell papers. I am grateful to Martin Klein for his helpful comments, and to the discussants when I presented the paper at Johns Hopkins in April 1985.

Notes

1. Sylvester Papers, Library of St. John's College, Cambridge.

2. Lord Kelvin, *Baltimore Lectures on Molecular Dynamics and the Wave Theory of Light* (Cambridge University Press, 1904), p. vii. On Kelvin's work on electricity and magnetism see Ole Knudsen, "Mathematics and Physical Reality in William Thomson's Electromagnetic Theory," in *Wranglers and Physicists: Studies on Cambridge Physics in the Nineteenth Century*, ed. P. M. Harman (Manchester University Press, 1985).

3. P. M. Harman, *Energy, Force, and Matter. The Conceptual Development of Nineteenth-Century Physics* (Cambridge University Press, 1982), passim.

4. P. M. Harman, *Metaphysics and Natural Philosophy. The Problem of Substance in Classical Physics* (Totowa, N.J.: Barnes and Noble, 1982), pp. 127–150.

5. Sir William Thomson, *Notes of Lectures on Molecular Dynamics and the Wave Theory of Light*, stenographically reported by A. S. Hathaway (Baltimore, 1884), pp. 6, 132, 270–271.

6. Kelvin, *Baltimore Lectures* (note 2 above), p. 9.

7. See M. Norton Wise, "The Flow Analogy to Electricity and Magnetism, Part 1: William Thomson's Reformulation of Action at a Distance," *Archive for History of Exact Sciences* 25 (1981): 19–70; Daniel M. Siegel, "Thomson, Maxwell and the Universal Ether in Victorian physics," in *Conceptions of Ether. Studies in the History of Ether Theories 1740–1900*, ed. G. N. Cantor and M. J. S. Hodge (Cambridge University Press, 1981).

8. Kelvin attended a lecture course given by Meikleham at Glasgow; Maxwell attended one given by J. D. Forbes at Edinburgh. See David B. Wilson, "The Educational Matrix: Physics Education at Early-Victorian Cambridge, Edinburgh and Glasgow Universities," in *Wranglers and Physicists*, ed. Harman.

9. P. M. Harman, "Edinburgh Philosophy and Cambridge Physics: The Natural Philosophy of James Clerk Maxwell," in *Wranglers and Physicists*, ed. Harman.

10. Harvey Becher, "William Whewell and Cambridge Mathematics," *Historical Studies in the Physical Sciences* 11 (1980): 1–48.

11. P. G. Tait, "James Clerk Maxwell," *Proceedings of the Royal Society of Edinburgh* 10 (1879): 331–339.

12. For example: S. D. Poisson, "Mémoire sur l'équilibre et le mouvement des corps élastiques," *Mémoires de l'Académie Royale des Sciences* 8 (1829): 357–570, 623–627; G. Lamé and E. Clapeyron, "Mémoire sur l'équilibre intérieur des corps solides homogènes," *Journal für die reine und angewandte Mathematik* 7 (1831): 145–169, 237–252, 381–413; G. G. Stokes, "On the Theories of the Internal Friction of Fluids in Motion, and of the Equilibrium and Motion of Elastic Solids," *Transactions of the Cambridge Philosophical Society* 8 (1849): 287–319.

13. Poisson, "Mémoire sur … corps élastiques," p. 361.

14. *The Scientific Papers of James Clerk Maxwell*, ed. W. D. Niven (Cambridge University Press, 1890), vol. I, p. 31.

15. George Gabriel Stokes, testimonial dated February 25, 1856, in support of Maxwell's application for the Professorship of Natural Philosophy in the Marischal College and University of Aberdeen. Copy in Glasgow University Library, Y1-h.18.

16. Maxwell to William Thomson, February 20, 1854, in J. Larmor, "The Origins of Clerk Maxwell's Electric Ideas, as Described in Familiar Letters to W. Thomson," *Proceedings of the Cambridge Philosophical Society* 32 (1936): 699–702.

17. James Clerk Maxwell, " On Faraday's Lines of Force" (manuscript), University Library, Cambridge, Add. 7655/V/c/7.

18. Maxwell, *Scientific Papers*, vol. II, pp. 155–158.

19. Tait, "James Clerk Maxwell."

20. Maxwell to Thomson, September 13, 1855, in Larmor, "Origins," p. 711.

21. W. Thomson, "On the Uniform Motion of Heat in Homogeneous Solid Bodies, and its Connection with the Mathematical Theory of Electricity," *Cambridge Mathematical Journal* 3 (1842): 71–84; Thomson, "Notes on Hydrodynamics. On the Equation of Continuity," *Cambridge and Dublin Mathematical Journal* 2 (1847): 282–286.

22. James Clerk Maxwell, *A Treatise on Electricity and Magnetism* (Oxford: Clarendon, 1873), vol. I, p. 98.

23. Maxwell, *Scientific Papers*, vol. I, p. 156.

24. Maxwell, *Scientific Papers*, vol. I, p. 158–160.

25. Maxwell, "On the motion of fluids" and "On the steady motion of an incompressible fluid," University Library, Cambridge, Add. 7655/V/c/4 and 5.

26. Maxwell, *Scientific Papers*, vol. I, p. 160.

27. Maxwell to Thomson, September 13, 1855, in Larmor, "Origins," p. 711.

28. Harman, "Edinburgh Philosophy and Cambridge Physics," in Harman, *Wranglers and Physicists.*

29. Michael Faraday, *Experimental Researches in Electricity*, (London: Taylor and Francis, 1839–1855), vol. I, pp. 60, 284. See P. M. Heimann [Harman], "Faraday's Theories of Matter and Electricity," *British Journal for History of Science* 5 (1971): 235–257 for discussion of the electro-tonic state in Faraday's electrical theory.

30. Maxwell, *Scientific Papers*, vol. I, p. 203; J. C. Maxwell, [Abstract of "On Faraday's Lines of Force"], *Proceedings of the Cambridge Philosophical Society* 1 (1856): 165.

31. Maxwell to Thomson, November 13, 1854, in Larmor, "Origins," pp. 702–703. See M. Norton Wise, "The Mutual Embrace of Electricity and Magnetism," *Science* 203 (1979): 1310–1318.

32. Maxwell, *Scientific Papers*, vol. I, p. 206.

33. G. G. Stokes, *Mathematical and Physical Papers* (Cambridge University Press, 1880–1905), vol. V, p. 320; Thomson to Stokes, July 2, 1850, University Library, Cambridge, Add. 7656/K 39. See J. J. Cross, "Integral Theorems in Cambridge Mathematical Physics, 1830–55," in Harman, *Wranglers and Physicists*.

34. Maxwell, *Scientific Papers*, vol. I, pp. 188, 205.

35. Maxwell, *Scientific Papers*, vol. I, pp. 188, 451.

36. W. Thomson, "On a Mechanical Representation of Electric, Magnetic, and Galvanic Forces," *Cambridge and Dublin Mathematical Journal* 2 (1847): 61–64.

37. Thomson to Faraday, June 11, 1847, in S. P. Thompson, *The Life of William Thomson, Baron Kelvin of Largs* (London: Macmillan, 1910), vol. I, p. 203.

38. Maxwell to Faraday, November 9, 1857, in L. Campbell and W. Garnett, *The Life of James Clerk Maxwell*, second edition (London: Macmillan, 1884), p. 204.

39. W. Thomson, "Dynamical Illustrations of the Magnetic and Heliocoidal Rotatory Effects of Transparent Bodies on Polarized Light," *Proceedings of the Royal Society* 8 (1856): 150–158.

40. Maxwell, *Scientific Papers*, vol. I, pp. 505, 502.

41. Maxwell to Thomson, December 10, 1861, in Larmor, "Origins," p. 728. See Ole Knudsen, "The Faraday Effect and Physical Theory," *Archive for History of Exact Sciences* 15 (1976): 235–281.

42. Maxwell, *Scientific Papers*, I, 464, 476–479.

43. Maxwell, *Scientific Papers*, I, 486.

44. Maxwell, *Scientific Papers*, I, 502–504. See Daniel M. Siegel, "Mechanical Image and Reality in Maxwell's Electromagnetic Theory," in Harman, *Wranglers and Physicists* (especially pp. 191–194).

45. Maxwell, *Scientific Papers*, I, 505; Maxwell, *Treatise*, II, 416. See P. M. Heimann [Harman], "Maxwell and the Modes of Consistent Representation," *Archive for History of Exact Sciences* 6 (1970): 171–213.

46. Maxwell to Tait, December 23, 1867, in C. G. Knott, *Life and Scientific Work of Peter Guthrie Tait* (Cambridge University Press, 1911), p. 215.

47. Maxwell, *Treatise*, II. 416–417.

48. Maxwell to Thomson, October 15, 1864, Glasgow University Library, Kelvin Papers M 17.

49. Maxwell, *Scientific Papers*, I, 532–533, 563.

50. Maxwell, *Scientific Papers*, I, 536–538.

51. James Clerk Maxwell, "A Dynamical Theory of the Electromagnetic Field" (manuscript), Royal Society, London, PT 72/7.

52. Maxwell, *Scientific Papers*, I, 564.

53. Maxwell, *Treatise*, II, 417.

54. Harman, *Energy, Force, and Matter*, passim.

55. Poisson, "Mémoire sur ... corps élastiques," p. 361.

56. Thomson to Stokes, March 15, 1865, Royal Society, Referees' Reports, 5/137.

57. Knudsen, "William Thomson's Electromagnetic Theory," in Harman, *Wranglers and Physicists*, p. 178.

58. Maxwell, *Treatise*, I, vi.

59. Maxwell, *Scientific Papers*, II, 216.

60. *The Mathematical Papers of Sir William Rowan Hamilton* (Cambridge University Press, 1931–1967), vol. III, pp. 355–359. See also Thomas L. Hankins, *Sir William Rowan Hamilton* (Baltimore: Johns Hopkins University Press, 1980), pp. 283–325.

61. Maxwell to Tait, November 7 and 14, 1870, in Knott, *Life of Tait*, p. 144.

62. [J. C. Maxwell], "Quaternions," *Nature* 9 (1873): 137; Maxwell, *Treatise*, I, 8–9.

63. Maxwell, *Treatise*, I, 28; *Scientific Papers*, II, 265.

64. Maxwell to Tait, November 7, 1870, in Knott, *Life of Tait*, p. 143.

65. Maxwell, *Treatise*, II, 238.

66. Maxwell, *Treatise*, I, 12–13.

67. Maxwell, *Treatise*, second edition (Oxford: Clarendon, 1881), vol. I, p. 30.

68. Maxwell to Thomson, undated postcard, in Larmor, "Origins," p. 748.

69. Maxwell, *Treatise*, II, 416.

70. Maxwell to Stokes, February 6, 1879, in *Memoir and Scientific Correspondence of the late Sir George Gabriel Stokes*, ed. J. Larmor (Cambridge University Press, 1907), vol. II, p. 42.

71. Maxwell, *Scientific Papers*, I, 189–192.

72. Maxwell returned to his original term, "intensities," in the second edition of the *Treatise*.

73. Maxwell, *Treatise*, first edition, I, 10–11; second edition, I, 12.

74. Maxwell to Stokes, January 11, 1871, in *Memoir and Correspondence*, II, 31.

75. P. G. Tait, "On Green's and Other Theorems," *Transactions of the Royal Society of Edinburgh* 26 (1870): 75.

76. Maxwell to Tait, April 4, 1871 [Cavendish Laboratory, Cambridge]; Add. 7655/I/b/21, University Library, Cambridge: photocopy of the original.

77. Poisson, "Mémoire sur ... corps élastiques"; Maxwell, *Scientific Papers*, I, 32. See Cross, "Integral Theorems," in Harman, *Wranglers and Physicists*, p. 120.

78. Tait to Maxwell, April 5, 1871, University Library, Cambridge, Add. 7655/I/a/10. (In the passage quoted, Tait referred to himself and Thomson as "the Archiepiscopal pair"; their names were the same as those of the archbishops of Canterbury and York.)

79. Maxwell, *Treatise*, I, 25–27.

80. Maxwell, *Treatise*, II, 214–216.

81. Maxwell, *Treatise*, II, 151, 159.

82. Maxwell to Tait, May 8 and 12, 1871, University Library, Cambridge, Add. 7655/I/b/23 and 25.

83. Maxwell, *Treatise*, I, 24n.

84. Maxwell, *Treatise*, I, 24; Maxwell to Tait, May 12, 1871, University Library, Cambridge, Add. 7655/I/b/25.

85. Maxwell, *Treatise*, II, 403–404, 407–408.

86. J. B. Listing, "Der Census räumliche Complexe," *Abhandlungen der Gesellschaft der Wissenschaft zu Göttingen, Mathematische Classe* 10 (1861): 97–182.

87. Hermann Helmholtz, "Über Integrale der hydrodynamischen Gleichungen, welche den Wibelbewegungen entsprechen," *Journal für die reinen und angewandte Mathematik* 55 (1858): 25–55; tr. P. G. Tait, *Philosophical Magazine* 33 (1867): 485–511. Kelvin's paper (note 88 below) was first presented to the Royal Society of Edinburgh on April 29, 1867, and was recast and augmented in the following year.

88. W. Thomson, "On Vortex Motion," *Transactions of the Royal Society of Edinburgh* 25 (1869): 217–260.

89. The paper by B. Riemann, "Lehrsätze aus der analysis situs für die Theorie der Integrale von zweigliedrigen vollständigen Differentialen," *Journal für die reine und angewandte Mathematik* 54 (1857): 105–110, is most relevant.

90. Maxwell to Thomson, September 28 and October 7, 1868, Glasgow University Library, Kelvin Papers M 25 and 27.

91. Add. 7655/V/c/40 and V/d/10 and 12, University Library, Cambridge.

92. Maxwell, *Treatise*, I, 16–24.

93. Maxwell, *Scientific Papers*, II, 325.

94. Maxwell, *Treatise*, II, 185.

95. Maxwell, *Treatise*, II, 184—185.

96. D. F. Moyer, "Energy, Dynamics, Hidden Machinery: Rankine, Thomson and Tait, Maxwell," *Studies in History and Philosophy of Science* 8 (1977): 251—268.

97. Maxwell, *Treatise*, II, 193—194.

98. Maxwell, *Treatise*, II, 435; C. F. Gauss, *Werke* (Göttingen, 1863—1933), vol. V, p. 529. See Heimann [Harman], "Maxwell and the Modes of Consistent Representation."

99. Maxwell, *Treatise*, II, 184, 194.

100. Add 7655/V/e/9, University Library, Cambridge.

101. Maxwell, *Treatise*, II, 189.

102. Maxwell, *Treatise*, II, 185, 194.

103. Thomson, *Notes of Lectures*, 270—271 (Lecture XX).

104. Maxwell, *Treatise*, II, 202.

105. Maxwell to Stokes, February 6, 1879, in Stokes, *Memoir and Correspondence*, II, 42—43.

106. Thomson, *Notes of Lectures*, 6.

107. Thomson to FitzGerald, April 9, 1896, in Thompson, *Life of Thomson*, II, 1065.

Bruce J. Hunt

"How My Model Was Right": G. F. FitzGerald and the Reform of Maxwell's Theory

"I never satisfy myself until I can make a mechanical model of a thing," Kelvin declared in the twentieth of his Baltimore lectures. "If I can make a mechanical model I can understand it. As long as I cannot make a mechanical model all the way through I cannot understand...." For Kelvin, and for a whole generation of British physicists, mechanical models had an importance they have since largely lost. These men regarded the construction of a model as the real test of one's understanding of a phenomenon, and as one of the main ways in which one might seek to extend that understanding. Kelvin's Baltimore Lectures, like his other writings, are filled with designs for intricate assemblages of weights, rods, and springs with which to mimic the behavior of molecules and the ether, and similar efforts can be found in the works of many of his contemporaries and successors. To dismiss the construction of such models as no more than a "playschool" activity, as a recent writer has done, would be a grave mistake.[1] Mechanical models were in fact an integral and characteristic part of the British approach to physical theorizing in the nineteenth century, and any attempt to understand Victorian physics must take careful account of the role they played.

Most historical studies of models have focused on the work of Kelvin and Maxwell, in particular on the crucial role Maxwell's "vortices and idle wheels" model played in the genesis of his electromagnetic theory.[2] But Kelvin and Maxwell were far from the only Victorian physicists to devise and use models, and to obtain a full picture we must examine the work of others as well. An especially interesting case is that of George Francis FitzGerald (1851–1901) and the "wheel and band" model he devised in 1885.[3] FitzGerald's exploration of the workings of this model led him to revise Maxwell's treatment of electromagnetic propagation in a subtle but important way, and so helped change the way Maxwell's theory was interpreted in the crucial period in the late 1880s when it was put into its definitive form. FitzGerald's trust in his model, and his willingness to modify Maxwell's equations to make them fit it better, played a key part in this reform. A look at the way FitzGerald used his model will enable us to fill in an important and hitherto largely ignored chapter in the evolution of Maxwell's theory and to see more clearly how British physicists of the nineteenth century used mechanical models as genuine research tools.

Maxwell's Model

The roots of FitzGerald's model went back to a discovery of Faraday's and to the implications that were drawn from it by Kelvin and Maxwell. In 1845 Faraday found that when a beam of polarized light was sent through a piece of glass in a magnetic field, its plane of polarization was turned slightly to one side. This became known as the "Faraday effect." In 1856, Kelvin deduced from this that there must be tiny "molecular vortices" spinning around along the lines of force in a magnetic field, and in 1861–62 Maxwell made these vortices the basis of his well-known "vortices and idle wheels" model of the ether.[4]

Maxwell supplemented Kelvin's basic vortex array with moveable "idle wheel" particles which passed rotation from one layer of vortices to the next and whose translatory motion represented an electric current. The result was a remarkably detailed and comprehensive model of the electromagnetic field that not only depicted known phenomena but provided clues to new ones as well. For instance, when Maxwell made the vortices elastic and identified the strain pattern around displaced idle wheels with an electrostatic field, he found that the motion of the wheels into their new positions constituted a "displacement current"—and so introduced the revolutionary hypothesis of electric currents in seemingly empty space. Moreover, Maxwell found that this newly elastic medium could carry waves whose speed, calculated from purely electrical and magnetic constants, was almost exactly that of light.[5] That the model he had devised for electricity and magnetism turned out to account for light as well was a pleasant surprise to Maxwell and a great triumph for his model-building approach.

Despite this great success, Maxwell did not regard his model as a realistic picture of the ether. The vortices were real enough, he thought, but the connecting idle-wheel mechanism was just a contrivance. "I do not bring it forward as a mode of connexion existing in nature," he said, but merely as one that was mechanically definite and readily investigated.[6] Since many of his results were independent of the details of the "mode of connexion," Maxwell thought that choosing a specific mechanism and exploring it exhaustively was the best way to make progress at this early stage of theory construction. Once he had hit upon his electromagnetic theory of light, however, he wanted to show explicitly that this important result rested on a more secure and general basis than the idle-wheel hypothesis. He did this in "A Dynamical Theory of the Electromagnetic Field,"[7] presented to the Royal Society in 1864, in which he used Lagrangian methods to derive his electromagnetic equations from very general dynamical considerations without regard to the underlying structure of the ether.

It was in "A Dynamical Theory" that Maxwell first made his theory of light fully electromagnetic. Whereas in his 1862 paper he had derived the wave equation from the mechanical variables of his model, in 1864 he derived it from the purely electromagnetic variables of the field. The wave equation Maxwell found was perfectly valid, but the derivation he used in 1864, and repeated in his *Treatise* nine years later, was far from flawless. He based his equations on the electric and vector potentials, and his discussion of their propagation ran into difficulties over what would today be called the question of "gauge choice."[8] This caused considerable confusion for Maxwell and his successors, much of which turned on the question of the physical significance of the potentials.

It is sometimes said that Maxwell abandoned his vortex model after 1862, and that "A Dynamical Theory" marked his decisive turn away from a search for ether mechanisms and toward an exclusive concern with the formulation of electromagnetic laws.[9] In fact, however, Maxwell continued to pursue both laws and mechanisms to the end of his career. He saw these two pursuits as complementary, not antagonistic; the construction of mechanisms could suggest new phenomena and theories, much as the vortex model had led to the electromagnetic theory of light, while the formulation of laws helped lay down the constraints without which the construction of models was empty speculation. A concern with laws was dominant in Maxwell's *Treatise*, as was proper in a work of that type, but even there Maxwell continued to use the vortex hypothesis in connection with the Faraday effect. Indeed, he ended his chapter on the subject by stating again that "we have good evidence for the opinion that some phenomenon of rotation is going on in the magnetic field," and that "this rotation is performed by a great number of very small portions of matter"—that is, by molecular vortices.[10] Maxwell's goal, here as elsewhere, was to integrate laws with mechanisms; just as he had helped integrate the gas laws with the kinetic theory of gases, he hoped to integrate the electromagnetic laws with a theory of the structure of the ether.

Maxwell's two-pronged attack, shifting back and forth between models and equations, was characteristic of British physics in the Victorian era. He and Kelvin were the great masters of this technique, but it had many other practicioners, of whom G. F. FitzGerald was among the most adept. He combined a mechanical turn of mind and mathematical skill to an unusual degree, and in 1885 he brought the two together in his wheel-and-band model of the electromagnetic field.

Wheels and Bands

FitzGerald's model was similar to Maxwell's vortices and idle wheels, but mechanically it was much simpler. It was simple enough, in fact, for FitzGerald to have a working version of it built—"rather pretty on a mahogany board with bright brass wheels"[11]—with which he could illustrate Maxwellian ideas to his classes at Trinity College Dublin. But our focus here will be not on this important pedagogical use but rather on the way FitzGerald used the model in his own research; in the late 1880s it was the key to his reform of Maxwell's theory.

The path that led FitzGerald to his model began late in 1884, while he was pondering J. H. Poynting's recent theoretical discovery of the paths of energy flux in an electromagnetic field, (a discovery that was to have far-reaching effects on the way Maxwell's theory was understood). Oliver Heaviside, who discovered it independently a few months after Poynting, made the flux theorem the key to his recasting of Maxwell's theory into the now familiar form of "Maxwell's equations," and to a concomitant reform of Maxwell's treatment of propagation phenomena.[12] The flux idea played a similar role in FitzGerald's own revision of Maxwell's theory. But whereas Heaviside essentially passed straight from the flux theorem to his new set of propagation equations, Fitz-Gerald followed a less direct and in some ways more interesting route. In keeping with Kelvin's dictum that the best way to understand a phenomenon was to construct a mechanical model of it, FitzGerald set out to depict Poynting's energy flux with a model. It was through this model—the wheels and bands—that the flux theorem eventually led him to revise his whole approach to electromagnetic propagation.

FitzGerald hit upon his arrangement of wheels and bands at the very end of 1884, and he immediately sent an account of it to his friend Oliver Lodge. Lodge had already devised several ether models of his own, and FitzGerald wanted to be sure that his new model did not too closely resemble one of his friend's. "I have been constructing a model ether," he told Lodge, "and if it is the same as yours I want to give you credit for it, as I want it to illustrate Poynting's great discovery that the energy of an electric current must come in at its sides and is not carried along with the current."[13]

FitzGerald's model was very simple. "I propose," he told Lodge, "a series of wheels of course on fixed axes all connected in pairs by indiarubber bands." (See figure 1.) The wheels, when set spinning, represented a magnetic field, just as the vortices did in Maxwell's model; their rotational inertia represented self-induction. Instead of using idle wheels to convey motion from one vortex to the next, as Maxwell had, FitzGerald simply connected each wheel to its four neighbors with rubber bands. If all the wheels in a region were spinning at the same rate, there was no consequent strain on the bands. But if some wheels were turned through different angles than their neighbors—if, for instance, the wheels in one region were held fixed while those in another were turned—the bands would be strained. The elements of the medium would then be "polarized," FitzGerald pointed out, with one side of each band tightened and the other loosened. This represented an electric field, with the elasticity of the bands corresponding to the "dielectric constant" of the medium.

Figure 1
FitzGerald's ether model as viewed from above. Each wheel is connected by rubber bands to its four neighbors; all can spin freely together at the same rate without straining the bands. This spinning corresponds to a magnetic field.

All this can best be seen by examining how a specific phenomenon, such as the charging and discharging of a condenser, could be depicted with FitzGerald's model. Imagine two regions from which the bands have been removed; these represent perfect conductors. Then imagine them to be connected by a line along which the bands have also been removed; this represents a conducting channel, or a wire. If we now turn the wheels above this channel one way—clockwise, say—and those below it the other, the two conductors will become "charged" oppositely. (See figure 2.) Note that this "charge" appears purely as a reflection of conditions in the medium—in this case, the strain of the stretched rubber bands. This was very characteristic of the Maxwellian approach, which focused entirely on the field and looked upon charges and currents as purely derivative phenomena.

Because of the way the wheels are connected to one another throughout the region around the charged conductors, all the bands will be strained and will try to turn the wheels back toward their original positions; it is only by continuing to pull along the connecting channel with an "impressed force" (corresponding to a battery or a dynamo) that the charge can be maintained. But now imagine the bands along the channel to be replaced; this corresponds to insulating the two conductors by removing the connecting wire. The tension in the surrounding bands will now be partly spent in straining the replaced bands, and a self-locked system of strain will be established throughout the region between and around the conductors. (See figure 3.) Each rubber band

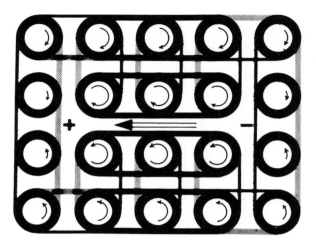

Figure 2

If a battery is connected between two conducting bodies, it will charge them oppositely and so set up an electrostatic field in the space between them. This figure shows
how this can be depicted on FitzGerald's model. The regions without bands represent
conductors—a small plate on each side and a channel or "wire" connecting them.
The large arrow in the middle represents the "impressed force" of the battery; when
applied along the wire, it causes the surrounding wheels to turn in the directions and
amounts shown by the curved arrows. Since neighboring wheels turn by different
amounts, the rubber bands connecting them are strained; the narrow lines correspond
to tightened bands and the shaded lines to loosened ones. Notice that all of the
bands around the left plate are loose, while those around the right plate are tight.
This corresponds to a positive "charge" on the left plate, as shown, and a negative
"charge" on the right. The pattern of strained bands between the two plates corresponds to an electrostatic field. Since the strained bands will try to turn the wheels
back to their original positions, the opposite charges of the plates will persist only as
long as the impressed force is applied.

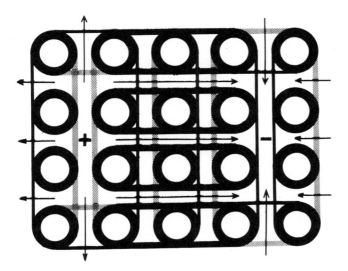

Figure 3

If the impressed force in figure 2 is removed and the missing bands along the conducting channel replaced—that is, if the battery wire is cut and the two plates insulated from each other, so that they form a condenser—the result is the self-locked system of strain shown here. The bands that were strained by the application of the impressed force will try to return to their original unstrained state, and in doing so will turn the wheels slightly and stretch the replaced bands until the tensions balance. The plates will retain their "charge" and will be surrounded by a pattern of strained bands corresponding to the electrostatic field around a charged condenser. The arrows indicate the lines of electric displacement, which run from the loose side of each band (shaded lines) to the tight side (narrow lines).

will be loose on the side toward the first conductor and tight on the side away from it. A vector drawn from the loose side to the tight side represents the electric "displacement" of each element; the pattern of lines thus generated is precisely that of the lines of force around two charged bodies. Notice that nothing is actually "displaced" along such a line of displacement; there is only a rearrangement of the tension, or "change in structure," of the elements along it. This was an important point, and FitzGerald later made much of it.

Now consider how FitzGerald's model depicts the discharge of a condenser and the accompanying flow of current and energy—the energy flow that had prompted FitzGerald to devise his model in the first place. Imagine the bands along a line connecting the two charged regions to be partly loosened, so that they slip slightly over the wheels and dissipate part of their energy in friction. This represents a conducting path, but one with some resistance; low friction corresponds to a good conductor, high friction to a bad one, and the heat generated by the rubbing of band against wheel represents the Ohmic heating of a current-carrying wire. With such a conducting path available, the strain in the medium is no longer self-locked; the bands along the conducting path will slip, representing a conduction current, while the wheels on either side will begin to turn in opposite directions, representing the accompanying magnetic field in the surrounding space. The energy stored in the strained bands will gradually be dissipated as heat in the conducting "wire"—that is, the region of slipping bands—and the medium will return to its original unstrained and motionless state. The energy comes in from the surrounding medium along the length of the strained bands and enters the "wire" from its sides, just as in the Poynting flux that FitzGerald had originally devised his model to illustrate. In the model, just as in an electromagnetic field, the energy does not flow along the line of the current, but rather in paths perpendicular both to the axes of the wheels (the direction of the magnetic force) and to the "polarization" line from the loose to the tight sides of the bands (the electric displacement vector).

The model could also illustrate such other characteristic Maxwellian phenomena as the generation of electromagnetic waves. If the slippage along a conducting path like that described above is not very great—if the "resistance" is low—then when the impressed force is removed, the inertia of the wheels will carry them past the equilibrium point and the wheels and bands will "bounce" from one state of strain to its opposite in oscillations of gradually decreasing energy and amplitude. Such an oscillatory discharge will cause the wheels in the surrounding field to bounce as well; they will turn slightly from side to side while the connecting rubber bands alternately tighten and loosen, and a wave of such changing motion and strain will spread out from the discharging wire. The motion of the wheels and the changing strain of the bands correspond to oscillating magnetic and electric fields, respectively; together they represent a train of electromagnetic waves spreading out from a discharging condenser in just the way FitzGerald had first described in 1882.[14] FitzGerald even showed how a sudden slippage of the bands would mimic a spark discharge. "Could anything be more complete!" he exclaimed to Lodge. "The whole thing seems a very fair representation of the ether."[15]

Indeed, the representation was more than "very fair"; it was exact. As early as 1885, in his first publication on the subject, FitzGerald showed that "the equations representing the energy of the model are of the same form as those of Maxwell representing the energy of the ether."[16] The potential energy of the strained bands was proportional to the square of their "displacement," just as in Maxwell's term for the electric energy; the kinetic energy of the spinning wheels was proportional to the square of their rotational velocity, just as the magnetic energy was to the square of the magnetic force. The Hamiltonian governing the behavior of the model was identical to that of the electromagnetic field, so that mathematical reasoning could be transferred freely between Maxwell's theory and FitzGerald's model.

Strictly speaking, the correspondence between theory and model was exact only when Maxwell's equations could be restricted to two dimensions; the wheels and bands became tangled if one tried to extend them to three dimensions. But FitzGerald devised another model, consisting of paddle wheels pumping fluid between elastic partitions, which, though far too complicated to be built, could be shown to yield Maxwell's energy terms for three dimensions. From this equivalence, FitzGerald pointed out, "it follows at once that all the results deduced from this form of the energy can be reproduced on a model," including all the phenomena of Maxwellian electrodynamics and of optics.[17] The correspondence between FitzGerald's paddle-wheel model and Maxwell's ether seemed to extend as far as one wished to take it.

The Vortex Sponge

Despite the success of his models, FitzGerald, like Maxwell, regarded them as no more than analogies, and he drew a careful distinction between such analogies and what he called "likenesses." The stretching of the bands was related to the turning of the wheels in the same way that electric force was related to magnetic force, but this was a similitude only of relations, not of things. "I do not in the least intend to convey the impression that the actual structure of the aether is a bit like [that of his model]," FitzGerald declared; if the ether could somehow be examined at great magnification, it would certainly not be found to be "actually made up of wheels and india-rubber bands, nor even of paddle-wheels, with connecting canals."[18] While a good model might be "so far analogous to [the ether] that we may in many respects reason safely from one to the other," FitzGerald warned against uncritically accepting such a model as a true depiction of reality—a "likeness." "The danger," he said, "is that we may be satisfied with an analogy, and mistake it for a likeness," and so leap to premature and erroneous conclusions about the nature of reality.[19] Models were useful, but they had to be used with care.

This did not mean that FitzGerald despaired of ever finding a true likeness of the ether, or believed physical reality to be at some level opaque to human understanding. Far from it. He believed the ether and the rest of the physical universe to be fully explicable in purely mechanical terms. Immediately after describing his own models as mere analogies, he declared: "What physicists ought to look for is such a mode of motion in space as will confer upon it the properties required in order that it may exhibit electromagnetic phenomena. Such a mode of motion would be a real explanation of the phenomena. I have only given a description of them."[20] FitzGerald even had a candidate for such a true likeness of the ether: the vortex sponge. He hit upon this picture of the ether as a turbulent liquid simultaneously with his invention of the wheel model; indeed, he seems to have regarded the vortex sponge as a physically plausible way to give the ether the same rotational elasticity as the wheels and bands. FitzGerald ended the letter to Lodge in which he first described his wheel model with a postscript on the vortex sponge. "I am just elaborating a theory of the ether," he said, "that requires it to be a perfect liquid chock full of vortices.... It works a long way without a hitch." As a fundamental hypothesis it was, he said, "simplicity itself."[21]

The hypothesis was indeed simple, though there was much room for complication in working it out. It started from the assumption that there was only one ultimate substance in the universe, a perfect continuous liquid pervading all space. This liquid was filled with energetic turbulent motion, and all physical objects and phenomena resulted from its states of motion—a strongly mechanist view that FitzGerald and many other British physicists found attractive on philosophical grounds.[22] The randomly distributed vortex rings and filaments in the liquid gave it an elasticity almost like that of a solid body, much as the random motion of particles makes a gas elastic. Moreover, FitzGerald's initial investigations pointed to a close parallel between the mechanics of the vortex sponge and those of Maxwell's field theory: the sponge could support rotations and polarizations corresponding to magnetic and electric forces, and could perhaps even propagate waves like those of light.[23] The same day he wrote to Lodge, FitzGerald sent a long letter about the sponge idea to J. J. Thomson, known then as an authority on vortex motion, and asked if there were any obvious objections to it.[24] Thomson apparently found none; FitzGerald published an account of the sponge hypothesis a short time later, and he continued to pursue it with unflagging enthusiasm to the end of his life.[25]

FitzGerald's faith in the vortex sponge came out clearly in a remark he made to Heaviside in 1893. "It is a pity," he said, "that someone with a store of patience and energy for indefinite calculation does not work on the average motion of liquids. There is evidently a field quite comparable to that of the average motion of particles which is the kinetic theory of gases and I feel horribly confident that it is *the* theory of the ether. I have a sort of feeling in my bones that it must be so...."[26] The vortex sponge was meant as a *likeness*; FitzGerald felt in his bones that the physical world really *was* just a turbulent liquid, and that "ether, matter, gold, air, wood, brains, are but different motions."[27] The vortex sponge went far beyond the analogies of FitzGerald's wheel model, or Maxwell's vortices and idle wheels, and offered a truly ultimate and unified mechanical theory of physical reality. Though eventually abandoned, it was perhaps the highest and most ambitious expression of the mechanical program that dominated late Victorian physics.

Changing Displacement

FitzGerald invented the vortex sponge to provide a plausible likeness of Maxwell's ether, and the paddle-wheel model to give a three-dimensional analogy to it, but he found the simpler wheel-and-band model adequate for most purposes, and he used it for many years as an aid to his own understanding. "I . . . am working more or less on my own model ether for concrete ideas,"[28] he told Heaviside in 1893 in the course of a discussion of the inertia of currents, and the model had its greatest value as such a testing ground for new ideas. FitzGerald could reason more vividly and concretely about the workings of his wheel model than about disembodied equations or the very complicated motions of the vortex sponge, while the demonstrated identity between the equations governing the model and those of Maxwell's theory ensured that insights gained from the model could be reliably applied to electromagnetic questions.

FitzGerald's model influenced his understanding of Maxwell's theory in several important ways. For one thing (as mentioned above), it helped convince him that Maxwell's term "electric displacement" should not be taken literally, and that "the word 'displacement' was unfortunately chosen."[29] Maxwell and his immediate followers had generally treated "displacement" as an actual motion of the ether, an idea that fit with the older forms of the wave theory of light.[30] But FitzGerald found that this literal interpretation led to contradictions, and he thought it "much more likely that what [Maxwell] called 'electric displacements' are changes in structure of the elements of the ether, and not actual displacements of the elements."[31] He used his wheel model to illustrate this point, showing how in it "electric displacement" arose from a change in the tension of a band rather from its actual displacement. The move away from a literal interpretation of "displacement" that FitzGerald was advocating marked a major step in the evolution of Maxwell's theory and helped free it from some of the inconsistencies of its original formulation.

The question of the true nature of displacement became an especially lively issue early in 1885 in the wake of Kelvin's Baltimore Lectures on the wave theory of light. Kelvin had essentially endorsed the elastic-solid theory of the ether, with light waves as real bodily vibrations in a "jelly." In one of his American lectures he went so far as to state flatly that "the luminiferous ether is an elastic solid," and to point to a bowl of jelly as "the nearest analogy [to the ether] I can give you."[32] But the elastic solid had serious shortcomings even as a purely optical ether; it was plagued by unwanted longitudinal waves, and it failed altogether to account for electromagnetic phenomena. If an elastic solid ether were not to interfere with the free motion of planets and other bodies, it would, Kelvin said, have to be a very soft substance like pitch or wax, able to sustain rapid vibrations but giving way with little resistance before a steady force.[33] But an ether that gave way so easily could not be the medium by which strong and steady electromagnetic forces were exerted; as FitzGerald noted, it was difficult to see "how such a soft material could be the means by which tramcars are driven by shearing stresses."[34] Kelvin's well-known rejection of Maxwell's electromagnetic theory of light in the Baltimore

Lectures is largely traceable to his faith in the elastic solid ether and his realization that Maxwell's theory was incompatible with it. But whereas Kelvin took this as grounds for rejecting Maxwell's theory, FitzGerald saw it as a strong argument in favor of abandoning the elastic solid ether and the literal interpretation of "displacement" that went with it.

FitzGerald brought this new view of the ether into sharp contrast with Kelvin's more orthodox position in two letters to *Nature* in the spring of 1885. When an account of the Baltimore Lectures had first appeared, FitzGerald had thought it must be mistaken; he could not believe that Kelvin had misunderstood Maxwell's theory of electromagnetic waves in the way the quotations seemed to indicate.[35] But Kelvin replied that he had indeed been quoted accurately, prompting FitzGerald to draft a longer letter to *Nature* in which he showed in some detail that the supposed waves in telegraph wires that Kelvin had described were quite different from the waves in the ether that were the focus of Maxwell's theory.[36] FitzGerald's real point, however, came in his last paragraph: "I cannot conclude without protesting strongly against [Kelvin's] speaking of the ether as *like* a jelly. It is in some respects *analogous* to one, but we certainly know a great deal too little about it to say that it is *like* one." The jelly ether was no more than a misleading analogy, he said, which had to be abandoned before Maxwell's theory could be properly appreciated.

"My letter to *Nature*," FitzGerald wrote to Kelvin, "is not intended to inform *you* of anything as I might as well teach my grandmother to suck eggs."[37] But he was convinced that Kelvin's continued advocacy of the jelly theory was a great mistake, and he feared that the acknowledged leader of British physics— a man for whom he had great respect—was "lending his overwhelming authority to a view of the ether which is not justified by our present knowledge, and which may lead to the same unfortunate results in delaying the progress of science as arose from Sir Isaac Newton's equally guarded advocacy of the corpuscular theory of optics."[38]

It was as a way to keep physicists' minds open to alternatives that FitzGerald thought his wheel-and-band and paddle-wheel models could be especially useful. "It is worth while considering these models," he said, "because in them the disturbance which represents light is not the same as the vibrations of an elastic jelly, for what represents an electric displacement is a change of structure of an element, and not a displacement of the element; and it seems almost certain that, notwithstanding the very high authority with seems to support the view that the aether is *like* an elastic jelly, nevertheless its vibrations are much more of the nature of alterations of structure than of displacements."[39] FitzGerald wanted the true nature of "displacement" to be left an open question; it should, he believed, be settled by further exploration and experiment rather than by appealing to its correlate in a possibly inappropriate model, such as a jelly. It is interesting that he used his own very specific model of the ether to make this point, and to show that electromagnetic and optical phenomena could be represented by a mechanism quite different from that of the orthodox elastic-solid theories. FitzGerald turned his wheel-and-band model into a powerful tool in his campaign in the 1880s to change the way "displacement" was interpreted in Maxwell's theory, and used it to help complete the break between the Maxwellian electromagnetic ether and the old jelly theories.

The "Murder of ψ"

FitzGerald's models had an even more striking influence on his understanding of electromagnetic propagation. Maxwell, in his *Treatise*, had used forms of the potential functions that were ill suited to treating propagation problems; his equations implied, for example, that the electric or scalar potential ψ was propagated instantaneously.[40] This seemed much like direct action at a distance, and it led to considerable confusion throughout the 1870s and the 1880s about the nature of propagation in Maxwell's theory. FitzGerald had himself been led astray on this question in his first papers on electromagnetic waves in 1879–1880, and though his introduction of a retarded form of the vector potential **A** in 1882 had corrected part of the problem, it did nothing to alter the presumption that the electric potential ψ was propagated instantaneously.[41]

The discovery of electromagnetic waves by Hertz and Lodge in 1888 focused new attention on electromagnetic propagation and touched off what FitzGerald dubbed the "murder of ψ debate" at the September meeting of the British Association at Bath. FitzGerald, who was president of the physics section that year, announced that Henry Rowland would give a paper in which he would "murder ψ"—that is, show that the electric potential had no place in the basic propagation equations. According to the report in *Engineering*, Rowland arrived late and "stated his case briefly and with hesitation, as Professor FitzGerald knew more about it than he did himself."[42] What Lodge described as "a sort of battle or impromptu discussion" about the fate of ψ ensued.[43] "Many deaths, and many bringings to life, this unfortunate symbol suffered in the course of the week," Lodge said.[44] The early stages of the debate were inconclusive, and according to *Engineering* "everybody expressed regret at the absence of Mr. Heaviside, and kept on his guard."[45] But the disputants, led by FitzGerald and Rowland, gradually came to agree that the form of the electric potential given by Maxwell offered only a very cumbersome way to treat changing fields. Despite some dissent from Kelvin, they decided to drop ψ from the basic propagation equations—a step that had far-reaching effects on the way Maxwell's theory was understood and expressed.

That Heaviside's name came up at Bath was significant. He had been working to reform Maxwell's treatment of propagation since 1885, mostly in connection with telegraph theory, and he followed reports of the debate closely from his home in London. In October 1888 he sent the *Philosophical Magazine* a note,

"On the Metaphysical Nature of the Propagation of Potentials," in which he argued that the confusion could be traced to Maxwell's mistake of taking the potentials **A** and ψ as the physical basis of his theory.[46] In Heaviside's view, **A** and ψ were merely mathematical artifices, occasionally convenient in calculations but "considerably remote from the vectors which represent the state of the field."[47] The physically significant vectors, he said, were the electric and magnetic forces **E** and **H**; it was **E** and **H** that determined such physical states of the field as the energy density and stress. Since "it is perfectly obvious that in any case of propagation ... it is a physical state that is propagated," it was just as obvious that "it is **E** and **H** that are propagated," not **A** and ψ.[48] Heaviside's 1884 discovery that the energy flux depended solely on **E** and **H** solidified his belief in the primacy of the forces and, in 1885, led him to make them the basis of a new way of expressing Maxwell's theory. He dropped the equations in **A** and ψ that Maxwell had given in his *Treatise* and recast the theory into the four symmetrical vector equations in **E** and **H** now universally known as "Maxwell's equations," which in free space take the form

$$\mathrm{div}\,\mu\mathbf{H} = 0, \quad -\mathrm{curl}\,\mathbf{E} = \mu\dot{\mathbf{H}},$$

$$\mathrm{div}\,c\mathbf{E} = \rho, \quad \mathrm{curl}\,\mathbf{H} = c\dot{\mathbf{E}}$$

(where μ is the permeability, c the permittivity, and ρ the charge density).[49] With these new equations, the problem of the propagation of potentials disappeared; the question simply "does not present itself as one for discussion," Heaviside noted, once **E** and **H** are installed as the fundamental quantities.[50] Reinforced by the parallel and partly independent work of Hertz,[51] Heaviside's equations became the standard form of Maxwell's theory by the mid 1890s; they continue to be the most widely used form of the field equations today.

The acceptance of Heaviside's new version of Maxwell's theory was delayed, however. Published early in 1885, Heaviside's equations remained almost unknown until late in 1888. A former telegrapher with no academic training or position, Heaviside was an outsider in the scientific community. It was only when the dramatic experiments of Hertz and Lodge forced physicists to search for theoretical tools appropriate to electromagnetic waves that Heaviside's papers began to attract attention. Even then, his work would not have been assimilated as quickly as it was after 1888 had it not been for the receptive attitude of a few leading physicists. Lodge, Hertz, and Kelvin all helped draw attention to Heaviside's work after 1888, but on the question of propagation theory his most important supporter was FitzGerald. In an influential review of Heaviside's *Electrical Papers* in 1893, he declared that "since Oliver Heaviside has written, the whole subject of electromagnetism has been remodelled by his work," and pointed in particular to the way Heaviside's equations in **E** and **H** clarified the treatment of propagation.[52] This endorsement, combined with FitzGerald's long-standing privately expressed support, helped greatly to make

Heaviside's innovations respectable to other physicists and to establish his new version of Maxwell's theory by the mid 1890s as the standard form.

Model Research

FitzGerald was especially receptive to Heaviside's work because his own thinking had evolved along a roughly parallel course. Indeed, it seems likely that he would have eventually carried through the substance of many of Heaviside's reforms on his own, though not in quite the same way. By 1888 he too had come to regard the potentials (especially ψ) as mere analytical tools with no independent physical significance, and from the Bath meeting on he played a leading role in promoting a shift away from regarding the potentials as the physical basis of Maxwell's theory and toward giving that role to the electric and magnetic forces and the energy flux. FitzGerald was first led to this new position by reasoning based on his wheel-and-band model, and he continued to rely on the model as one of his main guides through the intricacies of propagation theory.

The connection between FitzGerald's model and his revision of Maxwell's theory can be seen most clearly in a notebook of his that was recently brought to light at Trinity College Dublin.[53] In a forty-page section apparently dating from the summer of 1888, FitzGerald attacked propagation questions along two parallel paths, one purely mathematical and one based on the workings of his model. The model was well suited to the treatment of such questions, since it had been designed from the first so that it could clearly depict the propagation of energy. In an effort to trace a detailed correspondence between the equations of his model and those of Maxwell's theory, FitzGerald filled his notebook with sketches of wheels and bands and mathematical analyses of how they would work under various circumstances. He then used the insights he drew from the model to modify and correct Maxwell's electromagnetic equations. Much as Maxwell had done with his own vortices and idle wheels, FitzGerald turned his wheel-and-band model into a research tool, which he used not just to illustrate electromagnetic theory but also to improve it.

The fundamental insight FitzGerald drew from his model was quite simple, and he stated it early on in his notebook. Just three pages into his discussion, he interrupted some calculations on the relation between ψ and **A** to remark: "Still there remains that the wheels seem to give a correct representation of the ether and that there can be no instantaneous propagation of anything in them."[54] If his model indeed gave "a correct representation of the ether," as FitzGerald believed it did, then the equations that had led Maxwell to conclude that the electric potential was propagated instantaneously must contain some error. FitzGerald set out to find this error, and if possible to correct it.

It was at this point that FitzGerald began to sketch arrays of tiny wheels connected by rubber bands and to analyze their motions mathematically. He

found that these motions could be decomposed neatly into terms corresponding to the vector and scalar potentials only in the special case of a steady
field. When the fields varied, as when electromagnetic waves were passing
through, the potentials could not adequately represent propagation phenomena as they appeared on the wheel model. "There is nothing exactly like F,
G, H [the components of the vector potential $\mathbf{A}$] in my wheel system," Fitz-
Gerald noted at one point; elsewhere he remarked: "My coordinates $\alpha\beta\gamma$ are
much nicer ones to use as they represent something in my wheel medium."[55]
The vector (α, β, γ) represented the turning of the wheels and corresponded to
the time integral of the magnetic force; FitzGerald had used it before in this
latter sense in his purely theoretical work, as had Heaviside.[56] It was an important addition to the representation of the electromagnetic field (especially its
energy) in terms of potentials. Moreover, as Heaviside in particular came to
realize, by completing the symmetry of the potential equations it pointed the
way to the simpler scheme in which the potentials were dispensed with and
the field treated purely in terms of the electric and magnetic forces.[57]

FitzGerald's study of the workings of his model eventually led him to the
source of Maxwell's difficulties about instantaneous propagation. In deriving
his wave equation, Maxwell had introduced an auxiliary quantity J, defined as
the divergence of the vector potential: $J = \nabla \cdot \mathbf{A}$.[58] In the modern form of
classical field theory, the vector potential is regarded as having no direct
physical significance, and since its divergence drops out of the equations
connecting $\mathbf{A}$ to the observable forces, $\nabla \cdot \mathbf{A}$ is simply assigned whatever value
is most convenient for the problem at hand. This is called choosing a "gauge,"
and within broad limits it is a matter of free assumption.[59] But Maxwell regarded
the vector potential as far more than a calculational convenience; he called it
"the fundamental quantity in the theory of electromagnetism" and invested it
with great physical significance.[60] He thus felt compelled to treat its divergence with similar seriousness. Rather than simply assign it a value arbitrarily,
he sought to deduce its value from physical considerations. He was able to
show that in free space

$$\frac{d^2 J}{dt^2} + \frac{d(\nabla^2 \psi)}{dt} = 0.$$

He then made a very important assertion, for which he provided no real
justification: "$\nabla^2 \psi$, which is proportional to the volume-density of the free
electricity, is independent of t"—that is, the electric potential at a given
time and place depends only on the spatial distribution of charge at that
moment.[61] This was the assumption usually made in electrostatics, and Maxwell simply extended it to general electromagnetic theory without alteration or
explanation. Time independence implied that the electric potential adjusted
instantaneously across all space to any changes in the position of the charges;
it also implied that $d^2 J/dt^2 = 0$, so that "J must be a linear function of t,

or a constant, or zero, and we may therefore leave J and ψ out of account in considering wave disturbances."[62] In practice Maxwell took $J = 0$, and so he worked in what would now be called the "Coulomb gauge"—a gauge well suited to electrostatic problems but with the serious drawback in treating changing fields that it required electric potential to be propagated instantaneously.[63]

FitzGerald had already found that there was "no instantaneous propagation of anything" in his wheel model. In tracing the correspondence between his model and Maxwell's equations he found it very inconvenient, even contradictory, to assume, as Maxwell had, the $d(\nabla^2 \psi)/dt = 0$ and $J = 0$. "The great difficulty," he said, "is the $dF/dx + dG/dy + dH/dz = 0$,"[64] i.e., $\nabla \cdot \mathbf{A} = J = 0$. This condition had no counterpart in the wheel model, and it was on this point that FitzGerald broke decisively with Maxwell's treatment of propagation. "My wheel motion ought to succeed," he declared, but only if Maxwell's assumption that J and ψ were independent were replaced by the relation $J = -d\psi/dt$, or, expressed another way, $\nabla \cdot \mathbf{A} + d\psi/dt = 0$.[65] Later to become known as the "Lorentz gauge," this was much better suited to treating propagation phenomena than was Maxwell's Coulomb gauge, $J = 0$. FitzGerald's condition, when combined with Maxwell's equation $d^2 J/dt^2 + d(\nabla^2 \psi)/dt = 0$, implied that in free space

$$\nabla^2 \psi - \frac{d^2 \psi}{dt^2} = 0$$

and

$$\nabla^2 \mathbf{A} - \frac{d^2 \mathbf{A}}{dt^2} = 0.$$

These homogeneous wave equations imply that the potentials, particularly ψ, are propagated in a way completely different from that assumed by Maxwell. Rather than adjust instantaneously across all space, ψ is now seen to propagate, along with $\mathbf{A}$, $\mathbf{E}$, and $\mathbf{H}$, in waves traveling at the speed of light.[66] The new "gauge" thus eliminated any question of the instantaneous propagation of electric potential.

FitzGerald saw how crucial this change was. During the "murder of ψ" debate at the Bath meeting he went back through his notebook and circled his derivation of the new equation for J. He scrawled beside it: "very important. 9.9.88. Must be all in O. Heaviside."[67] This was apparently the first time FitzGerald had taken notice of Heaviside's work, and it helped inaugurate the very fruitful scientific, and later personal, relationship between the two men. But FitzGerald was always rather slow to publish his work, and he did not present his results publicly until two years later, at the 1890 Leeds meeting of the British Association. In a short paper, "On an Episode in the life of J (Hertz's Solution of Maxwell's Equations)," he pointed out the source of Maxwell's difficulties and explained the advantages of taking $J = -d\psi/dt$ rather than $J = 0$.[68] He showed how the revised equations could be used to treat Hertzian waves and other propagation phenomena, and pointed out explicitly how the new equations in **A** and ψ could be used to calculate the electric and magnetic forces **E** and **H**, which he recognized as the variables of real physical interest. In the meantime, Heaviside and Hertz had developed equivalent methods independently. The real lesson from the work of all three men was that it was best to regard the potentials as no more than auxiliary functions in propagation problems, and to focus attention instead on forces and fluxes.

Conclusion

Beyond the intrinsic importance of his revision of Maxwell's equations, the point of most interest in FitzGerald's treatment of propagation questions is the use he made of his model. "I would now explain how my model was right," he wrote in remarks intended for the discussion at Bath,[69] and this confidence that his model represented the electromagnetic field more faithfully than did Maxwell's original equations was the key to FitzGerald's whole investigation. The wheel-and-band model was more concrete and familiar than a set of bare equations, and so could be followed more easily; in using it one was not as liable to get lost in tracing the implications of one's premises and land in an inconsistency, as had happened to Maxwell in 1864 and again in his *Treatise* when he had tried to derive the wave equation directly from the electromagnetic equations. "In the case of a mechanical model of the ether," FitzGerald observed, "we have before us a structure which we may easily conceive, and with the method of whose working we are familiar; and so we can reason about it and discover what it should do without being troubled at every turn to realise our analysis, for the realisation is so easy and familiar that it gives our minds no trouble."[70] The workings of a mechanical model could be followed easily and in detail in one's mind's eye, or, if one went so far as to build it, on a lecture table; the relation of its parts was explicit and concrete. Even if it were not an actual likeness of the ether, a good model could bring out analogies and relations that might remain hidden if one worked with equations alone.

Thus, Maxwell was led to his hypothesis of displacement currents and so to the electromagnetic theory of light simply by following out the workings of his idle-wheel model, and FitzGerald was led to revise the propagation equations after a close examination of his wheel-and-band model. As aids to clear thought, mechanical models were complementary to mathematical equations. Indeed, it should be borne in mind that the equations of physics are not themselves phenomena but only mathematical models of them, and in an age in which reality was believed to be ultimately mechanical, it was only natural for physicists to construct mechanical models of phenomena as well, and to regard them as legitimate and valuable steps toward a true physical theory.

Martin J. Klein has aptly described Kelvin's Baltimore Lectures, with their profusion of intricate mechanical models, as an example of "the High Baroque phase of the mechanical world view."[71] In the work of such men as Lodge, the exuberance of Victorian model building may even have verged on the rococo.[72] But all these mechanical models, even those that may now seem merely odd or fanciful, were originally devised for the serious purpose of advancing scientific understanding, and they fulfilled that purpose with considerable success. The advances made with the help of mechanical models have long since been put into different dress; no one today would depict the electromagnetic field in terms of brass wheels and rubber bands. But if, like Maxwell, we "wish to study the growth of ideas as well as the calculation of forces,"[73] we must recognize that Victorian physicists looked upon mechanical models as tools not just of exposition but also of discovery, and that, as the work of Maxwell and FitzGerald amply demonstrates, they were indeed able to draw from these models valuable insights into the workings of the world.

Acknowledgments

I would like to thank the following libraries and archives for permission to use manuscripts in their possession: Department of Physics, Trinity College Dublin; Royal Dublin Society; University College London; Institution of Electrical Engineers; and University Library, Cambridge. I would also like to thank the Smithsonian Institution for fellowship support while this paper was being written, and the Cornell University Press for permission to publish here material that will appear in revised form in my book *The Maxwellians*.

References

1. Jed Z. Buchwald, *From Maxwell to Microphysics: Aspects of Electromagnetic Theory in the Last Quarter of the Nineteenth Century* (University of Chicago Press, 1985), p. 68.

2. See the very full bibliography in M. Norton Wise, "The Maxwell Literature and British Dynamical Theory," *Historical Studies in the Physical Sciences* 13 (1982): 175–205.

3. For a fuller discussion of FitzGerald's career see Bruce J. Hunt, The Maxwellians, Ph.D. dissertation, Johns Hopkins University, 1984.

4. Ole Knudsen, "The Faraday Effect and Physical Theory, 1845–1873," *Archive for the History of the Exact Sciences* 15 (1976): 235–281.

5. James Clerk Maxwell, "On Physical Lines of Force," *Phil. Mag.* 21 (1861): 161–175, 281–291, 338–348 and 23 (1862): 12–24, 85–95, repr. Maxwell, *Scientific Papers*, ed. W. D. Niven (Cambridge University Press, 1890), vol. 1, pp. 451–513. For an analysis of this paper see Daniel M. Siegel, "The Origin of the Displacement Current," *Historical Studies in the Physical Sciences* 17 (1986): 99–146.

6. Maxwell, "Physical Lines," *Sci. Papers*, vol. 1, p. 486.

7. Maxwell, "A Dynamical Theory of the Electromagnetic Field," *Phil. Trans.* 155 (1865): 459–512, repr. *Sci. Papers*, vol. 1, 526–597.

8. These difficulties are discussed in Alfred M. Bork, "Maxwell and the Electromagnetic Wave Equation," *American Journal of Physics* 35 (1966): 844–849.

9. See, for example, Joseph Turner, "A Note on Maxwell's Interpretation of Some Attempts at Dynamical Explanation," *Annals of Science* 11 (1956): 238–245, esp. p. 240; L. Pearce Williams, *The Origins of Field Theory* (New York: Random House, 1966), pp. 133–135.

10. James Clerk Maxwell, *Treatise on Electricity and Magnetism* (Oxford: Clarendon, 1873), vol. 2, art. 831.

11. FitzGerald to Oliver Lodge, March 8, 1894, Lodge Collection, University College London. The model was displayed at the Royal Dublin Society for the FitzGerald Centenary in 1951 (*Scientific Proceeding of the Royal Dublin Society* 26: 2). According to E. T. S. Walton it was at Trinity College Dublin as late as the 1960s, but a search of the attics and storerooms of the Trinity College Physics Building in 1982 failed to locate it.

12. Oliver Heaviside, *Electrical Papers* (London: Macmillan, 1892), vol. 1, pp. 429–451 (first publ. 1885). See also Hunt, Maxwellians (ref. 3), pp. 76–121.

13. FitzGerald to Lodge, January 1, 1885, Lodge Collection, University College London. See also FitzGerald, "On a Model Illustrating some Properties of the Ether," *Scientific Proceedings of the Royal Dublin Society* 4 (Jan. 19, 1885): 407–419, repr. in *Scientific Writings of the Late George Francis FitzGerald*, ed. J. Larmor (Dublin: Hodges and Figgis, 1902), pp. 142–156.

14. On FitzGerald and waves, see Hunt, Maxwellians, pp. 49–75.

15. FitzGerald to Lodge, January 1, 1885, Lodge Collection, UCL.

16. Abstract of FitzGerald, "On a Model ..." (ref. 13), *Nature* 31 (March 26, 1885): 498–499.

17. FitzGerald, "On a Model ..." (ref. 13), repr. in *Sci. Writings*, p. 149.

18. FitzGerald, "On the Structure of Mechanical Models illustrating some Properties of the Aether," *Phil. Mag.* 19 (June 1885): 438–443, repr. in *Sci. Writings*, pp. 157–162 (see p. 162).

19. FitzGerald, "Foundations of Physical Theory: Function of Models," *Sci. Writings*, pp. 163–169, at 167.

20. FitzGerald, "On the Structure ..." (ref. 18), p. 162.

21. FitzGerald to Lodge, January 1, 1885, Lodge Collection, UCL.

22. On FitzGerald's radical mechanist leanings and their relation to the philosophical views of his uncle G. J. Stoney, see Hunt, Maxwellians, pp. 169–175.

23. FitzGerald, "On a Model...," *Sci. Writings*, pp. 154–156; "A Hydrodynamical Hypothesis as to Electromagnetic Actions," *Scientific Proceedings of the Royal Dublin Society* 9 (December 21, 1898): 50–54, repr. in *Sci. Writings*, pp. 472–477.

24. FitzGerald to J. J. Thomson, January 1, 1885, Thomson Collection, Cambridge.

25. FitzGerald, "On a Model...." One of FitzGerald's last and strongest endorsements of the sponge came in an address delivered on February 22, 1900: "The Applications of Science. A Lesson from the Nineteenth Century," *Journal of the Institution of Electrical Engineers* 29 (1900): 394–408, repr. in *Sci. Writings*, pp. 487–499; see esp. pp. 489–490.

26. FitzGerald to Heaviside, August 25, 1893, Heaviside Collection, IEE.

27. FitzGerald, "Electromagnetic Radiation," *Proceedings of the Royal Institution* 13 (March 21, 1890): 77–84, repr. in *Sci. Writings*, pp. 266–276 (see p. 276).

28. FitzGerald to Heaviside, June 21, 1893, Heaviside Collection, IEE.

29. FitzGerald, "Sir Wm. Thomson and Maxwell's Electro-magnetic Theory of Light," *Nature* 32 (May 7, 1885): 4–5, repr. in *Sci. Writings*, pp. 170–173 (see p. 173).

30. This point is discussed in E. T. Whittaker, *History of Theories of the Aether and Electricity*, second edition (New York: Philosophical Library, 1951), vol. 1, 266n. For an explicit statement by one of Maxwell's former students that "displacement" was understood literally, see R. T. Glazebrook to FitzGerald, March 5, 1881, FitzGerald Collection, Royal Dublin Society.

31. G. M. Minchin found in 1885 that a literal interpretation of "displacement" led to a contradiction between Maxwell's stress formula and the laws of statics; see Minchin to Lodge, September 5, 6, and 30, 1885, Lodge Collection, UCL. In the last of these letters Minchin reported discussing the matter with FitzGerald, who, "if hard pressed," escaped the contradiction by saying "that 'the electric displacement is not a displacement of the Ether.'"

32. Sir William Thomson (Lord Kelvin), "The Wave Theory of Light" (lecture at Philadelphia, September 29, 1884), *Nature* 31 (November 27, December 4, 1884): 91–94, 115–118 (see p. 115).

33. Ibid., and Lecture I of the present volume.

34. FitzGerald, "The Relations between Ether and Matter" (review of J. Larmor, *Aether and Matter*), *Nature* 62 (July 19, 1900): 265–266, repr. in *Sci. Writings*, pp. 505–510 (see p. 508).

35. George Forbes, "Molecular Dynamics" (account of Kelvin's Baltimore Lectures), *Nature* 31 (March 19, 1885): 461–463; 31 (April 2, 1885): 508–510; 31 (April 30, 1885): 601–603; FitzGerald, "Molecular Dynamics," *Nature* 31 (April 2, 1885): 503.

36. FitzGerald, "Sir Wm. Thomson..." (ref. 29).

37. FitzGerald to Kelvin, April 25, 1885, Kelvin Collection, Cambridge.

38. FitzGerald, "Sir Wm. Thomson...."

39. FitzGerald, "On the Structure..." (ref. 17), repr. in *Sci. Writings*, p. 162.

40. Maxwell, *Treatise*, vol. 2, art. 783; see also Bork, "Wave Equation" (ref. 8), pp. 846–848.

41. See Hunt, Maxwellians, pp. 49–67.

42. "The British Association," *Engineering* 46 (October 12, 1888): 352.

43. Oliver Lodge, *Advancing Science* (London: Benn, 1931), p. 99.

44. Lodge, "Sketch of the Electrical Papers in Section A at the recent Bath Meeting of the British Association," *Electrician* 21 (September 21, 1888): 624–625.

45. "The British Association" (ref. 42).

46. Heaviside, "On the Metaphysical Nature of the Propagation of Potentials," *Phil. Mag.* 27 (January 1889): 47–50, repr. in *Elec. Papers* (ref. 12), vol. 2, pp. 483–485.

47. Heaviside, *Electromagnetic Theory* (London: Electrician Co., 1893–1912), vol. 1, p. 127.

48. Heaviside, *Elec. Papers*, vol. 2, p. 483.

49. Ibid., vol. 1, pp. 429–451.

50. Ibid., vol. 2, p. 483.

51. See Hertz, "Ueber die Grundgleichungen der Electrodynamik für ruhende Körper," *Annalen der Physik* 40 (1890): 577–624, repr. as "On the Fundamental Equations of Electromagnetics for Bodies at Rest" in Hertz, *Electric Waves*, tr. D. E. Jones (London: Macmillan, 1893), pp. 195–240. Hertz acknowledges Heaviside's earlier work; see pp. 196–197. See also Hunt, Maxwellians, pp. 245–246, 280–283, and 307–312.

52. FitzGerald, review of Heaviside's *Electrical Papers*, *Electrician* 31 (August 11, 1893): 389–390, repr. in *Sci. Writings*, pp. 292–300.

53. Five of FitzGerald's notebooks are held by the Department of Physics at Trinity College Dublin, where I saw them in 1982. They have not been cataloged, but they are briefly described and my system for references to them explained in a "Note on Manuscript Sources," in Hunt, Maxwellians, p. 341.

54. FitzGerald notebook 4, p. 10, Physics Dept., Trinity College Dublin.

55. Ibid., pp. 32, 17.

56. FitzGerald, "On the Electromagnetic Theory of the Reflection and Refraction of Light," *Phil. Trans.* 171 (1880): 691–711, repr. in *Sci. Writings*, pp. 45–73 (see p. 47); Heaviside, *Elec. Papers*, vol. 1, pp. 466–468.

57. Heaviside, *Elec. Papers*, vol. 2, pp. 172–175.

58. Maxwell, "Dynamical Theory" (ref. 7); *Sci. Papers*, vol. 1, pp. 578–582; *Treatise*, vol. 2, arts. 616 and 783.

59. For a modern discussion of gauge conditions see Melba Phillips, "Classical Electrodynamics," in *Encyclopedia of Physics*, vol. 5, ed. S. Flügge (Berlin: Springer Verlag, 1962), pp. 46–48. Some effects in quantum physics depend directly on the vector poten-

tial; see R. P. Feynman, *The Feynman Lectures* (Reading, Mass.: Addison-Wesley, 1965), vol. 2, pp. 15-8–15-14.

60. Maxwell, *Treatise*, vol. 2, art. 540, quoted in Alfred M. Bork, "Maxwell and the Vector Potential," *Isis* 58 (1967): 210–222.

61. Maxwell, *Treatise*, vol. 2, art. 783.

62. Ibid. See also Bork, "Wave Equation" (ref. 8), pp. 847–848.

63. Phillips, "Classical Electrodynamics" (ref. 59), p. 48.

64. FitzGerald notebook 4, p. 32, Physics Dept., Trinity College Dublin.

65. Ibid., p. 28.

66. Ibid., pp. 44–45.

67. Ibid., opp. p. 37.

68. FitzGerald, "On an Episode in the life of J (Hertz's Solution of Maxwell's Equations)," *British Association Report* (1890): 755–757. Note a misprint on 755; it should read $J = -d\psi/dt$, not $J = d\psi/dt$. This paper was not reprinted in *Sci. Writings*, but it is discussed in Lodge, *Advancing Science*, pp. 136–137.

69. FitzGerald notebook 4, p. 47, Physics Dept., Trinity College Dublin.

70. FitzGerald, "Foundations..." (ref. 19), pp. 168–169.

72. Martin J. Klein, "Mechanical Explanation at the End of the Nineteenth Century," *Centaurus* 17 (1972): 58–82, at p. 73.

72. See Oliver Lodge, *Modern Views of Electricity* (London: Macmillan, 1889).

73. Maxwell to William Thomson (Kelvin), November 13, 1854, in *Origins of Clerk Maxwell's Electric Ideas as Described in Familiar Letters to William Thomson*, ed. J. Larmor (Cambridge University Press, 1937), p. 7.

M. Norton Wise
Crosbie Smith

The Practical Imperative: Kelvin Challenges the Maxwellians

My letter to "Nature" is not intended to inform *you* of anything as I might as well teach my grandmother to suck eggs.

George Francis FitzGerald to Kelvin, April 25, 1885

When Lord Kelvin published his *Baltimore Lectures*, in 1904, the era of dynamical theories of the ether had long since passed. Joseph Larmor expressed the consensus of the British physics community when he wrote in his review:

In this chain of simple, yet brilliant and attractive, ideas, Lord Kelvin has gradually forged a reconciliation between fact and theory that would probably have been received with universal acclaim thirty years ago. Nowadays, as regards most people, the need has ceased to be so strongly felt; for better for worse most of us are now wedded to the electric theory of light, the creation of Lord Kelvin's most famous disciple [Maxwell], which forms a consistent scheme of the relations of electricity and radiation. . . .

Larmor recognized that Kelvin "would perhaps say that it is a successful description rather than an explanation," the true explanation lying rather in dynamics. "And here," he insisted, "we are at the parting of the ways. Is it incumbent on us to treat the aether as strictly akin to the material bodies around us? or may we assign to it a constitution of its own, to be tested by its success in comprehending the complex of known relations of physical systems?" [1]

This change in goals, from dynamical models to mathematical structures, is of immense importance in the history of classical physics and in the emergence of the modern theories of relativity and quantum mechanics. Like Larmor, historians have often taken the difference between Kelvin and Maxwell as symptomatic and have sought a concrete grasp of the larger movement in an analysis of their styles.[2] Within this genre of studies, no problem has received more attention than the status of the displacement current ("Maxwell's only real discovery," as G. F. FitzGerald put it[3]). This mysterious electric current supposedly derived from displacements in the ether that occured when the electrical tension through it was increasing or decreasing. From

Kelvin's perspective, the displacement current constituted not a discovery, but an unfortunate invention, possessing neither empirical nor dynamical foundation. Nevertheless, it made possible Maxwell's symmetric picture of the propagation of light waves as a reciprocal induction of electric and magnetic waves in the ether. Maxwellian modernists accepted it for the beauty and success of the theory that surrounded it, despite its otherwise weak physical grounding.

Kelvin saw the issue as one of loyalty to the true goals of natural philosophy. The new breed were defectors who had lost touch with reality, having adopted a "nihilistic" attitude to physical theory that threatened to sink them in the metaphysical mire of Hegel and the German *Naturphilosophen* who conflated the realms of ideas and matter. Writing to Larmor in 1906 with advice for an article on energetics, Kelvin complained: "Young persons who have grown up in scientific work within the last fifteen or twenty years, seem to have forgotten that enery is not an absolute existence. Even the Germans laugh at the 'Energetikers.' I do not know if even Ostwald knows that energy is a capacity for doing work; and that work done implies mutual force ... between two bodies or two pieces of matter, or between two atoms of matter, or ... between two electrons."[4] Any concept that defied explanation within the dynamics of matter had no place in Kelvin's physics. The same proscription applied to mathematically defined entities such as vectors. Concerning a remark by Rayleigh, for example, critical of G. G. Stokes' ether theorizing, Kelvin said: "That belittlement is a consequence of the nihilistic statement in the so-called electromagnetic theory of light as to 'magnetic vector' and 'electric vector' to which Rayleigh long ago succumbed."[5]

Kelvin's position can be fruitfully discussed from metaphysical, methodological, and institutional points of view. All of those perspectives, however, must incorporate an overriding interest, an interest in the practicality and utility of science, for Kelvin held firmly to the view that practical reality provided the only valid source of scientific knowledge. His own professional career epitomizes that position. It illustrates also how the twin goals of industrialization and empire served as generative and legitimizing ideals for some parts of Victorian science. Kelvin's views on mathematical and experimental physics incubated in the industrial city of Glasgow and touched its needs at every point. He won his knighthood for his work on the Atlantic telegraph and first acquired wealth by supplying the instruments that made it practicable. Recounting in 1890 the early history of that project, he proudly united scientific enterprise with empire, noting that even the failed cable of 1858 had transmitted "some exceedingly important public messages that saved many thousands of pounds, as countermanding the moving of two regiments ordered over from Canada to help to quell the Indian Mutiny."[6] He attained the peerage partly by supplying the British naval and merchant fleets with advanced navigational instruments and partly by his active support of empire at home in the liberal unionist cause. With self-conscious consistency he carried the values that informed his social-political position, and that underlay his economic success, deep into the theoretical structures with which he would explain the properties of matter and ether.[7]

To construct physical theory was, for Kelvin, to practice the engineering of nature. Indeed, he drew the model for his conception of trans-ether light signals directly from his work on trans-ocean telegraphy. Understanding Kelvin's use of this analogy will illuminate his views on displacement current. Maxwell, by contrast—descendent of the Scottish landed gentry and in happy receipt of an independent income—had not the same personal stake in industry and empire. Fascinated with the workings of mechanical systems, he both derived inspiration from them and elaborated the principles of their action. But they did not serve the essential function, either in his science or his life, that they did for Kelvin.[8] Although we shall not develop the thesis here, it may be that the "Maxwellization" of British physics in the late nineteenth century should be understood in part as the emergence of an academic profession separated from the immediate needs of industry, one that identified more nearly with the gentrified interests of Cambridge than with those of commercial Glasgow.[9] This is not to say that industrialization ceased to matter in physics, but that physicists learned to split off pure from applied research and to identify their professional interests with purity. The unity of theory and practice that Thomson always advocated, and that his entire career represented, did not find expression in the values of the newly emerged profession.

Telegraph Theory: The (Non-) Relation of Induction and Conduction

In 1854 the young Professor William Thomson worked out his theory of the electric telegraph. It was designed to explain the retardation of signals in underground and undersea cables, which threatened the profitability of long-distance lines. Such cables contained a central conductor of copper surrounded by gutta-percha insulation and sheathed in iron. Thomson had already calculated the enormous electrostatic inductive capacity of such cables, treating them as very long Leyden jars. This calculation rested on an analogy between Faraday's theory of electrostatic induction and Fourier's theory of steady-state heat conduction that in 1845 had established Thomson's scientific reputation.[10] That analogy made thermal conductivity stand for inductive capacity.

Stimulated in 1854 by a paper on retardation by Faraday and by a letter on the subject from Stokes, Thomson realized that transmission of a signal along the wire would occur in the same way as conduction of heat along a rod. In this dynamic analogy, one could treat inductive capacity as analogous to thermal capacity, and electrical conductivity as analogous to thermal conductivity, to obtain Fourier's well-known diffusion equation for the linear conduction of heat. Just as the temperature in any small section of the rod would increase only as fast as heat conducted to that section could fill its capacity for storing heat, so the potential of the central conductor would increase at a rate proportional to the rate at which electricity entered the section and inversely proportional to the capacity of the section,

$$\frac{\partial v}{\partial t} = \frac{1}{ck}\frac{\partial^2 v}{\partial x^2},$$

where k is resistance per unit length, c capacity per unit length, v potential, and x the distance along the cable. In letters to Stokes, which he soon published, Thomson concluded that the retardation of signals in a wire of length l would be proportional to kcl^2, which converted the question into one of economics: "It will be an economical problem, easily solved ... to determine the dimensions of wire and covering which, with stated prices of copper, gutta-percha, and iron, will give a stated rapidity of action with the smallest initial expense."[11] Thomson's entrepeneurial activities began with this calculation.

Two features of Thomson's theory require notice. First, he employed two different heat analogies, one for static induction and another for dynamic conduction. Conduction required time; induction did not. Faraday had often insisted on time for induction, and Stokes raised the issue in his letter, but Thomson reduced induction to a static parameter: capacity. Second, although Stokes had explicitly regarded the charging of the cable by a battery as a closed-circuit problem (consisting of conduction down the cable, induction through the insulator, return conduction through the sheath or sea, and conduction through the battery), Thomson's analysis ignored the return circuit. Charging became simply diffusion down the wire. By making these two choices he had eliminated the two factors that would form the basis of Maxwell's alternative theory of charging. Displacement current, for Maxwell, would close the circuit through the insulator and would require time for propagation.

Thomson's approach was essentially practical. He ignored resistance in the sea and in the battery because his knowledge of the magnitudes involved told him that the retardation caused by these factors would be negligible in comparison with that due to resistance in the wire. Similarly, no one had ever measured a time for induction. Thus he eliminated a potential conceptual problem that three of Thomson's close compatriots found pressing and that one of them turned into the electromagnetic theory of light. Thomson had attacked the problem of retardation. Having explained it theoretically, he immediately set out to patent his discovery and to capitalize on it, thereby turning theory into material reality of the most concrete variety. The implications of his practical approach to physical theory are much more subtle than this directly economic imperative would imply, but one must not suppose those subtleties, which we now examine, to have been independent of the industrial ideal.

Thomson's treatment of telegraph signaling displays a feature characteristic of his analogical method in physical theory. He did not allow new physical entities to be introduced by analogy. That stricture was apparent in 1845 when he employed the heat-flow analogy to reconcile Faraday's experimental discovery of inductive capacity in dielectrics, and his associated theory of propagated electrical action, with the reigning mathematical theory of electrical forces acting at a distance. Both Faraday's theory of lines of force and Fourier's theory of heat flow contained a distinction between an *intensity* (tension or temperature gradient, measured longitudinally along the lines of propagation) and a *guantity* (number of lines or flux, measured laterally over a cross section). Electrical action at a distance, however, involved no distinction between quantity and intensity of force. Force was potential gradient, an intensity. To effect his reconciliation, Thomson showed that change in inductive capacity could always be replaced by new sources of force (induced electricity due to polarization of the dielectric). That replacement eliminated the distinction between intensity and quantity of force by reducing their proportionality to an equality. Rather than

Quantity = Intensity × Inductive capacity,

Thomson had

Quantity = Intensity.

He could then speak freely—though ambiguously—of the analogy of force to flux without having to introduce into electrostatics any physical distinction between intensity and quantity of force.[12]

By 1854 Thomson had employed analogies in this restricted fashion to relate electric and magnetic action to both fluid flow and elastic deformation of solids, each time either reducing a quantity-intensity pair to an identity or choosing one member of the pair as the analogue of force while disregarding the other. It is not at all surprising, therefore, that in applying the analogy of steady-state heat flow to electrostatic induction in telegraph cables Thomson did not employ a quantity-intensity distinction, even though Faraday had made that duality the centerpiece of his own analysis of retardation and even though Stokes had reminded Thomson of the closed-circuit picture of charging current. In other words, it is not surprising that Thomson did not invent Maxwell's duality of electric force and electric displacement, with displacement current (changing displacement) closing the charging circuit.

This is not to say that Thomson should have invented Maxwell's device, but to indicate why he never would have and why he always rejected that device. His unified vision of theory and practice would not allow invention by analogy of nonempirical, and thus nonpractical, physical entities. The identity suggested here of "practical" and "empirical" points toward Thomson's well-known ideal of measurement: "I often say that when you can measure what you are speaking about and express it in numbers you know something about it; but when you cannot measure it, when you cannot express it in numbers, your knowledge is of a meagre and unsatisfactory kind." Thus, electricial measurement was a "practical science" in the best sense, for "the life and soul of science is its practical application." [13]

One understood a thing when it could be measured, not before. Thomson dedicated his laboratory at Glasgow to measuring the properties of matter, often aiming to apply the measurements commercially. He devoted great effort, for example, to measuring the conductivities of copper wires under various conditions of temperature, tension, and twist, partly to develop the empirical basis for a theoretical understanding of those properties and partly to perfect telegraph cables. Both interests were, in his vocabulary, "practical." No one, on the other hand, had ever measured, or even contemplated measuring, a "quantity" of electrostatic force distinct from its "intensity." Maxwell's displacement current was pure fantasy.

Thomson's notes and letters contain a variety of remarks on Maxwell's use of quantity-intensity pairs (or flux-force pairs, as Maxwell also called them). In his copy of Maxwell's *Treatise*, for example, Thomson simply wrote next to the listings of both electrostatic and magnetic pairs: "not usefully a 'pair'." [14] In a well-known letter to FitzGerald in which he complained generally about the nihilism involved in accepting Maxwell's symmetric treatment of electric and magnetic pairs, Thomson was more strident:

It is not the equations I object to. It is the being satisfied with them and with the pseudo-symmetry (pseudo, I mean, in respect to the physical subject) between electrostatics and magnetism. I also object to the damagingly misleading way in which the word "flux" is often used, as if it were a physical reality for electric and magnetic force, instead of merely an analogue in an utterly different physical subject for which the same equations apply [citing here his original heat flow analogy of 1842].[15]

Displacement of some kind there had to be, he acknowledged, perpendicular to the direction of propagation of light waves, but what relation it bore to electricity was a question with no hint of an answer.

Thomson's rejection of Maxwell's displacement current left him with no theory of the electromagnetic nature of light. In that circumstance he turned repeatedly to the practical reality of the telegraph, extracting ideas directly and by analogy. In fact, he often suggested that he had come very close in the telegraph theory to Maxwell's discovery that electromagnetic waves would propagate at the velocity of light. In the fourth of his Baltimore Lectures he remarked:

I am quite conscious ... of what has been done in the so-called electro-magnetic theory of light. I know the propagation of electric impulse along an insulated wire surrounded by gutta percha, which I worked out myself about the year 1854 and in which I found a velocity comparable with the velocity of light. We then did not know the relation between electrostatic and electro-magnetic units. If we had, that [velocity] might have been obtained [,] in the way that Maxwell has brought out so beautifully [,] from the proper coefficients of capacity for the gutta percha. If we work that out for the case of air instead of gutta percha, we get practically the same v, I think, for the velocity of propagation of the impulse.

Thomson's telegraph equation involved the ratio of electrostatic units (for induction) to electromagnetic units (for conduction). To him, Maxwell's real accomplishment lay in establishing that ratio as the velocity of light. Like all other fundamental discoveries, this one emerged from a precise measurement of the properties of matter. It was a practical discovery, one that he would himself have made had he been able to obtain better data.[16]

Thomson's remarks are surprising because his theory was entirely different from Maxwell's. To compare them we must first extend Thomson's telegraph theory in the way he outlined in his 1860 article "Velocity of Electricity"[17] three years before Maxwell's great discovery. He there conceptualized the basic "properties of electricity" as parameters in fluid flow. The three properties "concerned in the transmission of an electric signal along an insulated conductor" were charge (characterized by capacity for electricial accumulation), electromagnetic induction (characterized by resistance of a current to changes in strength), and resistance to conduction. These properties, he said, were "in the present state of science not understood except as quite distinct from one another," but water flowing in a rubber tube exhibited analogous properties: compressibility of the water (very small) and expansibility in the tube, inertia, and viscosity.

Taking the compressibility and expansibility and the viscosity as primary factors in signal transmission, Thomson described qualitatively his previous telegraph equation. He regarded expansibility as the analogue of high capacity in submarine cables insulated with gutta-percha, and compressibility as the analogue of low capacity in air lines. Submarine cables acted like highly elastic rubber tubes, and air lines like perfectly rigid tubes. Thomson knew inertia to be insignificant in long submarine cables, but in short cables and in air lines the frequency of signaling (i.e., the rate of change of current) had only to be high enough for electromagnetic induction to dominate over the kind of capacity associated with expansibility. In this case the telegraph equation became essentially a wave equation rather than a diffusion equation.

Thomson's view made electric signals in air wires analogous to longitudinal pressure pulses (sound) in a rigid tube of water, propagating at a definite velocity depending only on compressibility and inertia. Available data indicated a velocity considerably greater than that of light, but Thomson pointed out that the inertial effects were as yet completely uncertain, thus justifying, in a literal sense, his later claim to near priority in establishing the link between electrical units and the velocity of light.

Convinced Maxwellians such as FitzGerald, however, were horrified lest anyone should suppose Thomson's conception to be at all like Maxwell's.

FitzGerald wrote to *Nature* in 1885 to protest what he thought must have been a mistaken report of Thomson's Baltimore Lectures. Upon perusing a copy of them, as well as the "Velocity of Electricity" article, he wrote again, sending the letter through Thomson, who complained of misrepresentations. Apologizing, with the excuse of being "so anxious to prevent what I find is a very common mistake namely its being supposed that the velocity of transmission of signals is the same thing as Maxwell's velocity of light," and insisting that he was not trying to "teach [his] grandmother to suck eggs," FitzGerald nevertheless insisted also on the fundamental issues. He pointed out that, "in order to introduce anything at all *analogous* to Maxwell's wave propagation in non-conductors," Thomson would have had to include tubes with infinitely thick rubber walls, to propagate waves into the surrounding space rather than along the tube. "This is just the difference between Sir Wm. Thomson's and Maxwell's views," wrote FitzGerald. "According to Maxwell's view *there is a great deal more going on outside the conductor than inside it*, and it is evident that the inertia of the water is a very bad analogue to electromagnetic induction, for the latter depends essentially upon the form of the circuit, and not only upon its section and length." [18]

Whereas Thomson's theory treated electromagnetic induction as an effect between different parts of the conducting wire, or between the wire and adjacent conductors, without considering how the effect was transmitted, Maxwell's theory made that effect depend on displacement currents propagated by electromagnetic induction in the surrounding insulator. A current in any section of the wire generated magnetic force in circles around it, which in turn produced displacement currents in the insulator opposite to the current in the wire, which induced further magnetic force, and so on—a propagation outward from the wire of transverse electric and magnetic inductions moving at the speed of light. The apparent inertia of the current in the wire was actually the inertia of this electromagnetic field outside it.

Concomitantly, electrostatic induction, whether in the gutta-percha of a cable or in the air surrounding above-ground lines, was always an induced displacement in the surrounding insulator, measured by the charge, or electrical density, on the wire. While FitzGerald might have agreed that expansibility in a rubber tube could somehow represent inductive capacity, he could accept nothing like compressibility in the water: "According to my view of Maxwell's theory there is no known phenomenon in electricity exactly *analogous* to the compressibility of the water in the tube." [19]

The battle between Thomson and the Maxwellians had thus been joined in 1885. It had changed little when he died in 1907.

Electricity and Ether

Thomson's stake in the telegraph theory as the practical reality of electrical knowledge helps to explain his lifelong view that Maxwell's transverse displacement, however essential for light waves, had nothing to do with electrostatic force, which propagated longitudinally. This practical commitment, however, requires integration with his equally strong commitment to a particular form of ether theory, for the two shared a common root and expressed a common goal.

The representation of electrostatic force as a pressure or a tension goes back to Thomson's earliest work in electricity. In 1843 he treated the ponderomotive force on an electrified conductor as the result of changes in pressure in the surrounding air produced by the electricity. He did not discuss the cause of the same force in vacuum (ether). Tension in the air became a constant theme in his writings during and after 1845, when he produced his reconciliation, via the heat flow analogy, of Faraday's theory of induction with the theory of action at a distance. Apparently at that time, if not earlier, he began to think of air and ether as modifications of the same substance, in line with Faraday's experiments showing that all gases had the same inductive capacity as vacuum.[20] Then in 1847, in his seminal "Mechanical Representation of Electric, Magnetic, and Galvanic Forces," he portrayed electrostatic force as a state of longitudinal strain in an elastic-solid ether (with strain equal to stress). That paper is known for its representation of magnetic forces as states of rotational strain in order to explain Faraday's 1845 discovery of the rotation of the plane of polarized light during propagation through magnetized bodies.[21] Thomson often claimed that he had ever afterward been convinced that a real rotation would be found to constitute magnetic force, but that he regarded the electrostatic representation as merely a mathematical analogy.[22] Yet he never gave up the idea that longitudinal electrostatic stress in the ether would accompany light waves. We focus on the electrostatic problem here.

Thomson's earliest surviving discussion of tension in air/ether in relation to light occurs in lecture notes taken in his Natural Philosophy class at Glasgow in 1852–53. Discussing transverse and longitudinal vibrations and the differences of sound and light, he remarked: "It is probable that vibrations like those of sound are propagated at the same time [as those of light]." Partly he meant that any material medium that would transmit transverse light waves would probably transmit longitudinal waves as well. But he intended more: "An ether has been assumed—a luminiferous ether. No proof has been given of the sudden ceasing of air. It is more probable that it [ether] is matter than that it is not." Not only was ether probably material, it was probably continuous with air.[23]

This opinion would seem quite puzzling were it not for related lectures on electricity. Thomson described three theories: a single-fluid theory, a double-fluid theory, and Faraday's tension theory. "Faraday," Thomson noted, "says that [electricity] is the air which is put into a state of tension—bodies are pulled asunder by air." Thomson shared part of this view ("An electrified body puts the air around it in tension & transmits this tension all round towards all points"), but he also claimed that "the one hypothesis of one electric fluid seems far the most probable." The "hypothesis of one electric fluid" might, Thomson said, "be perhaps reconciled to probability by Faraday's theory of tension. It may be that there is an elastic fluid pervading all space, and causing light and heat. This is very highly probable—we get both light and heat from electricity.... Repulsion may be when the tension on one side is greater than the other."

This reconciliation required splitting off the existence of electricity (a material substance) from the existence of electrical tension. Discussing electrostatic induction in dielectrics, Thomson remarked: "A plate of sulphur acts like a solid resister of electricity filled with [polarized] conductors throughout. [Faraday] supposes that there is some such polar action in air. It appears that there is probably no such polar electrification in air." Induction in dielectrics thus involved a true motion of electricity within very small conducting parts, polarizing them, so that, on the boundary between a dielectric and air or between two dielectrics of different capacities, a real density of electricity appeared. Air, on the other hand, probably possessed no true inductive capacity: no flow, polarization, or strain—no "displacement"—associated with tension through it. "The 2 electricities for all we know, cannot be separated in air."[24]

If not, air and all gases behaved exactly as free ether behaved and transmitted electrical force in the same way: as tension in an incompressible medium. In conductors the same tension was relieved or neutralized by motion of electricity: "All the electrical forces give a resultant equal to 0 on any interior point. Faraday & Snow Harris say that electr^y dont operate through conductors. Faraday considered that it was a tension of the air on the outside entirely. But electrical force acts everywhere, in every direction."

This "probable" theory translated the terms of action at a distance into those of field theory, replacing electrical force at a point by tension and instantaneous action at a distance by instantaneous propagation to a distance. It has sometimes been doubted that Thomson quite understood that Faraday's theory involved nothing like a real electrical density, or that it required a quantity-intensity duality for propagating forces. However, his lectures show unequivocally that he understood the theory and rejected it.[25] They also show that his use of the term "flux" in drawing the analogy between transmission of force and heat conduction was purely metaphorical, for his flux was not continuous across interfaces between different media, as flux of heat would be for changes in conductivity. There could be no continuity of current, therefore, in the circuit that charged a Leyden jar. The current carried electricity to and from the plates, but between them no "quantity" like current flux existed. In air only a tension ("intensity") existed, while in a true dielectric polarization produced a real negative charge on the side of the positive plate and a real positive charge on the negative side. In short, no electric "quantity" like Faraday's and no "displacement" like Maxwell's could complete the circuit.

If we ask now why Thomson insisted on employing the material view of electricity, two answers present themselves. First, since all measurements yielded the same inductive capacity for air at all densities down to the best attainable vacuum, the supposition of polarizability in either air or ether, and thus the tension theory of electricity, was pure speculation. No lack of positive evidence existed, however, for electrical density, which could be manipulated and measured at will. Second, Thomson had recently generated, after three years of struggle, a general argument for discriminating matter from motion in the question of heat. "Are magnetism, electr. heat and light matter?," he asked his students. "We must consider general principles of work and mechanical effects." The rhetorical question introduced the "principle of mechanical effect," the "axiomatic" doctrine that all physical effects are mechanical and that mechanical energy is conserved. Since work could produce heat, heat had to be energy and not matter, for only God could create or destroy either energy or matter. Thomson regularly applied that reasoning to Davy's ice-rubbing experiment to make of it a classic argument for the dynamical theory of heat. "What is the nature of heat? ... Sir Humphry Davy rubbed two pieces [of ice] so as to melt them.... Now ice cannot be melted without heat. Heat was made by Davy, but he can't make matter—Ergo—What can the rubbing do? Only inertia overcome or matter set in motion." Heat was a state of motion. By contrast, no one had ever created a net electrical density, which suggested that it was matter.[26]

Both arguments for electricity as matter should be seen as variations on the theme of knowledge as knowledge of the practical and measurable (or, in classical terms, "we know what we can make"). Thus, one sees immediately an important reason for Thomson's tenacity in modeling electrostatic force as tension in the ether. Only in that way could he obtain a mechanical model that exhibited the properties that one could actually produce directly. By extension, Thomson did not even entertain the idea of developing models that one could not literally make (i.e., non-mechanical systems, mathematical or otherwise). Such errors vitiated *Naturphilosophie*. "The Germans," he warned his students on their first day, "apply the term Nat[ural] Phil[osoph]y to meta-physical science." On the second day, he said: "Force cannot be conc[eive]d as having any independent existence without matter."

In line with his opinions on matter and motion, Thomson developed views on the interrelation of physical phenomena. For example, he stated: "What-ever electricity is it seems quite certain that electricity in motion IS *heat*; and that a certain alignment of axes of revolution in this motion IS *magnetism*."[27] It was the dynamical theory of heat that first brought him to public expression of these speculations. In elevating Sir Humphry Davy to the status of first knight of dynamical theory, Thomson had also adopted Davy's view that the elasticity of matter depended not on a repulsive fluid (caloric) but on the "*repulsive motions*" constituting heat.[28] And he told his class: "Probably in the friction of 2 solids, there is change of shape, and the heat produced is very probably on account of electrical currents."[29] Any adequate theory of electricity, therefore, had also to be a theory of heat and elasticity. But what of the ether and light?

Thomson led his undoubtedly bewildered students into this complex of issues from a discussion of the empirical necessity of treating the compressibility and the rigidity of solids as independent properties, ridiculing the unrealistic theory of Poisson that required a fixed proportion between these properties. He suggested, in essence, that by regarding elasticity as due to the molecular motions constituting heat one could account for the independent variations of compressibility and rigidity with temperature. Such independence would allow an elastic medium of the sort required for the ether, incompressible but not completely inflexible: "An incompressible elastic solid is the kind of medium that we must conceive as that in which the vibrations of light and [radiant] heat are considered. These vibrations are principally vibrations of distortion & not compressions."

Now the motions constituting heat that Thomson had in mind were not primarily translations or vibrations of entire molecules, but rotations. Rankine's 1851 paper "Laws of Elasticity of Solid Bodies," in which he extended his hypothesis of molecular vortices, provided Thomson's primary inspiration, although he had begun by 1847 to consider that elasticity might be a mode of motion.[30] Unhappy with Rankine's assumption of forces acting at a distance between the nuclei of his vortices and with the repulsive power in their rotating atmospheres, Thomson suggested to his students replacing those forces with motions. He already envisioned a hydrodynamical theory of matter, though only vaguely. No doubt the rotating substance responsible for elasticity was to be electricity, accounting also for magnetized matter and the magnetic rotations of light.

These are the basic assumptions, scattered through his lectures of 1852–53, that Thomson began to develop in 1856 and 1858 into a general theory of the properties of matter.[31] Into that theory went the following results of measurement: *Heat* is molecular motion and is reponsible for elasticity, both of figure and volume; *light* is a transverse vibration in a medium that is rotated by magnetism in all bodies and that probably has associated with it longitudinal vibrations traveling at infinite velocity; *magnetism* is a rotational motion that can be produced by currents of electricity; *electricity* behaves like a material substance that can move through certain bodies called conductors, producing heat as well as magnetism, and that accumulates on their surfaces, producing tension in the surrounding space.

Thomson's private notebook for 1858 contains a sketch of how these bits of knowledge might be assembled into a unified theory. Free interpretation yields the following picture. An electrical fluid exists throughout space, as a universal plenum, possessing primitive properties of inertia and contractility. Within the fluid swim motes or molecules, which are atomic (in the sense of permanent) but not indivisible. Ideally the motes would be "a particular form & order of motions or eddies" in the fluid, but Thomson had not succeeded in establishing the permanence of such structures of motion (e.g. vortices). Ideally also, rotations of the motes would produce a repulsive pressure between them, but Thomson had "not *yet (!!!) succeeded*" in proving the necessary hydrodynamical theorem. Nevertheless he assumed that the motes, with their repulsive forces, formed a kind of lattice in space, having the rigidity necessary for propagating the transverse vibrations of light, and almost infinitely incompressible. Thus, from the compressible fluid there arose an incompressible lattice. On that basis he hoped "to conceive that all the phenomena of matter might be explained by the consequences of contractility in a universal fluid constituting the material world, and created with such a distribution as to density ... that the present and all past phases of dead matter may have followed from it in accordance with constant mechanical laws."[32]

This scheme assumed two different sorts of bodies, conductors and nonconductors, both constructed from complex systems of motes (the molecules of normal matter). In conductors the electric fluid could flow between the motes, causing them to take on the additional rotations of heat: "It seems certain that any motion ... of the liquid itself while the motes are held back by a strainer or sieve enclosing the space under consideration, will tend to generate ... rotating motions of the motes in general ... [, illustrating] the generation of heat in a liquid or solid by a current of electricity through it." Such electric currents were also to generate, in an unexplained manner, rotational motion of the motes in the nonconducting space, with an alignment of axes in closed curves surrounding the currents and constituting magnetic lines of force.[33] To charge a capacitor in circuit with a battery (which Thomson did not discuss here) would involve pumping electrical fluid from one conducting plate to the other, creating a deficiency on the first (negative) plate and an excess on the second (positive) plate. In the nonconductor between the plates the lattice of motes would somehow prevent continuous fluid flow. The fluid pressure would therefore create a tension in the incompressible lattice, constituting electrostatic stress. True dielectrics presumably contained very small regions within which flow could occur, producing an electrical polarization; free space, and all gases, possessed no such polarizability.

This sketch reproduced Thomson's "Mechanical Representation" of 1847 while incorporating his dynamical theory of heat and looking toward the vortex theory of atoms and molecules that he finally reached in 1867. Meanwhile, Maxwell had in 1863 taken over the basic idea of magnetic lines of force as vortex filaments produced by electric currents, added his displacement current (to complete open circuits and to transmit magnetic rotations through space), and thereby created a highly successful structural theory of light. In doing so he of course abandoned Thomson's concept of electricity as a fluid,

adopting Faraday's view instead, and abandoned concomitantly the propagation of electric tension with infinite velocity. But since Maxwell had no adquate
mechanical construction to represent displacement, much less any basis for it
in measurements, Thomson could hardly regard the reformulation of his own
best ideas as an improvement. It violated both of his criteria for legitimate
physical theory.

Electricity and Ether in Thomson's Later Models

The basic presuppositions about the relation of electricity and light that went
into Thomson's heuristic model in the 1850s figured in the Baltimore Lectures
and subsequent presentations as well. The later models, however, illustrate
more clearly than the earlier ones the nature of Thomson's commitment, for in
the 1880s he began seriously to seek an alternative to Maxwell's perversion. He
began also to annotate his copy of Maxwell's *Treatise*, dating every entry.
Numerous annotations date from August 1888, shortly before Thomson presented to the British Association his "Simple Hypothesis for Electro-magnetic
Induction of Incomplete Circuits, with Consequent Equations of Electric Motion in Fixed Homogeneous Solid Matter." The style of this critique was fixed
from its genesis on August 13, when Thomson wrote beside Maxwell's vector
equations for electromagnetic induction that they merely "translated into
gibberish" the full Cartesian component equations. On September 4, on the
train to Bath to attend the B. A. meeting, Thomson was going over Maxwell's
analysis for at least the fifth time, ensuring that he had located the physical
gibberish as well.[34]

Thomson's critique derives directly from the 1860 paper "Velocity of Electricity"—denounced by FitzGerald in 1885—in which Thomson had attempted
to generalize his theory of telegraph signals into a full description of the
factors affecting the motion of electricity and had illustrated the properties of
electricity as properties of water: compressibility, inertia, and viscosity. (We
ignore for the moment expansibility in the containing tube, the main analogue
of inductive capacity.) Now he sought to justify that picture by showing that
Maxwell's doctrine of closed currents, taken apparently in the only form in
which it made sense, would lead to an inconsistency that could be removed
by the adoption of his own view. Thus, he began by assuming that electricity
behaves like an *incompressible* fluid, derived a contradiction, and ended with
the hypothesis of compressibility.[35] The derivation followed closely Maxwell's
chapters 9 ("General Equations of the Electromagnetic Field") and 20 ("Electromagnetic Theory of Light"). The former contains the remark that total current
"is subject to the condition of motion of an incompressible fluid, and ... must
necessarily flow in closed circuits." But Thomson omitted from his conception
Maxwell's further remark: "This equation is true only if we take [total current
as] that electric flow which is due to the variation of electric displacement as

well as to true conduction."[36] Instead, he assumed all currents to be closed conduction currents, and demonstrated that, taken in conjunction with the known equation of electromagnetic induction, the assumption led to four equations in three unknowns. Although consistent for homogeneous media, the equations were inconsistent for inhomogeneous media. The difficulty, Thomson ingeniously showed, could be ascribed to the fact that the closed-circuit assumption eliminated net electrical densities that might arise from inhomogeneities. Compressibility, on the other hand, would produce the net densities and remove the contradiction.[37]

Had he included displacement currents, Thomson would not have obtained a contradiction for inhomogeneous media, but of course he had set out to rid the theory of that "curious and ingenious, but ... not wholly tenable hypothesis." He can only have looked with dismay on Maxwell's method of introducing such gratuitous concepts; e.g., "We have deduced everything from purely dynamical considerations, without any reference to quantitative experiments in electricity or magnetism. The only use we have made of experimental knowledge is to recognise, in the abstract quantities deduced from the theory, the concrete quantities discovered by experiment, and to denote them by names which indicate their physical relations rather than their mathematical generation." Contrast Thomson's remarks on compressibility: "An interesting and important *practical* conclusion is, that when currents are induced in any way, in a solid composed of parts having different electric conductivities (pieces of copper and lead, for example, fixed together in metallic contact), there must in general be changing electrification over every interface between these parts. This conclusion was not at first obvious to me; but it ought to be so by anyone approaching the subject with mind undisturbed by mathematical formulas."[38]

The hypothesis of compressibility led Thomson to a wave equation for electrical density in an infinite homogeneous conductor, with a propagation velocity equal to the famous ratio between electromagnetic and electrostatic units. "It is interesting and helpful to remark," he observed, "that this agrees with the equation for the density of a viscous elastic fluid, found from Stokes's equations for sound in air with viscosity taken into account."[39] An infinite conductor would propagate "sound" waves at the velocity of light. That result confirmed Thomson's 1860 generalization of the telegraph theory; with appropriate boundary conditions to convert the infinite conductor into a telegraph wire in air (with no true inductive capacity), the equation agreed with his old analogy to water in a perfectly rigid tube. The perceptive FitzGerald had recognized immediately that, according to Maxwell's theory, "there is no known phenomenon in electricity exactly *analogous* to the compressibility of the water in the tube,"[40] but Maxwell's theory was exactly the issue. With the attained reality of the telegraph as his guide, Thomson would undo Maxwell's damage: "I find simple and natural solutions, with nothing vague or difficult to understand, or to believe when understood, by their application to *practical* problems, or to conceivable ideal problems, such as the transmission of ordinary or telephonic signals along submarine telegraph conductors and land lines, electric oscillations in a finite insulated conductor of any form, transference of electricity through an infinite solid, &c. &c."[41] Unfortunately for Thomson, he presented his critique of Maxwell at Bath to an enthusiastic group of Maxwellians: FitzGerald, Rowland, Lodge, and others. They had little sympathy for his practical theory of electricity, for Hertz had recently published experimental confirmation of Maxwell's electromagnetic waves.[42]

We have left out of account whatever notion of dielectric inductive capacity Thomson may have entertained in 1888. Having attacked Maxwell's displacement, he required an alternative. Several brief remarks in his papers of the succeeding two years indicate that he maintained hope for his original idea of electrical fluid within an elastic but incompressible ether lattice. The idea appears most succinctly in an abstract of January 1890, "On Electrostatic Stress," which reads in its entirety as follows:

A complete dynamical illustration of electro-dynamic action may be had in an elastic solid, homogeneous in so far as rigidity is concerned, permeated with pores of unalterable size containing liquid. These pores may be in part in communication with each other, and in part closed by elastic partitions. These cases correspond to conductors and non-conductors respectively. Electrostatic stress depends on the curvature and extension of the partitions. The law of capacity in the model is identical with that in [non?]conductors.[43]

Presumably capacity would be proportional to the density of closed pores.

Thomson had described these ideas a year earlier to the most appropriate audience: the Institution of Electrical Engineers, formerly the Society of Telegraph Engineers. Speaking on "Ether, Electricity, and Ponderable Matter," but considering primarily ether, he had discussed the essential differences between static rotational displacements in a "jelly" and the absolute rotations required for magnetism: To reproduce the proper boundary conditions at interfaces between media of different capacity for magnetization, and to explain magneto-optic rotation, the absolute rotations were essential. Thomson had even discovered how to model them in a rotationally elastic ether, using his famous gyrostatic molecules, but how was the viscous flow of electricity in conductors to produce molecular magnetic rotations throughout a surrounding medium? Were it not for this obstacle, Thomson said, he "should be perfectly satisfied with the problem of electro-magnetic induction, by taking the electricity as a viscous fluid, and ether an elastic solid, porous in some places, and continuous or non-porous elsewhere." For if static rotational displacement in the porous solid would suffice, viscous flow would produce that displacement in exactly the required manner. Without viscous flow, he had no conception at all of how electromagnetism worked. He could only say: "Whatever the current of electricity may be, I believe *this* is a reality: *it does pull the ether round....*"[44]

In spite of its deficiencies, therefore, Thomson continued to operate with his telegraph conception of electrical motion, looking for the key that would "extend the analogy to include the complete problem of the submarine cable, in which electromagnetic and electrostatic induction are both taken into account."[45] The abstract quoted above shows that he hoped to replace Maxwell's notion of displacement for interrelating the two forms of induction with his own model of viscous fluid in pores. Just as a pressure gradient in a conducting wire would produce flow through the channels between pores, so in dielectrics the pressure gradient would produce elastic deformation of the membranes blocking the channels, and thus a small net motion of electricity. This is the same idea of electrostatic polarization that Thomson had employed since 1845. But now he intended to include polarization by electromagnetic induction as well. In whatever manner the rotations of magnetism affected the pores in conductors to produce flow, they would act in the same way on those in dielectrics to produce elastic deformation. One could not separate the two forms of induction, any more than in Maxwell's theory.

In one sense, nothing substantive distinguished Thomson's polarization from Maxwell's displacement except the absence of polarization from free ether, which contained no pores. In another sense, however, they viewed the problems from different worlds. Maxwell took light waves as his foundation; Thomson took telegraph signals. Both showed that the ratio of electromagnetic units to electrostatic units was a velocity equal to that of light, but their velocities had completely different meanings. Thomson's was the velocity of longitudinal pressure pulses in a fluid, which for unexplained reasons happened to be very near the velocity at which the rotations of magnetism propagated transverse to their axes. Maxwell's was the velocity of propagation of transverse displacement currents by electromagnetic induction, and it had no relation to pressure pulses.

Electrostatic Stress and Longitudinal Ether Waves

We have focused our attention so far on Thomson's view of electrical motions rather than of ether motions. That has been possible because conceptually his model separated pressure gradients in free electrical fluid from tensions of reaction in the lattice that contained the fluid. The separation meant that tension in the lattice between charged objects, and electrostatic force between them, would require a somewhat different treatment than conduction and induction. In fact, Thomson bemoaned the lack of even "the slightest inkling of how a fragment of paper jumps to rubbed sealing-wax."[46] Aside from this ultimate deficiency, however, if the ether lattice behaved like an incompressible elastic solid, it would propagate electrostatic stress at infinite velocity, while pressure pulses in continuous portions of the electrical fluid (i.e. in conductors) would travel at the velocity of light.

In the fourth of his Baltimore Lectures, Thomson had stated, with all the authority he could command, that "there are excessively strong probabilities that there will be waves of condensation and rarefaction of the luminiferous ether." That opinion had already appeared in his lectures of 1852–53, in relation to his views on the nature of electricity. By the 1880s, however, if our interpretation is correct, Thomson's opinions on electricity were grounded in the practical success of the telegraph. Indeed, the preceding quotation follows immediately on his connection of the velocity of light with the telegraph. It was that connection that most annoyed FitzGerald (in 1885) and Larmor (in 1904) in their responses to the respective publications of the Baltimore Lectures. The existence of longitudinal electrostatic waves in ether, furthermore, followed immediately from Thomson's telegraph theory of electricity. If the electrical fluid in the telegraph cable exerted pressure against its resisting lattice in the copper conductor and against its containing lattice in the gutta-percha, then that lattice would transmit a corresponding tension. The ether, continuing the lattice, would transmit this tension across vacuum.

To appreciate Thomson's commitment to the telegraph theory, then, is to appreciate his commitment to longitudinal electrostatic waves. This perspective illuminates an intense exchange of letters in 1896 between Thomson (now Lord Kelvin) and FitzGerald, in which Kelvin tenaciously insisted that Maxwell's theory required longitudinal electric waves and FitzGerald again and again tried to school him in Maxwellian theory. The exchange began when Kelvin suffered a fresh attack of the disease he labeled "ether dipsomania" brought on by Röntgen's discovery of x rays. Believing for a time that the long-sought longitudinal waves had been found (as Röntgen had suggested), Kelvin began again to market them.[47]

In a letter to *Nature*, Kelvin rejuvenated an old thought experiment from the Baltimore Lectures. The original argument (in Lecture IV) had clearly been devised to extend telegraph signaling to include effects in the surrounding space, especially those continuing longitudinally from the terminations of the line. He began:

Suppose that we have at any place in air, or in luminiferous ether (I cannot distinguish now between the two ideas) . . . two spherical conductors united by a fine wire, and that an alternating electromotive force is produced in that fine wire. . . . It is absolutely certain that such an action as that going on would give rise to electrical waves. Now it does seem to me probable that those electrical waves are condensational waves in luminiferous ether; and probably it would be that the propagation of these waves would be enormously faster than the propagation of ordinary light waves.

Kelvin then presented the relation of the velocity of light to telegraph signals in an air wire, noting that light presented a very different case, and returned to the thought experiment: " . . . [that] is a case of excitation of a kind that we know; we know the a, b, c of it, and the laws of it, and feel certain that if this operation be performed but fast enough there will be waves." The "a, b, c" and the laws of the excitation were those of the telegraph; it remained to describe their continuation in space.

Kelvin's renewal of these ideas in 1896 eliminated the wire between the spheres and concerned simply a spark discharge between them, which was to cause a secondary discharge between two other spheres located at some distance along the axis of the first two. He asserted: "The elastic solid theory restricted to the supposition of incompressibility (which is expressed by Maxwell's formulas) makes the difference of times between the two sparks infinitely small. The unrestricted elastic solid theory gives for the difference of times the amount calculated according to the velocity of the condensational-rarefactional wave."[48] FitzGerald responded: "Surely you are not right in your letter to 'Nature' last week in stating that Maxwell's ether would give an instantaneous action in the case you mention. Whatever properties it may have in respect of gravitation &c &c so far as electric actions are concerned they are *all* represented by the [wave] equation $\nabla^2 v = K\mu\, \partial^2 v/\partial t^2$ and so represent an action propagated with a velocity $1/\sqrt{K\mu}$ [the velocity of light]."[49]

As we saw, incompressibility in Maxwell's theory meant to Kelvin the hypothesis of closed electrical currents, which in ether implied closed displacement currents, with no accumulation of electricity. The result agreed with Kelvin's view that ether contained no free electricity and that the electrostatic force was transmitted instantaneously by the incompressible ether lattice. But since Kelvin did not recognize displacement currents at all, and certainly not as sources of magnetic force, he could never agree that the electrostatic stress associated with them was propagated only by electromagnetic induction (i.e., not like pressure in a solid). FitzGerald did his best to explain this interpretation, giving a detailed analysis of the two-sphere problem and pointing out that Hertz had dealt with it long ago, all of which made no impression on Kelvin.

FitzGerald tried again two days later, reiterating his previous remarks and including a diagram of displacements in the space surrounding an oscillating electric dipole. He insisted that "Maxwell never suggested that his ether was an elastic solid." But Kelvin merely responded by sending proof sheets of an article containing a new version of his argument—one he had concocted for a possible experimental confirmation that Röntgen's rays were longitudinal electrostatic waves. On the proofs he wrote: "Read mark learn & inwardly digest Maxwell paras. 610, 611, 612, 613, 616. I think when you have done so you will see that the statements marked // on the proof are correct, and will reconsider the whole of your letter to me of 14th." The paragraphs mentioned are those in which Maxwell presented his doctrine of displacement currents and closed total currents, the same paragraphs that formed the basis of Thomson's critique in 1888. The marked statements on Kelvin's proof no doubt included those asserting longitudinal waves and instantaneous transmission in an incompressible ether.[50]

Tirelessly, in eight long letters from February to May 1896 and in five more during the month of November 1898, FitzGerald expounded the electromagnetic viewpoint, while just as tirelessly Kelvin reiterated the requirements of incompressibility in an elastic solid (although many of his letters do not survive). FitzGerald isolated the problem in his third letter: "I feel pretty confident that you are overlooking the electric effects [displacement currents] due to the magnetic force being generated while the spark is starting."[51] These effects prevented the spark and all other longitudinal electrostatic actions from propagating instantaneously, according to Maxwell's theory. On Kelvin's theory they did not exist.

The correspondence between Kelvin and FitzGerald illustrates with painful repetition the incommensurability of their viewpoints and commitments.[52] At no point did either of these friends quite recognize the force of the other's arguments. They even meant different things by the same words— *displacement*, for example. For Kelvin it was an actual motion of matter or it was nothing: "Maxwell's expression, 'electric displacement' is, I believe, absolutely true so far as it indicates a true displacement of matter, as in the undulatory theory of light, but my difficulty is in respect to the electric quality concerned in this displacement."[53] For FitzGerald it was a structural element in a total theory, whose meaning was to be extracted from the theory and whose validity depended on the success of the theory as a whole: "Your faith that Maxwell's 'electric displacement' is 'absolutely true so far as it indicates a true displacement of matter' goes far beyond anything I can feel about its certainty."[54] This use of physical entities without "understanding" them, either as measurement or model, was what Kelvin called nihilism. In turn, FitzGerald regarded Kelvin's elastic-solid argument for longitudinal waves as a false analogy and "a pure waste of time" that imposed on an otherwise useful but partial analogy "an addendum for which there seems to me no corresponding phenomenon."[55]

We thus return to our opening divide, material reality versus mathematical structure. To FitzGerald's charge of false analogy Kelvin had a ready retort: " It is certainly not an allegory on the banks of the Nile. It is more like an alligator. It certainly will swallow up all ideas for the undulatory theory of light, and dynamical theory of E & M. not founded on force and inertia. I shall write more when I hear how you like this."[56] But FitzGerald had sharpened his own teeth: "I am not afraid of your alligator which swallows up theories not founded on force and inertia.... I am quite open to conviction that the ether is *like* and not merely in some respects *analogous* to an elastic solid, but I will . . . wait till there is *some* experimental evidence thereof before I complicate my conceptions therewith."[57]

The righteousness of empiricism now dogged the wrong party. But how did Kelvin come to this unhappy end? It might be attributed straightforwardly to his stubborn adherence to an outmoded concept of ether, to the mechanical philosophy, to a methodology that insisted on independent empirical evidence for each physical entity in a theory, or even to his revulsion for German idealism. All these explanations are correct. We have attempted to ground them, however, in Kelvin's view of reality as *practical* reality. We have suggested briefly that this view must be understood at the broadest level in the economic and political terms of industry and empire. In the case of Kelvin's electrical theorizing, those terms focus on the telegraph. We have therefore sought to relate his personal stake in the telegraph to his practical criteria for science, and to the theory of electricity that the telegraph came to represent for him. This set of interrelations illuminates the intensity and tenacity of Kelvin's views on electricity and his unwillingness (or even inability) to accept the Maxwellian orthodoxy of late-nineteenth-century physics. Conversely, it illuminates what the orthodoxy was not.

Acknowledgments

We thank Peter Galison and Mary Terrall for helpful comments, and the Cambridge and Glasgow University libraries for permission to quote from (respectively) the Kelvin Collection and the Kelvin Papers.

Notes and References

Abbreviations of Kelvin Sources

E & M: Reprint of Papers on Electrostatics and Magnetism (London, 1872)

MPP: Mathematical and Physical Papers (Cambridge, 1882–1911)

Pop. Lect.: Popular Lectures and Addresses, second edition (London, 1891)

ULC: University Library Cambridge, Kelvin Collection.

ULG: University Library Glasgow, Kelvin Papers.

Life: Sylvanus P. Thompson, *Life of Lord Kelvin*, second edition (New York, 1976)

1. Joseph Larmor, "Lord Kelvin on Optical and Molecular Dynamics," *Nature* (Supplement) 70 (1904): iii–v.

2. See the excellent survey of Thomson's career by Ole Knudsen, "Mathematics and Physical Reality in William Thomson's Electromagnetic Theory," in *Wranglers and Physicists: Studies on Cambridge Physics in the Nineteenth Century*, ed. P. Harman (Manchester, 1985), pp. 149–201, and Jed Z. Buchwald's detailed treatment of Maxwellian electrodynamics, *From Maxwell to Microphysics: Aspects of Electromagnetic Theory in the Last Quarter of the Nineteenth Century* (Chicago, 1985). Knudsen emphasizes the essentially practical character of Thomson's physics, which we develop further here.

3. FitzGerald to Kelvin, February 17, 1896 (F123, ULC).

4. Kelvin to Larmor, October 9, 1906 (L37, ULC).

5. Kelvin to C. G. Knott, September 12, 1904 (K111, ULC).

6. W. Thomson, "Remarks on Retiring from the Presidential Chair," *Inst. Elect. Engrs. J.* 19 (1890): 2–6; *MPP* V: 577–580, at 578.

7. We develop this thesis at length in the biography we are preparing, *Energy and Empire: William Thomson, Lord Kelvin, 1824–1907* (Cambridge University Press, forthcoming).

8. See the biographical analysis by C. W. F. Everitt, "Maxwell's Scientific Creativity," in *Springs of Scientific Creativity: Essays on Founders of Modern Science*, ed. R. Aris et al. (Minneapolis, 1983).

9. The limitations of the industrial spirit among gentrified professionals (including many academics) are discussed convincingly in Martin J. Wiener's *English Culture and the Decline of the Industrial Spirit, 1850–1980* (Cambridge and London, 1981) and in Jack Morrell and Arnold Thackray's *Gentlemen of Science: Early Years of the British Association for the Advancement of Science* (Oxford, 1981). Contrast Robert H. Kargon's study of an industrial context, *Science in Victorian Manchester: Enterprise and Expertise* (Baltimore and London, 1977).

10. W. Thomson, "On the Electro-statical Capacity of a Leyden Phial and of a Telegraph Wire Insulated in the Axis of a Cylindrical Conducting Sheath," *Phil. Mag.* 9 (1855), Supplement: 531–535; *E & M*, pp. 38–41, intended as an 1854 appendix to "On the Elementary Laws of Statical Electricity," *Cambridge and Dublin Mathematical Journal* 1 (1846): 75–93; *Phil. Mag.* 8 (1854): 42–62; *E & M*, pp. 15–37.

11. W. Thomson, "On the Theory of the Electric Telegraph," *Proc. Roy. Soc.* 7 (1855): 382–399; *Phil. Mag.* 11 (1856): 146–160; *MPP* II: 61–76, at 68.

12. See M. Norton Wise, "The Flow Analogy to Electricity and Magnetism, Part I: William Thomson's Reformulation of Action at a Distance," *Arch. Hist. Ex. Sci.* 25 (1981): 19–70.

13. W. Thomson, "Electrical Units of Measurment," one of a series of lectures delivered in 1883 to the Institution of Civil Engineers on "The Practical Applications of Electricity," *Pop. Lect.* I: 80–143, at 80, 86.

14. Thomson's annotation, August 12, 1890, to p. 240 of volume II of his copy of James Clerk Maxwell's *Treatise on Electricity and Magnetism* (Oxford, 1873), now located in the Natural Philosophy Department of Glasgow University. The title page is signed and dated October 19, 1878. Extensive annotations span the period 1882–1907. We thank Rex Whitehead for calling our attention to this and others of Thomson's books now housed in his old department, and for making them accessible to us.

15. Kelvin to FitzGerald, April 29, 1896 (*Life* II: 1071).

16. For Thomson's initial concern with this ratio see "On Transient Electric Currents," *Phil. Mag.* 5 (1853): 393–405; *MPP* I: 540–553.

17. W. Thomson, "Velocity of Electricity," in J. P. Nichol's *Cyclopaedia*, second edition (1860); *MPP* II: 131–137.

18. G. F. FitzGerald, "Molecular Dynamics," *Nature* 31 (1885): 503; "Sir Wm. Thomson and Maxwell's Electro-magnetic Theory of Light," *Nature* 32 (1885): 4–5. Thomson to

FitzGerald, April 17, 1885 (*Life* II: 1038). FitzGerald to Thomson, April 25, 1885 (F118, ULC). Final quotation from "Sir Wm. Thomson" (emphasis added).

19. FitzGerald to Thomson, ibid. For an insightful and entertaining account of the conflict between Maxwellians and electrical engineers, see Bruce J. Hunt, "'Practice vs. Theory': The British Electrical Debate, 1888–1891," *Isis* 74 (1983): 341–355. "Practice vs. theory," however, oversimplifies the problem. Hunt treats "water-in-a-pipe" concepts as merely the practical intuitions of "half-educated electricians." His story would look somewhat different if it encompassed Thomson's sophisticated "water-in-a-pipe" theory and his relation to some of the practical men, such as his business partner Cromwell Varley. Varley's brother Samuel insisted that Maxwell's theory "rests solely on hypothesis"—a view Thomson largely shared.

20. W. Thomson, "On the Attractions of Conducting and Non-conducting Electrified Bodies," *Cambridge Mathematical Journal* 3 (1843): 275–276; *E & M*, pp. 98–99; "On the Elementary Laws" (ref. 10 above), p. 23; Thomson to Faraday, August 6, 1845 (*Life* I: 146–149).

21. See Knudsen, "Mathematics and Physical Reality" (ref. 2 above). Also see Knudsen's discussion of magneto-optic rotation: "The Faraday Effect and Physical Theory, 1845–1873," *Arch. Hist. Ex. Sci.* 15 (1976): 235–281.

22. See, e.g., Thomson to FitzGerald, April 29, 1896 (*Life* II: 1069).

23. Notes of Thomson's lectures by William Jack, 1852–53 (Ms. Gen. 130, ULG). The notes contain no page numbers, few lecture numbers, and few dates. The quotations precede the date November 25, 1852. See also Thomson's "Note on the Possible Density of the Luminiferous Medium and on the Mechanical Value of a Cubic Mile of Sunlight," *Trans. Edin. Roy. Soc.* 21 (1857): 57–61 (read May 1, 1854); *Comp. Rend.* 39 (1854): 529–534; *Phil. Mag.* 9 (1855): 36–40; *MPP* II: 28–33, where he said that it "appear[ed] to [him] most probable" that the ether was "a continuation of our own atmosphere." Related discussion occurs in correspondence with Stokes and Tait. D. B. Wilson has summarized the non-electrical aspects of Thomson's opinion in the introduction to his edition of the Stokes-Kelvin correspondence (unpublished); we are grateful to him for providing us with a copy.

24. Jack notes (ref. 23), several locations in the lectures, one shortly after the date November 25, 1852, and one following January 26, 1853.

25. In light of this evidence the analysis in Wise's "Flow Analogy" (ref. 12) must be stated in even stronger terms.

26. Jack notes, Lecture 2, first week of November 1852; also shortly before November 25. Thomson hit on this argument while drafting his "Dynamical Theory of Heat" in early March 1851. See Crosbie Smith, "William Thomson and the Creation of Thermodynamics: 1840–1855," *Arch. Hist. Ex. Sci.* 16 (1976): 231–288, especially pp. 261–268, appendix II, and pp. 285–286; also W. Thomson, "On the Dynamical Theory of Heat...," *Trans. Roy. Soc. Edin.* 20 (1853); *MPP* I: 174–210, at 174.

27. "Royal Institution Friday Evening Lecture" (May 18, 1860), in *E & M* pp. 208–225, at 224. Thomson first publicly expressed these views in "Dynamical Illustrations of the Magnetic and the Helicoidal Rotatory Effects of Transparent Bodies on Polarized Light," *Proc. Roy. Soc.* 8 (1856–57): 150–158; *Phil. Mag.* 13 (1857): 198–204; *Baltimore Lectures* (London, 1904), Appendix F, pp. 569–577.

28. "Dynamical Theory" (ref. 26); Smith, "Creation of Thermodynamics" (ref. 26), p. 285; *MPP* I: 174.

29. Jack notes, somewhat after the hypothesis of a single electrical fluid; maintained also in lecture notes by David Murray, 1862–63 (ms. Murray 325, ULG, p. 122).

30. W. J. M. Rankine, "Laws of the Elasticity of Solid Bodies," *Cambridge and Dublin Mathematical Journal* 6 (1851): 47–80, 178–181, 185–186. Thomson also referred in this connection to Davy and Herapath, as he would repeatedly in later writings. See Thomson to Stokes, October 20, 1847 (K21, ULC).

31. Thomson, "Dynamical Illustrations" (ref. 27). A much more speculative vision appears in Thomson's private notebook of 1858 (NB35, ULC), published in Ole Knudsen's "From Lord Kelvin's Notebook: Ether Speculions," *Centaurus* 16 (1971): 41–53. See Knudsen's remarks on Davy, p. 50, note a.

32. Knudsen (ibid.), pp. 47–49.

33. For a detailed discussion of how such rotation of the motes explained magneto-optic rotation, which Thomson said "brought on the whole attack," see Knudsen, "Faraday Effect" (ref. 21), pp. 273–281.

34. Thomson's annotations from this period are concentrated on pp. 233 and 385, with dates of August 13, 14, 18, and 27 and September 4.

35. W. Thomson, "A Simple Hypothesis for Electro-magnetic Induction of Incomplete Circuits, with Consequent Equations of Electric Motion in Fixed Homogeneous Solid Matter," *Brit. Ass. Rep.* (1888): 567–570; *Nature* 38 (1888): 569–571; *MPP* IV: 539–544.

36. Maxwell, *Treatise* II: 231.

37. Thomson's concern with net electrical density and inhomogeneity appears clearly in his annotations. They show him seeking simply to understand Maxwell's view or, more likely, to derive an absurdity from Maxwell's equation (H) for current (*Treatise* II: 233),

$$\vec{j}_t = \vec{j}_c + \frac{\partial \vec{D}}{\partial t},$$

where $\vec{j}_t$ is total current, $\vec{j}_c$ conduction current, and $\vec{D}$ displacement. He obtained, with $\vec{E}$ the electromotive force and $\vec{D} = (K/4\pi)\vec{E}$,

$$\mathrm{div}\vec{j}_t = \left(C + \frac{K}{4\pi}\frac{\partial}{\partial t} \right)\mathrm{div}\vec{E},$$

which would be correct only for homogeneous media (constant conductivity C and inductive capacity K), and concluded, emphasizing the hypothetical: "$\therefore$ div$\vec{j}_t$, if it *were* = 0 would give

$$0 = \left(C + \frac{K}{4\pi}\frac{\partial}{\partial t} \right)\left(\frac{\partial \vec{j}}{\partial t} - \nabla^2 \psi \right)$$

being § 783 (8)" [Maxwell's paragraph 783, equation 8, valid for homogeneous media]; and again: "$\therefore$ if div$\vec{j}_t$ = 0 [then] div$\vec{E}$ = 0 & $\therefore$ this = 0!," where "this" refers to Maxwell's equation (J) for electrical density, div$\vec{D}$ = ρ, or, for homogeneous media, $(K/4\pi)$ div$\vec{E}$ = ρ. Two years later (September 28, 1890), Thomson added a qualification:

"within each homog[eneou]s part. Through all space we have

$$\text{div}\left(C\vec{E} + \frac{K}{4\pi}\frac{\partial \vec{E}}{\partial t} \right) = 0."$$

38. Thomson, "A Simple Hypothesis" (ref. 35), pp. 542–543 (emphasis added); Maxwell, *Treatise* II: 229. Thomson wrote "quote" (undated) next to a similar passage (p. 231) referring directly to displacement: "We have very little experimental evidence relating to the direct electromagnetic action of currents due to the variation of electric displacement in dielectrics. . . ."

39. W. Thomson, "On the Transference of Electricity within a Homogeneous Solid Conductor," *Brit. Ass. Rep.* (1888): 570–571; *Nature* 38 (1888): 571; *MPP* IV: 545–546. We leave out of account in our analysis an entire series of papers on viscous flow in conductors and on the "skin effect" for rapidly alternating currents and waves, papers which also exemplify the importance of the telegraph.

40. FitzGerald to Thomson, April 25, 1885 (ref. 18).

41. Thomson, "A Simple Hypothesis" (ref. 35), p. 544 (emphasis added).

42. See Oliver Lodge's report of the meeting in *The Electrician* 21 (1888): 622–625, partially reprinted in *Life* (II: 1041 ff.).

43. W. Thomson, "On Electrostatic Stress," *Proc. Edin. Roy. Soc.* 17 (1890): 412; *Nature* 41 (1890): 358; *MPP* V: 482. No draft of the paper has been located.

44. W. Thomson, "Ether, Electricity, and Ponderable Matter," *Inst. Elec. Engrs. J.* 18 (1890): 4–37, corr. 18 (1890): 128 and 24 (1895): between 396 and 397; *MPP* III: 484–515, quotations on 502–503. A full mathematical treatment of the magnetic models in this popular lecture is "Motion of a Viscous Liquid; Equilibrium or Motion of an Elastic Solid; Equilibrium or Motion of an Ideal Substance called for Brevity *Ether*; Mechanical Representation of Magnetic Force" (*MPP* III: 436–465).

45. *MPP* III: 498. See also Thomson's contention (p. 488) that electromagnetic and electrostatic induction "cannot be separated," and his letter to Rayleigh on the viscous-fluid model (November 18, 1890; *Life* II: 1053).

46. Kelvin to FitzGerald, April 29, 1896 (*Life* II: 1069).

47. Kelvin to FitzGerald, April 9, 1896 (*Life* II: 1065).

48. Lord Kelvin, "Velocity of Propagation of Electrostatic Stress," *Nature* 53 (1896): 316 (February 6).

49. FitzGerald to Kelvin, February 12, 1896 (F121, ULC). Here K is inductive capacity and μ magnetic permeability.

50. FitzGerald to Kelvin, February 14, 1896 (F122, ULC). Kelvin copied his reply of the 16th on FitzGerald's letter. See "On the Generation of Longitudinal Waves in Ether," *Proc. Roy. Soc.* 59 (1896): 270–273, read February 13; *Nature* 53 (1896): 450–451.

51. FitzGerald to Kelvin, February 17, 1896 (F123, ULC). Three of Kelvin's letters to FitzGerald are reprinted in *Life* (II: 1064–1072).

52. Knudsen (ref. 2) has discussed the same point for Kelvin vs. Hertz.

53. Kelvin to FitzGerald, April 29, 1896 (*Life* II: 1069).

54. FitzGerald to Kelvin, May 11, 1896 (F128, ULC).

55. FitzGerald to Kelvin, April 17, 1896 (F127, ULC); November 25, 1898 (F136, ULC).

56. Kelvin to FitzGerald, November 28, 1898 (F18, ULG).

57. FitzGerald to Kelvin, November 29, 1898 (F137, ULC).

Lawrence Badash

Ernest Rutherford and Theoretical Physics

Albert Einstein considered Rutherford "one of the greatest experimental physicists of all time, and in the same class as Faraday."[1] Charles H. Townes wrote that "Rutherford could perhaps fairly be called the greatest experimental physicist of the day and the father of nuclear physics."[2] Evaluations of this sort are commonplace enough that there is no compulsion to document further the widespread recognition of Rutherford's experimental genius. He was, after all, the man who brought to its zenith the art of discovery using apparatus that was small enough to be hand-held, and which had been constructed under the sealing-wax-and-string philosophy of "make it no better than the experiment requires." With this bare-bones approach Rutherford collected the data that allowed him to argue (with Frederick Soddy, in 1902–03) that radioactivity is a manifestation of chemical atoms changing into other atoms (spontaneous transmutation: alchemy!), that the alpha particle is a charged helium atom (1908), that the mass of an atom is concentrated in its tiny nucleus (1911), and that bombardment of some atoms by alpha particles changes them into different elements (1919).

In the history of science there are many examples of scientists who combined great experimental skill with major theoretical prowess; Archimedes, Isaac Newton, James Clerk Maxwell, and Enrico Fermi are some of the peaks in this landscape. Does Rutherford fit into this picture? Almost certainly, he himself would have said no, although an argument can be made that he at least deserves to be in the foothills.

In his time Rutherford was notorious for his public disdain of theory, which not everyone took as a jest. "They [theorists] play games with their symbols," he noted, "but we, in the Cavendish, turn out the real solid facts of Nature."[3] When "the trend of modern physics" was suggested as the subject of a lecture Rutherford was to deliver in South Africa, he growled: "The trend of modern physics? I can't give a paper on that. It would only take two minutes. All I could say would be that the theoretical physicists have got their tails up and it is time that we experimentalists pulled them down again!"[4]

Charles Ellis, for many years one of Rutherford's senior assistants in the

Cavendish Laboratory, recalled an outburst at dinner in Trinity College over Arthur Stanley Eddington's derivation of the fine-structure constant: "How can a fellow sit down at a table and calculate something that would take me, me, six months to measure in the laboratory?"[5] "Just by thinking!," Rutherford muttered in bewilderment.[6] This apparent hostility to theory was displayed frequently at colloquia. To Werner Heisenberg: "We are all much obliged for your exposition of a lot of interesting nonsense, which is most suggestive."[7] To Niels Bohr, who had delivered the prestigious Scott Lectures, on the uncertainty principle: "You know Bohr, your conclusions seem to me to be as uncertain as the premises upon which they are built."[8] To most lecturers, as he probed to extract the physical picture behind their mathematical techniques: "I am a simple man and I want a simple answer."[9]

In private conversation, too, Rutherford's tongue could be sharp. Before George Gamow left Copenhagen to explain his quantum theory of nuclear transformation to Rutherford, Bohr, whose devotion to Rutherford was lasting and deep, warned the young Russian that the "old man" disliked innovations and believed that "any theory is good only if it is simple enough to be understood by a barmaid."[10] Two decades earlier, at the 1910 Radiological Congress in Brussels, Rutherford, Willy Wien, and A. S. Eve chatted about relativity while at lunch. Wien insisted that Newton's way of adding velocities must now be modified, and dispaired that any "Anglo-Saxon can understand relativity." "No!," replied Rutherford, "they have too much sense."[11]

At the height of his fame in the early 1930s—as a former president of the Royal Society, a member of the Order of Merit, and raised to the peerage as Baron Rutherford of Nelson—he used public platforms to continue joshing theoreticians. The *Daily Telegraph* quoted his rueful comment that "among the many reasons we have for envying our immediate ancestors is the lost security of science. The large scale universe has been made by the mathematicians notoriously incomprehensible."[12] And in a formal toast at the Royal Academy of Arts, Rutherford remarked:

Quite recently there has been much interest taken by the cultivated public in the metaphysical aspects of science, especially in those of theoretical physics. Some of our publicists have boldly claimed that the old ideas which served science so well in the past must be abandoned for an ideal world where the law of causality fails, and the principle of uncertainty, so valuable in its proper domain of atomic physics, is pushed to extremes. The great army of science in its march into the unknown discusses with interest and sometimes amusement, these fine spun disputations of what is reality and what is truth. But it still goes marching on calling out to the metaphysicians "there are more things in heaven and earth than are dreamt of in your philosophy."[13]

Sometimes caustic, sometimes bellicose, Rutherford's statements unfortunately fail to convey the atmosphere in which they were delivered. The printed word ignores the twinkle in his eye and the half-smile on his lips. His booming, intimidating voice was merely part of his personality, as much as his proverbial energy and enthusiasm. Secure in his own self-image and honestly appreciative of his colleagues' contributions, Rutherford nevertheless liked to tease those in other parts of physics and especially those in other scientific fields (physics was the only real science; chemistry was "stinks" and all the rest were "stamp collecting"). More of this bantering is apparent in Rutherford's 1934 letter to Enrico Fermi, who had begun his career as a theoretician and was well known for his theory of beta decay, but who had switched to experiment when the neutron was discovered and he found that it could make many elements artificially radioactive. Rutherford congratulated Fermi on his "successful escape from the sphere of theoretical physics." He continued: "You seem to have struck a good line to start with. You may be interested to hear that Professor Dirac is doing some experiments. This seems to be a good augury for the future of theoretical physics!" [14] Yet, beyond this anecdotal portrait of a joking, teasing, and even irritated experimenter, what can be said of Rutherford's true feelings about theoretical physics and, further, about his own contributions to and effect upon theory?

Rutherford's Real Attitude Toward Theory

For Rutherford physics was almost synonymous with experimental investigation: "The properly trained physicist has the power to form his judgment by performing experiments." [15] How could one understand nature except through direct testing? [16] Theory was valuable, to be sure, but the goal of science was a visual understanding, a picture of nature, and that was best accomplished by empirical methods. The bias is evident in Rutherford's early 1907 correspondence with Authur Schuster, whom he was to succeed at Manchester. Rutherford lamented that mathematical physicists "as a rule know a certain amount of theoretical physics but ... are almost entirely destitute either of experimental knowledge or instinct. Even the best of them have a tendency to treat physics as purely a matter of equations." Such specialization, he opined, was bad for physics. Rutherford recognized, however, that "the experimenter is inclined to drop his mathematics [because] it is extremely difficult to keep up ... when all your energies are absorbed in experimentation." [17] Clearly, specialization of this sort was justified.

This preference for experiment-based knowledge lasted throughout Rutherford's career. Samuel Devons, a Cavendish student in the early 1930s, recalled his professor's obsession with facts: "Facts were to be respected and treated quite differently from theory, which was, in a sense, opinion." [18] Opinion, and its esthetic underpinning, a sense of simplicity and order in nature, were not foreign to Rutherford. He did, in fact, have remarkable intuitive powers. Yet, while he could appreciate that a promising theory should not be rejected out of hand because it did not match experiment precisely, he would have roared his disapproval of P. A. M. Dirac's belief that "it is more important to have beauty in one's equations than to have them fit experiment"[19] or Einstein's indifference as to whether the measurement of the bending of starlight by Eddington's eclipse expedition confirmed his general-relativity calculation of that effect. If the data had not agreed, Einstein ventured, "Then I would have been sorry for the dear Lord—the theory *is* correct."[20]

All this is not to suggest that Rutherford for a moment supported the "Aryan" method of science, which had a degree of popularity in Germany in the 1930s under Philipp Lenard and Johannes Stark and which looked upon experiment as the only valid way to study nature.[21] Nor, of course, could he be sympathetic to the view of E. A. Milne and Eddington that nature might be derived solely from mind and mathematics.[22] Never an extremist, Rutherford preferred experiment but respected theory. In fact, he claimed to encourage the floating of ideas. Responding to W. H. Bragg's belief in the "suspended recombination" of ions, he wrote in 1906: "It doesn't sound too unreasonable and [it] seems to fit in with things. I am a great believer of hypotheses, and no great harm is done if they have to be modified. The facts remain, and the more attractive clothing they can be wrapped in the better."[23] But one's enthusiasm for ideas must be held captive to experiment, as Rutherford noted in 1913 when he sought to soothe Kasimir Fajans's anger that Soddy had plagiarized him: "I personally feel that the making of hypotheses of this kind [group displacement laws] is not nearly so important as the proof of them, for obviously any general deduction of this kind must depend on whether the chemistry of all bodies is known with sufficient certainty. I think it is not very difficult to formulate hypotheses, but the essential point in them is that they should be consistent with the known data and suggest lines of work."[24]

At times Rutherford could be eloquent and statesmanlike—especially when it was expected, as in his response to a toast by the president of the Royal Academy of Arts:

I think that a strong claim can be made that the process of scientific discovery may be regarded as a form of art. This is best seen in the theoretical aspects of Physical Science. The mathematical theorist builds up on certain assumptions and according to well understood logical rules, step by step, a stately edifice, while his imaginative power brings out clearly the hidden relations between its parts. A well constructed theory is in some respects undoubtedly an artistic production. A fine example is the famous Kinetic Theory of Maxwell.... The theory of relativity by Einstein, quite apart from any question of its validity, cannot but be regarded as a magnificent work of art....[25]

Rutherford's prose was equally ornamental in his 1923 presidential address to the British Association for the Advancement of Science, although he could not indulge in the luxury of speaking about something of which he knew so little as art:

Experiment, directed by the disciplined imagination [i.e., theory] either of an individual, or still better, of a group of individuals of varied mental outlook, is able to achieve results which far transcend the imagination alone of the greatest natural philosopher. Experiment without imagination, or imagination without recourse to experiment, can accomplish little, but, for effective progress, a happy blend of these two powers is necessary. The unknown appears as a dense mist before the eyes of men. In penetrating this obscurity we cannot invoke the aid of supermen, but must depend on the combined efforts of a number of adequately trained ordinary men of scientific imagination. Each in his own special field of inquiry is enabled by the scientific method to penetrate a short distance, and his work reacts upon and influences the whole body of other workers. From time to time there arises an illuminating conception, based on accumulated knowledge, which lights up a large region and shows the connection between the individual efforts, so that a general advance follows. The attack begins anew on a wider front, and often with improved technical weapons. The conception which led to this advance often appears simple and obvious when once it has been put forward. This is a common experience, and the scientific man often feels a sense of disappointment that he himself had not foreseen a development which ultimately seems so clear and inevitable.[26]

This alliance between theory and experiment was something which Rutherford could not deny—although on occasion he might temporarily have been reluctant to admit it, as during the heyday of theory that preceded the flurry of profound experimental discoveries in the early 1930s. In his opening remarks to the 1934 International Conference on Physics, in London, he called attention to the "most interesting and exciting stage" nuclear physics had reached, and urged "a close collaboration between the theoretical and experimental physicists ... for rapid progress in this most fundamental of problems."[27] And yet there was a nagging distrust of theoretical conclusions alone; they required empirical verification (or falsification). In 1921, in this spirit, Rutherford communicated to the *Philosophical Magazine* a paper in which J. A. Crowther described the J series of characteristic x rays from different elements, although Bohr's theory of the atom precluded their existence.[28] Similarly, although Rutherford was impressed with George Gamow's concept that particles did not have to climb over the nuclear potential barrier but could tunnel through it at lower energies, he hedged his bet by asking the electrical engineers for millions of volts when he spoke at the opening of the Metropolitan Vickers High Voltage Laboratory in February 1930.[29] Where either theory or experiment could do the job, Rutherford chose the technique he found most comfortable— even if it involved cut-and-try methods, as when he tested models to determine the shape of a large electromagnet he wished to construct in 1936. (He reasoned that the latter was faster and more certain than making elaborate and difficult calculations.[30])

Equations, to Rutherford, were the tail that should never wag the dog. They were necessary to explore the options of interpreting phenomena, but they should not have a life of their own. Physical explanation was the goal, not mathematical exotica. In Rutherford's ideal world, theoreticians would have some hands-on acquaintance with scientific apparatus and, even more important, would maintain contact with the laboratory's experimental program and direct their investigations toward the problems that arose there.[31] Departures from the ideal were rarely tolerated. Lancelot Law Whyte confessed to Rutherford in the early 1920s that he really did not belong in the Cavendish. While physics appealed to him deeply, he was neither a good experimenter nor a good mathematician. Rutherford asked him "What kind of physics interests you?" "Theories, principles, ideas," replied Whyte. "Then," said Rutherford, "go and join the Continentals!"[32] What Rutherford meant by this, and what was widely recognized in the early part of this century, is summarized in the postscript of a letter that Rutherford wrote to W. H. Bragg in 1911:

I was rather struck in Brussels [at the first Solvay Congress] by the fact that the continental people do not seem to be in the least interested in trying to form a physical idea of the basis of Planck's theory. They are quite contented to explain everything on a certain assumption, and do not worry their heads about the real cause of the thing. I must say that I think the English point of view is much more physical and much to be preferred.[33]

Rutherford continually dealt with mental images of what he hoped was reality, which is not surprising since he was a product of the great nineteenth-century British tradition of conceptual model-making. Physicists, Arthur Schuster wrote, "flet the necessity of having clear conceptions, and distrusted mathematical symbols that had no fully defined meaning." Maxwell, he added, was a practitioner of this school, and "devoted a good deal of time to the specification of electrostatic and magnetic strains in his medium."[34] It was obvious that, to understand the physics of a phenomenon, one had to have a picture of it. When Eddington (like Rutherford a longtime senior fellow of Trinity College) speculated one evening that electrons were perhaps only mental concepts and had no real existence—that our means of observation forced them to "appear"—Rutherford rose and, "with the air of one saying 'You have insulted the woman I love.'" exclaimed: "Not exist? Not exist? Why I can see them as plainly as I can see that spoon in front of me."[35] Not only did he distrust concepts that led to no clear model of nature, he felt it to be "unscientific and also dangerous to draw far-flung deductions from a theoretical conception which is incapable of experimental verification, either directly or indirectly."[36] Theory, to be judged worthwhile, had to yield sharp mental images *and* be capable of experimental test. As always, the proof of the pudding lay in experiment.

But although Rutherford always called himself a simple man, and this picture of a visual, experimental scientist with a distaste for mathematical abstractions seems quite straightforward, the real Rutherford was more complex. He did more than give lip service to the interaction of theory and experiment; he actively supported it. Niels Bohr spent four months in Rutherford's laboratory at the University of Manchester in the spring of 1912, and then returned to

Copenhagen. Rutherford characteristically read all his students' papers, since their work reflected upon the laboratory, and it was traditional for the professor to "communicate" such papers to the scientific periodicals (in some fashion serving as a referee). Bohr's series of articles the next year on the constitution of the atom demanded much from Rutherford in both editorial and scientific criticism—far more than Rutherford usually had to give—but he responded enthusiastically. The topic, of course, was of great interest, being based upon Rutherford's own recent (1911) nuclear model of the atom. In a famous letter one sees Rutherford trying to extract a picture of Bohr's ideas, and immediately focusing upon the points of greatest significance:

Your ideas as to the mode of origin of [the] spectrum [of] hydrogen are very ingenious and seem to work out well; but the mixture of Planck's ideas with the old mechanics make[s] it very difficult to form a physical idea of what is the basis of it. There appears to me one grave difficulty in your hypothesis, which I have no doubt you fully realise, namely, how does an electron decide what frequency it is going to vibrate at when it passes from one stationary state to the other? It seems to me that you would have to assume that the electron knows beforehand where it is going to stop.[37]

Despite his reservations about the "validity" and the "underlying physical meaning" of Bohr's assumptions, Rutherford recognized the importance of this contribution and became its chief British enthusiast.[38] Norman Feather, a Cavendish Laboratory student and staff member in the 1930s and an astute observer of his professor, sees here Rutherford's usual reflex of disbelief, followed by a gesture of tolerance.[39] It must be noted also that, by this time, Rutherford privately admired Max Planck and was sympathetic to the applications of his theory. In 1908, Rutherford and Hans Geiger had measured the charge on an alpha particle and from that deduced the electronic charge e. The figure was over 35 percent greater than the accepted value, and they were much relieved to learn that Planck had derived a number almost identical to theirs from the quantum theory of temperature radiation.[40]

In the late 1920s, Gamow's (and Gurney and Condon's) interpretation of alpha decay in terms of the new wave mechanics undoubtedly interested Rutherford, who regarded the alpha particle as almost a personal possession. While the details of wave mechanics would have been of little concern, the picture they gave of the phenomenon was revolutionary. Yet the Cambridge theoreticians A. H. Wilson and Ralph Fowler (Rutherford's son-in-law) soon found him convinced that the idea had great merit.[41] Since the logic applied to particles penetrating to the nucleus as well as to those emitted from it, accelerator builders had to take notice. Although their progress was far slower than Rutherford wished, John Cockcroft and E. T. S. Walton constructed their machine in the Cavendish and were the first to show the transformation of nuclei in their bombardment of lithium by protons.[42] Particle accelerators would have been built irrespective of Gamow's theory, but the successful test at energies below the potential barrier shows the warm reception given to new ideas—even theoretical concepts—in Rutherford's laboratory. Rutherford's openmindedness with respect to another conclusion drawn from Gamow's work was in evidence on February 7, 1929, when as chairman of a Royal Society discussion on "The Structure of Atomic Nuclei" he welcomed the young Russian by noting: "It will be seen that this theory makes the radius of the uranium nucleus very small, about 7×10^{-13} cm, and in this small nuclear volume 238 protons and 146 electrons have to be made room for. It sounds incredible but may not be impossible."[43]

Sometimes Rutherford learned of a new idea only after it had been successfully tested, but even here some credit may be given him for the atmosphere he created in his laboratories. If he himself was willing to try out almost any "damn fool idea," his colleagues adopted the same attitude. And if theory predicted an effect, the idea was probably not so foolish. Nevill Mott's early research involved showing that "Rutherford scattering" of projectiles by nuclei was explained quantitatively by wave mechanics. In the case of identical particles, however, Mott predicted that there should be twice the classically calculated amount at $45°$, because of interference phenomena. He was tremendously excited when James Chadwick and Patrick Blackett, both very senior members of the Cavendish, each undertook to bombard helium nuclei with alpha particles—an experiment no one would have thought to perform but for theory, because there had been no strange data that suggested a problem. When the tests confirmed the prediction, Chadwick took Mott to a surprised Rutherford, who bestowed great praise on him: "Well, Mott, if you think of anything else like that, come and tell me."[44]

If Rutherford thought the subject quizzical, he was practical nevertheless: "The theory of wave-mechanics, however bizarre it may appear—and it is so in some respects—has the astonishing virtue that it works."[45] Nothing succeeds like success, and Rutherford was too seasoned a scientist to reject contributions, even from theorists whose thought processes he found unfathomable. He presumably learned the lesson early, as a postgraduate student (1895–1898) of J. J. Thomson, a man who combined experimental and mathematical skills of a high order. Except for his years at McGill University (1898–1907), Rutherford had men who would have described themselves primarily as theoreticians attached to his laboratories, formally or informally. Their leavening influence prevented him from becoming a caricature of a tunnel-visioned experimentalist.

Rutherford's personality kept him from extremism. While he might make outrageous statements, his half-humorous approach deflected serious criticism of his positions. With rare exceptions he avoided controversy, and he made private efforts to settle disputes between others in the belief that the airing of dirty linen was bad for science.

Rutherford's orientation was basically nonphilosophical. He observed the great conceptual problems of the early twentieth century calmly from a distance, except when they intersected his own research; even then, he experienced no crises of conscience. When Bohr and H. A. Lorentz, together in Cambridge in 1923 to receive honorary degrees, had a heated debate at a meeting of the Cambridge Philosophical Society on how to reconcile quantum and classical theories, Rutherford found the dispute "amusing and instructive"[46]; however, it apparently did not raise his intellectual temperature.

In 1908 Rutherford and Geiger confirmed Egon von Schweidler's understanding that alpha particles were not emitted from radioactive sources uniformly in time but fluctuated according to the law of probability.[47] Since alpha emission was one path by which a radioelement naturally transmuted into another, this meant that radioactive decay followed statistical laws. A measured half-life of, say, 3.82 days (for radium emanation) meant not that every emanation atom decayed in 3.82 days from its moment of creation but that a large population of atoms gave this average figure. Any individual atom might transmute in a longer or a shorter time. In the early years of this century such concepts ran counter to the prevailing view of determinism in nature. For every effect there was a proximate cause; atoms of the same element should not differ and require treatment by statistical means. Rutherford's schooling, however, had included Maxwell's kinetic theory of gases, and he had at least heard of other statistical treatments of phenomena by physicists such as Ludwig Boltzmann and J. Willard Gibbs. An open mind was one of his strong suits. Perhaps even more important, Rutherford's view was entirely pragmatic; if something seemed to work, there might be truth in it. He might have an inchoate opinion about the deeper philosophical significance, but where hard evidence showed the need for a revolutionary interpretation he had little difficulty in adopting it.

Along with the determinism bred into so many classical scientists went a belief in continuum physics. Planck's quantum theory was a major step in altering the way in which phenomena were seen, and Rutherford's willing acceptance of it has been noted. A related change, involving the continuum of matter, was the age-old controversy over atomism. Even after 1900, physicists and chemists were far from unanimous that matter's ultimate resolution lay in the discrete particles called atoms.[48] For Rutherford the subject required no discussion at all. He knew atoms existed; he worked with them every day. Continuum physics could easily be jettisoned. As for the classical ideas that atoms were indestructible and homogeneous, Rutherford's own work in explaining the phenomenon of radioactivity and in creating the picture of the nuclear atom dealt them a fatal blow. Similarly, the belief that atoms were the smallest pieces of matter fell easily when electrons in various guises (sometimes as beta particles) were discovered. The point here is that Rutherford was not so introspective that he worried about the foundations of physics, even as he was rebuilding them.

Radioactive bodies, with their seemingly endless emission of alpha and beta particles and gamma rays, appeared to violate one of the cornerstones of nineteenth-century physics: the conservation of energy. Rutherford collected data rather than doubts, and he and Soddy, in their explanation of the phenomenon of radioactivity, published in several papers in 1902 and 1903, calmly suggested that atoms had enormous stores of internal energy.[49] Old beliefs might be hallowed, but not sacred. Even the non-intuitive behavior of nature could be digested easily, especially if backed by experiment. Before Einstein explained it by his theory of special relativity, a number of physicists, including J. J. Thomson, Oliver Heaviside, G. F. C. Searle, Max Abraham, and H. A. Lorentz, concluded that electrical charges in motion act as if they have an apparent mass which becomes large as the charges' velocity approaches that of light. Walther Kaufmann in 1902 detected just such a mass increase in beta particles from radioactive substances. Rutherford, who as a student became familiar with Maxwell's theory and J. J. Thomson's elaboration of it, probably would have accepted the concept of an electromagnetic mass increase on the strength of these theoreticians' arguments; Kaufmann's data, however, made it real.[50]

Most ideas that contradicted ordinary experience probably got little attention from Rutherford but did not receive automatic dismissal. So great was his stature in his specialty, however, that others often assumed that he knew far more than he said about topics beyond his own field. When his old friend Henry H. Dale asked him about the recent confirmation of Einstein's general theory of relativity (i.e., the demonstration that light bent when it passed near massive objects), Rutherford remarked: "I don't believe there are more than about six people in the whole world who really understand what that Einstein theory means." When Dale asked who were the other five, Rutherford looked puzzled for a moment and then, with a hearty laugh, exclaimed "Good lord! Dale. You don't suppose that I understand it, do you?"[51]

Another theory that could be observed from some distance was wave-particle duality. Charles Ellis met Louis de Broglie in Paris and responded enthusiastically to his ideas. When Ellis returned to Cambridge hoping to test them, Rutherford gave him little support, apparently because the Cavendish lacked the essential vacuum equipment and de-gassing techniques. Still, Ellis recalled that Rutherford's view was "a liberal one":

Radiation, which he was quite certain was radiation, behaved sometimes like a particle; therefore an electron, which he was finally and absolutely convinced was a particle, could, as far as he was concerned, have some of the properties of radiation.[52]

The asymmetry of nature generated some thought in Rutherford. Why should electrons be so abundant and positrons so rare? "It may be," he told the Institution of Electrical Engineers in 1935, "that in another part of our universe, the roles of positive and negative electrons are reversed, so that our outraged sense of fair play may be assuaged on the cosmic scale."[53]

On the question of the annihilation of matter Rutherford expressed skepticism,[54] but he was remarkably unopinionated about the majority of theoretical developments. His career spanned an astounding period of conceptual change, from classical to modern physics. He witnessed new ideas about the permanence and structure of atoms, the conservation of mass and energy, relativistic mass changes, space and time, the quantum and the statistical nature of some phenomena, wave-particle duality, the addition of nuclear forces to those of gravity and electromagnetism, the uncertainty principle, and even the inability to construct the necessary mental mechanical models of phenomena (mathematical models were necessary). He did not go out of his way to stay abreast of all the latest theoretical developments, and he rarely was interested in the internal logic or mathematics of the theory. What did arouse his attention occasionally were the speculative implications, and what invariably made him take notice were the predictions that could be tested experimentally. Robert A. Millikan is looked upon as an experimentalist who tried to put his finger in the dike to stop the flood of revolutionary change (although his work did just the opposite). Rutherford, without self-consciousness, was one of the engines of this change. Though he was content with the *ancien régime*, new discoveries demanded new interpretations, so revolution came to him with no doubts or regrets.

The popular image of Rutherford as hostile to theory, derived from the sort of anecdotes related above, must be revised in light of the evidence presented in this section. His real attitude ranged from indifference to keen interest, and behind the banter he clearly appreciated what his theoretical colleagues accomplished. Moreover, in addition to the factors of schooling, personality, and pragmatic attitude, which help to explain his relationship to theory, there is another: Rutherford was himself an occasional theoretician.

Rutherford as a Theoretician

Einstein once made the distinction that "a theory is something nobody believes except the person proposing the theory, whereas an experiment is something everyone believes except the person doing the experiment." [55] If we are to call Rutherford a theoretician, the term must be defined; this is not as simple as one might think, nor is Einstein's puckish humor of much help. Indeed, it seems that most practitioners of such work and most historians who write about them have a variety of incompletely or baroquely expressed explanations.[56]

An experimentalist is fairly easy to define as one who uses apparatus to perform tests upon nature. Experimentalists, of course, are not prohibited from using paper, blackboards, computers, or their minds, but there is universal agreement that they gather data about natural phenomena. It would be facile to define theoreticians as scientists who do not perform physical experiments, but a negative tells us too little. Theoreticians cover a broad spectrum, from those who analyze experimental data and look for regularities to those who ponder the most metaphysical and epistemological questions. Then try to discover the basic laws of nature and the consequences that flow from them. Depending upon the complexity of the subject, they may use logical argument or mathematical techniques to explain why a particular phenomenon occurs or to question the underlying assumptions of our view of the universe.

John Slater saw a distinction between two types of theoretical physicists: the "pragmatic, matter-of-fact sort" and the "magician [who] waves his hands as if he were drawing rabbits out of a hat and ... is not satisfied unless he can mystify his readers or hearers." He looked upon himself, Schrödinger, and Heisenberg as members of the first group, and Dirac as representative of the second. Understandably, Slater had little taste for what he regarded as "needlessly formal and general theoretical tools."[57] He also recognized that for students the greatest difficulty lay "in learning how to apply mathematics to a physical situation, how to formulate a problem mathematically, rather than in solving the problem when it is once formulated." Further, Slater stressed the value for theoreticians of some practical laboratory training:

The same ability to overcome obstacles, the same ingenuity in devising one method of procedure when another fails, the same physical intuition leading one to perceive the answer to a problem through a mass of intervening detail, the same critical judgment leading one to distinguish right from wrong procedures, and to appraise results carefully on the ground of physical plausibility, are required in theoretical and in experimental physics. Leaks in vacuum systems or in electric circuits have their counterparts in the many disastrous things that can happen to equations. And it is often as hard to devise a mathematical system to deal with a difficult problem, without unjustifiable approximations and impossible complications, as it is to design apparatus for measuring a difficult quantity or detecting a new effect.[58]

In addition to Slater's stylistic distinction between the minimum of straightforward mathematics adequate to do the job and a banquet of elegant formalism, one must distinguish between mathematical and theoretical physics. Max Born, in an obituary of Arnold Sommerfeld, called him a mathematical physicist because "his gift was not so much the divination of new fundamental principles from apparently insignificant indications or the daring combination of two different fields of phenomena into a higher unit" as the "logical and mathematical penetration of established or problematic theories and the derivation of consequences which might lead to their confirmation or rejection."[59] Sommerfeld, Born noted, had been especially gifted in uncovering the relations among experimental data. Others see almost a mirror image and associate theoretical physics more closely with experiment or physical intuition,[60] as did Slater. For the purposes of this essay the latter view seems more appropriate; both may use mathematics, but the theoretical physicist relies on physical insights gleaned from problems whereas the mathematical physicist depends more heavily on mathematical intuition. Lewis Pyenson has made the same distinction, pointing out that Felix Klein and Hermann Minkowski were mathematicians above all, and behaved as if they were applying mathematics, whereas Lorentz, Boltzmann, and their successors in the Leiden and Vienna chairs of theoretical physics, though using complicated mathematics, nevertheless sought primarily to make the physical principles clear.[61]

Though the above discussion leaves us with no precise capsule definition of a theoretical physicist, or even a fully reliable sieve to separate the theoretical from the mathematical species, we may extract from these ideas some points pertinent to Rutherford's case: Some theoreticians analyze data and search for regularities or laws therein; they look for the consequences of their hypotheses; logical argument may serve their purposes as well as mathematics; and simplicity rather than elegance is considered a virtue. If these characteristics are accepted, it may be argued that Rutherford occasionally wore the hat of a theoretician.

In the following, no attempt is made to be exhaustive. The purpose merely is to show something of the range of Rutherford's conceptual forays, which ran the gamut from his almost casual speculative remarks to his more common carefully reasoned arguments that seemed to be driven inexorably by the experimental data.

When Rutherford went to the Cavendish Laboratory as a research student in 1895, he brought with him from New Zealand a radio-wave detector with which he had made some interesting discoveries. However, his professor, J. J. Thomson, soon recognized his abilities and invited him to join in the laboratory's main line of investigation: the conduction of electricity in gases. Further, Thomson asked the young Rutherford to collaborate on a report to the 1896 meeting of the British Association for the Advancement of Science—an honor Rutherford could not decline. The result was a paper that went far toward specifying a theory of ionization. Charged particles were shown to account for conduction in a gas exposed to x rays, and an analogy was made to a dilute solution of an electrolyte.[62] Almost certainly, Rutherford ran the experiments; just a little less certainly, Thomson provided the interpretation. The work was rather straightforward, and the experience probably was a lesson to Rutherford that important ideas can flow from good data.

Rutherford was appointed to a professorship at McGill University in 1898, and there he conducted the research in radioactivity that brought him a Nobel Prize. One of his first significant discoveries was of thorium emanation. So uncertain was he at first of its nature—gas, vapor, or particulate matter from the thorium—that he deliberately chose a term (*emanation*) that had not been used much since the time of Robert Boyle and was thus suitably vague. Before long, Rutherford became convinced that emanation was a radioactive gas of the inert argon family. Pierre Curie objected to this interpretation, believing emanation to be "centers of energy condensation" entrained between gas molecules. Rutherford, in perhaps the only instance of his speaking of "my theoretical views," strongly defended his conclusion (which was based upon diffusion experiments that gave a high molecular weight to the gas, on chemical tests that showed its nonreactivity, and on the condensation of the gas at liquid air temperature).[63] One may quibble that this is more the explanation of experimental results than what we normally call theory. However, ideas, interpretations, and theories may come in all sizes. In this particular case, the recognition that emanation changed into another material (then called active deposit) set the stage for a full-fledged "proper" theory that explained the phenomenon of radioactivity. This was the idea most commonly called the transformation theory, occasionally called the disintegration theory, and sometimes even called the transmutation theory. Rutherford and Soddy developed it in a series of papers published in 1902 and 1903, and it was mathematized a little later. The central concept is that radioactivity is the manifestation of atoms of certain elements spontaneously emitting "radiations" (what, how, and why remained to be more fully explained) and transmuting themselves into atoms of other elements. This iconoclastic idea was based upon a wealth of highly competent experiments and correlated a great amount of data. However, inevitable as it looks today, it was not the only interpretation; the Curies, for example, championed for another year their own theory of undetectable aetherial waves that stimulated the heavy elements to emit alpha, beta, and gamma radiations. These radiations, particularly the alpha, carried an enormous amount of energy. Not a few people in that period cautiously suggested that this apparent violation of the law of conservation of energy might signify that the law was invalid. Rutherford and Soddy took pains to state that their theory did not threaten this bedrock of nineteenth-century physics, because the energy of radioactive decay must come from hitherto unexpected, large stores of internal energy within the atoms. Indeed, they specified that all atoms possess such energy. In addition, they speculated that radioactivity might be the means by which stars shine for eons, and that the inactive product at the end of each decay series was accumulating and its quantity might be detectable—an idea that was to lead to a method of dating ancient rocks. The Rutherford-Soddy theory, moreover, had the desirable characteristic of being "pregnant" with new ideas to test. By claiming that there were parent-daughter relationships between pairs of radioelements arranged in decay series, Rutherford and Soddy stimulated the radiochemists to detect and identify the numerous active bodies in these series.[64] In a real sense the transformation theory is analogous to Mendeleev's construction of the periodic table of the elements, which also predicted the existence of new elements that subsequently were found.

Concerning the theory's mathematization, mentioned above, Rutherford used the occasion of the Royal Society's distinguished Bakerian Lecture in 1904 to describe several cases in which a radioactive body was removed from its daughter product and from the daughter's descendents. Then he presented the equations by which the quantity of any product at any time could be calculated. He also derived the equations needed to calculate the radioactivity at any time of a sample containing several radioelements, each with a different half-life.[65]

Rutherford had had a solid mathematical training at Canterbury College in Christchurch, New Zealand, and he occasionally (as in the above-mentioned case) used simple integrals and differential equations in his work. His theories were not the result of educated yet arbitrary approximations or boundary conditions necessary to allow a calculation to proceed. By no means could the moderate amount of mathematics found in his publications be called sophisticated or elegant, and he did not always use the most efficient approach. He used what was "available," and it was effective. In 1910 Harry Bateman "cleaned up" this particular piece of Rutherford's work and provided a less laborious and more symmetrical means of deriving the transformation formulas.[66]

John A. Wheeler writes:

I think that Rutherford deserves some credit for one of the key ideas of quantum theory. In the ... theory of radioactive decay he insisted that the concept of the radioactive decay constant had to be accepted. You could not understand it; you could hardly believe it. It was so incredible. It did not matter how old an atom ... was. But however old it was, there was the same decay constant. And just simply to insist that the experiments told you this and you had to take it was really the key idea of transition probabilities in the quantum theory.[67]

Wheeler is correct that Rutherford was impressed with the very unsettling realization that an atom of, say, radium emanation, with a half-life of just under four days, might disintegrate in seconds or years. But, as mentioned above, Rutherford was part of a generation that increasingly came to accept statistical phenomena, and he also followed Galileo's approach of asking how and not why changes occur.

When Rutherford moved to the University of Manchester in 1907, he continued his occasional forays into theory. Radioactivity could be detected by the darkening of the emulsion of a photographic plate, but this technique was more qualitative than quantitative and Rutherford rarely used it. At McGill he had preferred to measure the accumulated charges produced by the alpha and beta particles, using electrometers and electroscopes. As the science grew more sophisticated, and after Sir William Crookes popularized the observation of scintillations of light caused by a speck of radium placed near a zinc sulphide screen, Rutherford was able to show that each flash was produced by a single alpha. The counting of individual decays was now possible, and Rutherford and the superb experimenter Hans Geiger perfected the technique. The two men went on to develop an electrical method of counting events (which evolved into the famous Geiger-Müller tube many years later), but at Manchester scintillation counting reigned supreme. Because of its importance to his work, Rutherford had his student Ernest Marsden study the decay of the luminosity excited by alpha rays in the minerals most commonly used for screens, and then derived a theory to describe the process. One of the consequences of this work was the recognition that the active centers producing the scintillations were of molecular size, which negated the speculation that the light arose when small crystals were mechanically cleaved by the alphas.[68] This small theory wrapped up a line of work.

Another contribution from Manchester was a major theory that opened an enormous line of investigation. This, of course, was the nuclear theory of the atom. Marsden, as an undergraduate, was given a little research project: to see if alpha particles were reflected from a metal surface. Unexpectedly, he found that some alphas were deflected from their path by 90° and more when a beam of them was directed upon a thin gold foil.[69] In 1909 Marsden and Geiger counted about one alpha in 8,000 experiencing such scattering, whereas the most probable deflection was less than one degree.[70] This was done during a decade when there were several atomic models, including Lenard's dynamids, Nagaoka's Saturnian atom, and (most important) J. J. Thomson's plum-pudding model. Since Thomson had shown the existence of the electron in 1897, it was clear that atoms were not homogeneous "billiard balls" but had a structure. Thomson himself took the lead in trying to devise quantifiable schemes to account for experimental evidence, and these models generally consisted of an atomic-size sphere of positive electrification with negative electrons arranged in a geometric pattern within the volume. Alpha and beta scattering were explained by the electrostatic attraction and repulsion of numerous encounters with the electrons. This was called small scattering, because for alphas the amount of deflection in a single encounter was tiny. However, even with a succession of twists and turns as one alpha was influenced by a multitude of electrons, Thomson's theory could not account for large-angle scattering.[71]

Rutherford was thereby led in 1911 to conceive of an atom that was mostly empty space. In its final form, electrons moved through a void, while the atom's mass and positive charge resided on a nucleus tens of thousands of times smaller than the atomic diameter. It was the strong electric field of this "point charge" that was able to turn an alpha, with its great energy and momentum, through such a large angle in a single collision or near miss.[72] Rutherford used analytic geometry to describe the hyperbolic orbits followed by the alpha particles. This is more mathematics than is commonly found in his papers, and it is math of a sort that he probably learned at Canterbury College. (Today it is taught in many high schools.) It was adequate and effective, even if it was not the simple integral that more mathematically sophisticated theoreticians would have chosen.[73] The idea of an atom composed like a solar system was radical, despite Nagaoka's somewhat similar model, which never was accorded much interest. It violated the conventional wisdom of solid atoms. Yet Rutherford scarecely promoted his own theory, because he lacked proof that it could be applied to phenomena besides radioactivity. The few others who discussed it likewise did not recognize its revolutionary potential. It took two more years until Niels Bohr showed how the nuclear model could explain x ray, optical, and chemical properties of the atom and pointed out that radioactivity was a nuclear phenomenon.[74]

Not long after Rutherford moved to Cambridge to succeed J. J. Thomson as Cavendish Professor, he was invited to deliver the Royal Society's Bakerian Lecture for a second time. Just as the 1904 lecture caught the crest of interest in the transformation theory of radioactivity, the one in 1920 followed hard upon Rutherford's spectacular work in deliberately disintegrating certain nuclei. In some subsequent experiments he detected what he believed to be a particle of mass 3. According to the ideas of nuclear constituents at that time, this meant that one electron bound three protons. Extrapolating from this, Rutherford pointed out that

If we are correct in this assumption it seems very likely that one electron can also bind two H nuclei and possibly also one H nucleus. In the one case, this entails the possible existence of an atom of mass nearly 2 carrying one charge, which is to be regarded as an isotope of hydrogen. In the other case, it involves the idea of the possible existence of an atom of mass 1 which has zero nucleus charge.... Such an atom would have very novel properties. Its external field would be practically zero, except very close to the nucleus, and in consequence it should be able to move freely through matter. Its presence would probably be difficult to detect by the spectroscope, and it may be impossible to contain it in a sealed vessel. On the other hand, it should enter readily the structure of atoms, and may either unite with the nucleus or be disintegrated by its intense field, resulting possibly in the escape of a charged H atom or an electron or both.[75]

In this famous passage Rutherford predicted the existence of the deuteron and the neutron, based on his belief that he had found helium 3. His premise happened to be wrong; the "light helium" was an ordinary alpha particle of mass 4, which had an unusually long range and had not been found before. Nevertheless, all the predicted particles were ultimately discovered. This says something about Rutherford's intuition, but even more about general attitudes at that time. Isotopes of the elements were accepted as real bodies, and, within unspecified limits, many were anticipated. The neutron prediction, for which this Bakerian Lecture is best known, also was an idea whose time had come. In 1920, Orme Masson in Australia and William D. Harkins in the United States independently described the neutron too.

Despite his prediction of the neutron, Rutherford was in no hurry to incorporate it into his conception of the nucleus. From his McGill days onward, he had been concerned with the constituents of first the atom and then the nucleus. Alphas and betas, being the only particles emitted from radioelements, were naturally the prime candidates. In 1906, Rutherford wrote to his good friend W. H. Bragg: "You will have seen J. J. T. finds only one electron for one atomic weight. I believe he will make a theory of matter with the α particle the basis next time. At any rate, I intend to do so later."[76]

In various subsequent papers, Rutherford did indeed advance and modify a theory of matter. In 1914, although he reported a failure to find protons in radioactive decay, he took protons as the fundamental unit of all atoms. Recent evidence that helium's mass was less than that of four hydrogen atoms indicated that four protons and two electrons were especially closely packed to constitute the alpha particle. Perhaps, therefore, the alpha was a secondary unit.[77]

When Rutherford in 1919 showed protons to be emitted under certain conditions of alpha bombardment.[78] and experiments were conducted upon many elements, he was able to be more specific in his modeling. Nuclei with masses of $(4n + 2)$ and $(4n + 3)$, where n is an atomic mass unit, gave off protons, whereas none were detected from $4n$ nuclei. This meant, according to Rutherford, that alphas constituted a central core, and less tightly bound protons existed as satellites within the nuclear system. During the 1920s Rutherford's nuclear model grew more complex, with both positive and negative satellites.[79] Even after 1928, when George Gamow advanced his quantum theory of alpha decay, which accounted for the phenomena that Rutherford's theory was designed to handle, Rutherford found it hard to let go of his theory. As Roger Stuewer points out, the famous 1930 text by Rutherford, Chadwick, and Ellis, *Radiations from Radioactive Substances*, contains both Gamow's and Rutherford's theories, but the satellite model was followed by "highly critical comments ... written by Chadwick, which were included over Rutherford's grunts."[80]

Conclusion

A case has been made that Rutherford should rank among theoretical as well as experimental physicists. Is this, however, a distinction without a difference? No, because this terminology, which originated during Rutherford's lifetime, is still in use. Do we learn anything by such categorization? The answer emphatically is yes, because it clarifies our understanding of the manner in which physicists worked in various countries at other times. Such insight is especially useful now, when students are streamed into one path or another fairly early in their training and have far less familiarity with the other approach.

Michael Faraday, with whom Rutherford often is compared for his intuitive experimental skill, likewise advanced theoretical concepts (notably the lines of force in space), but suffered personal duress because his contemporaries failed to appreciate this aspect of his work.[81] Rutherford is not known to have been discouraged or frustrated when an idea met with indifference. Part of the contrast between the two men lies in their personalities; Rutherford was enormously self-confident. Another part lies in the dissimilarity of their careers;

Faraday was self-taught in science, knew little mathematics, and held no university appointment (although his position at the Royal Institution and his standing among scientists of his day were of great distinction). Rutherford was more than conventionally educated, knew enough mathematics for whatever he chose to do, and was in the thick of the activities of his profession. He knew that he was part of the "establishment" in the organizational structure of science and, even more, in the leadership of ideas.

Rutherford's feeling of ease or comfort with what he was doing is notable. His lack of concern about overturning long-held concepts has been mentioned above. To that can be added the comfort he derived from doing what he believed was expected of him. In Great Britain a tradition in physics had developed, of which James Clerk Maxwell, Lord Kelvin, Lord Rayleigh, and J. J. Thomson were the most prominent figures. Trained in rigorous analytical techniques for Cambridge's mathematical tripos examination, they combined, in their careers, extensive mathematical investigations with superb experimental work. By Rutherford's time the natural-sciences tripos had supplanted the mathematical tripos as the more common vehicle for admission to research in the Cavendish Laboratory,[82] but Rutherford took neither, since he had completed his undergraduate studies elsewhere. Still, he fits into this British tradition of designing experiments, conducting them, interpreting the results, predicting consequences, and testing for them. In an earlier age this would have been called natural philosophy; the scientist was expected to sit at both desk and workbench. Only during the latter part of the nineteenth century did physics bifurcate, and then far more on the Continent than in the British Isles. In the twentieth century, the experimental and theoretical lines slowly split in the United Kingdom, and then the division was especially pronounced because theory was taught far more as part of an applied-mathematics curriculum than as part of physics.[83]

Thus, when we label Rutherford a (part-time) theoretical physicist we are using a distinction that evolved during his lifetime and which by now is quite pronounced. In his generation, and particularly in Great Britain, a physicist was expected to possess a considerable range of talents. In the following generation, owing in large part to the success of quantum theory and relativity,[84] specialization was already noticeable, even in Rutherford's laboratories. At Manchester, Schuster endowed, for Rutherford's benefit, a mathematical readership within the physics department; Harry Bateman and Niels Bohr held this post, and C. G. Darwin's presence in the laboratory also must be mentioned. At Cambridge, although formally based in the mathematics department, Ralph Fowler in reality was resident in the Cavendish. The other theoretical personnel, comprising staff members, students, and visitors of longer or shorter duration, included G. I. Taylor, Nevill Mott, Alan Wilson, George Gamow, J. Robert Oppenheimer, John Slater, Hans Bethe, E. C. Stoner, E. J. Williams, Max Born, and Rudolf Peierls. Rutherford often encouraged them to go through an experimental "kindergarten," and even to undertake an experimental investigation, on the reasonable assumption that theorists would work to greater effect with some hands-on experience with apparatus and a closer understanding of current problems in the laboratory. For some (Bohr and Oppenheimer, notably), the workbench days were short; they soon pursued their primary interests. For others (Stoner, for example), the experimental period was long and unhappy; Stoner persevered because that was the route to his doctoral degree. The point here is that this next generation already perceived the dichotomy in physics and, while still students, aspired to follow either an experimental or a theoretical approach, but not both. That Enrico Fermi achieved great success in each line is, today, looked upon as awesome.

For Great Britain this specialization was costly, if inevitable. Rutherford was widely perceived as the leader of experimental physics (though the point of this essay is that he was more), and Cambridge was the Mecca for the Commonwealth's best physics students. To an unfortunate degree, these students sought to emulate Rutherford; indeed, training in the Cavendish was pointedly experimental. The consequence was a relative paucity of British theoreticians. Many of the posts in British universities went to foreigners, especially when Hitler's racial policies made a number of brilliant candidates available. These comments are in no way meant to be a criticism of Rutherford's behavior; in fact, he frequently tried to expand the field of theoretical physics in England. Rather, the circumstance was in large part due to the institutional separation of applied mathematics (i.e., theory) and experimental physics.

Rutherford, without question, was a great physicist. Some have gilded the lily and claimed that he never made a mistake. Of course he made mistakes, though they were relatively few and he often was the one to rectify them. The other revision necessary in evaluating his career is the topic of this essay: that he was not only an experimentalist. We can better understand physics in Rutherford's day and his own contributions if we recognize that he was overpraised and underestimated.

Acknowledgments

I wish to thank Alfred Romer, Paul Hoch, Judith Goodstein, and Lewis Pyenson for sharing with me their ideas and information on this topic. My appreciation is extended also to the John Simon Guggenheim Memorial Foundation for a fellowship, during which this essay was written.

Notes and References

1. Carl Seelig, *Albert Einstein. A Documentary Biography* (London: Staples, 1956), p. 188. I am grateful to Helen Dukas and to Alan Beyerchen for this reference.

2. C. H. Townes, "Quantum Electronics, and Surprise in Development of Technology," *Science* 159 (February 16, 1968): 702.

3. E. N. da C. Andrade, *Rutherford and the Nature of the Atom* (Garden City, N.Y.: Doubleday Anchor, 1964), p. 210. The same story is told in P. M. S. Blackett's paper "The Old Days of the Cavendish," *Rivista del Nuovo Cimento* 1 (1969), special number, p. xxxvii.

4. Andrade (note 3), p. 209.

5. N. F. Mott, "Rutherford and Theory," *Notes and Records of the Royal Society of London* 27 (August 1972): 65–66.

6. C. D. Ellis, interview by the author, June 3, 1970.

7. M. L. Oliphant, *Rutherford: Recollection of the Cambridge Days* (Amsterdam: Elsevier, 1972), p. 28.

8. Ibid., pp. 28–29.

9. Ibid., p. 28.

10. W. B. Lewis, "Some Recollections and Reflections on Rutherford," *Notes and Records of the Royal Society of London* 27 (August 1972): 61.

11. A. S. Eve, *Rutherford* (Cambridge University Press, 1939), p. 193.

12. *Daily Telegraph*, March 24, 1934.

13. Royal Academy of Arts toast, April 30, 1932, item 145 in Rutherford collection, Cambridge University Library. Published, with some errors, in Eve (note 11), p. 354.

14. Rutherford, letter to Fermi, April 23, 1934, reproduced in Edoardo Amaldi's "Neutron Work in Rome in 1934–36 and the Discovery of Uranium Fission," *Rivista di Storia della Scienza* 1 (February 1984): 9, and in Emilio Segrè's *Enrico Fermi: Physicist* (University of Chicago Press, 1970), pp. 74–75.

15. Article on Rutherford's talk at the Institute of Physics, *Nature* 129 (June 4, 1932): 823.

16. C. D. Ellis, interview by the author, June 3, 1970.

17. Rutherford, letter to A. Schuster, January 27, 1907, Schuster collection, Royal Society Library, London. I am grateful to Rutherford's grandchildren and to the Royal Society for permission to quote from this letter.

18. S. Devons, "Recollections of Rutherford and the Cavendish," *Physics Today* 24 (December 1971): 38–45, at 41.

19. P. A. M. Dirac, "The Evolution of the Physicist's Picture of Nature," *Scientific American* 208 (May 1963): 45–53, at 47. Dirac, in fact, was arguing that inadequate data could undercut a valid theory.

20. Quoted from a manuscript by Ilse Rosenthal-Schneider, Reminiscences of Conversation with Einstein, dated July 23, 1957, published in Gerald Holton's "Mach, Einstein, and the Search for Reality," *Daedalus* 97 (spring 1968): 653.

21. Alan Beyerchen, *Scientists Under Hitler: Politics and the Physics Community in the Third Reich* (New Haven: Yale University Press, 1977); J. Stark, "The Pragmatic and the Dogmatic Spirit in Physics," *Nature* 141 (April 30, 1938): 770–772.

22. Max Born, *Experiment and Theory in Physics* (New York: Dover, 1956), pp. 1, 36–44.

23. Rutherford, letter to W. H. Bragg, February 24, 1906, Bragg collection, Royal Institution, London.

24. Rutherford, letter to K. Fajans, April 2, 1913, Fajans collection, University of Michigan Library.

25. Rutherford, Royal Academy of Arts toast, April 30, 1932, item 145 in Rutherford collection, Cambridge University Library, quoted on p. 353 of Eve, *Rutherford* (note 11).

26. Rutherford, "The Electrical Structure of Matter," presidential address, *BAAS Report, 1923*, pp. 23–24.

27. Rutherford, "Opening Survey," *International Conference on Physics, London 1934* (London: Physical Society, 1935), vol. 1, pp. 4–16, at 16.

28. R. Stuewer, *The Compton Effect* (New York: Science History Publications, 1975), pp. 198–200. On the tradition of communicating papers, see below.

29. T. E. Allibone, letter to the author, May 7, 1969.

30. Rutherford, letter to P. Kapitza, August 5, 1936, Rutherford collection, papers section, Cambridge University Library.

31. Rutherford, letter to A. Schuster, January 27, 1907, Schuster collection, Royal Society Library, London.

32. L. L. Whyte, *Focus and Diversions* (London: Cresset, 1963), p. 47. I am grateful to L. Pearce Williams for this reference.

33. Rutherford, letter to W. H. Bragg, December 20, 1911, Bragg collection, Royal Institution, London.

34. A. Schuster, *Progress of Physics During 33 Years (1875–1908)* (Cambridge University Press, 1911; repr. New York: Anno, 1975), p. 43.

35. Andrade (note 3), pp. 208–209. C. D. Ellis (interviewed by the author, June 3, 1970) confirms this attitude.

36. Rutherford, letter to Herbert Samuel, 1933, quoted in Samuel, *Belief and Action* (London: Cassell, 1937), pp. 300–301.

37. Rutherford, letter to N. Bohr, March 20, 1913, Niels Bohr Institute, Copenhagen.

38. Rutherford, "The Structure of the Atoms," *Phil. Mag.* 27 (March 1914): 488–498, at 498. Reprinted in *The Collected Papers of Lord Rutherford of Nelson*, ed. J. Chadwick (New York: Interscience, 1962, 1963, 1965), vol. 2, pp. 423–431. The *Collected Papers* will henceforth be referred to as *CPR*.

39. N. Feather, "Some Episodes of the Alpha-Particle Story, 1903–1977," in *Rutherford and Physics at the Turn of the Century*, ed. M. Bunge and W. Shea (New York: Dawson and Science History Publications, 1979), p. 81.

40. Rutherford, note in honor of Max Planck, *Naturwissenschaften* 17 (June 28, 1929): 483.

41. N. F. Mott, "Rutherford and Theory," *Notes and Records of the Royal Society* 27 (August 1972): 65; Harrie Massey, "Nuclear Physics Today and in Rutherford's Day," ibid. 27 (August 1972): 26.

42. Guy Hartcup and T. E. Allibone, *Cockcroft and the Atom* (Bristol: Hilger, 1984), pp. 40–57.

43. Rutherford, "Discussion on the Structure of Atomic Nuclei," *Proc. Roy. Soc.* A 123 (April 6, 1929): 373–390, at 379.

44. N. F. Mott, On His Personal Experiences of the Development of Quantum Physics (unpublished notes, dated March 18, 1962), in Archive for the History of Quantum Physics, University of California, Berkeley; Mott, interview by the author, October 10, 1969; Mott (note 41); Blackett (note 3), p. xxxv.

45. Andrade (note 3), p. 209.

46. Rutherford, letter to T. H. Laby, June 18, 1923, Rutherford collection, Cambridge University Library.

47. E. Rutherford and H. Geiger, "An Electrical Method of Counting the Number of α-Particles from Radioactive Substances," *Proc. Roy. Soc.* A 81 (1908): 141–161 (*CPR* 2: 89–108).

48. Mary Jo Nye, *Molecular Reality* (New York: American Elsevier, 1972).

49. See, e.g., E. Rutherford and F. Soddy, "Radioactive Change," *Phil. Mag.* 5 (May 1903): 576–591, esp. 591 (*CPR* 1: 596–608).

50. See Rutherford, *Radio-Activity* (Cambridge University Press, 1904), pp. 108–112.

51. H. H. Dale, "Some Personal Memories of Lord Rutherford," in *An Autumn Gleaning* (London: Pergamon, 1954), pp. 179–180.

52. C. D. Ellis, letter to Thomas H. Kuhn, April 18, 1963, Archive for the History of Quantum Physics, University of California, Berkeley.

53. Rutherford, address at annual dinner of North Western Centre of the Institution of Electrical Engineers, Manchester, January 22, 1935, item 160 in Rutherford collection, Cambridge University Library.

54. James Jeans, letter to Oliver Lodge, August 1, 1931, Lodge collection, University College Library, London.

55. A. Einstein, quoted in Robert Resnick, "Misconceptions about Einstein," *J. Chem. Educ.* 57 (December 1980): 855.

56. That the debate over definitions continues and is often quite philosophical may be seen in the following: Helier J. Robinson, "A Theorist's Philosophy of Science," *Physics Today* 37 (March 1984): 24–32; letters to the editor, ibid. 37 (September 1984): 13–15, 114–117. For some comments about definitions and their change over time, see *Exploring the History of Nuclear Physics*, ed. C. Weiner (New York: American Institute of Physics, 1972), pp. 53–56.

57. H. Ehrenreich, review of J. C. Slater's *Solid State and Molecular Theory: A Scientific Biography, Science* 188 (May 23, 1975): 839–840. The first two quotations are Slater's words; the third is the reviewer's.

58. J. C. Slater and N. H. Frank, *Electromagnetism* (New York: McGraw-Hill, 1947), p. vii.

59. Max Born, "Arnold Johannes Wilhelm Sommerfeld," *Obituary Notices of Fellows of the Royal Society* 8 (1952): 275–296, at 282.

60. William G. Pollard, "Physics," *McGraw-Hill Encyclopedia of Science and Technology* (New York: McGraw-Hill, 1982), vol. 10, p. 277.

61. Lewis Pyenson, "La réception de la relativité généralisée: Disciplinarité et institutionalisation en physique," *Revue d'Histoire des Sciences* 28 (1975): 61–73.

62. J. J. Thomson and E. Rutherford, "On the Passage of Electricity through Gases Exposed to Röntgen Rays," *Phil. Mag.* 42 (November 1896): 392–407 (*CPR* 1: 105–118).

63. Rutherford, "Some Remarks on Radioactivity," *Phil. Mag.* 5 (April 1903): 481–485 (*CPR* 1: 576–579).

64. E. Rutherford and F. Soddy, "The Radioactivity of Thorium Compounds," *Trans. Chem. Soc.* 81 (April 1902): 321–350 (*CPR* 1: 376–402); ibid. 81 (July 1902): 837–860 (*CPR* 1: 435–456); "The Cause and Nature of Radioactivity," *Phil. Mag.* 4 (September 1902): 370–396 (*CPR* 1: 472–494); ibid. 4 (November 1902): 569–585 (*CPR* 1: 495–508); "Radioactive Change," ibid. 5 (May 1903): 576–591 (*CPR* 1: 596–608). For a valuable analysis of this work see Thaddeus J. Trenn, *The Self-Splitting Atom: The History of the Rutherford-Soddy Collaboration* (London: Taylor & Francis, 1977).

65. Rutherford, "The Succession of Changes in Radioactive Bodies," *Phil. Trans.* A 204 (1904): 169–219 (*CPR* 1: 671–722). Abraham Pais calls the decay equations one of Rutherford's two contributions to theoretical physics, the other being the picture of the nuclear atom: "Radioactivity's Two Early Puzzles," *Rev. Mod. Phys.* 49 (October 1977): 925–938, esp. 935. Rutherford first used the decay equations in "A Radioactive Substance Emitted from Thorium compounds," *Phil. Mag.* 49 (January 1900): 1–14 (*CPR* 1: 220–231).

66. H. Bateman, "The Solution of a System of Differential Equations Occurring in the Theory of Radio-active Transformations," *Proc. Cambridge Phil. Soc.* 15 (1910): 423–427.

67. J. A. Wheeler, in Weiner (note 56), p. 55. In this same source Philip A. Morrison calls the decay constant a statistical rather than a quantum idea—a view that additionally avoids charges of whiggish history. My point in presenting their comments, however, is to support my own contention—that Rutherford made theoretical offerings—with the words of theoretical physicists.

68. Rutherford, "Theory of the Luminosity Produced in Certain Substanced by α Rays," *Proc. Roy. Soc.* A 83 (1910): 651–672 (*CPR* 2: 182–192).

69. Rutherford, "Forty Years of Physics: The Development of the Theory of Atomic Structure," in *Background to Modern Science*, ed. J. Needham and W. Pagel (Cambridge University Press, 1938), p. 68.

70. H. Geiger and E. Marsden, "On a Diffuse Reflection of the α-Particles," *Proc. Roy. Soc.* A 82 (July 31, 1909): 495–500.

71. John Heilbron, "The Scattering of α and β Particles and Rutherford's Atom," *Arch. Hist. Exact Sci.* 4 (1968): 247–307; "J. J. Thomson and the Bohr Atom," *Physics Today* 30 (April 1977): 23–30.

72. Rutherford, "The Scattering of α and β Particles by Matter and the Structure of the Atom," *Phil. Mag.* 21 (May 1911): 669–688 (*CPR* 2: 238–254).

73. Hans Bethe, letter to the author, November 24, 1970.

74. J. Heilbron, "Rutherford-Bohr Atom," *Am. J. Phys.* 49 (March 1981): 223–231.

75. Rutherford, "Nuclear Constitution of Atoms," *Proc. Roy. Soc.* A 97 (July 1, 1920): 374–400 (*CPR* 3: 14–38, at 34).

76. Rutherford, letter to W. H. Bragg, August 1, 1906, Bragg collection, Royal Institution.

77. Rutherford, "The Structure of the Atom," *Phil. Mag.* 27 (March 1914): 488–498 (*CPR* 2: 423–431, esp. 428).

78. Rutherford, "Collison of α Particles with Light Atoms. IV. An Anomalous Effect in Nitrogen," *Phil. Mag.* 37 (June 1919): 581–587 (*CPR* 2: 585–590).

79. E. Rutherford and J. Chadwick, "The Artificial Disintegration of Light Elements," *Phil. Mag.* 42 (November 1921): 809–825 (*CPR* 3: 48–62); "Scattering of α-Particles by Atomic Nuclei and the Law of Force," ibid. 50 (November 1925): 889–913 (*CPR* 3: 143–163).

80. R. H. Stuewer, "The Nuclear Electron Hypothesis," in *Otto Hahn and the Rise of Nuclear Physics*, ed. W. Shea (Dordrecht: Reidel, 1983); "Rutherford's Satellite Model of the Nucleus," forthcoming; P. I. Dee, "The Rutherford Memorial Lecture, 1965," *Proc. Roy. Soc.* A 298 (April 4, 1967): 103–122. For further insights to Rutherford's concept of mass as electrical or electromagnetic in character (which ties into his nuclear models), see Thaddeus J. Trenn, "Rutherford's Electrical Method: Its Significance for Radioactivity and an Expression of His Metaphysics," *Actes du XIIIᵉ Congres International d'Histoire des Sciences, 1971* (Moscow, 1974), sec. 6, pp. 112–118; Daniel M. Siegel, "Classical-Electromagnetic and Relativistic Approaches to the Problem of Nonintegral Atomic Masses," *Historical Studies in the Physical Sciences* 9 (1978): 323–360.

81. Joseph Agassi, *Faraday as a Natural Philosopher* (University of Chicago Press, 1972).

82. David B. Wilson, "Experimentalists among the Mathematicians: Physics in the Cambridge Natural Sciences Tripos, 1851–1900," *Historical Studies in the Physical Sciences* 12 (1982): 325–371.

83. Paul K. Hoch, "The Reception of Central European Refugee Physicists of the 1930s: U.S.S.R., U.K., U.S.A.," *Annals of Science* 40 (1983): 217–246; "Transmission of a New Mathematical Physics from Germany to Britain and America," presented at International Congress of History of Science, University of California, Berkeley, 1985; Paul Forman, John Heilbron, and Spencer Weart, "Physics circa 1900: Personnel, Funding, and Productivity of the Academic Establishments," *Historical Studies in the Physical Sciences* 5 (1975): 1–185.

84. Hoch (note 83).

Howard Stein

After the Baltimore Lectures: Some Philosophical Reflections on the Subsequent Development of Physics

Kelvin appears to be associated, in the historical mythology of science, with the opinion that the science of physics was nearly at an end—that there only remained some small clearing up of difficulties, to be followed by a process of determining physical constants to increasing numbers of decimal places. James Clerk Maxwell alluded to this view—of course without the slightest thought of implicating his friend Sir William Thomson—in his Inaugural Lecture at the Cavendish Laboratory; Maxwell remarked: "If this is really the state of things to which we are approaching, our Laboratory may perhaps become celebrated as a place of conscientious labour and consummate skill, but it will be out of place in the University, and ought rather to be classed with the other great workshops of our country, where equal ability is directed to more useful ends." How far Kelvin stood from the complacent view that Maxwell deprecated is immediately apparent from the first page of the preface he supplied when the *Baltimore Lectures* were published by the Cambridge University Press in 1904:

Having been invited by President Gilman to deliver a course of lectures in the Johns Hopkins University after the meeting of the British Association in Montreal in 1884, on a subject in Physical Science to be chosen by myself, I gladly accepted the invitation. I chose as subject the Wave Theory of Light with the intention of accentuating its failures; rather than of setting forth to junior students the admirable success with which this beautiful theory had explained all that was known of light before the time of Fresnel and Thomas Young, and had produced floods of new knowledge splendidly enriching the whole domain of physical science. My audience was to consist of Professorial fellow-students in physical science; and from the beginning I felt that our meetings were to be conferences of coefficients, in endeavours to advance science, rather than teachings of my comrades by myself. I spoke with absolute freedom, and had never the slightest fear of undermining their perfect faith in ether and its light-giving waves: by anything I could tell them of the imperfection of our mathematics; of the insufficiency or faultiness of our views regarding the dynamical qualities of ether; and of the over-whelmingly great difficulty of finding a field of action for ether among the atoms of ponderable matter. We all felt that difficulties were to be faced and not to be evaded; were to be taken to heart *with the hope of solving them if possible*; but at all events with the certain assurance that there is an explanation of every difficulty though we may never succeed in finding it.

Among the "coefficients" at the lectures were Rowland, Michelson, and Morley. The eighty-year-old Kelvin, in his preface, after noting with satisfaction that the difficulties of 1884 had (in his opinion) all been resolved in the ensuing two decades, wrote with a certain relish of pride that "two of ourselves, Michelson and Morley, have, by their great experimental work on the motion of ether relatively to the earth, raised the one and only serious objection against our dynamical explanations." (It should not be supposed that this one remaining serious objection represented, in Kelvin's view, the last step that would need to be taken; on the contrary, he immediately proceeded to indicate what advances he expected toward "the grand object ... of finding a comprehensive dynamics of ether, electricity, and ponderable matter, which shall include electrostatic force, magnetostatic force, electromagnetism, electrochemistry, and the wave theory of light.")

Yet for all the restless energy of his mind, and despite his generous attitude toward the work of colleagues, it cannot be denied that Kelvin's view of physics was, in a way, curiously stiff and hidebound. Indeed, the advances he celebrated in the preface of 1904 (in which he asserted that the difficulties posed in his lectures of 1884 had all been resolved) have left no trace on the physics of our century, whereas the work that from some years before those lectures were delivered to the time of their publication had achieved the greatest advance in the subject to which the lectures were themselves devoted—work that had laid the groundwork for the profound developments of the next two decades—was passed over almost entirely in both the original lectures and published volume. There is in this something of more than merely biographical interest.

The full title of Thomson's course (it was not until some eight years later that he was baronized as Lord Kelvin) was "Lectures on Molecular Dynamics and the Wave Theory of Light"; the subject, in other words, was to be the problem of the dynamical interrelations of ordinary matter—treated as molecular in its structure—and that extraordinary matter the ether (the carrier of the waves constituting light). To resolve this problem completely, one would require a theory that gave a satisfactory representation of the propagation of light in free space (the "free ether"), in the interior of ordinary transparent or partially transparent matter, and at the surface of separation between two ordinary bodies of different constitution, or between free ether and an ordinary body. Beyond this, one would require an account of the dynamical interactions involved when bodies *move through* the ether. This inventory of desiderata essentially defines what deserves to be called "the classical ether problem for light." In Kelvin's opinion, by 1904 he had satisfactorily resolved this problem, except for the difficulty posed by the Michelson-Morley experiment. There was, in the latter part of the nineteenth century, a further and more intricate set of issues concerning the ether, which may be called "the classical ether problem for the electromagnetic field." Kelvin fully recognized this; in the preface from which I have already quoted, he goes on to say:

My object in the Baltimore Lectures was to find how much of the phenomena of light can be explained without going beyond the elastic-solid-theory. We have now our answer: *everything non-magnetic; nothing magnetic.* The so-called "electromagnetic theory of light" has not helped us hitherto: but the grand object is fully before us of finding a comprehensive dynamics of ether, electricity, and ponderable matter, which shall include electrostatic force, magnetostatic force, electromagnetism, electro-chemistry, and the wave theory of light.

Kelvin had already adverted to this object in 1889, in a paper on ethereal dynamics:

... to give anything like a satisfactory material realisation of Maxwell's electro-magnetic theory of light ... essentially involves the consideration of ponderable matter per-meated by, or imbedded in ether, and a *tertium quid* which we may call electricity, a fluid go-between, serving to transmit force between ponderable matter and ether.... I see no way of suggesting properties of matter, of electricity, or of ether, by which all this, or any more than a very slight approach to it, can be done, and I think we must feel at present that the triple alliance, ether, electricity, and ponderable matter is rather a result of our want of knowledge, and of capacity to imagine beyond the limited present horizon of physical science, than a reality of nature.[1]

Some eleven years before those words were written, H. A. Lorentz published a paper on the propagation of light in a material medium, treated on the basis of the electromagnetic theory of light.[2] Although an abridged German version was published in 1880 in a prominent journal,[3] this work appears to have remained largely ignored; Helmholtz, for instance, was evidently unaware of it in 1892, when he published a paper on the electromagnetic theory of disper-sion, and it is accorded only passing reference, with inadequate appreciation of its contents, in Whittaker's *History of the Theories of Ether and Electricity*.[4] In fact, Lorentz's paper of 1878, which contains the first development of a theory of dispersion on the basis of the electromagnetic theory, initiated the line of research that eventuated in what we now know as Lorentz's theory of electrons, and it is this theory that provided the framework for all progress on the question of molecular dynamics and the theory of light.

It is of interest, for the understanding of Lorentz's point of view and pro-cedure, and also for the sake of certain general lessons I wish to preach, to go back a little further—to Lorentz's doctoral thesis of 1875, which itself has as background and foundation the critical discussion of electrodynamic theories that Helmholtz had published in 1870. Helmholtz's purpose was to examine the relationships among, and to lay the basis for comparing the merits of, the several electrodynamic theories then in competition, with somewhat special concern for the status of Maxwell's theory. To this end, Helmholtz followed one of the leading ideas of Maxwell: that the ether—presumed to fill otherwise free space (as the carrier of light waves)—was essentially a polarizable dielectric medium. But, unlike Maxwell, Helmholtz treated this medium after the manner of the Poisson-Mossotti theory, according to which polarization consists in a separation of charges resulting from the ordinary action at a distance of external exciting charges. Thus, it may be said that Helmholtz's version of the theory employs an ether whose fundamental constitution is intrinsically electrical rather than elastic.

The Helmholtz electrodynamics contains two parameters—let me call them κ and λ—whose values serve to discriminate a spectrum of possible specific theories. The parameter κ appears in the expression for the electrodynamic interaction at a distance between two current elements; $\kappa = 1$ yields the electrodynamic law of Franz Neumann, $\kappa = -1$ yields the once-celebrated law of Weber, and, according to Helmholtz, $\kappa = 0$ corresponds to the theory of Maxwell. This last statement, however, is not right, as Poincaré seems to have been the first to point out.[5] The parameter λ may be regarded as defining the dielectric coefficient of the ether; more exactly, λ is the ratio of the electrostatic force between two given charges at given distance apart *in an ideal nonpolarizable medium* to the force between the same charges at the same distance in ordinary empty space (i.e., "in the ether"). Thus the value $\lambda = 1$ gives the "pure" action-at-a-distance theories, in which the ether itself is nonpolarizable, whereas any value greater than 1 makes the ether electrically polarizable. The very interesting results of Helmholtz's investigation were the following:

- that any polarizable medium will be a carrier of electromagnetic waves;
- that there will occur in general both longitudinal and transverse waves, propagated with different velocities;
- that the velocity of the transverse waves will depend upon the polarizability of the medium, that of the longitudinal waves upon both the polarizability of the medium and the value of the parameter κ; and
- that, on the other hand, for $\kappa = 0$ the velocity of longitudinal waves will be infinite—hence, there will be no such waves—and, on the other hand (in view of the value found by measurement for the ratio of the electromagnetic unit of charge to the electrostatic unit), in order for the propagation of transverse electromagnetic waves in a given medium to have the velocity of light the polarizability of this medium must be practically infinite.

In the last case (more strictly, in the limit in which the polarizability of the ether goes to infinity) one arrives at Maxwell's theory in full detail: the infinite polarizability of the ether corresponds to Maxwell's hypothesis that "all electric currents are closed" (that is, that the current distribution is solenoidal); and the parameter κ, if one supposes only that it is held bounded during the limiting process, drops out of the theory altogether—once again the longitudinal waves disappear.

It was, then, upon Helmholtz's formulation that Lorentz based his treatment of reflection and refraction (the subject of his thesis) and his treatment of the dependence of the optical properties of bodies on their physical constitution (the subject of his paper of 1878). In doing so, he did not commit himself to the strictly infinite polarizability of the ether, or to the strict nonexistence of longitudinal waves, but only to the very high polarizability of the ether and the very great velocity of longitudinal waves. The thesis concluded with an indication of problems that remained to be investigated and that seemed ripe for investigation from the vantage point already attained; in this connection, Loretz remarked: "If it is true that light and radiant heat are constituted by electric vibrations, it is natural to admit that the molecules of bodies, which give rise to such vibrations in the ambient medium, are likewise the seat of electric oscillations, whose intensity increases with the temperature. This conception, which is not new but which derives from the electromagnetic theory a high degree of probability, seems to me very fruitful."[6] Just that conception is employed in the paper of 1878, in which material dielectric bodies are regarded not (like the ether) as continuous polarizable media, but as molecular in structure, the space between the molecules being pervaded by ether. In the chapter on dispersion,[7] Lorentz finds it necessary to specialize his assumption about the molecules by positing that they contain particles possessing both charge and mass. This is the first characteristic assumption of the theory of electrons. Although introduced very quietly, in the midst of a painstaking piece of work that can be seen as a paradigm of what Thomas Kuhn calls "normal science,"[8] the assumption in its context is momentous. Lorentz, in addition to opening his scientific career with an investigation devoted, in however cautious a fashion, to the theory of Maxwell (whose status among physicists in the 1870s was as insecure as was that of the theory of relativity in the years immediately following 1905), had combined that theory with an element superficially quite foreign to it: the *tertium quid* referred to so unenthusiastically by Kelvin—electricity—not in its old guise as a fluid, but in the form of intrinsic charge of a particle of "ordinary" matter, which charge serves to mediate the interaction of the particle with the ether.

The problem of the nature of electric charge deserves to be regarded as the main crux of Maxwell's theory; thus Poincaré wrote in 1890 in the preface to his published lectures on Maxwell's theory: "It is in electrostatics that my task has been the hardest; it is there above all in fact that precision is wanting [in Maxwell]. One of the French scientists who have most deeply studied the work of Maxwell said to me one day: 'I understand everything in his book except this: what is a charged sphere?'"[9] Lorentz's postulating of intrinsically charged particles can be regarded as a cutting of the Gordian knot. But such cutting is in itself a crude act: remember Kelvin's suggestion that "the triple alliance, ether, electricity, and ponderable matter is rather a result of our want ... of capacity to imagine beyond the limited present horizon of physical science, than a reality of nature."

The postulate of charged massive particles in Lorentz's 1878 paper does, in a sense, lack depth; it serves simply to tie charge and mass together, to provide something that can resonate or fail to resonate with the oscillations of the electric field in a light wave. But in Lorentz's subsequent work a radically new thing comes to pass. Put simply, it is the more far-reaching assumption that the charges of such particles constitute the *sole* mediators of interaction between ordinary "ponderable" matter and the ether. This extended assumption requires, to make it precise, a fuller analysis of the nature of the interaction in question; that is, in the terms that prevailed at the time, an analysis of the mechanical interconnection of an intrinsically charged particle and the ether.[10]

Just such an analysis was provided by Lorentz in his famous memoir of 1892 on the electrodynamics of moving bodies;[11] but what Lorentz gave was a mechanical analysis with a difference. The introductory section of the memoir begins as follows:

In one of the most beautiful chapters of his *Treatise on Electricity and Magnetism*, Maxwell shows how the principles of mechanics can serve to elucidate the questions of electrodynamics and the theory of induced currents, without the need to penetrate the secret of the mechanism that produces the phenomena. The illustrious scientist limits himself to a small number of hypotheses, which are known to all physicists and of which I may be permitted here to recall the principal ones.

Lorentz then reviews the chief assumptions upon which Maxwell, in part IV, chapter VI of the *Treatise* ("Dynamical Theory of Electromagnetism") had based his own subsumption of his theory under the principles of Lagrangian mechanics; he continues:

After having posited the principles I have summarized, Maxwell applies the equations of Lagrange; he arrives thus at well-known formulas for the electrodynamic forces and for the induction of currents....

[But t]he equations that determine the motions of electricity in three-dimensional bodies do not follow, in the book of Maxwell, from a direct application of the laws of mechanics; they rest upon the results that have been obtained for linear conductors.

Lorentz next adverts to the work of Heaviside and of Hertz in simplifying the form of Maxwell's field equations, and remarks that Hertz had hardly concerned himself with a mechanical account of processes in the field. He adds:

Needless to say, this method has its advantages.

Nevertheless, one is always tempted to revert to mechanical explications. That is why it has seemed useful to me to apply directly to the most general case the method of which Maxwell has given the example in his study of linear circuits.

There follows a crucial statement:

I had yet another motive to undertake these researches. In the memoir in which M. Hertz treats of bodies in motion, he takes the ether they contain to move with them.

Now, optical phenomena have long since demonstrated that this is not always so. I therefore wished to know the laws that govern electrical motions in bodies that traverse the ether without its entrainment; and it seemed to me difficult to achieve this aim without having as guide some theoretical idea. The views of Maxwell may serve as foundation for the theory sought.

The exposition of the results here adumbrated occurs in chapter IV of Lorentz's memoir, which begins: "It has seemed to me useful to develop a theory of electromagnetic phenomena based upon the idea of a ponderable matter completely permeable to the ether, and able to move without communicating the slightest motion to the latter." But how, Lorentz asks, is one to form a precise idea of a body that, moving within the ether and thus traversed by it (in effect, by an "ether wind"), is at the same time the seat of an electric current or a dielectric process?

To overcome the difficulty as far as it was possible for me, I have sought to reduce all the phenomena to a single one, the simplest of all, namely just to the motion of a charged body. It will be seen that, without a deeper analysis of the relation between ponderable matter and the ether, one can establish a system of equations suited for the description of what happens in a system of such bodies.

Lorentz's fundamental hypotheses (presented, with the greatest simplicity and clarity, in section 75 of the memoir[12]) are these:

(i) All charges are carried by particles of "ponderable matter"—i.e., bodies in the ordinary sense, having mass and subject to the application of "ponderomotive" forces. (I shall give these particles the name later adopted for them by Lorentz: *electrons*.) For convenience, it is assumed that the distribution of charge is given by a charge-density function that is, at each instant, a smooth function on space. Further, it is assumed that the individual electrons are rigid, and that the charge density of each one is rigidly associated with the particle.

(ii) Electrons and ether mutually penetrate one another. Ether, in particular, is strictly ubiquitous; not only does it occupy all spaces *between* charges, it occupies all of space *simpliciter*.

(iii) Charge density is connected with the "components of the dielectric displacement in the ether"—that is, with the vector function of electric field intensity—by the divergence equation of Maxwell.

(iv) The "total electric current," in the sense of Maxwell, is the sum of the current of convection (determined by the motions of the electrons) and the "displacement current" (determined by the rate of change of the electric intensity). Lorentz remarks that, in consequence of these two assumptions, the distribution of total current is solenoidal. In comparison with the theory of Helmholtz, this result is now obtained in a quite different (and in a sense more abstract) way: there is no longer any question of the circuits' being completed by complementary motions of charges in the dielectric medium; the ether itself—which has now become, from a fundamental point of view, the only dielectric medium—is not assumed to be a Mossotti-type polarizable body or to contain any electric charges other than those of the "electrons."

(v) The total current is connected with the vector function of magnetic field intensity by the pertinent curl equation of Maxwell (a postulate that is consistent with the foregoing, in view of the solenoidal character of the current). The magnetic field itself is source-free. The electromagnetic energy associated with the total current—given by an energy-density function proportional to the square of the intensity of the magnetic field—is, following Maxwell, to be regarded as *kinetic energy of the ether*.

(vi) The *configuration* of an electromagnetic system (in the sense of generalized dynamics) is determined by the configuration in the ordinary sense of the system of the electrons involved, together with the vector function that defines the dielectric displacement. The potential energy of the ether is distributed with a density proportional to the square of the electric field intensity.

The application of d'Alembert's principle to these hypotheses leads, in the first place, to the remaining set of Maxwell equations: those for the curl of the electric intensity. (An analogous result had been established, so far as the free ether is concerned, by G. F. FitzGerald some thirteen years earlier.[13]) Beyond this, however, Lorentz obtained the new basic result he sought: the formula that gives the force per unit charge in the electromagnetic field, and so completes the account of the role of electricity as intermediator between ponderable matter and the ether. Indeed, the influence of ordinary matter upon the ether is now expressed, through the divergence equation of the electric field and the curl equation of the magnetic field, as determined by the distribution of charge carried by material particles and by the convection of charge as those particles move, and the influence of the ether upon ordinary matter is expressed as the "ponderomotive force" exerted upon the charged particles. A notable circumstance in this result is the asymmetry it introduces in the treatment of the "displacement current" of Maxwell; the latter is on a par with the convection current, so far as its role in influencing the magnetic field is concerned, but it is not involved (as the convection current is) in determining ponderomotive forces. The theory of a polarizable ether makes no such distinction.

I have called Lorentz's account a mechanical analysis "with a difference." By this I meant to refer not just to the fact that the mechanics involved is of the generalized kind that allows one to avoid detailed hypotheses about microstructure (Maxwell himself had invoked such a procedure twenty-eight years before), nor just to the fact that no elastic medium is posited (that is equally the case in the Helmholtz theory to which Lorentz's earlier investigations had attached themselves). The profound difference lies, rather, slightly beneath the surface of Lorentz's characterization of his theory as "based on the idea of a ponderable matter perfectly permeable to the ether and able to move without communicating the least motion to the latter."

In one sense, of course, this phrase explicitly contains a notion that stands in striking contrast with traditional ideas; impenetrability has often been considered one of the essential properties of body, so the notion of two sorts of matter that are literally co-present at the same points of space is rather a radical departure. It is therefore of some interest that such a noted traditional "mechanist" as Kelvin subscribed, toward the end of his life, to a theory of the ether that assumed mutually penetrable forms of matter. It is this theory that constitutes his proposed solution of the problems of his Baltimore Lectures in the 1904 edition.[14] But Kelvin's ether, although it does not interact *by contact* with the particles with which it interpenetrates, does interact with these particles by *distance* forces, so that motion is communicated between the ether and ponderable matter. In Lorentz's theory, no such effect occurs; instead, the ether appears as a system with mechanical attributes; it is the seat of energy (a part of which is formally identified as kinetic), and it is able to exert forces on bodies, yet is itself incapable of motion.

In putting the case so, I am perhaps open to a legitimate objection. Lorentz himself certainly does not present the conception he has introduced as "revolutionary." He characterizes his attempt as one "to surmount the difficulty as far as it was possible for me," and "without a deeper analysis of the relation between ponderable matter and the ether"; he leaves us free to envisage the possibility that such a deeper analysis might yet identify the energy of the magnetic field as genuine energy of motion within the ether and reconcile small-scale motion within the latter with its immobility in the large. On the other hand, it is quite clear that Lorentz was aware of the formidable difficulties in the way of any such classically mechanical theory.[15] What in effect he did deserves comparison with such other cuttings of Gordian knots as that by Einstein in his 1905 quantum paper and that by Bohr in his 1913 papers on the constitution of atoms and molecules: in his quiet way, he had the boldness and the insight to combine those parts of existing physical theory that could lead to definite and interesting results, while simply ignoring—as in the cases of Einstein and Bohr, tentatively and in the spirit of heuristic inquiry—what he saw no way to incorporate consistently. Perhaps the chief difference between Lorentz and the other two in this respect is that in the later period there was in the background far more definite evidence of an out-and-out contradiction between the classical notions and natural phenomena, so that the expectation of an ultimate clarification on strictly classical principles was more clearly excluded. This may be enough to render Lorentz "normal" and the other two "revolutionary," but if one attends to the objective consequences of the respective investigations the distinction appears of limited interest.

Furthermore, in his next large work on the subject Lorentz made his departure from classical principles quite definite:

Why should we, having once assumed that the ether does not move, ever talk of a force acting upon this medium? It would be simplest to assume that no force ever acts upon a volume element of the ether, considered as a whole; or even to refrain from applying the very concept of force to such an element, which indeed never moves from its place. To be sure, this conception would conflict with the theorem of the equality of action and reaction—since we have grounds for saying that the ether *exerts* force upon ponderable matter; but, so far as I see, nothing compels us to elevate that theorem to a fundamental law valid without restriction.[16]

Perhaps, after all, we should recognize Lorentz as a *quiet* revolutionary.

The consequences that flowed from Lorentz's treatment of the electromagnetic ether problem may be divided into two classes. The first of these surely deserves the epithet "normal": rapid and impressive progress, partly at the hands of Lorentz himself and partly at the hands of others, in the theoretical representation—both "explanation" and "prediction"—of optical, electrical, magnetic, and thermal properties of bodies. One exemplar of this class is the work for which, in 1902, Lorentz shared with Pieter Zeeman the second Nobel Prize in physics: the theory of the so-called "normal Zeeman effect." This theory, which Lorentz furnished immediately upon Zeeman's discovery, allowed both the prediction of the finer structure of the phenomenon (doublet when emission is in the direction of the applied magnetic field, triplet when emission is perpendicular to the field; circular polarization of the emitted light) and the determination of the charge-to-mass ratio of the particles responsible for it; in effect, then, it led to the discovery of the electron, in the current sense of that word.[17]

This "discovery of the electron" by Zeeman and Lorentz was independent of, and essentially contemporaneous with, the corresponding discovery made by J. J. Thomson in his work on cathode rays; together with the latter, it constitutes the beginning of the acquisition of genuine information about subatomic particles. In calling this work an exemplar of the class of "normal" consequences of Lorentz's theory, I had in mind not only that it illustrates the power of that theory in advancing our understanding of natural processes, but also its noteworthy limitation: The "normal" Zeeman effect is in point of fact exceptional. What is called the anomalous effect is far more common, and of this Lorentz's theory proved unable to give a satisfactory account. This case is typical, in that the processes the theory of electrons deals with all lie on or over the borderline that separates the domain of competence of classical physics from the domain in which recourse to the quantum theory is essential; thus, the "normal science" involved tends to arrive rapidly at "anomalies" in the sense of Kuhn, and may in this respect be called pre-revolutionary. If the terminology appears frivolous—and I have already indicated my reservations about any such sharp distinction as Kuhn has suggested—there is nevertheless a quite serious point involved: The power of a scientific theory does not consist solely in its success in explanation and in prediction; the *failures* of a theory can be of the greatest importance when they serve to bring deep problems into sharp focus.[18]

The second class of consequences I have mentioned is related to the genuine Lorentzian revolution. That is to say, these consequences involve directly the new fundamental conception Lorentz had introduced: that of the immobile ether that nevertheless has attributes of a mechanical system interacting with charged particles. Here the most obvious development is the one Lorentz is best remembered for in the folk tradition: his reconciliation of the postulate of immobility with the failure of all attempts to detect the motion of ordinary bodies—in particular, the earth—relative to the ether. As is well known, this achievement was attained rather painfully, step by step, starting from the introduction in 1892 of the hypothesis of a contraction of bodies in the direction of their motion relative to the ether.[19] The near-final stage was reached by Lorentz in 1904; the definitive formulation was given by Poincaré in 1905–06 and—with a difference—by Einstein in 1905.[20] But parallel to that development, and not altogether separate from it, was another, whose result was to consolidate into the conceptual frame of physics the notion of the ether as what I shall now call a *non-Newtonian mechanical system*. In this connection, let me mention especially the introduction into the theory of the notion of *electromagnetic* (linear and angular) *momentum*, first as a kind of *jeu d'esprit* by Poincaré in 1900[21] and then, on the basis of Poincaré's suggestion (but now treated as having serious and far-reaching import), by Max Abraham in 1903.[22] This move, which served to rescue the conservation principles of momentum and angular momentum despite the absence of any "force of reaction" to the Lorentz force on the electrons, constituted the first extension of a specific concept of Newtonian mechanics to a domain strictly outside its original home; one now spoke of momentum with no necessary reference to any mass in motion.

I fear I may have dwelt too long upon historical details that are by now fairly familiar; but I wished to present them in a particular form. In the remainder of this essay I shall depart from that detailed historical mode in order to make some more general comments on the matters I have spoken of and their sequel. First I want to say how I construe the difference that (as everyone now appears to agree) subsists between the Lorentz-Poincare theory and that of Einstein.

In the perfected from given it by Poincaré, Lorentz's theory may be described as follows:

(1) All the postulates of the theory of electrons, as formulated in the memoir of 1892, remain in effect when (as in that memoir) the coordinate system is taken

as one with respect to which the ether is at rest—except that the assumption of the rigidity of the electron is replaced by the assumption of a distribution of stresses that just equilibrate the mutual repulsions of its charge elements.

(2) If one introduces, for a system of reference in uniform translatory motion through the ether, an array of auxiliary spatial and temporal coordinates and of electrodynamic variables related to those for the system at rest by the set of transformations given by Poincaré, then exactly the same postulates hold in the moving system. This result is a purely mathematical theorem; it is what Hermann Weyl, in his splendid exposition of the theory of relativity, calls Lorentz's Theorem of Relativity.[23]

(3) If (as Lorentz does) one postulates further that, in Poincaré's words, "all forces, of whatever origin, are affected by a translation in the same manner as the electromagnetic forces,"[24] then the spatio-temporal, electromagnetic, and dynamical quantities introduced formally under (2) will be just the ones determined in empirical measurement by an observer in a state of uniform translatory motion.

The physical content of the theory, therefore, is determined by the postulates I have mentioned in (1) and (3). The theorem cited in (2), with the further remark in (3), yields the conclusion that uniform motion with respect to the ether is in principle undetectable by any physical means.

Now, in comparing this theory with that of Einstein, I want to say emphatically that the physical contents of the two—as I understand the notion of "physical content"—are identical. The distinction between them resides, as I see it, not at all in their physical content but in two aspects of the point of view with which they are presented by their respective authors—aspects that, since they do not belong to the *objective content* of physics and yet have important bearings upon our *understanding* of the physics, may appropriately be called "meta-physical."

The first difference lies simply in the degree of conviction with which the theories are presented. In Lorentz's case, one finds an entirely characteristic expression of caution: "It need hardly be said that the present theory is put forward with all due reserve. Though it seems to me that it can account for all well established facts, it leads to some consequences that cannot as yet be put to the test of experiment. One of these is that the result of Michelson's experiment must remain negative, if the interfering rays of light are made to travel through some ponderable transparent body."[25] Lorentz made analogous statements on many subsequent occasions; what they come to is that he was not entirely convinced that phenomena would never be found that are sensitive to uniform translatory motion through the ether, and he therefore considered the theory *tentative.*

Poincaré's case is rather different. For some ten years he had been setting forth, with increasing emphasis, the conviction that motion relative to the ether was a strictly fictitious concept, and that a really satisfactory physical theory should reveal this in all rigor.[26] He could not, therefore, have shared Lorentz's doubts about *this* aspect of the theory. His misgiving, rather, was that the desired objective had been obtained by artificially complicated means: "We cannot be contented with formulas that are simply juxtaposed, and which agree only by a happy chance; we require these formulas to come as it were to penetrate one another mutually. The mind will not be satisfied until it believes itself to perceive the reason for this agreement, to the point of having the illusion that it could have foreseen the same."[27] Moreover, he suspected that this artificial complication was connected with an inappropriate relation of the concepts of the theory to methods of measurement: "In this theory, two equal lengths are, by definition, two lengths that light takes the same time to traverse. Perhaps it will suffice to abandon this definition, for the theory of Lorentz to be overthrown as completely as was the system of Ptolemy by the intervention of Copernicus."[28] It was, therefore, with an expression of considerable diffidence that Poincaré introduced his paper; the passage I have just quoted continues: "If this occurs one day, that will not prove that the effort made by Lorentz was useless; for Ptolemy, whatever one thinks of him, was not useless to Copernicus. Therefore I have not hesitated to publish these partial results, even though at this very moment the whole theory may seem placed in jeopardy by the discovery of the magneto-cathodic rays."[29]

Einstein, it need hardly be said, had no such misgivings; he possessed in full the "illusion" desiderated by Poincaré, and it is a famous fact that the conviction thereby engendered was strong enough to sustain his confidence through a period in which serious experimental counterevidence—nothing so vague as the mysterious "magneto-cathodic rays" mentioned by Poincaré—seemed to have refuted his theory.

The second difference I have in mind is suggested by the contrast between Poincaré's reasons for doubt and Einstein's for conviction. It has, of course, to do with what has often been called the *interpretation* put upon the theory. The word is apt enough; but I think it demands some elucidation, and with this aim I want to compare the case at hand with a famous hypothetical example offered, with a philosophical purpose, by Poincaré.

In chapter IV of *Science and Hypothesis*, Poincaré describes a fictitious world, concerning which he makes the claim that beings essentially like us who are native to it would be led to describe the "space" in which they live as non-Euclidean; but that we ourselves, should we find ourselves there someday, would continue to use Euclidean goemetry in our representation of its physical phenomena. The example is based upon the conformal mapping—due to Poincaré—of Bolyai-Lobachevsky space onto the interior of a Euclidean sphere; Poincaré in effect just supposes that solid bodies move so as to remain congruent with themselves in the non-Euclidean metric and that light is propagated with constant "non-Euclidean velocity" along non-Euclidean "straight lines," and he argues that this very description of the hypothetical world shows that we, if we entered it, would be able to describe it in Euclidean language.

Now, there is an interesting oversight in Poincaré's account. His "world," by his own assumption, can be described by a physics that admits the Bolyai-Lobachevsky congruence group as a group of symmetries. The Euclidean description he invokes must postulate systematic deformations of physical bodies and a systematic variation, over space, of the velocity of light. Reichenbach had Poincaré's discussion in mind when he tried to formulate a principle of "ruling out universal forces," but this, I think, was a misstep, although a sound motive can be seen behind it. The point is that the deformations required for the Euclidean description break the symmetry of the world. But again, this can be misconstrued. What is important is not the "esthetic" loss from breach of symmetry; it is, rather, a serious implication which that breach has for the practice of physics. Because the Euclidean description breaks the symmetry, it requires that one postulate a difference in physical character among the points or regions of the space itself. In Poincaré's model, this difference is expressed in terms of the distance from the center of the (Euclidean) sphere. But because the physics is symmetric, this immediately implies that the Euclidean description is far from unique; any point could have been chosen as "center," and so there are infinitely many ways to choose a Poincaré-style Euclidean description. (The case is in fact still worse. Poincaré's own alternative conformal model for Lobachevskian geometry, using a half-space instead of the interior of a sphere, would entail another form of the law of deformation of bodies and velocity of light. This model *has* no center. And if one renounces conformality of the mapping, the possibilities grow outlandishly.)

How, then, is our "Euclidean" physicist to proceed? He may begin by choosing the Poincaré representation and declaring some place he has reason to prefer to be "the center"; but then he will be forced to keep track of his spatial relation to that center (which may be hard to do), wherever he happens to wend his way in that world, and to use that relation constantly to correct the account of his experiences—and his experiments—for the postulated distortions (which will be onerous in the extreme). Our physicist will also be in an awkward position in relation to his sister and brother Euclideans, who agree with him in their ideology but have elected different realizations thereof. In the end, surely, this would be an intolerable state of affairs; Poincaré surely ought to regard the theory as "Ptolemaic," and as based upon an inappropriate relation of concepts to methods of measurement.

The parallel to the case of Poincaré versus Einstein is exact. The place of the non-Euclidean symmetry is taken by the "Lorentz Theorem of Relativity"—proved, in strictness, not by Lorentz but by Poincaré. The place of the artificially elected center of the world is taken, not (of course) by a point, but by a (space-time) direction—a state of motion: the postulated state of motion of the immobile ether.

The most far-reaching consequence of the acceptance of Einstein's point of view is its consequence for physical investigation—its "heuristic" significance, as one says. Having recognized Lorentz symmetry as (to use Poincaré's words once more) not "a happy chance," but something for which there is a fundamental "reason," we shall expect all physical laws yet to be discovered to conform to this symmetry until reasons to the contrary are found. Notice that this heuristic principle is already implied by the Lorentz-Poincaré theory, if it is taken in all strictness; the difference is that Lorentz and Poincaré were free not to take their theory strictly—to leave open the door (as Lorentz deliberately chose to do) for the eventuality that motion relative to the ether might yet be made manifest. Einstein's form of the theory definitely closed that door.

There is an issue of methodology here. In general, leaving doors open when possible ought certainly not be regarded as a vice. For Lorentz, that was his procedure of choice on all occasions and at all stages of his career. In this, he stands in rather instructive contrast to Kelvin, although a superficial view might regard both men simply as "scientific conservatives." We have seen the cautious young Lorentz choosing to base his first investigations into the electromagnetic theory of light, not on the more decisive formulation of its creator, Maxwell, but on the more "open" formulation of Helmholtz (a formulation that has been attacked as "rather heavy-handed" and as "spoiling the subtle harmony" of Maxwell's conceptions[30]). Clearly, his caution did not prevent Lorentz from making fundamental contributions to Maxwell's theory and fundamental advances beyond it. What is still more striking is that as late as 1903, when Lorentz wrote the article on Maxwell's theory for the *Enzyklopädie der mathematischen Wissenschaften*, he included in it a brief discussion of the distance-action theory of Helmholtz and its relation to the "field-action" theory.[31] This ended, to be sure, with the comment that "no one will dispute that the merit of simplicity and of greater perspicuity [*Anschaulichkeit*] and intelligibility [*Verständlichkeit*] in physical regard lies with Maxwell's theory,"[32] but it also contained the statement that "should the need ever appear of dropping the law of the solenodial distribution of current, further developments could attach themselves to the Helmholtz theory with a finite value of ε_0 [i.e., of the constant I called λ above]."[33]

The point I want to make about this is that Lorentz's conservatism *was* just caution. He did not commit himself obstinately to "received" views, any more than he rushed to embrace "revolutionary" views (even his own). It is sometimes urged that such caution is a drawback to the energetic pursuit of a program. So it can be; and so can the deep commitment to a program be a drawback to the open-minded entertainment of fruitful alternative suggestions. The last is what Kelvin surely lacked. He was willing to consider the most radical conceptions, and the most startling combinations of structure (mutually penetrating kinds of matter; an ether engaged in fundamental action at a distance, a "rotationally elastic" ether, a "quasi-labile" ether with negative compressibility, an ether that was in part quasi-labile and in part nearly incompressible[34]), so long as these fell within the scope of classical mechanical constructions. He was also quite capable, as his pathbreaking work in thermodynamics shows, of working fruitfully outside that mechanical framework. However, for a certain species of theory—one that involved fundamentally dynamical conceptions but that transcended, or even just tentatively abstracted from, the strict Newtonian frame—his mind seems to have been simply closed; he did not *pay attention* to the successes of such theories.

That one cannot say of Lorentz. The literature contains some really grotesque statements about his attitude toward Einstein's theory, and even about his lack of capacity to understand that theory. But Lorentz's remarks on the subject in his *Theory of Electrons*[35] ought to be enough to convince anyone that in 1909 Lorentz both understood and fully appreciated Einstein's work, although he maintained his characteristic attitude of caution (judiciously moderated, although still not abandoned, in the notes he added for the second edition of 1915). And if those remarks about the special theory of relativity do not suffice, the case is put beyond the cavil of a doubt by the circumstance that Lorentz made significant contributions to the *general* theory of relativity—and did so in the period from his sixty-first to his sixty-ninth year.[36]

I rehearse these facts not merely in the interest of justice to Lorentz (although I do believe that justice should be done, to the best of one's ability, in historical commentary) but for the sake of a larger point about the dialectic of scientific investigation. It seems to me that too many clichés, and too many oversimplifications, are propagated both in the historical and in the historically oriented philosophical literature about science—clichés and oversimplifications about programmatic divisions, about national divisions, about generation gaps. All these divisions (and others as well) are really present, and have their influence on research and on the combat of theoretical views; they are an inescapable part of the human comedy; but exactly *what* influence they have is a subtler question than it is often taken for.

To add a dimension to the evidence I have already cited, let me make a brief comment on the issue of national differences. One encounters over and over again the statement that Maxwell's views were long neglected on the Continent, and especially in Germany. This is undoubtedly true to a certain, not insignificant, extent. But there are *local* nuances. Berlin was far more sympathetic than was Göttingen; in fact the first significant experimental testing of, and evidence for, the theory of Maxwell came from Helmholtz's laboratory in Berlin (and from Boltzmann in Vienna). Again, the Scotsman Kelvin—a compatriot and friend of Maxwell—was cool to his theory and rather uncomprehending of its content. This is explained in terms of the famous British penchant for "mechanical models" (a penchant, however, shared in superlative degree by the Austrian Boltzmann, who was one of the earliest and most enthusiastic of Maxwell's adherents). And finally we have our Dutchman Lorentz, who pursued the theory of the Scot Maxwell, initially under the banner of the German Helmholtz; who developed his own theory of the electrodynamics of moving bodies with important stimulation and constructive criticism from the Frenchman Poincaré; and whose work found one of its culminations in what during the time of infamy was called "Jewish physics."

On the subject of these clichés, I cannot omit to mention one particular tidbit: In 1906, Planck delivered an address to the Congress of German Scientists and Physicians on Kaufmann's measurements of the deflection of β-rays by electric and magnetic fields and the significance of those measurements for the dynamics of the electron.[37] Planck's conclusion was that Kaufmann's results did not suffice to decide between the Lorentz-Einstein theory of the deformable electron and the theory of Max Abraham, which held the electron to be a rigid sphere. Abraham's theory made it possible to suppose that the electron mass was exclusively electrodynamic in origin, whereas Lorentz's theory did not; the former was accordingly preferred by the adherents of the "electromagnetic world-view." In the discussion following Planck's talk, Arnold Sommerfeld said: "On the question of principle formulated by Herr Planck I should like to express the conjecture that the gentlemen under forty years old will prefer the electrodynamic postulate [i.e., the rigid electron of Abraham] and those over forty the mechanical-relativistic postulate. I give the preference to the electrodynamic."[38] The report of the proceedings indicates amusement in the audience; Sommerfeld was thirty-eight at the time. In the previous decade Sommerfeld had worked on mechanical models of the ether; in the next decade (when, it must be conceded, he was past forty) his application of the special theory of relativity to Bohr's atomic model led to the explication of the fine structure of spectral lines—a success that served to establish the theory of relativity as one of the most firmly grounded parts of physics.[39]

Besides suggesting that the sociological contemplation of science might profit from caution in respect of large conclusions (and from a more relaxed enjoyment of the human comedy), the example of Sommerfeld, when set beside that of Lorentz, illustrates another important fact about the role in science of "commitment" to theories or to programs. I have argued before, from Lorentz (and Helmholtz provides another notable instance here), that a principled attitude of caution is no necessary bar to the pursuit of fruitful work. The case of Sommerfeld illustrates the complementary point: that enthusiastic commitment to a program need not be an insuperable bar to the serious entertainment of alternatives; for such commitment can be revised.

I have called the difference in point of view between Lorentz and Einstein "meta-physical." I should like now to say something about a matter that may deserve that epithet without the hyphen: the question whether the ether has been eliminated from physics. A few years ago, Larry Laudan, citing a statement by Hilary Putnam to the effect that the ether does not exist (or, in current jargon, that the term 'ether' occurring in nineteenth-century physical theories does not genuinely "refer"), used this in an argument against the position, then held by Putnam, that "terms in a mature science typically *refer*,"[40] and more generally against a certain version of "theoretical realism." In the same connection, Putnam had used the term 'atom' as an example of one that did "refer." If we replace 'atom' here by 'molecule', we have the two elements of Kelvin's subject in Baltimore: the dynamical relations of molecules and the ether.

I am not going to take sides on the issue of realism, which I do not consider to be well posed; however, I want to quote Lorentz on the question of the ether, from an article he wrote on Maxwell's theory and the theory of electrons for the volume on physics in the series *Die Kultur der Gegenwart*, published in 1914:

In conclusion I should like just to say something about the significance of the ether for theoretical physics. Although, as the foregoing exposition shows, the role of this medium has continually gained in significance, on the other hand the attempts to penetrate further into its nature have fallen increasingly into the background. Since the development of the theory of relativity many physicists have indeed gone so far as to speak no longer of an ether at all, but just of the electromagnetic field propagating itself in space. The question, to what extent this is expedient, must here be left undecided. In part it reduces itself to a verbal question; if one does not want to say that all forces are transmitted *through the ether*, one will nonetheless have to explain, according to the theory of relativity, that they are all propagated in space with the speed of light.[41]

The "explanation" Lorentz calls for here, applied specifically to the force of gravity, led in the year or two following his remark to the definitive formulation of the general theory of relativity. In the physics of our own time, the problem of the transmission of force has received one partial solution in that theory of Einstein and another partial solution in the special-relativistic quantum theory of fields and their associated particles. How to resolve the still outstanding difficulties of the latter, and beyond this how to find a framework in which both of these partial solutions have their place, is certainly the greatest issue facing theoretical physics today. It seems to me entirely appropriate to say, with Lorentz, that whether or not one calls this "the problem of the ether" is a merely verbal question.

Something else ought not to be lost sight of, however. In this transformation of the ether problem, the presuppositions of Kelvin have undeniably been left behind; it would be absurd to raise the question of whether the ether, as anything like an elastic medium, "really exists." But it is equally true, although often ignored, that the old notion of "space," that empty and quiescent container within which bodies exist and forces are propagated, has also been left behind. First, with special relativity, we were led to space-time as the frame whose structure constrained the form of all interactions; then, with general relativity, we were led to the view that space-time is not a quiescent container but is itself interactive; finally, with quantum electrodynamics, we have been led to the view that even "empty" regions of space-time are seething with—I almost said "physical activity," but I suppose it would be more correct to say physical possibilities that have to be reckoned with. It is therefore, in an important sense, more accurate to say that space and time themselves have been assimilated to the ether than that the ether has been eliminated from our view of the world.

And what kinds of properties does this ether have? The answer is that it has the properties of a dynamical system—in the generalized and transformed sense that term itself has acquired in general relativity (where the curvature tensor and the mass-energy tensor play the central role) and in the quantum theory (where Lagrangian densities and Hamiltonian operators are the recognizable descendants of the concepts which Lorentz was one of the first to generalize to the electromagnetic field).

As to the other pole of Kelvin's pair, the atoms or molecules, it is at least arguable that their status has undergone a quite parallel transformation, and that the sense in which they have been retained in our own physics is quite analogous: Some of what had seemed their most fundamental properties have fallen away, but their recognizable conceptual descendants have continued to play a basic role in our theories.

Finally, a word as to the character of this "recognizable conceptual descent": What is in fact "recognizable" is a distinct relationship, from older to newer theory, of *mathematical forms*—not a resemblance of "entities." This has always seemed to me the most striking and important fact about the affiliations of scientific theories. I do not suggest a philosophical "explanation" of this fact; I cite it, on merely historical evidence, just as a fact. But I think that, in its turn, this fact helps to "explain" why such a "conservative" as Lorentz, who was willing to borrow the mathematical structures suggested by older theories and to explore their application in contexts where the presumed "substrates" of those structures were lacking (Should one call this "realism," or should one call it a purely "instrumentalist" use of theory?), was able so greatly to advance our understanding of the world.

Notes

1. W. Thomson, "Motion of a Viscous Liquid; Equilibrium or Motion of an Elastic Solid; Equilibrium or Motion of an Ideal Substance Called for Brevity Ether; Mechanical Representation of Magnetic Force," in *Mathematical and Physical Papers*, vol. III (London: C. J. Clay and Sons, Cambridge University Press Warehouse, 1890), pp. 436–465, at 465.

2. H. A. Lorentz, *Collected Papers*, vol. II (The Hague: Martinus Nijhoff, 1936), pp. 1–119: "Concerning the Relation Between the Velocity of Propagation of Light and the Density and Composition of Media." (This is an English translation of the Dutch original.)

3. "Über die Beziehung zwischen der Fortpflanzung des Lichtes und der Körperdichte," *Wiedemanns Annalen der Physik und Chemie* 9 (1880): 641–665.

4. See H. Stein, "'Subtler Forms of Matter' in the Period Following Maxwell," in *Conceptions of Ether*, ed. G. N. Cantor and M. J. S. Hodge (Cambridge University Press, 1981), pp. 325–326.

5. H. Poincaré, *Electricité et optique*, vol. II, *Les théories de Helmholtz et les expériences de Hertz* (Paris: George Carré, 1891), pp. 50, 103, 111. In the single-volume second edition (Paris: Gauthier-Villars, 1901), the corresponding pages are 275, 329, and 337–338.

6. Lorentz, *Collected Papers*, vol. I, p. 383.

7. Ibid., vol. II, pp. 70–87; see especially pp. 79 ff.

8. Indeed, the work of Lorentz *tout ensemble* is so characterized by Kuhn himself. He remarks, in commenting on the views of Popper, that the latter's emphasis upon the "occasional revolutionary parts" of the scientific enterprise is "natural and common": that "the exploits of a Copernicus or Einstein make better reading than those of a Brahe or Lorentz." See Thomas Kuhn, "Logic of Discovery or Psychology of Research," in *Criticism and the Growth of Knowledge*, ed. I. Lakatos and A. Musgrave (Cambridge University Press, 1970), p. 6; also in Kuhn, *The Essential Tension* (University of Chicago Press, 1977), p. 272.

9. Poincaré, *Electricité et optique*, vol. I (Paris, 1890), pp. xvi–xvii.

10. In the 1878 paper, Lorentz dispensed with such a fuller account and simply postulated that the particle is subject to a force equal to the product of its charge and the intensity of the electric field at its location—or, more strictly, to the volume integral of the product of charge density and field intensity. There was no issue of motion through the ether, because he located the charged particles at the centers of cavities in the ether and considered only oscillations small enough so that each particle stayed inside its cavity. (In the theory of dispersion, contained in chapter III of the work, no magnetic forces were taken into account.)

11. "La théorie électromagnétique de Maxwell et son application aux corps mouvants," in Lorentz, *Collected Papers*, vol. II, pp. 164–343.

12. Ibid., pp. 230–233. (In stating Lorentz's assumptions, I have deviated slightly from the arrangement he gives to them.)

13. G. F. FitzGerald, "On the Electromagnetic Theory of the Reflection and Refraction of Light," *Philosophical Transactions of the Royal Society*, 1880; reprinted in *The Scientific Writings of the Late George Francis FitzGerald*, ed. J. Larmor (Dublin and London, 1902). See pp. 45–49 and 41–42. See also Stein, "'Subtler Forms of Matter'" (note 4 above), pp. 312–315.

14. The assumptions of this theory are stated in Lecture 19 (rewritten 1903) of Kelvin's *Baltimore Lectures* (Cambridge University Press, 1904). See also what appears to be Kelvin's last published paper, "On the Motions of Ether Produced by Collisions of Atoms and Molecules, Containing or not Containing Electrons," in *Mathematical and Physical Papers by the Right Honorable Sir William Thomson, Baron Kelvin*, vol. VI, ed. J. Larmor (Cambridge University Press, 1911), pp. 235–243, in particular p. 237, §8.

15. The following remark is from a paper of some eight years later (1903): "The physicists who have endeavoured, by means of certain hypotheses on the mechanism of electromagnetic phenomena, to deduce the fundamental equations from the principles of dynamics, have encountered considerable difficulties, and it is best, perhaps, to leave this course, and to adopt the equations (I)–(VII) [that is, the Maxwell-Lorentz equations]— or others, equivalent to them—as the simplest expression we may find for the laws of electromagnetism." ("Contributions to the Theory of Electrons," in *Collected Papers*, vol. III, p. 136.)

16. Lorentz, *Versuch einer Theorie der elektrischen und optischen Erscheinungen in bewegten Körpern* (Leiden: E. J. Brill, 1895); reprinted in *Collected Papers*, vol. V (The Hague: Martinus Nijhoff, 1937). See, in the latter, p. 28.

17. See P. Zeeman, "On the Influence of Magnetism on the Nature of the Light emitted by a Substance," *Philosophical Magazine*, 5th series, (January–June 1897): 226–236. Note in particular the following (pp. 230 ff.):

§15. The train of reasoning ... by which I was induced to search after an influence of magnetism, was at first the following:—If the hypothesis is true that in a magnetic field a rotatory motion of the aether is going on, the axis of rotation being in the direction of the magnetic force (Kelvin and Maxwell), and if the radiation of light may be imagined as caused by the motion of the atoms, relative to the centre of mass of the molecule, revolving in all kinds of orbits, suppose for simplicity circles; then the period ... will be determined by the forces acting between the atoms, and then deviations of the period to both sides will occur through the influence of the perturbing forces between aether and atoms.... The deviation will be the greater the nearer the plane of the circle approximates to a position perpendicular to the line of force.

§16.... [T]he above-mentioned considerations are at most of ... value as indications of somewhat analogous cases....

§17. A real explanation of the magnetic change of the period seemed to me to follow from Prof. Lorentz's theory.

... Prof. Lorentz, to whom I communicated these considerations, at once kindly informed me of the manner in which, according to his theory, the motion of an ion in a magnetic field is to be calculated, and pointed out to me that, if the explanation following from his theory be true, the edges of the lines of the spectrum ought to be circularly polarized. The amount of widening might then be used to determine the ratio between charge and mass, to be attributed to a particle giving out the vibrations of light.

The above-mentioned extremely remarkable conclusion of Prof. Lorentz relating to the state of polarization in the magnetically widened lines I have found to be fully confirmed by experiment (§20).

. . . .

§23. The experiments 20 to 22 may be regarded as a proof that the light-vibrations are caused by the motion of ions, as introduced by Prof. Lorentz in his theory of electricity. From the measured widening (§14) ... the ratio e/m may now be deduced. It thus appears that e/m is of the order of magnitude of 10^7 electromagnetic C.G.S. units.

18. Maxwell remarked that "the scientific or science producing value" of a piece of work is to be measured by its effect in stimulating investigation. See *The Scientific Papers of James Clerk Maxwell*, ed. W. D. Niven (Cambridge University Press, 1890; New York: Dover, 1965), vol. II, p. 486. The passage occurs in Maxwell's article "Attraction," written for the ninth edition of the *Encyclopaedia Britannica*. Lorentz's own far-seeing comment on the unresolved problems connected with Zeeman's phenomenon, from his Nobel Lecture of 1902 (*Collected Papers*, vol. VII [1934], p. 84) is worth quoting:

I am convinced that the theory will not make significant advances, until it directs attention not merely upon a single spectral line, but upon the totality of all lines of a chemical element.

If we once succeed in understanding theoretically the constitution of the spectra, then, and not sooner, one will be able to tackle the more complicated forms of the Zeeman phenomenon with success. Or to express my thought better: in the future the investigations concerning the regularities in the spectra and those concerning the Zeeman effect will have to proceed hand-in-hand; so they will be able someday to lead to a theory of the emission of light, to attain which is one of the finest goals of contemporary physics.

19. This hypothesis is the only aspect of Lorentz's work to which I have found a reference in Kelvin: he cites it, from Lorentz's monograph of 1895, in two papers of 1900, reprinted as Appendices A and B in *Baltimore Lectures* (1904). (The second of these is the celebrated article on the "two clouds" over the dynamical theory of heat and light.) The two passages, verbally identical (see *Baltimore Lectures* [1904], pp. 485 and 492), refer to "a brilliant suggestion made independently by FitzGerald and by Lorentz of Leyden" that may serve to reconcile with the Michelson-Morley experiment the hypothesis of the "free motion of ether through space occupied by the earth." No mention is made by Kelvin of the general theoretical context into which Lorentz incorporated this suggestion.

20. The chief stages were the following:

• the introduction, in the *Versuch* of 1895 (see note 16 above), of spatial coordinate transformations (§23), of associated transformations of the electric and magnetic field vectors and the charge-density (§§20, 23), and of a "local" time coordinate (§30), serving to simplify the electrodynamic equations when these are referred to a system moving relative to the ether; the use of the resulting formulation to show the undetectability of the earth's motion by optical means, in so far as the square of the ratio of the earth's velocity to that of light can be regarded as negligibly small; and the indication (§§90–92) of the way in which the contraction hypothesis, which extends the nondetectability to the case of the Michelson-Morley experiment, can be plausibly incorporated into the theory

• a modification (or, rather, two successive modifications) of the previous transformations, in a paper of 1899 ("Théorie simplifiée des phénomènes électriques et optiques dans des corps en mouvement," in Lorentz, *Collected Papers*, vol. V, pp. 139–155; see §§4 and 10); the most important new feature, introduced toward the end, is the incorporation into the "local time" of a factor that represents, in effect, the "time dilatation," which here makes its first appearance

• the incorporation, in a 1904 paper ("Electromagnetic Phenomena in a System Moving with any Velocity Smaller than that of Light," in Lorentz, *Collected Papers*, vol. V, pp. 172–197), of the second set of transformations of 1899—with one change respecting the components representing (or, rather, formally replacing) the velocity of an electron in the equations referred to a moving system—into a theory showing the close correspondence of all electromagnetic processes in such a moving system with those in a system at rest

• the perfection of the theory by Poincaré, who achieved the definitive form of the transformations (in particular, of those for charge-density and velocity) and exhibited the strict symmetry that then results between the representation of processes in systems at rest and those in uniform motion (H. Poincaré, "Sur la dynamique de l'électron," in *Oeuvres de Henri Poincaré*, vol, IX [Paris: Gauthier-Villars, 1954], pp. 494–550; see pp. 499–503).

The paper of Poincaré was published in 1906, but an abstract containing the equations of transformation appeared the year before (ibid., pp. 489–493). Einstein's contribution, of course, is to be found in his famous paper "Zur Elektrodynamik bewegter Körper," *Annalen der Physik* 17 (1905): 891–921.

21. Poincaré, "La théorie de Lorentz et le principe de réaction," *Oeuvres*, vol. IX, pp. 464–488.

22. M. Abraham, "Prinzipien der Dynamik des Elektrons," *Annalen der Physik*, 4th series, 10 (1903): 105–179; see p. 110.

23. Weyl, *Space—Time—Matter*, tr. H. L. Brose (New York: Dover, 1950), pp. 160–166. I have praised Weyl's account of the theory; unfortunately, this translation is badly defective, and the reader interested in reading his exposition must be referred to the German original, preferably its fifth or sixth edition: *Raum—Zeit—Materie: Vorlesungen über allgemeine Relativitätstheorie*, sixth edition (Berlin: Springer-Verlag, 1970).

24. Poincaré, *Oeuvres*, vol. IX, p. 491.

25. Lorentz, *Collected Papers*, vol.·V, p. 190.

26. See, e.g., Poincaré, *Oeuvres*, vol. IX, pp. 381–382, 412–413; and *Rapports du Congrès de Physique de 1900*, vol. I, pp. 22–23.

27. Poincaré, *Oeuvres*, vol. IX, p. 497.

28. Ibid., p. 498.

29. The phenomenon referred to by Poincaré was first described by André Broca (*Comptes Rendus, Paris* 126 (1898): 736–738, 823–826), who reported that in a sufficiently intense magnetic field there are produced two distinct sorts of cathode rays: those "of the first kind," which wind about the lines of magnetic force, and those "of the second kind," which follow the field lines. The subject was taken up again in 1904 by Paul Villard (*ibid.* 138 (1904): 1408–1411), who concluded that the rays "of the second kind," to which he gave the name "magneto-cathodic rays," were *uncharged* (p. 1410: "Les rayons magnéto-cathodiques ne sont pas électrisés.... Leur charge, si elle existe, est ... incomparablement moindre que celles des premiers. Il est fort probable que cette charge est nulle et que ces rayons sont autre chose qu'électrisés.") and that their properties were "inverse to those of the rays of Hittorf" (i.e., the ordinary cathode rays): "Le champ électrique agit sur les premiers comme le champ magnétique sur les seconds, et reciproquement." It is clearly to Villard's account, with its suggestion that the phenomenon represents a fundamentally new kind of process, that Poincaré refers. The same volume of the *Comptes Rendus*, however, contains a discussion by Charles Fortin in which it is pointed out that Villard's observations are consistent with the behavior to be expected of cathode rays of the ordinary kind if these wind about the magnetic lines of force in helices of very small radius.

In the *Comptes Rendus* for the years 1909 and 1911 (vols. 148 and 152), discussion of the magneto-cathodic rays was continued by G.-L. Gouy, who, finding bright and dark fringes at positions for which the lengths of the rays were integral multiples of a certain length (itself inversely proportional to the intensity of the magnetic field), suggested that an appropriate modification of the Newtonian theory of periodic "fits" might have to be invoked for the rays. I find no subsequent reference to the phenomenon in the *Comptes Rendus* through 1914; the whole matter has by now lapsed into such obscurity that in the English translation of Poincaré's paper by H. M. Schwartz, which appeared in the *American Journal of Physics*, (39 [1971]: 1287; 40 (1972): 862), Poincaré is made to say that the whole theory is put in jeopardy by the discovery of cathode rays! (These had of course been discovered several decades before, and they fit very well indeed into the structure of Lorentz's theory; but clearly the translator did not know what to make of the "rayons magnéto-cathodiques.")

30. See L. Rosenfeld, "The Velocity of Light and the Evolution of Electrodynamics," *Nuovo Cimento*, supplement to vol. IV, series X, no. 5 (September 1956): 1664.

31. *Enzyklopädie der mathematischen Wissenschaften mit Einschluss ihrer Anwendungen*, vol. V, part 2 (Leipzig: B. G. Teubner, 1904–1922), article 13: "Maxwells elektromagnetische Theorie," by H. A. Lorentz, §§44–45.

32. Ibid., p. 144.

33. Ibid., p. 143.

34. See the references given in note 13 above; also Stein, "'Subtler Forms of Matter,'" pp. 320–321 and 329–330.

35. H. A. Lorentz, *The Theory of Electrons and Its Applications to the Phenomena of Light and Radiant Heat* (reprint of second edition, 1915; New York: Dover, 1952); see preface and pp. 223–230; also (added in second edition) notes 72*–74, 75*–76 (pp. 321–327, 328–334). In particular, compare with the passage earlier cited from Poincaré about the chance agreement of formulas the following remarks of Lorentz's.

• on the last page of the main text of 1909: that Einstein "may certainly take credit for making us see in the negative result of experiments like those of Michelson, Rayleigh and Brace, not a fortuitous compensation of opposing effects, but the manifestation of a general and fundamental principle" (p. 230)
• in a note added in 1915: "If I had to write the last chapter now, I should certainly have given a more prominent place to Einstein's theory of relativity ... by which the theory of electromagnetic phenomena in moving systems gains a simplicity that I had not been able to attain. The chief cause of my failure was my clinging to the idea that the variable *t* only can be considered as the true time and that my local time *t'* must be regarded as no more than an auxiliary mathematical quantity." (p. 321)

36. See a series of articles reprinted in Lorentz, *Collected Papers*, vol. V, pp. 229–245, 246–313, 330–355, 363–382.

37. Planck, "Die Kaufmannsche Messungen der Ablenkbarkeit der β-Strahlen in ihrer Bedeutung für die Dynamik der Elektronen," *Physikalische Zeitschrift* 7 (1906): 753–761.

38. Ibid., p. 761.

39. Note the following passage in Sommerfeld's book *Atomic Structure and Spectral Lines*, tr. (from third edition, 1922) H. L. Brose (New York: Dutton, 1923), p. 531: "In this chapter we have seen how the theory of relativity, just as it has remodelled all our physical thought and ideas, has also been able to help forward spectroscopy in a decisive manner. Conversely, we note that, in return, spectroscopy is in a position to lend support to one of the main pillars of the theory of relativity and to decide in its favor the question of the variability of mass of the electron."

40. L. Laudan, "A Confutation of Convergent Realism," *Philosophy of Science* 48 (1981): 19–49; see in particular p. 24 and p. 21, n. 1.

41. Lorentz, "Die Maxwellsche Theorie und die Elektronentheorie," in *Physik*, ed. E. Warburg (Berlin: B. G. Teubner, 1915), pp. 311–333; the quotation is from the closing sentences of the article.

Abner Shimony

The Methodology of Synthesis: Parts and Wholes in Low-Energy Physics

1 Aspects of the Problem

One of the most pervasive features of the natural world is the existence of reasonably stable systems composed of well-defined parts which are to a large extent unchanged by entering into composition or leaving it. The problem of parts and wholes is to understand with the greatest possible generality the relation between the components and the composite system.

The parts-wholes problem has an ontological aspect, which concerns the properties of the components and the composite system without explicit consideration of how knowledge of them is obtained. Among the ontological questions are the following: Is there an ultimate set of entities which cannot be subdivided and which are therefore "atomic" in the etymological sense? If the properties of the components are fully specified, together with the laws governing their interactions, are the properties of the composite system then fully determined? In particular, are those properties of composite systems which are radically different from those of the components, and which might properly be characterized as "emergent," also definable in terms of the latter? Do composite systems belong, always or for the most part, to "natural kinds"? Is the existence of natural kinds explicable in terms of the laws governing the components? Are both the possible taxonomy and the actual taxonomy of natural kinds thus explicable? Is there a hierarchy of "levels of description"— i.e., microscopic, macroscopic, and possibly intermediate—such that laws can be formulated concerning a coarser level without explicit reference to the properties at a finer level of description?

The parts-wholes problem also has an epistemological aspect. Suppose that the most precise and best-confirmed laws turn out to govern relatively simple systems—as is indeed mostly the case in physics—but that the systems of interest are enormously complicated combinations of simple components. Then there will be insuperable experimental difficulties in gathering knowledge about all the initial conditions of the parts, and insuperable mathematical difficulties in deducing from the basic laws the properties of the composite system.

To what extent can the composite system be said to be understood in terms of the laws governing its parts? And if there is independent phenomenological knowledge of laws on a coarse level of description, how do we know that these are in principle derivable from the laws on a finer level?

Aside from the ontological-epistemological distinction, there is a subdivision of the parts-wholes problem according to domains of investigation. There is the domain of inorganic systems (the physical sciences), that of organic systems (biology), that of systems endowed with minds (psychology), and that of groupings of human beings (social sciences). There is no *a priori* reason to believe that the questions listed above should be answered in the same way in all these domains, and it may not even be appropriate to pose the same questions in all of them. Of course, even the separateness of these domains is an aspect of the parts-wholes question. If the properties of an organism can in principle be defined in terms of those of the constituent molecules, and biological laws can in principle be derived from those of physics, then the domain boundary between the inorganic and the organic loses its fundamental significance and is maintained only for reasons of professional specialization. Likewise for the other domain boundaries.

The parts-wholes problem is so vast and ramified that no apology is needed for restricting attention to a small part of it. I shall consider almost nothing outside of physics, and within physics almost exclusively those branches in which the energies involved are typically of the order of an electron volt per particle or less: atomic physics, molecular physics, solid-state physics, and in general the physics of bulk matter. Low-energy physics is roughly Democritean. Electrons and stable nuclei have properties which can be studied in isolation, and to a very good approximation they keep these properties unchanged when they enter atoms, molecules, fluids, and solids. Hence electrons and nuclei behave as "building blocks," even though high-energy investigations reveal a rich internal structure of the latter. Democritus's characterization of his atoms as infinitely hard, indivisible, and immortal certainly does not apply without drastic reservations to these building-blocks. Nevertheless, they preserve their integrity in low-energy interactions to such an extent that Democritus's vision of explaining the full variety of the natural world in terms of the combinations of a very few kinds of unchanging bodies is realized to a remarkable extent. By contrast, high-energy physics—according to which particles are created and annihilated and the nucleon components (quarks) are removed from confinement either with great difficulty or not at all—is radically non-Democritean.

This essay consists largely of summaries of the treatment of parts and wholes in five areas of low-energy physics, concerning atoms, molecules, fluids, infinite Coulomb systems, and spin systems. Roughly Democritean answers are given in each area to most of the ontological questions concerning parts and wholes (if the divisibility of nuclei is set aside). I wish to call the attention of

philosophers of science to the great variety and subtlety, and the often surprising nature, of the derivations of properties of composite systems from
those of the components. Ontologically, the theory of the wholes is reduced
to the theory of the parts, but the pejorative overtones of "reductionism"—a
suggestion of flattening and loss of distinctive features—is certainly inappropriate. Epistemologically, the reduction of the theory of the wholes to that
the parts has to be understood with serious qualifications. In the derivations
that will be mentioned, the first principles of low-energy physics are usually
supplemented with secondary principles, which can reasonably be regarded
as consistent with the former but rarely rigorously inferrable from them. The
methodology of the physical understanding of complex systems turns out
to have intricacies that are seldom discussed and are difficult to formulate
clearly.

This essay does not pretend to be authoritative, for the simple reason that I
am not an expert in any of the branches of physics from which my examples
are drawn. I nevertheless hope to convey my enthusiasm for inquiry in the
territory that is reconnoitred.

2 First Principles

The first principles of low-energy physics fall into three groups: those of nonrelativistic quantum mechanics, those concerning the dominant forces among
the building blocks, and the principles of statistical mechanics.

A. Quantum mechanics is a framework theory that at present is commonly
believed to apply to every physical system from elementary particles to the
whole of the cosmos. My skepticism about the universal validity of quantum
mechanics is largely irrelevant to the present discussion, since there is little
doubt that quantum mechanics holds with great accuracy within the domain
of low-energy physics. As a framework theory, quantum mechanics has to be
supplemented by detailed information about the constitution of any particular
system to which it is applied before inferences can be made about the behavior of the system and before experimental results can be predicted. Nevertheless, the framework which quantum mechanics provides is rich.

(i) It prescribes the general characteristics of the space of states of any system,
most commonly summed up by saying that there is a one-to-one correspondence between the pure states and the rays of a Hilbert space. Implicit in this
characterization of the space of states is the superposition principle.

It prescribes the general structure of the class of observables (as they are
commonly called, though a less anthropocentric name such as "dynamical
variables" would be preferable).

(iii) It prescribes the rule for calculating the probability distribution of possible
values of an observable, contingent upon its actualization (by measurement or
possibly by other means), when the state is known.

(iv) Nonrelativistic quantum mechanics offers a general law for the temporal evolution of the pure state of an isolated system, provided that a sufficient characterization of the system is given (namely, its Hamiltonian operator).

(v) Quantum mechanics itself has a remarkable principle of composition. If A and B are two systems, associated respectively with the Hilbert spaces H_A and H_B, then the Hilbert space associated with the composite system $A + B$ (provided that their dynamical interaction does not change them internally) is the tensor-product space $H_A \otimes H_B$. Only in special cases is it possible to write a vector ϕ belonging to $H_A \otimes H_B$ in the simple-product form $u \otimes v$, where u belongs to H_A and v to H_B, and only in these special cases is it correct to say that the pure state represented by ϕ is equivalent to the attribution of pure states to each of the components A and B. In general, A and B have an "entangled" state (to use Schrödinger's expression), so that there is no complete characterization of A without reference to B and vice versa. It is evident that quantum-mechanical entanglement requires a partial retrenchment from the Democritean conception of the parts-wholes relationship, but it leaves open the possibility of realizing other aspects of this conception. The quantum-mechanical ground state of a many-particle system is determined by its Hamiltonian operator, which consists of kinetic-energy contributions for each particle and potential-energy contributions for each pair of particles. Such a Hamiltonian is in the spirit of Democritus—*mutatis mutandis*—since it contains no contributions from the whole that cannot be traced explicitly to the parts. If the quantum-mechanical ground state then turns out to account for the physical properties of the system at a temperature of absolute zero (assumed for the present in order to avoid complications due to thermal excitations), then it is reasonable to say that a Democritean description of the many-particle system has been realized, even though the ground state is entangled.

(vi) Finally, there are symmetrization principles; i.e., if there are n identical bosons (integer-spin particles) in the system, then any vector in the Hilbert space representing a physically allowable state must be invariant under the interchange of two of these bosons, and if there are n identical fermions (particles with half-integer spin), then any vector representing a physically allowable state must change sign under exchange of two of these fermions. The Pauli principle, which prohibits two fermions to occupy the same single-particle state, is a direct consequence of this anti-symmetrization. Both the symmetrization rule and the anti-symmetrization rule are modifications of the tensor-product principle, since each restricts the space of states of a composite system to an appropriate subspace of the full tensor-product space.

B. One of the great simplifications of the restriction to low-energy physics is that the dominant forces among the constituents are electromagnetic. Furthermore, these forces are treated essentially classically, by potentials, rather than by postulating a quantized electromagnetic field which interacts with matter fields. Nevertheless, much complexity remains, for electromagnetic forces are exhibited in electron-nucleus interaction, electron-electron repulsion, nucleus-nucleus repulsion, orbit-orbit interaction, spin-spin interaction, electron spin–electron orbit coupling, electron orbit–nuclear spin coupling, and interaction with external electromagnetic fields. The potential terms in the total Hamiltonian can be of any of these types. In addition, of course, there are kinetic-energy terms. When a specified set of nuclei and a specified number of electrons are given, usually together with certain constraints upon the configuration, then the electromagnetic-force laws and the general principles of quantum mechanics yield a definite Hamiltonian. A large set of physical questions (What are the stationary states of the system? What is the ground state? What is the energy of the ground state? At what energy above the ground state does dissociation occur? What are the degeneracies of the various allowable energies? What are the geometrical properties of the system in the ground state? and so on) then become answerable in principle.

C. Statistical mechanics is a science of systematically extracting a relatively small amount of reliable physical knowledge of statistical matters from an ocean of ignorance. It treats mainly of two classes of systems. Systems of the first class consist of very many parts, typically falling into a small number of types (e.g., a homogeneous or a heterogeneous gas, or an alloy). Because of the very large number of parts, it is practically unfeasible to gather exact information about the initial conditions of the system. And even if, *per impossibile*, a complete knowledge of the initial conditions were given, it would be humanly impossible to solve the equations of motion (classical or quantum mechanical) in order to infer exactly the state at a later time. The other class of systems considered by statistical mechanics consists of open systems, interacting with an environment the exact constitution of which is not known. (If the constitution were known in detail, then the system plus the environment might constitute a single system of many parts that would fall in the first class just mentioned.) The reliable information that one wishes to extract from the immense background of ignorance consists of probability distributions of especially interesting quantities concerning the system. When the system consists of a large number of parts, then typically the probability distributions are very sharply peaked, so that interesting quantities concerning the system as a whole, such as energy, state of condensation, and magnetization, can be predicted with virtual certainty. Statistical mechanics thus provides a powerful instrument for making inferences about wholes from the properties of parts, though it is an instrument that usually requires supplementation by secondary principles. The enterprise of statistical mechanics is greatly expedited if the environment with which the system exchanges energy is negligibly affected by the exchange, so that it can be considered a reservoir. When the system is in equilibrium with the reservoir—which is a concept that can be defined either phenomenologically or probabilistically—then the probability distribution over the space of states can reasonably be shown to be the canonical distribution,[1] which can be written as follows for classical systems:

$$\rho(x) = \frac{1}{Z(\beta)} e^{-\beta E(x)},$$

where x represents the state in an appropriate space of states, $E(x)$ is the energy of the system when it is in the state represented by x, β is the inverse of the product of Boltzmann's constant k and the absolute temperature T, and $Z(\beta)$ is a normalizing factor defined in such a way that the integral of $\rho(x)$ over the space of states is 1. Explicitly,

$$Z(\beta) = \int e^{-\beta E(x)} \, dx.$$

If $Z(\beta)$ is considered as a function of β, it is called the partition function; some indication of its importance and utility will be conveyed below. There is also a quantum-mechanical version of the canonical distribution, which will not be needed explicitly in this essay even though it is implicit in much of the discussion of low-energy physics.

3   Atoms

The structure of atoms is one of the best-understood kinds of composition of wholes from parts in all of physics. Some important features of this composition can be understood quantitatively from first principles, but other features are described only by semi-empirical rules, for which there is some explanation but by no means a rigorous general derivation.[2]

Well established is the possibility of describing the ground state of any atom by a configuration, which tells how many electrons are in single-electron states of given principal quantum number n and given orbital angular momentum quantum number ℓ. The ascribability of definite configurations to each species of atom is the essence of the shell model of the atom. For example, for aluminum the configuration is $1s^2 2s^2 2p^6 3s^2 3p$, where the letters s, p, and d conventionally stand for the values 0, 1, and 2, respectively for the quantum number ℓ; the number preceding a letter prescribes the value of n; and the superscript after the letter gives the number of electrons with the specified values of n and ℓ. The wave function ψ_{al} (a vector in the Hilbert space of square-integrable functions), representing the ground state of aluminum, is a superposition of a number of vectors, each of which is a product of single-particle wave functions, of which two have $n = 1$, $\ell = 0$; two have $n = 2$, $\ell = 0$; six have $n = 2$, $\ell = 1$; two have $n = 3$, $\ell = 0$; and one has $n = 3$, $\ell = 1$. In each term in the superposition, spin states for each of the 13 electrons are given. Furthermore, the superposition is so contrived that ψ_{al} is antisymmetric under exchange of any two electrons, as required by principle A(vi) above. This is already an enormous amount of information, and it is derived almost rigorously from first principles, together with the specification of the charges of the electrons and nuclei and the spin $\frac{1}{2}$ of the electron. The value $\frac{1}{2}$ for the spin not only implies that the electron is a fermion but also fixes the dimensionality of the spin space associated with each electron to be 2. Consequently, when a fixed one-particle spatial wave function is given—characterized by n and ℓ together with one other quantum number, the magnetic quantum number m—there is a further option of making the spin either up or down along a specified axis, thereby choosing between two orthogonal directions in the two-dimensional spin space. Hence, a pair of electrons can be characterized by n, ℓ, and m without violating the exclusion principle. Another ingredient in the derivation of the configuration is the approximate spherical symmetry of the effective potential which each electron feels, due to the small size of the nucleus in comparison with the atomic radius and due also to the effective "smearing" of the charge distributions of all the other electrons. This treatment of the electronic charge distribution is an instance of a mean-field approximation, which will recur below. Because the effective potential is nearly spherically symmetric, the angular momentum of the electron is a conserved quantity and hence ℓ is a good quantum number. Further exploitation of spherical symmetry yields rigorous information about angular momentum: for example, that for fixed n a given value ℓ of the angular momentum quantum number is compatible with exactly $2\ell + 1$ values of the component of the angular momentum along any specified axis, and hence $2\ell + 1$ values of the magnetic quantum number m (i.e., $m = -\ell, \ldots, \ell$). That fact was implicit in the configuration stated above for aluminum—e.g., there are six electrons in the one-electron state 2p, because the p is an abbreviation for $\ell = 1$, which permits three possible values of m $(1, 0, -1)$, for each of which there are two possible spin orientations. That is why the closed shell 2p has six electrons. To be sure, the multiplicity of 3p is only 1 in aluminum, but that is because the 3p shell is incomplete; in all the atoms from argon onward there are six 3p electrons.

Although the configuration provides much information about the ground state, there is more to be determined, and almost everything else is much harder to extract from first principles. One wishes to know, for example, the radial dependence of the single-particle wave functions, which in turn determines such important properties as the average size of the atom. To find the radial dependence, one must know the effective potential felt by a single electron and use it to solve the differential equation in the radial variable r which is obtained from the time-independent Schrödinger equation. But the potential depends upon the wave functions of all the electrons. How does one emerge from a maze in which the potential depends upon the wave functions and the wave functions depend upon the potentials—even if one sets aside all the mathematical difficulties of solving the differential equations? A way out is provided by the self-consistent field method of Hartree and Fock. A sequence of successive approximations is applied in practice, and it is reasonable to expect rapid convergence to a situation in which wave functions and potentials fit each other self-consistently.[3]

Except in the simplest atoms (through beryllium), the configuration does not determine how the orbital angular momenta ℓ_i and the spins s_i of the individual electrons are combined to yield a total orbital angular momentum quantum number L for the whole set of electrons, a total spin S, and also a total angular momentum J. In principle the possible combinations could be checked to see which yields the lowest-lying energy, but that requires detailed knowledge of the wave functions, which—as just noted—is a formidable obstacle. There are, however, Hund's rules based on spectroscopic evidence, which are quite reliable. The first two of Hund's three rules are the following:

1. The LS with the largest S compatible with the configuration has the lowest energy.

2. In the case where the largest S is associated with several different allowable values of L, the largest L has the lowest energy.

There is no generally valid proof for these rules, though they have been confirmed for a number of atoms by detailed calculations. But much plausibility can be given to them, especially to the first, by general arguments from first principles. The larger S is, the more parallel are the spins. When the spins are parallel, they make a symmetrical contribution to the complete wave function, and hence the antisymmetrization principle for electrons implies that the spatial part of the wave function is antisymmetric. But that has the general effect of keeping the electrons, on the average, farther from one another than would be the case with symmetrical spatial wave functions, and therefore the potential energy of electrostatic repulsion among the electrons is diminished, thus lowering the energy.[4]

The mathematical complications of determining the ground-state wave function of an atomic species should not make us lose sight of a fundamental Aristotelian fact: that the species is a natural kind. All atoms of a given species (i.e., a given number of neutrons and protons in the nucleus, and as many electrons as protons for electric neutrality) have the same ground state when there are no external perturbations, and they have the same array of excited states. The *identity* of all atoms of a given species ensures that they all emit and absorb electromagnetic radiation at the same frequencies and have the same size, shape, and internal motion. Weisskopf (1979, p. 87) emphasizes two other features of atoms: stability ("The atoms keep their specific properties in spite of heavy collisions and other perturbations to which they are subjected") and regeneration ("If an atom is distorted and its electron orbits are forced to change by high pressure or by close neighboring atoms, it regains its exact original shape and orbits when the cause of distortion is removed"). Atoms thus exhibit form in Aristotle's sense, and even have the tendency to maintain this form, which phenomenologically is like his final cause. But the Aristotelian form is achieved by Democritean means—by interactions among the electrons and the nucleus, which leave these building blocks intact. Of course, one additional element is required for the achievement of form: the principles of quantum mechanics. Classical physics did not have the resources to explain natural kinds of composite systems, even when natural kinds of the indivisible building blocks were postulated, whereas Aristotelian science accounted for natural kinds uneconomically, by postulating an irreducible principle for each. In contrast to both, quantum mechanics ensures that the formal properties of very many kinds are implied by the properties of a small number of kinds of components.

4   Molecules

On the whole, molecular structure is more difficult to understand than atomic structure, primarily because molecules lack the spherical symmetry of the single-nucleus atom. Nevertheless, there is one principle of molecular composition that was extracted from laboratory experience and is simple enough to be taught in elementary chemistry: the principle of valence.[5] The valence of an element is defined phenomenologically as the number of atoms of hydrogen that one atom of the element can combine with or take the place of in forming compounds. From H_2O, HCl, NH_3, and CH_4 we infer that O, Cl, N, and C, respectively, have valences 2, 1, 3, and 4. Very soon, however, one finds that valence is not an intrinsic characteristic of an element, since in CO carbon behaves as if it has valence 2, and iron would have to be assigned valence 2 because of FeO and valence 3 because of Fe_2O_3. Clearly, valence is a crude and complicated principle of composition—and there are wonderful monsters like xenon-fluoride to increase the complication.

The hope of physical chemists is to put the valence concept on a firm basis, with all appropriate qualifications, by using the principles of composition of low-energy physics, especially quantum mechanics and the electromagnetic forces. Even before the new quantum mechanics, G. N. Lewis proposed to interpret chemical bonding in terms of the sharing of a pair of electrons by each of two atoms in a molecule, and the valence of an atom was interpreted as the number of electrons available for sharing. W. Heitler and F. London later gave a quantum-mechanical treatment of this idea in which the valence electrons are those that are not paired in the atom's configuration, where pairing means having the same spatial wave function (or "orbital") but opposite spins. In the bonding of two atoms in a molecule, a valence electron from one of the atoms is free to combine with a valence electron of the other in such a way that their combined spin state is the singlet state

$$|\uparrow\rangle_1|\downarrow\rangle_2 - |\downarrow\rangle_1|\uparrow\rangle_2.$$

This is a vector in the two-electron spin space which obviously changes sign under the exchange of electrons 1 and 2. Conseqently, the overall antisymmetrization of electrons 1 and 2 is achieved by a symmetric spatial wave function for the two electrons. The effect of the symmetry of the spatial wave function can best be seen if one considers a coordinate r along an axis passing through the nuclei of the two atoms, utilizing the fact that the nuclei are more massive than the electrons and hence more precisely localizable without violating the uncertainty principle. An antisymmetric spatial wave function will have an amplitude close to zero halfway between the nuclei, and a symmetric spatial wave function will usually have a large amplitude and even be peaked near the midpoint. Therefore the symmetric spatial wave function makes the two electrons, on the average, closer to each other than does a comparable antisymmetric one, and hence the contribution of the repulsive electrostatic potential of the two electrons (a positive energy) is larger in the symmetric case than in the antisymmetric; but this positive contribution is more than compensated by the fact that, on the average, the electrons spend more time close to the nuclei in the symmetric case, thereby increasing the attractive contribution. Hence the singlet spin state is favored for the achievement of the lowest possible energy. This is an excellent argument, except for some hand-waving concerning the relative magnitude of the competing repulsive and attractive contributions. In the case of the H_2 molecule, Heitler and London were able to perform a quite accurate quantitative calculation from first principles to confirm the hand-waving argument. The extrapolation to more complex molecules is always somewhat risky, however.

Without any consideration of more detailed calculations it is evident that some modification of the foregoing account of valence is needed. The ground-state configuration of carbon is $1s^2 2s^2 2p^2$, which provides two unpaired electrons in the 2p shell (unpaired because there are three mutually orthogonal spatial wave functions available in this shell—$m = 1$, $m = 0$, and $m = 1$). Usually, however, carbon exhibits a valence of 4, though it was noted that in CO it exhibits valence 2. Pauling (1931) offered a resolution of this and similar difficulties by arguing that in molecular bonding one must consider not only the atomic ground-state configuration but also the lower-lying excited configurations. The reason is that the energy expended to raise some of the electrons to excited levels is more than compensated for by the availability of additional unpaired electrons. This suggestion is a fruitful new secondary principle of composition in molecular physics, known as "the hybridization of molecular orbitals." It is not merely an *ad hoc* device to save the Heitler-London theory of spin pairing, since good calculations of molecular geometry can be made by means of hybridization theory. Hybridization has to be used with tact, and sometimes must be refined by additional secondary principles. For example, Pauling and Keaveny (1973, p. 93) answer as follows an argument that hybridization theory breaks down for iron-group elements: "It has been recognised that the orbital occupied by a bonding electron may be either expanded or contracted considerably with respect to the orbital in the isolated atom. This expansion or contraction is accompanied by a contraction or expansion of other orbitals occupied by unshared electrons. Expansion or contraction of one orbital decreases or increases its shielding of nuclear charge for another orbital." Machine calculations and empirical data seem to support this refinement of hybridization theory.

Machine calculations of wave functions are not imperturbable, objective judgments independent of all appeal to secondary principles, because the choice of initial guesses of wave functions in an iteration procedure is crucial for the rate at which converges occurs. If a principle such as hybridization of molecular orbitals is a good approximation to the truth, then the initial wave function that it suggests will lead to rapid convergence. If not, the convergence will be too slow to be useful.

5   Coulomb Systems in the Thermodynamic Limit

A vast range of phenomena can be described quite well by the formalism of thermodynamics. Essential to this formalism is the possibility of characterizing a system at a macroscopic level in terms of a small number of extensive variables: V, U, and $N_1, \ldots, N_k$, where V is the volume, U is the total energy, and N_i is the mole number of the ith chemical constituent. By *extensive* is meant that when the system is spatially subdivided into subsystems (not too small) the value of the extensive variable of the system as a whole is the sum of the values of that variable in all the subsystems. Furthermore, it is assumed that there is an entropy S, which is a linear homogeneous function of the extensive variables and which therefore is extensive too. The maximization of S relative to constraints determines the thermodynamic state of the system.[6]

Extensiveness as a constructive principle is so often taken for granted in ordinary applications (for example, the energy necessary to heat a kilogram of milk one degree is 1,000 times the energy needed to heat a gram of it one degree, largely independent of the shape of the container) that it takes some reflection to see that it is thoroughly nontrivial. Extensiveness depends, for example, on the negligibility of surface forces, which of course is not always the case. In order to avoid the complications of the surface, the standard procedure is to idealize by going to the thermodynamic limit, in which there is infinite volume and infinitely many particles, but to take the limit in such a way that

$\lim(N/V) = \text{density}$

exists. It remains to be established, however, that U and S are extensive variables in the thermodynamic limit, so that $U = Vu$, where u is the internal energy per unit volume, and $S = Vs$, where s is the entropy per unit volume.

There is no *a priori* assurance that the constructive principle of thermodynamics is consistent with the constructive principles of atomic and molecular physics, in which the properties of the composite are determined by the properties of component particles and their interactions. One might properly worry, for example, that if the fundamental forces among particles are long-range, then putting together more and more particles will not determine an internal energy proportional to volume, even if the density is fixed.

These questions turn out to be remarkably difficult. Using classical mechanics, one can prove the existence of the thermodynamic limit, provided that the intermolecular forces are repulsive at short distances and attractive at long distances but that the attractiveness falls off very rapidly as the distance increases.[7] Coulomb forces, which are by far the most important of electromagnetic forces at low velocities, do not have these characteristics, and the physically important question is whether the thermodynamic limit holds for a neutral Coulomb system (equal densities of positive and negative charges). The answer is no if classical mechanics rather than quantum mechanics is taken as the framework theory. But, most remarkable, the answer is yes if quantum mechanics is used. The proof is extraordinarily difficult, but hinges upon two simple facts: that the exclusion principle keeps fermions from coming too close to one another, thereby providing an effective repulsive core if the negatively charged particles are electrons (or, more generally, if either the positively charged particles or the negatively charged particles are fermions of one kind); and that electrostatic shielding in a neutral Coulomb system counteracts the long-range character of the Coulomb force.[8] The very existence and stability of bulk matter with ordinary macroscopic properties depends crucially upon an intimate quantum-mechanical principle that underlies the first of these two facts. In 1519–1522 Magellan proved that the Pacific Ocean existed by sailing across it, but 450 years had to elapse before Dyson, Lenard, Thirring, Lebowitz, and Lieb proved by quantum mechanics that something like the Pacific Ocean *could* exist. Thus science progresses.

6   Normal Fluids

In a very large, spatially extended system, it is trivial to define macroscopic physical variables by summing or averaging over a very large number of microscopic variables. The local matter density $\sigma(\mathbf{x})$ of a fluid can be defined classically by summing the masses of the particles in a small region $R_\mathbf{x}$ centered about x, which is large enough to contain on the average many particles but is small in comparison with the macroscopic dimensions of the system, and then dividing by the volume of $R_\mathbf{x}$. An analogous definition can be given quantum-mechanically, but it makes use of expectation values of positions of particles. Likewise, a local mean velocity $\mathbf{v}(\mathbf{x})$ of the fluid can be defined quantum-mechanically as well as classically. And similarly for other macroscopic quantities such as pressure and temperature, which also may be local quantities, varying with **x**. Enough has been said to make it clear that *local* can be *macroscopic.* What is nontrivial, and clearly not simply a matter of definition, is whether there are any laws of nature that can be formulated solely in terms of a set of macroscopic quantities. If there are such laws, then the microscopic level is not the only level of description at which laws are exhibited. Actually, we are quite confident that there are laws at a macroscopic level of description, because of the great empirical success of such disciplines as fluid dynamics. The difficult problem is to explain why this is so.

An outstanding example is the Navier-Stokes equation, which is the dynamical law for a viscous fluid in nonturbulent motion:

$$\sigma(\mathbf{x})[\dot{\mathbf{v}}(\mathbf{x}) + (\mathbf{v}(\mathbf{x})\, \mathrm{grad})\mathbf{v}(\mathbf{x}) = -\mathrm{grad}\, p(\mathbf{x}) - \eta_1 \,\mathrm{curl\, curl}\, \mathbf{v}(\mathbf{x}) + \eta_2 \,\mathrm{grad\, div}\, \mathbf{v}(\mathbf{x}).$$

The matter density $\sigma(\mathbf{x})$ is the mass per unit volume in the neighborhood of **x**; $\mathbf{v}(\mathbf{x})$ is the local velocity; $p(\mathbf{x})$ is the local pressure; and η_1 and η_2 are friction constants. There have been many derivations of this equation, but I find that of Fröhlich (1973) impressive for its generality and lucidity. The derivation begins with the exact time-dependent Schrödinger equation for a system consisting of N particles of specified mass, but restates this equation in terms of the von Neumann density operator Ω instead of the wave function, as is appropriate where there is imperfect knowledge of the true quantum state. In other words, Ω is the quantum-mechanical analogue of the distribution function of classical statistical mechanics. Ω contains information about the correlated behavior of all the N particles, but for practical purposes most of these correlations are of no interest. It therefore is useful to define reduced density operators, Ω_1, Ω_2, Ω_3, etc., concerning (respectively) single particles, pairs of particles, triples of particles, etc. From the original dynamical equation for Ω, a hierarchy of equations for Ω_1, Ω_2, Ω_3, etc. follows easily. The complication of this hierarchy lies in the fact that the equation in Ω_1 contains a term dependent on Ω_2, the equation for Ω_2 contains a term dependent on Ω_3, and so on. The macroscopic quantities in the Navier-Stokes equation depend only upon Ω_1 and Ω_2, and with appropriate additional assumptions the equation itself can be derived. First, it is assumed that in equilibrium there is rotational and translational invariance, so that the pair-correlation function $P(\mathbf{x}, \mathbf{y})$, which is essentially the probability of jointly finding particles located at **x** and **y**, becomes at equilibrium a function $P_e(|\mathbf{x} - \mathbf{y}|)$ which depends only on the distance between the two points. Second, it is assumed that the interaction between particles is short-range, so that large changes of σ, **v**, and p do not occur over distances in which the interaction is non-negligible. And third, attention is restricted to situations in which the deviation of $P(\mathbf{x}, \mathbf{y})$ from the equilibrium pair-correlation function $P_e(|\mathbf{x} - \mathbf{y}|)$ is a function of **x** only via the distance $|\mathbf{x} - \mathbf{y}|$ and via linear dependence upon the macroscopic fields $\sigma(\mathbf{x})$ and $\mathbf{v}(\mathbf{x})$. Since these assumptions are quite mild, Fröhlich's derivation has great generality. In principle, values of η_1 and η_2 can actually be calculated from Ω_2, though to do so the hierarchy of equations for Ω_i must be terminated at $i = 2$ by an approximation, and the resulting dynamical equation for Ω_2 must be solved. For practical purposes it usually is sufficient to know that the form of the Navier-Stokes equation is legitimate and to depend upon measurement to supply the friction constants.

7   Phase Transitions in Spin Systems

Most of the discussion of composition so far has drawn only upon the first two classes of principles summarized in section 2, those of quantum mechanics and of electromagnetism. These are the principles that one would expect to dominate at low temperature, when thermal fluctuations are negligible. At elevated temperatures, however, the thermal fluctuations become more and more important as a disordering influence. The derivative principles of composition that one expects to find at finite temperatures are likely to be the result of competition between the ordering tendencies implicit in the rules of quantum mechanics and the disordering tendencies due to thermal excitations. Splendid instances of such competition are provided by phase transitions, such as from a solid to a fluid phase, from a superfluid to a normal fluid, or from a ferromagnetic to a nonferromagnetic arrangement of spins in a crystal.

Below 1,043°K (the Curie temperature), a single iron crystal exhibits net magnetization in the absence of an external magnetic field; above this temperature it does not do so. This phenomenon is interpreted as the preferential alignment of the unpaired electronic spins (each of which behaves as a little magnetic dipole) of the atoms below the Curie temperature, whereas no preferred direction survives above it. Qualitatively, this interpretation is in accordance with the foregoing remark that elevation of temperature has a disordering effect. It is nevertheless remarkable that there is a sharp transition from the ferromagnetic phase (in which magnetization occurs in the absence of an external field) to the nonferromagnetic phase, and that discontinuities occur in the specific heat and in other quantities.

A more detailed explanation requires a statistical-mechanical calculation.[9] First a reasonable expression is written for the energy of a specified configuration of spins:

$$E[\mu_1, \ldots, \mu_N] = \sum_{1 \leq i < j \leq N} \phi(|\mathbf{r}_i - \mathbf{r}_j|)\mu_i\mu_j - H \sum_{i=1}^{N} \mu_i,$$

where each μ_i can be either $+1$ or -1, according as the ith spin is up or down with respect to a given direction (ordinarily that of the external magnetic field, if it is nonvanishing), H is the intensity of the field, and ϕ is a function that depends upon the distance between the locations of the ith and jth spins. In principle the calculation of the partition function

$$Z(\beta, H) = \sum_{\text{all configurations}} e^{-\beta E[\mu_1, \dots, \mu_N]}$$

can be carried out when the function ϕ is given, and from Z the magnetization, the specific heat, and other macroscopic quantities of interest can be calculated by differentiations and other mathematical operations. For almost all choices of ϕ the calculation of Z is extremely difficult. A great simplification is the mean-field assumption, according to which each spin effectively interacts with the mean field produced by all the other spins together, with the result that

$$\sum_{1 \leqslant i < j \leqslant N} \phi(|\mathbf{r}_i - \mathbf{r}_j|) \mu_i \mu_j \cong \frac{\text{const}}{N} \sum_{1 \leqslant i < j \leqslant N} \mu_i \mu_j.$$

This calculation can be performed quite easily, and it exhibits a phase transition: the spontaneous magnetization per spin, as a function of the absolute temperature T, has the following form:

$$m_0 = [3(1 - T/T_c)]^{1/2} \quad \text{for } T \text{ close to but less than } T_c,$$

$$m_0 = 0 \quad \text{for } T > T_c.$$

This derivation of a phase transition is an incomplete triumph. In the first place, the behavior of the magnetization, the specific heat, and other quantities near T_c does not agree with experiment. Furthermore, the mean-field assumption is entirely implausible physically, since a crystal can be enormously large in comparison with a single atom, and it makes no sense that the far-off spins should contribute as much to the total field felt by a given spin as the ones closer by.

A more plausible model is the Ising model, in which a given spin is assumed to interact only with its nearest neighbors:

$$E[\mu_1, \dots, \mu_N] = -J \sum_{\substack{i=1 \\ j=\text{nearest neighbors}}}^{N} \mu_i \mu_j - H \sum_{i=1}^{N} \mu_i.$$

The one-dimensional version of this model was analyzed by E. Ising in 1925 and shown not to exhibit a phase transition. (Ising erroneously argued that the same was true in two dimensions.) In 1944 L. Onsager performed a rigorous calculation for the two-dimensional Ising model and found that

$$m_0 = [1 - (\sinh 2v)^{-4}]^{1/8} \quad \text{for } T < T_c,$$

$$m_0 = 0 \quad \text{for } T > T_c,$$

where $v = J/kT$ and $\sinh 2v_c = 1$. An intuitive explanation can be given for the failure of a phase transition in one dimension and its occurrence in two. In one dimension, the occurrence of a single spin flip due to a thermal perturbation interrupts the long-range order; in two dimensions, each spin has four neighbors, and if one is deviant because of a thermal perturbation, the other three are likely to maintain the long-range order, except in the improbable event of two or more becoming deviant. (The political implications of this argument are obvious.) A beautiful variant of this reasoning explains how the periodic structure of a crystal is maintained over lengths many orders of magnitude greater than the lattice spacing, so that deviations from exact periodicity do not become cumulative with distance (Peierls 1979, pp. 85–91).

The spectacular work of the last two decades by K. Wilson, B. Widom, M. Fisher, and L. Kadanoff on phase transitions and critical phenomena[10] has exhibited the power of several further principles of composition. Their main results concern the values of critical exponents and relations among them, these exponents being numbers that characterize the singular behavior of thermodynamic quantities, such as susceptibilities and specific heats, in the vicinity of singular points in thermodynamic space (e.g., critical points and lines of phase transition). These critical exponents turn out to depend crucially upon a small number of parameters, notably the dimension d of the system and the "spin dimensionality" D or its analogue.[11] What is striking is that the values of the critical exponents are independent of many microscopic features that one might intuitively think to be important, such as the geometry of the lattice array and the strengths of the interactions. The explanation for the insensitivity of the critical exponents to these factors is that near the singular points fluctuations are very large. (In some cases the fluctuations can even in a sense be visible; near the critical point of a fluid there is "critical opalescence," in which giant fluctuations cause unusually large scattering of light.) But if the extent of the average fluctuation is much larger than the average spacing between atoms, then the details of the interactions become irrelevant or are swamped out. This is the essence of the hypothesis of universality for each class characterized by a few parameters. It is at first surprising, but then upon reflection quite reasonable, that general statements can be made about whole classes of macroscopic systems in virtue of giant fluctuations, which are disordering factors. The neglect of fluctuations is the chief respect in which the mean-field treatment of phase transitions fails. The rigorous exploitation of the hypothesis of universality requires the machinery of the renormalization group (a group of transformations in each of which a change of scale plays a crucial role); in a few cases, quantitative calculations of critical exponents have actually been achieved by this machinery.

Although the universality results just mentioned concern only phase transitions and critical phenomena, they are probably indications of a principle of much greater generality. This is the principle enunciated by P. Anderson (1984, p. 85) and called by him "the principle of continuation." Anderson writes: "In a very deep and complete sense we can refer back whole classes of problems involving interacting fermions to far simpler and fewer problems involving only noninteracting particles. While there can be great or small quantitative differences between a real metal, for instance, and a gas of noninteracting electrons, the essentials of their qualitative behavior—specific heat, T dependence of susceptibility, T dependence of various transport coefficients, etc.—are the same." Later, Anderson states the conjecture that the renormalization group "affords at least one way of putting mathematical teeth into the basic concept of continuation . . . , and perhaps it is even *the* most fundamental way of doing so" (p. 167).

8   Ontological Comments

On the basis of the foregoing summaries a number of propositions can be asserted with some confidence concerning the ontological aspects of the parts-wholes relationship in physics. All the ontological questions listed in section 1 will be addressed, though the first only briefly. Furthermore, a few additional matters not anticipated in the list of questions will be touched upon.

A. It is an open question in elementary-particle physics, which has been bypassed in this essay, whether there is an ultimate level of noncomposite particles. Electrons and other leptons may have no internal structure, since they exhibit pointlike behavior in scattering experiments;[12] the same may be true of quarks, which are the components of protons and other heavy particles. Whatever may be the answer to this question, the Democritean equating of noncompositeness with immortality is surely abandoned. Electron-positron pairs can be created and annihilated, even though each member of a pair is created or annihilated as a whole rather than by fabrication and dismantling.

B. The Democritean picture is transformed by the fact that, from the point of view of quantum field theory (the best framework of fundamental physics that we have), a particle is a quantum or excitation of a field, and a field in some sense is a holistic entity, given over all of space-time. Although this statement seems inconsistent with a Democritean answer to the parts-wholes problem, this *prima facie* judgment must be carefully qualified by locality considerations. The propagation of a field is presumably governed by relativity theory, which forbids direct causal connection between two points with spacelike separation. As a result, there is complete freedom to specify a classical field over a spacelike surface, constrained only by smoothness and good behavior at infinity. Except for these restrictions, a classical relativistic field on the whole spacelike surface is determined by the field values on an exhaustive set of disjoint subregions of the spacelike surface, and hence a residuum of the Democritean point of view concerning parts and wholes is found in classical field theory. Analogous (but suitably modified) statements can be made about a quantized relativistic field.

C. From the standpoint of fundamental physics, the low-energy domain to which we restricted attention in sections 3–7 is derivative. The elementarity of electrons may be an open question, but there is no doubt that all nuclei other than that of ^1H are composite, even if one does not probe the quark structure of the protons and neutrons. Nevertheless, there are several crucial physical facts which ensure that the level of low-energy physics has a certain autonomy, allowing it to be investigated with little attention to deeper and finer levels, and has a quite firmly Democritean character. One fact is the scale of energies in nature. Typically, the energy required to excite a nucleus is of the order of 100,000 electron volts, whereas the excitation energy of an atom is of the order of one electron volt and that of a molecule even less; excitations in solids can be much less. Consequently, in energy transactions typical of low-energy physics, the nuclei are not disrupted or even excited (though there are unstable or nearly unstable nuclei for which this statement must be amended). A second fact is that the lightest of the particles with nonvanishing rest mass, the electron, has (according to the relativistic equation $E = mc^2$) an energy equivalent of about 500,000 electron volts, and therefore the creation of an electron-positron pair has an energy threshold far beyond the domain of low-energy physics. A third fact is that even though there is no such hindrance to the production of low-energy photons, since photons have zero rest mass, their existence can largely be neglected in the low-energy domain, where the semi-classical radiation theory makes few predictions that differ from those of quantum electrodynamics.[13]

D. The properties of atoms, molecules, normal fluids, infinite Coulomb systems, and spin systems are derivable in principle from the properties of their components and the physical laws governing the interactions of the components. It is not strictly correct, however, to say that the laws underlying these derivations are none other than the principles of low-energy physics summarized in section 2. One reason for caution is that corrections from relativistic quantum mechanics and gravitational theory are required in order to obtain in full detail the properties of the systems surveyed; for instance, relativistic effects are already found in the hydrogen atom, and they become increasingly important with increasing atomic number. Furthermore, the rigorous derivations of phase transitions in spin systems did not assume electromagnetic interactions among the elementary spins, but rather mathematically concocted interactions that hold only between pairs of spins of a restricted class (e.g., nearest neighbors or next-nearest neighbors). The results of Onsager and his successors are not to be depreciated for this reason. What that work shows is that long-range order can be accounted for in terms of interactions with strictly short range, which is a more severe restriction than the falling off with distance of electromagnetic forces. Consequently, the motivation for postulating some kind of hitherto unobserved long-range interaction for the purpose of accounting for long-range order is removed. (Similarly, the success of the Bardeen-Cooper-Schrieffer theory of superconductivity showed that electromagnetic forces in the framework of quantum mechanics could account for that kind of long-range order, without the postulation of a new fundamental long-range force.)

E. Many of the properties of composite systems which are implied by the laws governing the components are emergent, in the sense that they are qualitatively radically different from the properties of the components. The long-range order of a ferromagnet is an example of an emergent property, as are the rigidity of a crystal and the viscosity of a fluid. The term *emergent* is a sometimes used to mean "underivable in principle from the properties of the components," but this is not the meaning that I adopt, nor would it be semantically economical for anyone maintaining a roughly Democritean point of view to do so.

F. The ontological reduction of the properties of the composite system to those of the components, as asserted in paragraph D, does not constitute a renunciation of the entanglement of states of composite systems, which was one of the peculiarities of quantum mechanics emphasized in section 2. Quite the contrary, in many instances the ontological reduction is possible only because of entanglement—e.g., in the role of hybridization in molecular bonding.

G. Composite systems fall into natural kinds, as the discussion of atoms pointed out explicitly and that of molecules tacitly. The same can be said for nuclei and the elementary particles composed of quarks, though these systems lie outside the domain of this essay. The stable configurations permitted by quantum mechanics and the force laws determine in principle the complete taxonomy of possible kinds of stable systems composed from the given components. In the case of relatively simple composite systems—the composite elementary particles, the nuclei, the atoms, and the smaller molecules—the actual taxonomy is also determined, provided that the environmental conditions are suitably specified. (In the case of the heavier nuclei, the requisite environmental conditions are satisfied only in sufficiently hot stars.) The reason is that one can reasonably assume enough trial encounters to have occurred among the components that the space of possible configurations has been well explored in the time interval under consideration. In the case of complex systems, such as macromolecules, there is certainly not enough time within the relevant intervals (e.g., the existence of the earth) to explore the space of possible configurations, and hence the actual taxonomy is in large part the result of the contingencies of evolutionary history. It should be added that within the framework that has been sketched, imperfections in natural kinds (such as flaws in crystals and errors in DNA) are as comprehensible as the near perfections. The Democritean treatment of the central Aristotelian doctrine of natural kinds must be reckoned as one of the great triumphs of modern physics.

H. One of the remarkable features of the physical world is the existence in many situations of several well-defined levels of description for the same physical system. Three levels were considered in the analysis of normal fluids and spin systems, but can be found in many other types of systems. The deepest level—that of relativistic field theory—was only briefly mentioned above, and its existence is acknowledged but not normally exploited by working condensed-matter physicists. The second level is the nonrelativistic quantum-mechanical many-body treatment of interacting electrons and nuclei. The third and coarsest level is that of macroscopic description, in which the fundamental variables for spin systems are temperature, applied magnetic-field strength, and magnetization, and those for fluids are local density, local velocity, and local pressure (all these local quantities being defined by averages over regions large compared to an atomic volume, and all of them possibly varying with position). The remarkable feature of a level of description is not that there are quantities at that level which are defined by summing or averaging over the quantities at a deeper level of description, but that there are laws which govern the specified level without supplementary information from a deeper level. The Navier-Stokes equation for fluid flow and the equation of state $M = M(H, T)$ for a spin system are formulated entirely in terms of appropriate macroscopic variables. To be sure, there are parameters entering these equations which are not supplied by the macroscopic theory and which must either be measured empirically or derived by resorting to a deeper level, but there are no physical variables in these equations except those of the macroscopic level.[14]

I. As pointed out in the summary of Fröhlich's derivation of the Navier-Stokes equation, the derivation of a macroscopic equation from the laws governing a deeper level is not absolutely general; it holds only in a range of circumstances which fortunately is very wide. Sometimes this failue of generality is cited as an objection against the thesis of reducibility of macroscopic physics to physics at a deeper level of description. The most common version of this objection is that classical thermodynamics is a nonstatistical theory, whereas statistical fluctuations require changes in the thermodynamics which Boltzmann and Gibbs purport to derive from statistical mechanics. My answer to this objection is that excessive rigidity concerning the concept of reducibility can mask the truly wonderful relations between levels of description. From an Olympian point of view, unclouded by the difficulties of mathematical inference, the laws at the deeper level imply, for each set of system constitutions, initial conditions, and boundary conditions, a definite set of values or possibly a definite set of probability distributions of the macroscopic quantities; the extent to which the putative macroscopic laws are satisfied is then made explicit. If the agreement is good (by a reasonable standard), in a set of circumstances that is wide (by a reasonable standard), then a striking relation holds which captures the intuitive and practical sense of "reduction." What constitutes "reasonable standards" obviously cannot be specified at the level of generality of the present discussion, but the treatment of fluctuations in statistical mechanics shows that sensible things can often be said about this question.

9   Epistemological Comments

The epistemological problem that has most preoccupied philosophers of science is the legitimation of scientific principles on the basis of empirical evidence. Some variant of the hypothetico-deductive method is recognized widely to be appropriate for this purpose, though, of course, there are many different opinions about the details of the method and the cognitive claims that can be justifiably made for it. The deductive steps in the hypothetico-deductive method are usually given less attention than the other steps; deduction is supposed "in principle" to be well understood, even though in practice it encounters enormous technical difficulties. In other words, there is a tendency to take an Olympian point of view for granted in discussing the deductive steps of the hypothetico-deductive method.

In the enterprise of reducing macrophysical theories to microphysical theories and, in general, explaining the properties of composite systems in terms of those of the components, the principles governing the components are taken as established, and therefore the problem of deduction emerges from the background and assumes a central role. The special epistemological problems concerning the relation between parts and wholes in physics arise just because humans are debarred by their intellectual limitations from taking an Olympian point of view. If there is not a satisfactory surrogate for Olympian deduction, then human beings are not entitled to assert that macrophysics is reduced to microphysics and that wholes are understood in terms of parts, no matter how matters appear to the deities. The five illustrations discussed in sections 3–7 contain a number of procedures, none of which by itself provides a satisfactory surrogate for Olympian deduction but which work together remarkably well to this end.

One procedure is systematic approximation to the exact solutions of difficult mathematical equations—for example, the Hartree-Fock iteration method mentioned in section 3. Approximations might be thought to be purely mathematical, carried out without reference to the physical problems that posed the difficult equations. In practice, however, the sublimation of the mathematics from the physics is incomplete, and physical considerations direct and supplement purely mathematical ones in several ways. First of all, the starting point of an approximation method is very important, and observational data together with intuitive understanding can suggest a good beginning. Conversely, a qualitatively wrong starting point—e.g., one with a different symmetry from that of the phenomenon of interest—will ensure that the approximation method will never converge to the true solution; a barrier of singularity separates a state with the wrong symmetry from those with the correct symmetry

(Anderson 1984, pp. 27–30). Even if the starting point is not radically wrong in this sense, a clumsy choice will entail convergence so slow that the outstanding features of the true solution will not be recovered by a computation of reasonable duration. Another way in which physical considerations guide and direct mathematics is by indicating that an approximation method has been carried far enough in a situation where no rigorous proof exists of convergence or where no rigorous bounds can be put upon the error of truncation. Example: In the ground state of a system with finitely deep potential wells, the energy is a minimum; if the energy computed with an approximate solution is very little above the experimentally measured energy, then it is reasonable to conclude that the approximation is good and has captured the major features of the true solution.

A second procedure is the performance of simplifying steps in the course of physical analysis of a problem. This procedure is quite different from the construction of a model, in which the simplifications are made initially in characterizing the constituents of the system and the forces among them. The procedure to which I am now referring is exemplified at several points in section 6—e.g., it was assumed that changes of σ, $\mathbf{v}$, and p are small over distances of the order of the interaction range between atoms, and that the dependence of the pair-correlation function $P(\mathbf{x} - \mathbf{y})$ on powers of $\mathbf{v}$ higher than the first is negligible. These simplifications are made quite naturally in the course of the analysis, but would be hard to insert into a rigorously treated model. Usually these simplifications are intuitively appealing, and when they seem particularly significant they are often dignified with the name of *Ansatz*.[15] Simplifications are made plausible not only by physical intuition but also by the fact that there is an empirically well-established *terminus ad quem*, a phenomenological equation that is known to work well for normal liquids that are not driven to the point of turbulence. This empirical evidence provides inductive support for the supposition that the simplifications are valid broadly and not merely in highly idealized special cases.

In the two procedures considered so far, informal inductions are made in order to remedy the shortcoming of human deductive powers. Disagreement with empirical evidence would then constitute *prima facie* evidence that the truncation was premature or the simplification unjustified. Conversely, agreement with empirical evidence provides inductive support for the legitimacy of the truncation or the simplification, though the informality and the lack of rigor must be acknowledged; it is certainly possible that good agreement is achieved, not because of the smallness of the terms neglected relative to those retained, but because of fortuitous cancellations among the neglected terms. Confirmation is always riskier than disconfirmation. As in inductive inference generally, however, when good agreement is found in a variety of experimental situations, and when the physical picture becomes more and more coherent, then confidence in the confirmation justifiably increases.

A third procedure is the construction of rigorously soluble models possessing some of the crucial qualitative features of real systems—for example, the short-range character of the microscopic magnetic moments that compose an actual ferromagnet. As noted in sections 7 and 8, the great significance of these results is the rigorous establishment of emergence; that is, the exhibition of macroscopic properties radically different from those of the constituents. Conceptually, though not technically, it is a larger step to establish the possibility of emergence at all than to show that real composite systems behave qualitatively in the same way as the models.

A fourth procedure is to implement what Anderson calls "the principle of continuation," thereby providing the missing linkage between models and real systems. For this purpose the renormalization group is very powerful, but an exposition of that is beyond the scope of this essay and its author. Evidently, however, Anderson's suggestion that the renormalization group is not merely a technique for studying a special body of phenomena but an instrument of great generality in many-body physics deserves very careful study.

The systematic deployment of these procedures for the purpose of understanding the properties of composite systems in terms of their components deserves the epithet "the methodology of synthesis," which has previously been applied to statistical physics (Toda et al. 1983, p. v). The epistemological significance of the physics of composite systems is summed up in this phrase.

One final epistemological comment is appropriate. Throughout this essay, physical first principles have been assumed to be given. Of course, they are not given, but have to be established by ingenious experimentation and profound analysis of the experimental results. In view of the great mathematical difficulties of determining what experimental prediction a theory makes concerning a many-body system, it is obviously desirable to carry out the critical experiments for assessing first principles upon the simplest possible systems.[16] But a reasonably successful execution of a physics of composite systems within a given framework of first principles, carried out in accordance with the "methodology of synthesis" sketched above, should be regarded as very impressive supporting and reconfirming evidence for those principles. Contrariwise, the failure to execute successfully a physics of composite systems within a proposed framework in spite of much effort (for which we now have a good standard of strenuousness) should be regarded as serious disconfirming evidence. The investigation of the properties of composite systems would usually be classified as "normal science," but it should be obvious from the foregoing discussion that "normal" cannot be equated with "mundane" or "routine."

Acknowledgment

I wish to thank Peter Achinstein, Howard Stein, William Klein, and Charles Willis for their stimulating comments and questions.

Notes

1. See, for example, Khinchin 1949.

2. See, for example, Bethe and Jackiw 1968.

3. A good treatment is found in Merzbacher 1970, pp. 535–539.

4. A good discussion is given in Baym 1969, p. 458.

5. See, for example, Baym 1969, pp. 486–498; Roby 1973.

6. See, for example, Tisza 1966, pp. 102–193.

7. See, for example, Thompson 1972, pp. 67–71.

8. A survey of the argument is given in Lieb 1976.

9. See, for example, Thompson 1972, chapters 4–6.

10. See, for example, Ma 1976.

11. As already pointed out, the qualitative behavior of the Ising model depends upon whether d is 1 (in which case there is no phase transition) or 2 or 3 (in which case there is a transition). D is 1 if the spin can be either up or down in a fixed direction; it is 2 if the spin can be a vector confined to a plane; it is 3 if it is a vector that can point in any direction in space. There are generalizations of this meaning of spin dimensionality when one is dealing with other types of transitions between order and disorder.

12. See, for example, Bransden et al. 1973, p. 210. For the view that the electron may be composite, see Greenberg 1985.

13. See, for example, Sargent et al. 1974, p. 97.

14. The Ω_i in section 6 could be considered to constitute a hierarchy of levels of description intermediate between the second and third levels listed here, except that the dynamical equation for Ω_i refers to Ω_{i+1}. With suitable assumptions, the hierarchy of equations can be truncated at the kth without serious error (this was in fact done for $k = 2$ in section 6), and then there is an autonomous intermediate level with its own dynamical law. Grad 1967 contains a fine discussion of levels of description.

15. A famous example is L. Boltzmann's *Stosszahlansatz* in the kinetic theory of gases, which says essentially that the velocity distribution of a molecule that has just undergone a collision is the same as that of a randomly selected molecule. As a matter of fact, this *Ansatz* is automatically satisfied by an appropriate model (Grad 1967, p. 56).

16. It may be objected that a many-body system supplied the data that inspired the beginning of quantum theory: the radiation in a black cavity. This objection is not decisive, however, because black-body radiation is in fact extremely simple, consisting of noninteracting excitations.

References

Anderson, P. W. 1984. *Basic Notions of Condensed Matter Physics*. Menlo Park, Calif.: Benjamin/Cummings.

Baym, G. 1969. *Lectures on Quantum Mechanics*. New York: Benjamin.

Bethe, H. A., and R. W. Jackiw. 1968. *Intermediate Quantum Mechanics*, second edition. Reading, Mass.: Benjamin/Cummings.

Bransden, B. H., D. Evans, and J. V. Major. 1973. *The Fundamental Particles*. London: Van Nostrand Reinhold.

Fröhlich, H. 1973. "The Connection between Macro- and Microphysics." *Rivista del Nuovo Cimento* 3: 490–534.

Grad, H. 1967. "Levels of Description in Statistical Mechanics and Thermodynamics." In *Delaware Seminar in the Foundations of Physics*, vol. 1, ed. M. Bunge. New York: Springer-Verlag.

Greenberg, O. W. 1985. "A New Level of Structure." *Physics Today* 38, no. 9: 22–30.

Khinchin, A. 1969. *Mathematical Foundations of Statistical Mechanics*. New York: Dover.

Lieb, E. 1976. "The Stability of Matter." *Reviews of Modern Physics* 48: 553–569.

Ma, Shang-keng. 1976. *Modern Theory of Critical Phenomena*. Reading, Mass.: Benjamin.

Merzbacher, E. 1970. *Quantum Mechanics*, second edition. New York: Wiley.

Pauling, L. 1931. "The Nature of the Chemical Bond. Application of Results Obtained from the Quantum Mechanics and from a Theory of Paramagnetic Susceptibility to the Structure of Molecules." *Journal of the American Chemical Society* 53: 1367–1400.

Pauling, L., and I. Keaveny. 1973. "Hybrid Bond Orbitals." In *Wave Mechanics*, ed. W. Price et al. New York: Wiley.

Peierls, R. 1979. *Surprises in Theoretical Physics*. Princeton University Press.

Roby, K. R. "Mathematical Foundations of a Quantum Theory of Valence Concept." In *Wave Mechanics*, ed. W. Price et al. New York: Wiley.

Sargent, M., M. O. Scully, and W. E. Lamb. 1974. *Laser Physics*. Reading, Mass.: Addison-Wesley.

Thompson, C. 1972. *Mathematical Statistical Mechanics*. New York: Macmillan.

Tisza, L. 1966. *Generalized Thermodynamics*. Cambridge, Mass.: MIT Press.

Toda, M., R. Kubo, and N. Saito. 1983. *Statistical Physics* I. Berlin: Springer-Verlag.

Weisskopf, V. F. 1979. *Knowledge and Wonder*, second edition. Cambridge, Mass.: MIT Press.

Paul Teller

Space-Time as a Physical Quantity

Traditional debates about the nature of space have been dominated by two opposing views. Substantivalism, associated with the name of Newton, maintains that space is a substance, that spatial points are substantival particulars,[1] and that this concrete system of particulars provides an objective frame of spatial reference. According to relationalism, associated with the name of Leibniz, there are only the spatial relations exhibited by objects. According to this view, substantival space and the objective locations which it is supposed to supply are both illusions.

The Theory of Relativity has changed the form of this debate. Relativity drives home the point that velocity is relative. So space is relative too. But these facts do not settle the question, for we can raise it again in terms of Relativity's four-dimensional space-time. One customarily discusses Relativity in terms of a collection of points, often called "space-time locations," which are supposed to serve as the "places" at which physical events occur. We can now ask about Relativity's four-dimensional space-time points: Are they to be understood substantivally, in the old Newtonian sense, or should we insist that, accurately speaking, there are only the space-time relations exhibited by physical events?[2]

I maintain that neither substantivalism nor relationalism provides the best way of thinking about space-time. Instead, I want to suggest, we should take space-time as a physical quantity differing from quantities such as mass and temperature only in details of structure, and we should take space-time points to be the properties that constitute particular values of the space-time quantity.

Throughout this essay I will express myself with the realist's voice, which takes there to be particulars, properties, and relations and which takes the problem to be one of figuring out which of these space-time "really" is. But the reader may mask out this realist stance. If one prefers, one may start from the view that we use thing-talk, property-talk, and relation-talk and take the problem to be one of determining which kind of language is most appropriate for the referential apparatus of space-time discourse. In other words, one may read my arguments as supporting the view that, in the respects I will indicate, space-time is rather more like mass and rather less like atoms. Indeed, it is only this more cautiously formulated conclusion that I take to be clearly established by my arguments. In yet other words, I recommend that one read the realist language of this essay in terms of what Fine (1984) calls the Natural Ontological Attitude, which attributes to assertions of existence a commonplace, practical significance without some heavier, traditional, metaphysical burden.

The Quantity View

Most people seem initially to find the idea that space-time points are properties extremely strange. So I start with a consideration that may help to make the idea seem less unnatural. Suppose you are a substantivalist, that is, you take space-time points to be substantival particulars. Now consider the question, "Are space-time points contingent or necessary particulars?" understood in the following sense: Given any specific space-time point, is it a contingent particular, in the sense that it might not have existed or might cease to exist; or is it a necessary particular, in the sense that it is impossible for it not to have existed or for it to cease to exist?[3]

I feel we have a contrast here between the substantivalist's particulars and more familiar cases. The substantivalist claims that space-time points are particulars that, like atoms, are hypothesized as part of our best available theory of the world. It makes perfectly good sense to say of a particular atom that it might not have existed. But can we say of a particular space-time point that *it* might not have existed? I have no argument showing that substantivalists must answer "no," but their conception hardly seems compatible with a view of space-time points as things that might or might not have been.

Let us try an even more painful question. Might a substantivalist believe that there could have been a physical world with no space-time points at all? How, except possibly by being a relationalist, could one maintain that there could be a world with physical events but no space-time locations at which they occur? Doesn't a physical event have to occur somewhere and at some time?

These considerations suggest that the substantivalist may be plausibly supposed to be committed to two theses: (1) that space-time points are necessary particulars (in the sense indicated), and (2) that, necessarily, each physical event occurs at some space-time point or other. But both of these theses suggest, in turn, that space-time points are abstract objects rather than concrete particulars. Properties are not contingent entities in the way that atoms and soccer games are. So the inclination to think of space-time points as necessary suggests thinking of them as more like properties than particulars. The second thesis actually makes the analogy stronger still. A quantity or determinable, such as mass, is something of which qualifying entities must have some specific value or other; something that can have mass must have some specific mass. An event (according to substantivalists) clearly *can* occur at some space-time point. So if, in addition, every event *must* have a space-time point at which it occurs, then space-time points look to be functioning in just the way that masses and other physical quantities or determinables do.

In sum, our considerations suggest a new view,[4] which I will call the *quantity view of space-time*. Space-time is a physical quantity or determinable, and space-time points are specific values of that quantity. All space-time points "exist" in whatever (perhaps not very good) sense in which all values of quantities such as mass, including unexemplified values, exist. In other words, space-time points are necessary existents in what ever sense all properties are necessary existents. They are exemplified in some possible worlds and not in others. But we think of them as equally "available" in all possible worlds, so that in a possible-world semantics we do not specify which exist in which possible world.

In suggesting that space-time points are properties ("location properties," as we might call them), I want to be sure that my proposal is not misunderstood in the following way. In struggling to get a grip on what a location property is supposed to be, one might say to oneself that it is the property of being located at such-and-such a space-time point. One might then take this to mean that there are substantival space-time points and that for an event to have a location property is just for the event to occur at one of these space-time points, that is, for the event to bear the *occurs at* relation to some space-time point. In other words, on this interpretation, location properties would be taken to be covertly relational, in just the way that the property of being married is really relational. This is emphatically not what I have in mind. Masses are not relational, at least not in the way that being married is relational. There are no concrete particulars, the masses, such that for an object to have some specific mass is for the object to bear some obscure relation ("having the mass of") to one of these particulars. (Mass may be relational in other ways, which I will presently consider in great detail.) Likewise, the location properties are not to be understood as covert relations to substantival space-time points.

My proposed location properties are unusual in one respect: No two distinct physical objects can share one of these properties. That is, if *a* and *b* are distinct physical objects as we ordinarily conceive of them, and if *a* has the location property *A* (occurring-at-space-time-location-A), then *b* cannot have that same property. But I see no reason why this fact should count against space-time occurrence's having the status of a property. Moreover, the same quirk does not hold for events. Distinct events (such as my becoming hot and my becoming tired) can share space-time occurrence. In addition, contemporary quantum field theory suggests that, strictly speaking, there are no physical objects as we ordinarily conceive of them, and, in particular, no objects that would have to be unique possessors of my space-time properties. I can clarify this claim with an analogy to what science tells us about solidity. There is a perfectly useful sense in which ordinary chairs and tables are solid, even though atomic physics has revealed to us that they are mostly empty space. Consequently, there is a stricter or at least alternative sense in which (with the possible exception of neutron stars) there are no solid objects. Analogously, quantum field theory indicates a reconception of what we usually conceive of, and usefully speak about, as ordinary physical objects. It does this by treating physical matter as a kind of field, the point being that fields (unlike ordinary physical objects) can coincide in space and time.

All told, the issue of space-time coincidence is in no obvious way an objection to the quantity view of space-time. There is no reason why certain properties might not apply uniquely. And although space-time location properties would apply uniquely to physical objects as we usually conceive of them, contemporary physics suggests that a more fundamental characterization of the physical world should be in terms of events and fields, for which the issue of unique application does not even come up.

I hope that these introductory remarks will give the reader confidence in the coherence of the quantity view. I expect there to be no knock-down argument for (or against) it. In what follows I want to compare the quantity view with its rivals, substantivalism and relationalism. I hope to show that the quantity view avoids the difficulties and also captures what seems right about both of these rivals.

The Quantity View Compared with Substantivalism

How does the quantity view fare in face of the arguments for substantivalism? Historically, the predominant argument (and the one Newton used) appealed to absolute acceleration. We experience the effects of acceleration all the time, as when we exhibit centrifugal force with a weight on a string and when we are pushed against the back of our seat in an accelerating car. These are called "inertial effects." Unlike relative uniform motion, inertial effects are absolute. If two systems are moving uniformly relative to each other, we can detect only the relative motion. An observer in each system sees the other system as the one that seems to move, and no experiment in either system will differentiate between the two, enabling us to put our finger on one of them as "really" moving. The situation is different when two systems are accelerating relative to each other. Relatively accelerating systems exhibit different inertial effects. In nice cases, one system will exhibit inertial effects and the other will exhibit none. (In your accelerating car you feel pushed against your seat. In my parked car I feel no such force.) Such differences enable us to draw an experimentally based distinction between systems that are really accelerating and those that are not. Traditionally, the substantivalist argues that the relationalist, with only space-time relations to which to appeal, cannot explain these differential inertial effects and the absolute accelerations they identify, whereas substantivalism can explain these differences in terms of differing motions through an objective space-time.[5]

This argument from inertial effects takes the form of a dilemma. It assumes that we have to choose between substantivalism and relationalism. On this assumption, if one of the two doctrines succeeds and other fails in accounting for some phenomenon, this fact automatically counts in favor of the doctrine that succeeds. But I have introduced a third alternative, the quantity view. I will argue that the quantity view accounts for inertial effects just as well (and in fact in the same manner) as does substantivalism.

To see this we need to look a little more carefully at the argument from inertial effects, and to set forth the ideas I will work for the moment in terms of old-fashioned three-dimensional absolute space. The way the argument is usually put is that the substantivalist can, while the relationalist cannot, *explain* inertial effects. But how is acceleration relative to a substantival space supposed to explain these effects? In what way does motion through absolute space tell us *why* inertial effects occur? Since this is unclear, we will be better off drawing the intended contrast in somewhat different terms.

As far as we can tell from our common observations, inertial effects correlate with accelerations, and with other things only insofar as these themselves correlate with accelerations. But on the relationalists' position, acceleration is a symmetric relation—there is only relative acceleration between objects, and relative acceleration is symmetric. However, inertial effects clearly do not apply symmetrically to relatively accelerating objects. So either relationalists must be neglecting some features of the world—features that in some way go beyond our common observations—or they must acquiesce in the conclusion that inertial effects are susceptible to no systematic characterization in terms of other facts about objects.

Substantivalists reject the latter conclusion as uncharacteristic of physical phenomena. Better, they say, to postulate a substantival space that will serve systematically to characterize those objects that exhibit inertial effects. So far this is only description. But within Newton's mechanics, the description is deeply systematic, providing detailed, lawlike links between inertial effects of various kinds and various kinds of acceleration through absolute space. If one is inclined to count such systematic, lawlike descriptions as explanatory, then this description also counts as an explanation of inertial effects.

Switching to the four-dimensional space-time point of view does not really change any of this. Events, or trajectories, do not themselves move relative to space-time, of course. So we have to rephrase. On the four-dimensional view, acceleration at a point on a trajectory is understood as the trajectory's being curved at the point. So in order to make absolute acceleration available to serve as a correlate of inertial effects we must determine absolute (i.e., observer-independent) conditions under which a trajectory can be said to be curved. Relationalists cannot do this. They can appeal only to actual physical events, actual trajectories made up of such events, and the space-time relations that hold between these events. They still have a notion of relative observed acceleration. But this notion is again symmetric. When you appear to me to be moving away from me faster and faster every second, I give the same appearance to you. The space-time substantivalist can, however, characterize absolute acceleration. Consider the trajectory of a soccer ball. Does the ball accelerate at any point on its trajectory? Yes, if there are two points on the trajectory such that along some alternative trajectory connecting the points the space-time separation would have been shorter.[6] But such an account requires alternative trajectories, and thus alternative space-time points, besides those at which events actually occurred. Denying these unoccupied points and the frame of reference which they supply (in the form of a general notion of space-time distance along an arbitrary trajectory), relationalists are stuck. They must say that some trajectories exhibit inertial effect, that others do not, and that nothing more can be said about the difference between those that do and those that do not.

Now I can make the needed point about the quantity view quite simply. Nothing in the foregoing account turns on the spatial or space-time points' being substantival particulars. Location properties can play the same role. Instead of saying that there are substantival points at which nothing is (though something could have been) located, we can say that there are location properties which are not (though they could have been) instantiated. Location properties can support a system of space-time relations every bit as much as substantial space-time points are supposed to. And the argument from inertial effects as I have detailed it requires only a frame of reference to supply locations which could be occupied (or instantiated) and which support the needed space-time relations. I can put the point even more simply if I call such a frame of reference "objective" — objective, in the sense of being independent of observers, of where objects happen to exist, and of where and when events happen to occur. Such a frame of reference is objective in the same way that the system of mass properties is objective: in being independent of what masses happen to be instantiated. The argument from inertial effects calls for space-time to be objective, in the sense indicated. Apparently substantivalists thought that only a substantial space-time could be objective in this sense. But there is no reason to believe this. A system of location properties can be objective in the same sense. Thus, the argument from inertial effects supports the quantity view against relationalism to whatever extent it supports substantivalism.

I call the second argument for substantivalism the "Field-theoretic argument" (pun shamelessly intended). Hartrey Field (1980, p. 35) argues that it is very hard to see how one could do modern physics without fields. And it is very hard to see how to do field theory except in terms of attributing field quantities to space-time points. Consequently, modern physics commits us to an ontology of substantival space-time-point particulars.

We can grasp the motivation for Field's attitude by outlining the way in which the field concept arises. Consider a medium, such as a block of iron, in an idealized way so that we attribute a temperature to each point of the medium. This illustrates the primitive idea of a field as the attribution of a value of a quantity to each point of a medium. Laws then relate the value of the quantity at one point to values at nearby points and govern the way that values change over time, for example in the form of waves propagating through the medium. Historically it was thought that this picture was mandatory, in that there always had to be a medium to possess the values of a field quantity. In particular, it was thought that there had to be an "ether" to serve as the medium for the electromagnetic field quantities that constitute light. We all thought it was a major conceptual advance when physics let go of the ether and realized that we can take electromagnetic waves to propagate without any ether in which to "wave." Still, if fields are attributable values of quantities, we must have something to which to attribute the values. Having rejected the ether as the substance of attribution, Field concludes that we must at least have space-time points as particulars to which the field quantities can be attributed.

Field's argument is really just the old argument for the ether all over again. (This fact, of course, does not in itself show the argument to be wrong.) The argument was that there had to be something substantial for field quantities to apply to, something for the electromagnetic waves to "wave in." Having abandoned all substances occurring in space-time, we must, says Field, appeal to a substantial space-time to supply the needed medium. We get around the relativistic problems with the old ether as a three-dimensional substance persisting through time by moving to a four-dimensional space-time substance as described by special or general relativity. But space-time is still playing just the role that the old ether played: It is the bearer of the field quantities.

This gloss on Field's argument poses the question of whether we can really let go of the ether. Indeed, the theories of electromagnetic and other fields developed in the (at least apparent) absence of the ether concept provide an alternative. According to contemporary physics, fields themselves carry momentum and energy, and so also mass. In short, contemporary physics enables us to view the fields as themselves substantial. These substantival fields are not particulars, like soccer balls, since fields superimpose. But fields provide a new kind of substance, a substance that still has properties and parts differentiated by the different properties possessed by the parts. In particular, parts of the field take on different values of the space-time quantity, which we misleading gloss as the field "occurring at various space-time points."[7]

The discussion to this point shows, but shows no more than, that none of the arguments for substantivalism demonstrates substantivalism to be a better view than the quantity view. Are there any reasons for positively preferring the quantity view to substantivalism? One might suggest that it is better to appeal to fewer and less mysterious things. Substantivalism must postulate space-time points and the *occurs at* relation, both of which must be taken as primitive and which are otherwise mysterious. The quantity view instead appeals to properties and to the instantiation of these properties by events (or by physical objects at times). Though it is undoubtedly a matter of intuition or taste rather than one of real argument, I find instantiation of space-time properties by events or things at times more palatable and less mysterious than a relation of occurrence holding between independently existing particular events and particular substantival space-time points.

However, these last considerations are pretty mealy-mouthed, the sort of argument that at most convinces the already converted. I don't put much weight on them. My main reason for preferring the quantity view to substantivalism turns in part on comparison with relationalism. We have seen that the quantity view captures what seems right or at least attractive about substantivalism. Both views characterize space-time as providing an objective frame of reference. Indeed, I suspect that one could make a case for the quantity view's being what at least some substantivalists always had in mind. But, as I will argue next, the quantity view also catpures what seems right about relationalism. It is because the quantity view encompasses what seems right about each of its apparently antithetical predecessors that I believe it should be preferred to both.

The Quantity View Compared with Relationalism

Historically, Leibnizian indiscernibility arguments have provided the mainstays of relationalism. Suppose that substantivalism is right. We can then describe a possible or counterfactual situation that is just like the real world except that everything has been "moved over" (that is, relocated) from the beginning, but in such a uniform manner as to preserve all space-time relations. All of us have a funny feeling about this case. It just does not seem to be a genuine alternative. But if substantivalism is right, it seems that there must be such an alternative way the world could have been; therefore substantivalism must be wrong.

No one is very moved these days by Leibniz's more specific formulation of this argument in terms of God's needing a sufficient reason to place things in space-time the way they are rather than in some alternative, relationally equivalent way. And a formulation in terms of the identity of indiscernibles threatens to beg the question. According to this second formulation, the actual and the alternative space-time placements are indiscernible and hence identical. Since substantivalism would have them be distinct, substantivalism must be wrong. But substantivalists are going to grant only that the two alternative ways of locating every thing in space-time may be indiscernible with respect to space-time relations and other non-spatio-temporal relations and properties. If substantivalism is right, the alternative placements differ with respect to where things really are. So one can bring the identity of indiscernibles to bear only by assuming that substantivalism is wrong, which is just what was in question.

Finally, the argument can be given a verificationist formulation according to which, if no "observable" difference distinguishes between two allegedly different situations, the situations are after all the same. Many will dislike this formulation because they find verificationism fatally flawed. But we can put this version of the argument much more strongly: Except for the alleged differences in space-time placement, there is *no* difference—observational, theoretical, or whatever—between the alternatives we are considering. Even inertial effects are the same in the two cases. Inertial effects show acceleration to be absolute, and we will have the same absolute acceleration in a world that is simply "moved over from the beginning." But if two descriptions agree in such a thoroughgoing way, surely sound methodology and good sense require us to count them as verbally different descriptions of the same situation.[8]

The last few years have seen another approach to relationalism in the form of representation theorems.[9] Suppose we have a system of relations holding between members of some collection of things (such as a system of space-time relations holding between events; a system of relations of less, equal, and greater mass holding between objects; or a system relations of comparative preference holding between an agent's desires). To get a representation theorem, we begin by characterizing the system of relations in terms of axioms describing the relations' lawlike features and, if one likes, specific descriptions specifying what individuals bear what relations toward what other individuals. (In what follows I will always allow "axioms") to include such specific descriptions.) Next we take a formal or mathematical structure (e.g., the positive real numbers for mass, or the real numbers for utility), which is going to represent the relations by means of a mapping of the original relata (massive objects, specific desires, etc.) into the structure. This mapping must be such that, whenever a relation holds between some of the original relata, an exactly corresponding relation holds between the corresponding formal or mathematical elements of the representing structure. (This relation is exactly corresponding in the sense that the induced system of relations between the formal elements satisfies the same axioms that characterized the original relational system.)

In a second step we prove that such a representation is "unique" in the following sense (Krantz et al. 1971, p. 12). Suppose that we are determined to use one mathematical structure to do the representing. In general there will be more than one way of mapping the original relata into the structure. For example, given any mapping that represents masses in the positive reals, with the adding of masses represented by $+$, that representing mapping multiplied by a positive constant will do the same job. And in representing desires we can arbitrarily pick a "zero point" as well. A uniqueness proof characterizes the complete set of mappings that adequately represent the original relational structure in the fixed mathematical structure. Typically, one gets this characterization by taking advantage of symmetries of the original relational system. The uniqueness proof then works by showing that any representation in the fixed mathematical structure can be obtained from any other by application of the mathematical correlate of these symmetries.

Relationalists can now use a representation theorem for space-time relations to attack substantivalism. Substantivalists claim that there exists a substantival space-time. But they must also describe their substantival space-time in a way that allows reference to at least some specific space-time points. This they do, for example, with a vector space. But, of course, if one vector-space description will do, so will any other obtained from the first by translations and other allowed transformations, since any of these descriptions states the facts about the alleged space-time as well as any other. It is only the statements common to all these descriptions that correctly describe the facts about space-time. Relationalists now refine and formalize these observations with a representation theorem. The needed theorem starts with objects or events and the space-time relations holding between them. The theorem provides axioms describing the space-time relations and then maps the objects or events into exactly the vector space (or other structure) used by the substantivalists to describe their space-time. With this mapping, the representing elements of the vector space represent the space-time relations uniquely "up to isomorphic transformations" in the manner explained above. This can be arranged by taking advantage of just those transformations that (as the substantivalists had to acknowledge) change one adequate description of their substantival space-time into another. But, once again, only the statements common to all these descriptions correctly describe the space-time facts. And these common descriptions can now be seen to be exactly what follow from the representation theorem's axioms (including, as mentioned, specific statements about space-time relations holding between specific real events). All consequences of the axioms will be (relational) facts about space-time which the substantivalist will acknowledge. Can there be further facts about space-time? Only if the axioms have non-isomorphic models only some of which, the substantivalists will claim, correctly describe space-time. This possibility is ruled out if the axioms are categorical, as is usually the case with the axioms obtained in these representation theorems; for then all models of the axioms are isomorphic to the models obtained from the original representing vector space.

In sum, the substantivalists must allow many equally adequate descriptions of their alleged substantival space-time, and, if the axioms for space-time relations are categorical, all the facts that are common to all the adequate descriptions already follow from these axioms. What happens if the axioms are not categorical? The tendentious assumptions will be an Archimedean axiom, roughly guaranteeing that there are no points "infinitely far away" or "infinitesimally close together," and a Cauchy completeness axiom, roughly guaranteeing that convergent sequences of points have limit points. If either or both of these axioms are omitted, the field over which the vector space is defined need not be the reals, and the axioms will not differentiate between descriptions of space-time that differ as to whether or not there are points "infinitely far away" or "infinitesimally close together," or as to whether there are as "many" space-time points as the reals, as "few" as the rationals, or something "in between." Substantivalists can now claim more, but at most that the facts about "real" space-time are those that flow from the relationalists' axioms plus facts that differentiate between such non-isomorphic models, e.g., facts about how "many" space-time points there are, and whether they are "infinitely far or close." But such facts will include nothing about objective space-time placement, and without facts about objective space-time placement substantivalism loses its point.

I hope that my summary of this recent use of representation theorems makes it plain that the representation-theorem argument really constitutes a formalization of the viable version of the traditional indiscernibility argument. The indiscernibility argument turned on the premise that there is no difference in the real facts presented by two descriptions, one of which we obtain from the other, figuratively speaking, by moving everything over, or by other permissible transformations. In other words, the only real facts are those common to all such descriptions. But the representation theorem simply identifies these facts common to all equally admissible descriptions as the consequences of the axioms for space-time relations.[10]

Seeing the representation-theorem argument as a formalization of the indiscernibility argument provides an important tool for clarifying the status of the indiscernibility argument in application to a world described by general relativity. If the curvature of space-time varies sufficiently irregularly, then no representation theorem can be brought to bear. Such cases lack the symmetries needed to characterize the family of representations; or, to give the same problem another description, there will be no family of (first- or second-order) axioms that finitely or recursively characterize the relational structure of space-time when the curvature is sufficiently complicated. In particular, there will be no axioms that express the sort of symmetries that give indiscernibility arguments their compelling quality. But if the representation-theorem argument is just a formalization of the indiscernibility argument, it should follow that the indiscernibility argument will not apply either in such cases. Indeed, just this conclusion was reached by Earman (1970, p. 303) and Weingard (1977, p. 289). Their argument can be epitomized by emphasizing with the extreme: If there were a different curvature at each space-time point, the curvature could be used to individuate objective space-time points.

Yet, in another form, an indiscernibility argument might nonetheless apply (Earman 1979, p. 272; Hellman, unpublished). The traditional indiscernibility argument assumes that it would make no difference if all objects or events were shoved over in space-time "from the beginning." In the substantivalists' language, the intuition would be expressed as follows: Space-time relations are "carried" by the space-time points in a uniform way; the space-time points embody the space-time relations with the same homogeneity displayed when they are represented in a mathematical structure. In other words, a transformation such as "moving everything over" by the same amount in the same direction does not affect the structural relations, in rather the same way that moving to a new isomorphic mathematical representation does not affect the structural relations. This line of argument breaks down in general relativity, as we have seen, because moving over takes objects to a new "region" in which there may be a different curvature pattern and hence a different structure of space-time relations. Earman's suggestion, however, is that when we move everything over we move *everything* over; that is, we not only move the physical objects or events over, but we let these take their space-time structure with them.

In the context of classical space-time theory this suggestion might not have seemed to make much sense. Space-time structure was considered fixed—an intrinsic feature of space-time, not something separable from it. But general relativity has liberalized this conception of space-time relations, or physical "geometry." According to general relativity, space-time relations in one area depend on the distribution of mass and energy throughout space-time, and also on initial conditions. Thus, in general relativity, space-time relations, or "geometry," is connected only contingently with space-time points; and it makes perfectly good sense, as Earman suggests, to think of transposing everything—physical events and the space-time relations holding between them—to a new setting in the substantivalists' hypothesized space-time arena.

In a number of respects one has to wonder about this argument. The representation-theorem argument seemed to capture so exactly what is involved in the indiscernibility argument. But in this new formulation there is no work left to be done by an axiomatization of the relations even if one were available. Since the representation-theorem argument seems clearly to formalize the indiscernibility argument, and since the former seems irrelevant here, one might wonder whether this new formulation really is an indiscernibility argument. Furthermore, the new argument seems suspiciously strong. Why should the same line of reasoning not apply to any subject matter, to show that there are really only relations and no objects or nonrelational properties?

At the moment I can give no more than my hunches about answers to these questions. I see nothing wrong with our earlier conclusion that the representation-theorem argument is a formalization of the traditional indiscernibility argument. And the indiscernibility style of argument in application to general relativity or to any subject matter is as good as the traditional original. But in most cases indiscernibility arguments cannot be formalized. More specifically, the axiomatization of the original representation-theorem version gives us a formal statement of the relational facts that are held constant during the permutation of the alleged nonrelational objects of the traditional substantivalist. But nothing in the original argument *requires* that these fixed relational facts admit of formal axiomatic description. It merely makes the argument more vivid. If these hunches are right, the issues for indiscernibility arguments are really the same as the issue of the inverted spectrum. And the same style of argument will have to work for any subject matter as well as it works for space-time points or color sensations. The claim will be that, given any subject matter, the objects and their nonrelational properties can be permuted, leaving the relational properties invariant. And this will be claimed to indicate that only relational properties are "real."[11]

Space does not allow complete examination of this speculation. But appreciation of the representation-theorem argument as a formalization of the traditional indiscernibility argument already shows that the repercussions reach shockingly far. For we can show in some detail that if the representation-theorem/indiscernibility argument is right, then all measurable quantities are really relational. The representation theorems were originally introduced to clarify the formal properties of measurable quantities, such as mass, temperature, and utility. So if the representation-theorem/indiscernibility argument really succeeds in showing that space-time is relational, it will equally apply to show that all measurable quantities are relational.

It is worth looking in some detail at how the representation-theorem argument applies to the case of mass. The representation theorem for mass makes explicit the system of relations that enable us to assign numbers to represent specific masses. For example, we can determine that an object has a mass of 1 kilogram by putting the object on one side of a balance with a standard 1-kilogram object on the other side, and then seeing that the balance balances. That is not to say that having a mass of one kilogram *means* balancing or being able to balance a standard 1-kilogram object. Mass might be established in indefinitely many other ways. But because the unit of measurement is arbitrary, all these ways must involve direct or indirect comparison with some object which serves as the standard unit of comparison. Determining other masses works by an extension of these processes of comparison. For example, to determine that an object has a mass of 2 kilograms we find one object that exactly balances the standard 1-kilogram object, and then ascertain that the two 1-kilogram objects exactly balance the 2-kilogram object. In one way or another, any measurement of mass turns on the fact that there is a system of relations holding between massive objects—relations of having the same, less, or greater mass. And it is by using these relations, together with an arbitrary standard of comparison, that we assign numerical values to the masses of objects.

The extensive theory of measurement seeks to capture the formal features of mass relations that enable this process to work. The theory characterizes the mass relations in terms of a system of axioms describing various order and richness assumptions. The theory then proves a representation theorem. In the now familiar way, we are given a mapping of objects into the positive real numbers. The numerical relations between the numbers representing the masses satisfy the axioms describing the mass relations, and the representation is shown to be unque up to an arbitrary choice of unit. That is, any other mapping will do as well if the second is obtained from the first by multiplication by some positive real number.

By this time the punch line should be pretty clear: If the representation-theorem argument works to show that space-time is relational, it equally shows mass to be relational. The mass relationalist will argue that we would not be able to distinguish, observationally or theoretically, between our world and a world in which everything had a thousandfold greater mass and in which there were compensating shifts in other quantities involving mass relations, such as force and the gravitational constant. (Such other quantities are also specified numerically with reference to the standard unit of mass.) Every physics student who learns to switch between systems of description involving grams and kilograms proves that this is so. Since these alternative worlds with differing objective masses are indistinguishable, we see that only the mass relations (and relations holding between mass and other quantities) are real facts about mass. Put formally in terms of the representation theorem, the mass relationalist argues that all real facts about mass are equally well represented by any of the representations. And, exactly as in the case of the space-time representation theorem, one can argue than any conclusion a mass absolutist might reasonably draw about mass already follows from the axioms for mass relations. Obviously the line of argument works as well for any quantity to which the formal theory of measurement applies.

Some people seem persuaded by this argument for mass relationalism. Already they had thought that since mass can be specified only relative to a perfectly arbitrary standard, mass must be relational. I was shocked. Mass seems such a clear-cut example of what we mean by a nonrelational property that if it should turn out to be relational, so, one fears, will all other properties. This fear reminds us of the conclusion that threatened to emerge from the attempt to extend the indiscernibility argument to general relativity: the conclusion that the same line of reasoning might apply to any subject matter to show that there are only relations.

Something seems to me to have gone grotesquely wrong. My first reaction was to say that alternative representations for mass are only alternative systems of nomenclature for a system of nonrelational properties. Again, to translate between nomenclatures is what the physics student learns when he learns to switch descriptions from grams to kilograms. But the availability of alternative nomenclatures for properties, systematic or otherwise, hardly shows the properties to be relational. In other words, it now seems that the permutation of nonrelational things and properties has been unmasked as a trivial renaming. What is special about (e.g.) mass represented by numbers is that not only do the numbers function as names of the masses, but they also represent the masses in the sense that the numerical relations between the number-names correspond exactly to the mass relations between the named masses. These numerical relations, in turn, admit of formal axiomatic specification—this is the point at which formalization enters the discussion, as was explained more generally above. But if the argument proper turns only on the availability of alternative nomenclatures, the representational role of the names does no work in the argument. The argument then is perfectly general, but it is also silly; the availability of whole-scale renaming hardly shows everything to be relational.

However, I believe that this initial diagnosis does not do justice to the intuition of those who feel that mass *is* somehow relational, or to the intuition of those who *do* see relationality in everything. Can we reconcile the recurrent suggestion that there may be no limits to be put on the relational with the obvious fact that we do distinguish between relational and nonrelational properties, however much difficulty we may have in drawing the distinction?

Our considerations to this point show that none of us have been as clear as we thought about how the representation-indiscernibility argument uses the specifically representational features of names of masses, space-time points, and the like, if the argument uses them at all. We must defer a more satisfactory examination of this question. But even without this needed clarification, I think we can see that the accusation of trivial renaming does not get to the bottom of what fuels the relationalists' point of view, nor does it do justice to the force we all feel in the indiscernibility argument.

I believe that the relationalists are onto something—something that has to do with the connection between dispositions and nonrelational properties, and something that comes out more clearly if one focuses on the case of mass rather than the case of space-time. Having a nonrelational manifest property, such as mass, involves having certain dispositions. Indeed, it may be that mass is in some sense exhausted by its attendant dispositions, though not in any way that would serve the purposes of a reductionist, let alone a phenomenalist. For example, the dispositions of a 1-kilogram object include its disposition to balance the standard 1-kilogram object and its disposition to outbalance one of two objects which balance each other and which together balance the standard 1-kilogram object. More generally, for objects to have mass involves their having dispositions to display just the system of mass relations on which the representation theorem and our system of measurement for mass turn.

There are actually (at least) two kinds of dispositions involved in having mass. First, take a case in which there are at least two objects, one with a mass of 1 kilogram and one with a mass of 2 kilograms. We appropriately say that the 1-kilogram object manifests the relation of having half the mass of the 2-kilogram object, where this manifest relation involves dispositions such as that of the 1-kilogram object to be outbalanced by the 2-kilogram object, to exactly balance one of two maximal mass balancing parts of the 2-kilogram object, and so on. But now consider a case in which there is a 1-kilogram object but no object with a mass of exactly 2 kilograms. Such a 1-kilogram object, in such a situation, still has a mass one-half that which a 2-kilogram object would have were one to exist; but this 1-kilogram object cannot be said to actually bear the relation of having one-half the mass to a 2-kilogram object because in the case we are considering there is no such second object to supply the second term of the relation. Nonetheless, in this situation the 1-kilogram object still has the disposition to bear the relation of having one-half the mass to a 2-kilogram object, were a 2-kilogram object is exist; and having this disposition is clearly part of what is involved in having a mass of 1 kilogram. I shall actually be most interested in this second kind of disposition to manifest mass relations.

Inasmuch as having mass is constituted by the dispositions to manifest mass relations, there is a sense in which the mass relationalist is right about mass. I should really state this conclusion more cautiously since I have not given a detailed statement of, let alone argued for, the view that dispositions to manifest mass relations fully constitute having mass: To the extent that having mass involves dispositions to manifest mass relations, there is a sense in which the mass relationalist is right about mass. Still, mass is not a relational property in the way that, say, being married or being as tall as anyone is relational. To be married, dispositions are not enough. There must *be* an *actual* person who is the married person's spouse. Being as tall as anyone turns on the relative distribution of heights among actual people. Again, what happens in merely possible situations is not enough. Thus there is a distinct sense (I think the conventional sense) in which these properties are relational in a way in which mass is not. Altogether, the mass relationalist is right about mass, in the sense and to the extent that mass boils down to a system of mass relations, where it is understood that this system need concern no more than massive objects' dispositions. The same may be true of all properties. At the same time, there is a more conventional sense of "relational," a sense in which the relations concern what holds in the actual world. In this more conventional sense, mass is nonrelational.

Some may be distressed by the lack of detail in my discussion of the relational aspects of mass. But for present purposes what really matters is the analogy I want to draw with the space-time quantity. Regardless of how the details are to be filled in, I want to say that space-time points are like masses. Differences, such as in complexity of representing structures, are completely irrelevant to present concerns. Let me draw the parallel between masses and space-time points as sharply as possible. What mass something has is a matter of what mass relations it bears to some perfectly arbitrary standard. That does not make mass relational in the way in which being married or being as tall as anyone is relational. In contrast with these cases, it is enough that the other terms of the mass relations could exist so that the relations would then actually hold. In this way it is enough that the mass relations be merely potential. In just the same way, where something is is a matter of where it is relative to a perfectly arbitrary reference point (for example, me). As in the case of mass, that does not make place relational in the way that being married and being as tall as anyone are relational. But it does make place relational in exactly the same dispositional way in which I have tried to explain mass to be relational.

This last characterization has to be corrected. I have put it in terms of place, intending that you hear 'place' as spatial position. I have proceeded in this way because only thus can one (or I, at least) firmly get hold of the idea. But once we have a grip on the conception, we must correct it; identification of spatial points at different times is arbitrary (the relativity of space), so we need to work with space-time occurrences rather than spatial position. Thus, what we need more accurately to say is that where an event occurs is a matter of where it occurs relative to four perfectly arbitrary reference space-time points, this statement being in one way relational and in another way nonrelational as before.

It will have dawned on some readers that my strategy for distinguishing ways in which mass and space-time properties are and are not relational will work only if we can make sense of identifying properties across possible worlds; and some will find such an identification absurd, at least for the case of my space-time properties. I can bring out the problem by giving an alternative statement of the way in which mass and space-time are nonrelational. Mass: Consider a 1-kilogram red ball. It has a mass of 1 kilogram because it is equal in mass to the standard 1-kilogram object (as might be shown by a balance experiment). Having a mass of 1 kilogram is nonrelational in the sense that the red ball could exist with the same mass it actually has without there being the standard 1 kilogram object, or any other 1 kilogram object. This is shown to be true by the fact that (speaking in terms of possible worlds), in a possible world in which the red ball existed with the same mass it has in the real world but in which the standard 1-kilogram object did not exist, it would still be true that if the standard 1-kilogram object did exist with the same mass it actually has it would balance the red ball. Or, avoiding talk of possible worlds, we say that if the red ball were to exist with its actual mass but without the standard 1 kilogram object, it would still be true that the red ball would exactly balance the standard 1-kilogram object if this standard were to exist with its actual mass. Space-time: Similarly, the place of my neighbor's house is fixed by the relation of being 100 feet north of my house. This is likewise nonrelational in the sense that if my neighbor's house were to exist (existed in a possible world) with its actual location but without my house's existing, it would still be true that were my house to exist after all, with its actual location, my neighbor's house would still be 100 feet north of mine. Now, the problem comes out most graphically if one is willing to talk in terms of possible worlds. Most people who will talk about possible worlds at all seem happy enough talking about objects with the same mass in separate worlds. But most balk at talking about the same place across possible worlds, and drawing the analogy as I just did clearly requires appealing to such cross-world identification of space-time locations.

Given our thoroughgoing analogy between mass and space-time, I don't see why there should be a difference with respect to mass and space-time location with respect to this question. One might well react here by saying "So much the worse for identifying masses across possible worlds." Consider a world in which there is only one object—the red ball. It patently makes no sense, one might suggest, to say that the ball is located at one place rather than another. Similarly, it makes no sense to say that the ball has one mass rather than another. With nothing else in the possible world, what difference could such an attribution possibly make? But if this is so for the property of having a mass of 1 kilogram, why should it not be so for properties generally? By following this line of reasoning, we have "discovered" the fact that there is every bit as much a problem of cross-world identification of properties as there is for objects.[12]

Let us look at another example of attempted cross-world identification of properties. If the problem holds generally, it should make no sense to say that a ball could exist in another possible world with the same color as the red ball. Indeed, it should make no sense to say that something in another possible world is red! Once again, something has obviously gone wrong. To make this clearer, let us start with familiar counterfactuals. For example, it makes perfectly good sense to say that there could be a red ball right here on the table in front of me. Now, I take talk about possible worlds to be just a handy way of reexpressing what we say using counterfactuals. Since the counterfactual makes perfectly good sense, so must the corresponding talk about possible worlds. Similarly, counterfactual talk about masses makes perfectly good sense. Consider: "If you had stuck to your diet for a month you would weigh ten pounds less than you do now." If possible-world talk is to serve its role in reexpressing and clarifying formal features of counterfactual discourse, we must find a way, corresponding to this example, to make sense of cross-world identification of masses. Space-time locations are no different. Consider: "If you were sitting where I am now you would have seen the right wing score that goal." Colors, masses, and space-time locations are all on a par, and any sort of possible-world analysis worth its salt will have to make sense of cross-world identification for all of them.

After twenty years of discussion of related issues, I don't think the solution to the problem proves very difficult. Indeed, already in "Identity and Necessity" (1971, p. 148) Kripke noted that the problem of cross-world identification is the same for objects and for properties, and that the solution is likewise the same. Here is Kripke's formulation: Possible worlds are not "foreign lands" which we might see through a telescope. Instead they are counterfactual situations. Rather than discover them, we stipulate which one or ones we are talking about by using normal referential discourse to pick out both things and properties, in exactly the same way in which we use this referential discourse to pick out things and properties when talking about actual situations. The same point can be made by remarking that we have constancy of reference across counterfactual contexts for proper names, natural-kind terms, and indexicals, indicating both objects and properties. Consider, for example, the constancy of reference of both 'Tom' and 'cold' in "Tom is drinking a cold beer. Tom would not drink it if it were not cold." and the constancy of reference of indicated object and property in "If *this* stamp were *that* color it would be worth a lot more."

I would like to amplify. I take counterfactual talk to be a way of asserting things about the real world. Thus, I am saying something true about the real world when I say "If I had been more polite I would not have shortstopped the catsup," or "If I had let go of the brick it would have dropped and hit my foot," or "If I had correctly multiplied 1,645 times 5,286 I would have gotten 8,695,470." Inasmuch as counterfactual talk is talk about the real world, there need be nothing mysterious about the fact that the referential apparatus that functions to pick out things when we engage in indicative discourse about the real world works, and works in the same way, when we talk counterfactually. We use names, natural-kind terms, indexicals, and so forth to pick out things and properties, and having picked them out we go ahead and make counter-factual assertions about or concerning *them*. In particular, for location properties, we pick out location properties of events in counterfactual situations by reference to the location of occurrence of events in the real world. Thus, when I say that if you had been sitting where I am now you would have seen the right wing score the goal, reference to the place (and time) you would have occupied in the counterfactual situation is fixed by reference to where I am now sitting in the real world. Even reference to the red ball's being located where I am now in the philosopher's otherwise empty counterfactual situation is securely anchored by the event of my actually being where I now am.

Space-time properties of location are fixed as nonrelational by where events occur in the real world, or, better, by where they actually occur. This has always seemed wrong, for it has always seemed that if there are objective real-world locations then there are possible worlds, or counterfactual situations, in which all events are displaced from ours; and such alternatives would be distinct, yet indistinguishable from the actual world. We can finally see that this is a mistake. The argument turns on the supposition that possible worlds or counterfactual situations have an independent existence. But when we appreciate that they are parasitic on the actual world, on the way things in fact are, we see that the supposed alternatives *are* distinguishable from the actual world. Reference in a possible world is established by reference in the actual world. When I pick out a specific thing to talk about and properties to attribute to it in a counterfactual situation or possible world, I do this by picking out the thing and the properties in the real world and then speaking about them counterfactually. In particular, if I talk about a counterfactual situation in which everything has been uniformly moved over from where it now actually is, I speak of a situation in which, for example, *I* have been moved over from where I actually now am. Reference to *where I actually now am*, just as reference to the mass and color of this red ball, stays constant across counterfactual contexts. Thus, the counterfactual case in which everything has been uniformly moved over is, we can now see, distinguishable from the actual case; for in the counterfactual case, I (to pick an arbitrary example) am now at a different place from the place I can identify in the counterfactual situation as the place I now actually occupy.

So Leibnizian alternatives are, after all, distinguishable. This distinguishability is no "deep" metaphysical fact. It is just a reflection of facts about how language works in counterfactual contexts.[13] Yet, consistent with this distinguishability, space-time location is in some way relational, inasmuch as any space-time location gets fixed or "measured" only relative to some perfectly arbitrary standard. The kind of relationality in question differs from conventional relationality, like that of being married, because a counterfactual situation can incorporate the "relational" location without including the standard that sets the relation. This differs from the relationality involved in being married; if someone is married in a counterfactual situation, the relation-creating spouse must co-occur in the *same* counterfactual situation. The kind of relationality that infects space-time location is more general than this—it equally infects all measurable quantities, and quite likely all properties whatsoever.

Conclusion

Many patient readers of a prepublication draft of this paper have remarked that the considerations that show Leibnizian alternatives to the distinguishable in no way turn on the difference between substantivalism and the quantity view. Consequently, if I am right about Leibnizian alternatives, the quantity view is proof against the indiscernibility argument; but so is substantivalism. What then, these readers have wondered, really is the difference between substantivalism and the quantity view? Isn't the quantity view really just substantivalism all over again, in disguise? In fact, Hilary Putnam tells me that for years he has been remarking that one could as well take space-time points to be properties. However, he has intended this remark not as a resolution of the Newtonian-Leibnizian debate but as a way of showing that at bottom there is no real difference between calling something a "property" and calling it a "concrete thing."

Nothing is this essay shows that Putnam's suggestion is wrong. Why, then, have I insisted on contrasting the quantity view with substantivalism? I think that both traditional substantivalism and traditional relationalism, as I am conceiving of them, make a mistake. Traditional relationalism wants to reject an objective frame of space-time reference (i.e., a frame of reference independent of where observers, objects, or events may happen to be situated in space-time). We have seen that the reason given, the indiscernibility argument, is fallacious. Moreover, the traditional substantivalists are right in maintaining the need to postulate an objective frame of reference to provide a systematic description (and insofar an explanation) of inertial effects. But traditional substantivalism also makes a mistake. It suggests that an objective frame of reference could only be substantival, could only consist in a substance. Substantivalists give no argument for this supposition, and once the alternative is suggested we quickly see that description in terms of space-time as an objective physical quantity will work as well. In addition, a number of considerations speak in favor of the quantity description.

How these considerations look depends somewhat on whether or not one is inclined to agree with Putnam. If Putnam is right and there is no "ultimate" distinction between "things" and "properties," we still have the fact that we group things differently in language. For whatever (perhaps practical) reasons, we use thing-talk in speaking about some things and property-talk in speaking about others. The analogy to mass and other quantities shows that in many respects talk about space-time is more naturally grouped with a particular kind of property-talk: talk about physical quantities. This is just the minimal conclusion I advertised in the introduction. Or Putnam is wrong, and there is a real thing/property distinction to be drawn. But then again the fruitful analogy to other physical quantities constitutes at least a *prima facie* reason for taking space-time to be a quantity and not a substance. Either way, we should count as substantial evidence the ways in which space-time and physical quantities such as mass are similar: similar with respect to the arbitrary unit or other parameters in setting up a system of linguistic reference, similar with respect to the connections between this arbitrariness and the ways in which the item is or is not relational or absolute, and similar with respect to the usefulness of these considerations in clarifying both what is wrong with the indiscernibility argument and what is still somehow right about relationalism. All these parallels and analogies suggest that we should group space-time with mass, temperature, charge, and the like, rather than with atoms, fluids, planets, and the like.

In bare outline the argument for the quantity view is extremely simple: Newton and Leibniz gave us two competing conceptions of space, which we apply now to space-time. After three hundred years of debate, the issue between these conceptions remains at a stalemate, Why? Because, after all, both views have got something right as well as something wrong. There is some way in which space-time is relational, a way having to do with the manner in which both real and possible things are located in relation to things in the real world. But since this way of being relational only requires dispositions to bear relations, there is another way in which space-time is not relational, a way in terms of which we can make sense of our intuition that there are objective space-time locations at which events occur, and a way that can then serve within a space-time theory as part of a systematic description of inertial forces. The quantity view helps us to see how both of these things can be true at the same time. It provides an objective sense to space-time location, and at the same time it leads us to see, as I have tried to explain, how space-time partakes of the kind of relationality that inheres in all measurable properties and possibly in all properties whatsoever. It is because the quantity view captures what is true about both relationalism and substantivalism that this view of space-time provides a better account than either of its predecessors.

Acknowledgment

So many people have helped me write this essay that it would seem silly to mention them individually. But I want each of them to know how much I appreciate their advice and all the time spent trying to explain to me so many things I didn't know.

Notes

1. A more cautious formulation might work in terms of spatial regions instead of points. Since nothing will turn on this refinement, I will speak exclusively in terms of points; this may be taken as an expository simplification.

2. In this form the issue has received considerable recent attention. See Sklar 1977, Friedman 1983, and the numerous references in these works.

3. In the present context, this question just drops out of the sky. In fact, the question arises naturally in a broader context of issues concerning relational and nonrelational properties and their role in theories. I spell out these connections in Teller (to appear).

4. The view is not entirely new. Horwich (1978) explicitly mentions the possibility of taking space-time points to be properties. In some respects this proposal is like Sklar's idea (1977, pp. 229–234) that relationalists could take acceleration to be nonrelational and primitive. Some passages in both Leibniz and Newton (or Clark) sound rather like the present proposal, which also might be seen in the medieval doctrine of the ubi. I will have to leave to historians the task of detailing these older occurrences of the quantity view.

5. See Friedman 1983, pp. 236–263, for a modern formulation with detailed attention to the methodological underpinnings of the argument.

6. "Shorter" would be right if the metric were Euclidean. The non-Euclidean metric of relativity actually calls for "longer" in the case of timelike trajectories. This technical point is of no relevance here.

7. Malament (1982, p. 532, n. 11) expresses the same response to Field's position.

8. My gloss on the various versions of the indiscernibility argument leans heavily on Sklar (1977, pp. 167–181), and I take my brief remark reconciling relationalism with inertial effects to be a summary expression of Sklar's suggestion that relationalists take absolute acceleration to be a nonrelational, primitive property (ibid., pp. 229–234).

9. Manders (1982) and Mundy (1983 and unpublished) offer explicit accounts, but one can find suggestions along these lines in Earman 1979 and Friedman 1983. The next few paragraphs are a gloss on the kind of arguments to be found in the literature. As will become clear further on, one may question what role the axioms and theoremhood actually play in the argument.

10. I should mention the connections between these observations and previous work. Manders's overall exposition (1982) also draws the connection between the indiscernibility argument and the representation-theorem argument pretty clearly. Though not quite as explicit, this relation of the argument is likewise suggested in Earman 1979 and on pages 216–223 of Friedman 1983. Mundy (1983) also conveys the point. This is also a good place to suggest the connections I see between my work and that of Manders and Mundy. Manders appeals to a network of possible configurations, interpreted dispositionally. If my remarks below on the interconnections between properties, relations, and dispositions are sound, differences between my scheme and that of Manders might be only a matter of detail. Mundy wants to avoid appeal to possibilia and to put the weight instead on the lawlikeness of the axioms describing space-time relations. But if lawlike generalizations support counterfactuals and if talk of possibilia is just a reformulation of counterfactual talk, I am not sure that the difference is important. I will outline my views on these connections below. Although I disagree with Mundy's final evaluation, his emphasis on the tools of measurement theory in spelling out the representation-theorem argument has played a vital role in the formation of my own position.

11. These speculations may also be connected with what W. V. O. Quine has in mind when he speaks about indeterminacy of translation and relativity of reference. But I leave to Quineans the task of examining this suggestion.

12. Curiously, there is an extensive literature on the latter problem (see papers reprinted in Loux 1979) but, with the exception mentioned below, there is apparently no discussion of the former.

13. The point is the same as the one to be made about Kripke's *a posteriori*, so-called metaphysical necessities. That Hesperus is identical with Phosphorus is *a posteriori* and is necessary, but this is merely a reflection of how rigid designators work in counterfactual contexts rather than a reflection of any "deep" metaphysics.

References

Earman, John. 1970. "Who's Afraid of Absolute Space?" *Australasian Journal of Philosophy* 48: 287–319.

Earman, John. 1979. "Was Leibniz a Relationist?" In *Midwest Studies in Philosophy,* vol. IV, ed. Peter French et al. Minneapolis: University of Minnesota Press.

Field, Hartrey. 1980. *Science Without Numbers: A Defense of Nominalism.* Princeton University Press.

Fine, Arthur. 1984. "The Natural Ontological Attitude." In *Scientific Realism,* ed. J. Leplin. Berkelely: University of California Press.

Friedman, Michael. 1983. *Foundations of Space-Time Theories.* Princeton University Press.

Hellman, Geoffrey. Location in Space-Time. Unpublished.

Horwich, Paul. 1978. "On the Existence of Time, Space, and Space-Time." *Nous* 12: 397–419.

Krantz, David H., Dunkan R. Luce, Patrick Suppes, and Amos Tversky. 1971. *Foundations of Measurement Theory,* vol. I New York: Academic.

Kripke, Saul. 1971. "Identity and Necessity." In *Identity and Individuation,* ed. Milton Munitz. New York University Press.

Lewis, David. 1983. "Extrinsic Properties." *Philosophical Studies* 44: 197–200.

Loux, Michael. 1979. *The Possible and the Actual.* Ithaca: Cornell University Press.

Malament, David. 1982. Review of *Science Without Numbers. Journal of Philosophy* 79: 523–534.

Manders, Ken. 1982. "On the Space-Time Ontology of Physical Theories." *Philosophy of Science* 49: 575–590.

Mundy, Brent. 1983. "Relational Theories of Euclidean Space and Minkowski Space-Time." *Philosophy of Science* 50: 205–226.

Mundy, Brent. Relational Theories of Space and Spacetime. Unpublished.

Teller, Paul. 1986. "Relational Holism and Quantum Mechanics." *British Journal for Philosophy of Science* 37: 71–81.

Teller, Paul. To appear. "Relational Holism in Modern Physics."

Sklar, Lawrence. 1977. *Space, Time, and Spacetime.* Berkeley: University of California Press.

Weingard, Robert. 1977. "On Cracking That Nut, Absolute Space." *Philosophy of Science* 44: 288–291.

John Earman

Locality, Nonlocality, and Action at a Distance: A Skeptical Review of Some Philosophical Dogmas

The philosophy of science as practiced earlier in this century showed a tendency to pose issues in terms of bogeymen and to read the latest developments in science as rescuing us from these scourges. A prime example is provided by Reichenbach's (1959) framing of the fundamental issues in the Newton-Leibniz debate about space, time, and motion in terms of the bogeyman of Absolute Space. Newton, we are told, "begins with precisely formulated empirical statements, but adds a mystical philosophical superstructure. Leibniz had to oppose a theory which regards space and time as autonomous entities existing independently of things, nor could he help but consider his own theory superior, feeling as one who had emancipated himself from the primitive notions of everyday life." (p. 53) And: "In their opposition to Newton, physicists of our day have rediscovered the answers which Newton's two contemporaries [Leibniz and Huygens] had offered in vain." (p. 47) We have, thankfully, passed beyond these crudities,[1] though the bogeyman of Absolute Space is still occasionally used to scare unsuspecting undergraduates.

We have not, unfortunately, passed beyond the crudities associated with another of Reichenbach's bogeymen, action at a distance. Again, Reichenbach took relativity theory to slay this specter by legislating in favor of local action by contact. And he claimed to draw from his analysis of these issues various methodological morals, including the principle of common cause (PCC). This principle, in its various guises, is often used in contemporary philosophy of science; in particular, it has been used by Salmon (1975, 1978, 1984) as part of an account of scientific explanation, and it is also used as a major premise in an argument for scientific realism, the idea being that only by resort to unobservable theoretical entities can we hope to satisfy the PCC in various domains.[2] As a would-be scientific realist I would like to embrace this argument, but, alas, I cannot endorse the PCC or the reasoning on which it rests.

Recently the focus of the discussion of action at a distance has shifted to the quantum domain, which seems to involve some disturbing nonlocal features. Attempts to explain away these features in terms of local hidden-variable interpretations, whether of a deterministic or a stochastic variety, have proved fruitless—that is the moral drawn from Bell's theorems and the experimental tests of the Bell inequalities. Now the tables seem to have been turned, and the fear has been expressed that the very relativistic structure of space-time, which was supposed to underwrite locality, is threatened.[3]

If we are going to discuss these issues, we should at least be able to say what we are talking about when we speak of locality and nonlocality; but the lexicon is missing, or at least I have never been able to find it, especially for the quantum case. In the first section of this essay I attempt to supply an important part of the missing lexicon for nonquantum theories, Newtonian and relativistic. I formulate a series of principles designed to capture various aspects of locality —localizability, contiguity, finite speed of action, etc. With these principles in hand it is possible to say something about the extent to which relativity theory demands or suggests locality. I then turn to a discussion of the underpinnings and status of the PCC. Also, in passing, I offer some comments on the vexed question of what constitutes a "field theory." The second section, which is devoted to locality and nonlocality in quantum physics, is less constructive and more critical. To be sure, there are substantive disagreements over the cases covered in the first section; however, the form of these disagreements is encouraging, since it implies agreement on what the issues are. In the quantum case no such agreement exists, for at present there exists no set of principles which are unproblematically applicable to quantum phenomena and which are recognizably and indisputably principles of locality in the way that the principles developed in the first section are. If you can't say it, then you can't say it, and you can't whistle it either. Nevertheless, I try to whistle up a tune about how to approach locality questions in quantum mechanics; more particularly, I try to see what lies behind Einstein's assertion that some interpretations of quantum mechanics imply "spooky actions at a distance"[4] and behind the more recent assertions that the Bell inequalities contain important new information on this matter.

Locality and Nonlocality in Newtonian and Relativistic Physics

Reichenbach on Normal Causality

First put forward in 1927 in *The Philosophy of Space and Time,* Reichenbach's ideas about action by contact versus action at a distance continue to influence contemporary discussions. In some cases the influence has been transmitted by contact through a chain of Reichenbach's students and disciples; in other instances the influence seems to act at a distance or by pre-established parallelism (as in the stochastic version of the principle of common cause, which turns up in recent discussions of hidden-variable interpretations of quantum mechanics).

According to Reichenbach, "normal causality" implies both that the transference of effects requires a time lapse and that the effects spread continuously, passing through all intermediate points. Action at a distance involves a "causal anomaly" since it violates normal causality. So far the point is purely terminological. But behind the terminology is a substantive claim and an implied methodology of scientific explanation. The claim is that the label "normal causality" is apt in that the laws of classical relativistic physics embody such a causality. The resulting methodology of explanation comes from reflecting on what we would say when confronted with a perfect synchronization of distant events. We might regard the synchronicity as a "parallelism of events which had 'by chance' the same initial conditions and since then have been running down like synchronized watches," but such an interpretation is possible only in deterministic worlds, "for otherwise the permanence of the strict parallelism would be infinitely improbable" (Reichenbach 1958, p. 65). Or, as Reichenbach puts it at a later juncture, strict parallelism without determinism is pre-established harmony, which is just a disguised way of postulating action at a distance and is, therefore, a "violation of the principle of [normal] causality" (1958, p. 277). Thus, the methodology: To explain a synchronization of distant events, first seek causes in the common causal past of the events; failing that, look for an explanation in terms of a single "chance" coincidence which is thereafter maintained by determinism; only after repeated failures to produce explanations of the first two kinds, consider the repugnant alternative of a causal anomaly.

This brief gloss of Reichenbach's ideas does not do justice to the richness of examples and analysis found in *Space and Time* and the posthumously published *The Direction of Time*. But it is this gloss that continues to influence and distort current discussions. What follows is an attempt to capture various aspects of normal causality and action by contact by means of a series of locality principles. Once these principles are set out, we can try to understand the role of locality in the methodology and the content of physics.

Locality: Bounds on Causal Propagation and the Possibility of Determinism

Reichenbach warns us that his version of the principle of action by contact does not entail the existence of a fixed finite limit to the speed of all causal propagation; rather, the principle "excludes only infinite velocities, whereas it is compatible with any arbitrarily high velocity" (1958, p. 132). To mark Reichenbach's distinction, let me state two locality conditions:

(L1) All causal propagation takes place with a finite velocity.

(L2) There exists a fixed finite limiting velocity for all causal propagation.

Another closely related principle is the following:

(L3) Determinism is possible in the sense that domains of dependence are nontrivial.

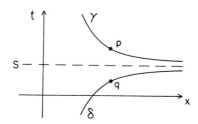

Figure 1

In explaining L3 I will use the term *curve* ambiguously to denote a differentiable map $\sigma: I \rightarrow M$, where M is the space-time manifold and I is an interval (open, closed, or half-open) of $\mathbb{R}$, or else to denote the image set $\sigma(I)$. A *causal curve* is a curve that is the space-time locus of some possible causal signal. For $S \subset M$, the *domain of dependence* of S, $D(S)$, is defined as the collection of all points $p \in M$ such that every causal curve that passes through p and has no endpoint meets S. That $p \in D(S)$ does not imply that the state on S determines the state at p but only implies that the spatio-temporal relation of S to p is favorable to such a determination in that every causal signal that can influence events at p must register on S. The relevant point to note here is that Reichenbach's L1 is apparently not strong enough to vouchsafe L3. To see why, suppose that L1 holds while L2 fails. Thus, the causal curves γ and δ in figure 1 represent possible causal processes whose velocities are everywhere finite in accordance with L1. However, for the time slice S (say, high noon on April 1, 1988), $p \notin D(S)$ and $q \notin D(S)$, since neither γ nor δ meets S even though they have no past or future endpoints. And, in fact, $D(S) = S$. The domain of dependence of S and every such time slice is trivial.

A little further reflection on this and similar examples leads to the worry that Newtonian physics, supposedly the source of paradigm examples of determinism at work, is not after all deterministic—does not Newtonian particle mechanics violate L2, and don't various Newtonian field theories violate L1 as well? The resolution of this "paradox" is that Newtonian physics is *not* deterministic. More properly, it is not deterministic without the imposition of various "boundary conditions at infinity" that rule out such rude past or future surprises as those pictured in figure 1. Whether the use of such boundary conditions amounts to an *ad hoc* rescue of determinism is a topic I will not broach here.[5]

Two further comments about L1–L3 are in order. First, by a causal signal I mean the propagation of a disturbance, whether by particle or by field. Although I can give no general and precise definition of this concept, it is, I claim, clear enough that the truth or falsity of L1–L3 is cleanly decidable for the Newtonian and relativistic particle and field theories studied below. I emphasize that I do not follow the tradition of explicating causal signal in terms of information transmission. A causal signal in my sense need not be capable of carrying information because, e.g., it cannot be modulated or because its emission cannot be controlled. In the other direction, information gains need not be explained in terms of a causal signal which carries the information. Reichenbach proposed to distinguish between genuine and pseudo-causal signals by means of the "mark method," which is essentially a test of information transmission. The distinction can be drawn without appeal to this apparatus. An advantage of avoiding the use or the mention of "information," "messages," etc. is that we are relieved of the pressure to move from the existence of correlations conveying information about relatively spacelike events to the violation of signal locality or the existence of action at a distance—a move that is illigitimate unless it is backed by special consideration, but a move that some philosophers nevertheless find irresistible. Second, L2, with the velocity of light as the maximum signal velocity, has been a tacit assumption of many formulations of relativity theory. In some presentations the assumption is made explicit, as when Hawking and Ellis declare that "our formulation excludes the possibility of particles such as tachyons, which move on spacelike trajectories" (1973, p. 60). Tachyon advocates hold that this is mere fiat, since, they maintain, the possibility Hawking and Ellis dismiss with a stroke of a pen is a live physical possibility. If the tachyon advocates are correct, then part of Reichenbach's normal causality and determinism as well are threatened. This is an important issue, but I will bypass it to concentrate on other, more neglected aspects of normal causality and locality.

Locality: Separability and Localization

Until otherwise noted, the discussion will be restricted to space-time theories, by which I mean theories whose intended models have the form $\langle M, O_1, O_2, \ldots \rangle$, where M is a differentiable manifold representing space-time and where the O_i are local geometric object fields[6] on M representing either the structure or the physical contents of space-time. The O_i need not be the only quantities in the theory, but they are assumed to be the basic ones; other, derivative quantities can be obtained as functionals of them.

Most of the familiar theories of Newtonian and classical relativistic physics can be cast in the form of space-time theories. However, the generality of this apparatus should not blind us to the fact that a significant amount of locality is being presupposed. In particular, the description of physical reality in terms of local geometric object fields rules out various forms of nonlocal holism for the state description of the world and demands semantic separability and semantic localizability (it implies that, given the state throughout M, it is meaningful to speak of the restriction of the state to proper subsets of M, and, conversely, the state throughout M can be seen as the "sum" of the states at all the points of M). Though this is commonsensical enough, there is no *a priori*

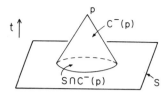

Figure 2

guarantee that nature must be nonholistic in this sense; indeed, quantum physics looms as a potential counterexample. Lest we forget the substantive locality assumptions built into our approach, let us list the following as an explicit principle:

(L4) T is a space-time theory.

For theories that conform to L4, we can make L1–L3 precise in various concrete instances. Thus, for the little theory consisting of the scalar wave equation for Minkowski space-time the intended models have the form $\mathcal{M} = \langle \mathbb{R}^4, \eta, \Phi \rangle$, where η is the Minkowski metric and Φ is a scalar field on $\mathbb{R}^4$ satisfying $\eta^{ij}\Phi_{,ij} = 0$. We can then proceed to prove statements about domains of dependence and speed of propagation; e.g., from the form of solutions to the wave equation we can see that the field $\Phi(p)$ at any point $p \in \mathbb{R}^4$ is uniquely determined by the values of th field and its normal derivative at those points where the past light cone $C^-(p)$ of p intersects a given spacelike hypersurface S (see figure 2). In fact, if space is assumed to be three dimensional, Huygens' principle holds and $\Phi(p)$ is uniquely determined by data on the boundary of $S \cap C^-(p)$ since then Φ propagates with exactly the speed of light, no more and no less.[7]

Without L4 it is hard to see how to make precise sense of L1–L3 and the other locality principles to be introduced below. Thus, I will refer to L4 as a *pre-locality*, noting that the semantic locality assumptions built into it seem just as necessary for paradigm cases of physical action at a distance as for local action by contact.

Pre-locality L4 demands semantic localizability. We might also want to demand physical localizability:

(L5) T permits localized states.

Suppose that the O_i's can be divided into those (S_j) that characterize the structure of space-time and those (P_k) that characterize the physical contents of space-time. This separation can be made for orthodox Newtonian and special-relativistic theories where the S_j are "absolute objects" (which are the same in every intended model) while the P_k are "dynamical objects" (which vary from model to model). We can then say that T permits localized states just in case there are models of T such that at every instant of time the P_k have support that is contained in a connected compact region of space.

The previous example of the wave equation can also be used here, since if Φ is localized at an instant then, by (L2), it is localized for ever after. It might seem then that any relativistically correct theory must fulfill L5. But not so; one-particle states in relativistic quantum mechanics cannot, on pain of contradicting L2, be localized at any instant as will be discussed below.

Locality: From Global to Local

The next principle to be considered expresses one aspect of the idea that local physics is independent of global considerations.

(L6) Every localization of a global model of T is again a model of T.

If $\mathcal{M} = \langle M, O_1, O_2, \ldots \rangle$ is a model in the above sense and $U \subset M$ is a connected open set, then the restriction $\mathcal{M}|_U$ of $\mathcal{M}$ to U is well defined;

$$\mathcal{M}|_U = \langle U, O_1|_U, O_2|_U, \ldots \rangle,$$

which is again a model in the intended sense. L6 demands that if $\mathcal{M}$ satisfies the laws of T then so does $\mathcal{M}|_U$ for any U. In special-relativistic physics it is sometimes taken for granted that all models be given on Minkowski space-time, and in general-relativistic theories it is sometimes assumed that any physically realistic space-time is inextendable. If these assumptions are made explicit postulates of T, then obviously L6 fails trivially. So for present purposes these assumptions must be taken to operate only on potential global models and not on local models.

Although L6 might seem mild enough, it captures a good part of the content of the action-by-contact principle. This claim is based on the observation that L6 divides paradigm cases in just the way that intuition suggests. For example, the special-relativistic version of a scalar field satisfying the wave equation and the source-free electromagnetic field satisfying Maxwell's field equations are clear cases of action-by-contact theories (if any such cases are clear), and they obviously conform to L6. More generally, any theory, Newtonian or relativistic, that is couched in terms of field quantities[8] obeying local partial-differential equations will be a local theory in the sense of L6. By contrast, particle theories that use only particle variables and yield intuitively clear cases of nonlocal action typically violate L6. Examples of Lorentz-invariant theories that violate L6 are provided by the Poincaré-type theories of gravitation (Whitrow and Morduch 1965), the Wheeler-Feynman (1945, 1949) formulation of classical relativistic dynamics in terms of direct particle action, the Currie-Hill instantaneous-action-at-a-distance formalism for classical relativistic particle mechanics (Currie 1966; Hill 1967a, b, 1982), and the Van Dam–Wigner (1965, 1966) theories. In the case of retarded-action-at-a-distance particle theories,

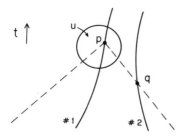

Figure 3

the laws of motion for two particles would specify the acceleration of particle 1 at event p (see figure 3) as a function of the behavior of particle 2 at event q, where the past lobe of the light cone from p intersects the world line of particle 2. The restriction to the neighborhood U of a solution of the equations of motion is not a solution, for particle 1 is accelerated at p even though in the restricted model it "sees" no other particle in U.

Particle theories (Newtonian or relativistic) that allow the particles to interact only by contact should count as local, and so they do by L6. On the other hand, field theories in which the field laws are in the form of nonlocal differential or integral equations may violate our ideas of contiguity; a violation of L6 will signal when this is so.

L6 is also useful for revealing and probing the fuzziness of our concepts of action at a distance and action by contact. Consider, for example, the classical heat equation, a partial-differential equation of the parabolic type in which the characteristics are the planes of absolute Newtonian simultaneity.[9] Heat is transmitted infinitely fast, violating Reichenbach's most basic locality requirement, L1. And yet the effects do not leap over spatial gaps without passing through intermediate points, as they do in pure particle theories; in short, L6 holds. Do we then have a case of action at a distance, or a case of action by contact? We have both. The dichotomy is not a dichotomy after all.

Consider next two versions of Newtonian gravitational theory. The first version employs only particle variables and posits direct action at a distance. The second version posits that the particles act not directly upon one another but through the mediation of a gravitational force field. This field is determined locally as the gradient of the gravitational potential φ which is assumed to obey Poisson's equation $\nabla^2 \varphi = \rho$, where ρ is the mass density. Does the introduction of these field variables succeed in replacing action at a distance by local action by contact? Locality in sense L1 is still violated. But contiguity in sense L6 seems to have been secured, at least at first glance. However, if the field version is to be predictively equivalent to the particle version in the minimal sense of yielding the same particle trajectories, then the field version must eschew source-free solutions to Poisson's equation; but L6 no longer holds if the true field law is not Poisson's but the expression of the field value at a point as an integral over sources. Furthermore, the combination of the fact that source-free gravitational fields are not allowed and the fact that momentum is not attributable to the field suggests that the field variables mediate the gravitational interactions of the particles in a merely descriptive sense that carries no ontological implications. But such a judgment about the status of the variables is beyond the power of my apparatus to resolve.

At this juncture our ideas about locality may follow a number of different but intersecting paths. One path traces out further aspects of the idea, partially captured in L6, that large-scale considerations do not constrain the local scene. Another path starts from the notion that the goings-on in relatively spacelike regions of space-time should be lawfully independent of one another. Still a third path starts with the premise that determinism, if it obtains, should be localizable so that what happens locally is determined locally. I will explore each of these paths in turn. But before we turn to these tasks, some remarks about the concept of a "field theory" are in order.

Digression: Fields and Field Theories

The history of field theories is such a tangled web of conflicting motivations, approaches, and results that any attempt to provide a neat and comprehensive classification of the multifarious senses of "field" is doomed to failure.[10] Nevertheless, the apparatus developed so far provides for a crude but useful initial sorting.

For a theory T to count as a field theory, we certainly want the variables to be field variables, which I take to mean both that they are geometric-object field variables $\mathbf{O}_i$ and that the allowed values O_i, construed as mappings from space-time M to $\mathbb{R}$ (say), are continuous on at least some open neighborhoods of M. This already rules out discrete-particle theories, but not continuous-fluid theories. And it is just here that we encounter one of the major conflicts of approach and usage between nineteenth-century field theories and present-day field theories.

On current usage, the field variables describe fields proper only if their values do not represent the state of a ponderable medium. This is in direct conflict with the usage of Faraday, Maxwell, and Kelvin, who are generally credited with introducing the modern field concept as well as the term 'field.' For them 'field' denoted the state of a subtle medium that transmits the action of ordinary matter. Note that, without further provisos, field theories in this sense need not be opposed to action-at-a-distance theories, for the medium could, for instance, be construed as a particulate ether with the ether particles exerting (say) short-range action-at-a-distance repulsive forces on one another. Maxwell stated the case for just such a view while rebutting the notion that it is an *a priori* truth that a body cannot act where it is not:

To this it was replied that we have no evidence that real contact ever takes place between two bodies, and that, in fact, when bodies are pressed against each other and in apparent contact, we may sometimes actually measure the distance between them, as when one piece of glass is laid on another, in which case a considerable pressure must be applied to bring the surfaces near enough to show the black spot of Newton's rings, which indicates a distance of about a ten thousandth of a millimetre. If, in order to get rid of the idea of action at a distance, we imagine a material medium through which the action is transmitted, all that we have done is to substitute for a single action at a great distance a series of actions at smaller distances between parts of the medium, so that we cannot even thus get rid of action at a distance. (Maxwell 1890a, pp. 485–486)[11]

This brings us to a second set of issues and a second set of conflicting usages revolving around the question whether the field is local. We have already distinguished three senses for this question: Is there action by contact (L6)? Is the speed of propagation finite (L1 and L2)? Is the field physically localizable (L5)? Many writers take positive answers to some or all of these questions to be part of the very definition of a field.[12] It is obvious why this should be so for those nineteenth- and twentieth-century field theorists whose motivation for introducing fields was to avoid direct action at a distance between gross bodies. But it is not at all obvious why there should be an *a priori* connection between locality and fields, either in the nineteenth-century sense or in the modern sense. Certainly field theories in the modern sense can violate L1, L2, L5, and L6, and Lorentz invariance does not demand or secure L5 or L6.

Locality attributes are sometimes also used to adjudicate the issue of whether the field is "real" as opposed to a "mere mathematical construct." Some of the other criteria also used to this purpose (Are source and/or receptor free fields permitted?[13] Is energy and/or momentum attributable to the field?) were noted above.

The locality principles studied so far were crafted broadly enough that they do not beg questions about particles versus fields or about the various conflicting senses of 'field' and 'field theory.' The other locality principles to follow will follow suit.

Locality: From Local to Global

The theme that local physics should not be influenced or constrained by global factors was explored above through the idea that localized pieces of global models of a local theory should be models of the theory as well. We can also explore this theme in terms of conditions that move in the opposite direction, from local to global. Thus, we might demand that local verifications of a local theory entail a global verification:

(L7) Let $\{U_i\}$, $i = 1, 2, 3, \ldots$, be an open covering of the space-time manifold in $\mathcal{M} = \langle M, O_1, O_2, \ldots \rangle$. Then if each of the local models $\mathcal{M}|_{U_i}$ is a model for T, so is $\mathcal{M}$.

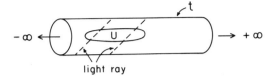

Figure 4

Or we might want to require of a local theory that every local solution to the laws of the theory on a space-time can be made global on that space-time. Since we will be conerned below with the implications of relativity for locality, I will formulate a version of this condition for relativistic theories. Let M,g be a relativistic space-time; i.e., g is Lorentz signature metric defined on all of M. (Minkowski space-time is, of course, a special case.) M,g is an *allowable space-time* for T just in case there is a model of T of the form $\langle M, g, O_2, O_3, \ldots \rangle$. Then the local-to-global condition states:

(L8) If M,g is an allowable space-time for T and $U \subset M$ is an open set and $\langle U, g|_U, O_{2U}, O_{3U}, \ldots \rangle$ is a model of T, then there is a model $\langle M, g, O_2, O_3, \ldots \rangle$ of T such that $O_2|_U = O_{2U}$, $O_3|_U = O_{3U}$, etc.

This condition may be violated in acausal space-times. Consider the cylindrical space-time formed by rolling Minkowski space-time up along the time axis (figure 4). Let T demand source-free solutions of Maxwell's equations of electro-magnetism on this space-time. One allowed solution on U has a single light ray entering and then leaving U. But this local solution cannot be made into a global solution on the cylinder. As Geroch and Horowitz (1979) note, L8 can be saved if the cylinder is cropped back enough on either end to ensure that the light ray will propagate off the "edge" of the resulting space-time before it has a chance to wrap around the cylinder. Thus, acausality in space-time structure need not always lead to a violation of L8. But it seems a plausible conjecture that a violation of L8 by a theory that is local in all the preceding senses is due to acausality or to some other objectionable global feature of space-time structure that provides grounds for saying that space-time is not allowable after all, thus preserving locality in the form L8. Of course, the conjecture is not useful until we have some independent characterization of what is to count as an "objectionable" feature of space-time structure. Nevertheless, L8 can be used to explore this theme.

Locality: The Independence of Relatively Spacelike Regions

Reichenbach's discussion of locality leaves the impression that, while his con-cept of locality may be compatible with some contingent forms of correlation between relatively spacelike events, it precludes purely lawlike correlations

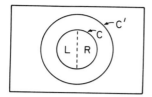

Figure 5

among such events. Whatever the exegesis of Reichenbach, the following locality principles seem worthy of consideration on their own merits.

We would like a condition that would guarantee that, consistent with the laws of T, we could hold fixed the state in one region of space-time while varying the state in a relatively spacelike region however we please. One way to capture this idea is to demand that any two local models of T on relatively spacelike regions can be made global within the same model:

(L9) If $\mathscr{M} = \langle M, g, O_2, O_3, \ldots \rangle$ and $\mathscr{M}' = \langle M, g, O_2', O_3', \ldots \rangle$ are models of T and if $U, V \subset M$ are relatively spacelike open regions, then there is a model $\mathscr{M}'' = \langle M, g, O_2'', O_3'', \ldots \rangle$ of T such that $\mathscr{M}''|_U = \mathscr{M}|_U$ and $\mathscr{M}''|_V = \mathscr{M}'|_V$.

Or, instead of dealing with open regions, we could formulate conditions similar to L9 for instantaneous states on spacelike hypersurfaces.

Trivial violations of L9 and its cousins that are not also violations of L1–L8 can be obtained in field theories whose laws are local partial-differential equations that allow an infinity of solutions but force the field quantities to be constant throughout space-time. Other trivial violations of L9 can be obtained in theories that place a fixed upper bound on some quantities when summed or integrated over all space. If these were the only counterexamples, one might be justified in thinking that L9 and company do capture a fundamental aspect of locality. But there are more interesting counterexamples to come.

Relativistic field theories sometimes entail lawlike constraints on instantaneous states. For example, Maxwell's field equations for electromagnetism and Einstein's field equations for gravitation entail divergence conditions (elliptic partial-differential equations) that constrain the allowed Cauchy-type initial data. When combined with the appropriate space-time geometry, these conditions lead to constraints on relatively spacelike events, in violation of L9. Consider the closed two-surface C on the spacelike plane pictured in Figure 5. Remove the boundary points where the left hemisphere joins the right, leaving the relatively spacelike regions L and R. Different solutions as restricted to L and R cannot always be joined in the same global solution. Any solution must satisfy $\nabla \cdot \mathbf{B} = 0$ so that in a solution encompassing both L and R the outward magnetic flux through L and through R must sum to zero, but there are obviously local solutions where, say, the outward flux through L is positive while the flux through R is zero (see Avishai et al. 1972).

If you do not like the fact that L and R are cheek by jowl, consider C', which may lie as far outside of C as you like. Add to Maxwell's theory the further axiom that only one kind of electric charge (say, positive) exists. Then in any global model where there is outward electric flux through C, there must also be outward electric flux through C'. This in turn places restrictions on any global model for the allowed states on relatively spacelike open annular regions surrounding the projections of C and C' a short way into the future. Consequently L9 fails.

Such examples show that the independence of relatively spacelike regions —in the sense of L9—is not entailed by L1–L8. Further, the examples also serve to show that a violation of L9 and kin need not entail pernicious action at a distance. Any tendency to see perniciousness is undercut by the fact that the large-scale constraints that violate L9 are derived by integrating purely local constraint equations ($\nabla \cdot \mathbf{B} = 0$ and $\nabla \cdot \mathbf{E} =$ charge density) and the fact that the two other Maxwell equations, themselves local dynamical laws, imply that if the constraints hold for any time then they hold for all times.

Yet the idea of the lawful independence of relatively spacelike regions remains an intriguing one. The only violations of L9 and kin of which I am aware have in common with the above counterexamples that they either use artificial "laws" or appeal to special geometric configurations. Is there then some principle that goes beyond L1–L8 and toward L9 and is a necessary part of locality? My attempts to isolate a pernicious subset of violations of L9 and kin have proved self-defeating in that when I try to explain what is pernicious about the example I find myself falling back on L1–L8.

In the other direction, note that L9 can be satisfied in action-at-a-distance particle theories that allow an indefinite number of particles. Thus, L9 by itself may do little to secure any fundamental sense of locality.

My tentative conclusion is that L9 and kin do not lie as close to the heart of the concept of locality as has sometimes been assumed.

Locality: Determinism and Local Determinism

It might be thought that the analogue of L9 should be routinely violated when the regions are relatively timelike rather than spacelike. But not so. Any two local solutions of the scalar wave equation on the regions R_1 and R_2 of Minkowski space-time pictured in figure 6 can be joined in a global solution. Intuitively, influences propagating from R_2 to R_1 can be canceled out by other influences that do not register on R_2. This cannot happen, of course, when R_2 is extended so that $R_1 \subset D(R_2)$.

The scalar wave equation also illustrates another form of locality:

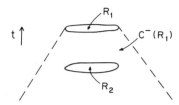

Figure 6

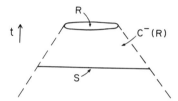

Figure 7

(L10) If T is Laplacian deterministic, then it is locally so.

To make L10 more precise we need to define our terms more carefully. For a relativistic space-time, the *future* (respectively, *past*) *domain of dependence* $D^+(R)$ ($D^-(R)$) of $R \subset M$ consists of all those points $p \in M$ such that every future directed nonspacelike curve that passes through p and has no past (future) endpoint meets R.[14] The total domain of dependence of R is

$$D(R) \equiv D^+(R) \cup D^-(R).$$

A *time slice* of M,g is defined to be a spacelike surface $S \subset M$ without edges.[15] Such a slice is said to be a *future* (respectively, *past*) *Cauchy surface* for M,g just in the case that all the points of M to the future (past) side of S are in $D^+(R)$ ($D^-(R)$). S is a Cauchy surface simpliciter just in case it is both past and future Cauchy. For special-relativistic theories based on Minkowski space-time, global Laplacian determinism means that if two models agree on appropriate initial data on a Cauchy surface then they agree everywhere. Both the meaning and the truth of determinism become more problematic in the general theory of relativity because cosmological models can fail to possess Cauchy surfaces; whether or not such a failure puts the models beyond the pale of the physically realistic is a matter of current controversy.[16]

Localization of Laplacian determinism demands that the state on a region R be determined by the state on any spacelike S such that $S \subset C^-(R)$ and $R \subset D^+(S)$ (see figure 7), and likewise with past and future interchanged. With this interpretation, L10 has been dubbed "Einstein locality" by Hellman 1982a.

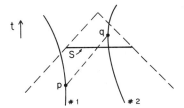

Figure 8

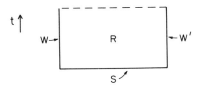

Figure 9

Intuitively, local determinism expresses action by contact through deterministic evolution, so one would expect that for deterministic relativistic theories L6 and L10 would stand or fall together. As we have seen, L6 is characteristic of local fields. If the fields propagate deterministically with a speed no greater than that of light, then the state of the fields on the local slice S of figure 7 will determine the state throughout $D(S)$—which is just to say that the domain of dependence has its intended meaning. In the converse direction, a failure of L6 is typically accompanied by a failure of local determinism. Thus, in the case of two particles interacting via retarded action at a distance (figure 8), the state on S does not suffice to determine the state at point q. The retarded action from particle 1 originating at p reaches particle 2 at q, but since the action is not transmitted by a field it does not register an S. (Thus, for action-at-a-distance theories the concept of domain of dependence as defined above is not a very useful concept.) However, it is not known to what extent counterexamples to L10 can be extracted from such cases. The equations of motion here are of the delay-differential type, which typically do not admit a well-posed Cauchy-type initial-value problem (see Driver 1977). But it has been shown that for the special case of two like-charged particles moving in one spatial dimension the retarded-action-at-a-distance equations of classical relativistic electrodynamics do admit a well-posed initial-value problem with Newtonian initial data (Driver 1969). Here, then, we have an example, though a somewhat artificial one, of how Einstein locality can fail in a relativistic theory.

For Newtonian theories where L1 of L2 fails, we can still hope for localization of determinism, or "Newtonian locality," meaning that the state on the local time slice S and on the walls W and W' (figure 9) uniquely determine the state in the interior region R. Newtonian field theories (e.g., the classical heat equation) typically display Newtonian locality, whereas Newtonian particle theories typically do not. Once again, the difference is reflected in which theories do and which do not satisfy L6.

Although philosophers have been almost totally preoccupied by the Laplacian variety of determinism, there are many other varieties worth considering. For instance, the future development of a system may not be uniquely determined by instantaneous data on a time slice, but it may be fixed by data given over a finite span of time. Differential-delay and/or differential-advance equations provide relevant illustrations (Driver 1969, 1977). These distinctions can be codified in a further locality principle that ranks deterministic theories in terms of the temporal locality of the initial-value problem:

(L11) T_1 is temporally more local than T_2 just in case the minimal initial data needed for a well-posed initial-value problem for T_1 are more temporally local than those needed for T_2.

A Laplacian deterministic theory has the highest locality ranking on L11; a theory requiring the specification of the entire past history of the system for a well-posed initial-value problem would rank at the bottom. There seems to be no natural connection between this sense of locality and the others studied above; viz., the Currie-Hill instantaneous-action-at-a-distance relativistic mechanics violates L6 and L10 but has the highest locality rank by L11.

Relativity and Locality

Does relativity theory require locality? The question is partly terminological, turning on what is to be counted as part of the theory of relativity and what is to be counted as locality. But it also engages substantive issues about the nature of scientific theorizing and the form and content of physical theories.

The special theory of relativity (STR) is not a theory in the usual sense but a second-level theory (i.e., a theory of theories) that constrains acceptable first-level theories. The constraint is Lorentz (or, more properly, Poincaré) invariance. In terms of the structure of space-time models, the demand is that the first-level theory be formulated in the arena of Minkowski space-time; the implicit understanding is that the models of the theory should not smuggle in any additional structure for space-time, such as distinguished reference frames. The general task of separating the O_i's into those that characterize the space-time structure and those that describe the contents of space-time is a delicate and difficult one.[17] But the vagaries of the general problem need not detain us here, since there are enough clear cases of Lorentz-invariant theories to serve our purposes.

If Lorentz invariance were the only constraint imposed by STR, the examples already presented would suffice to show that STR *per se* does not entail locality or action by contact in any of the senses L5–L10. However, it remains open that Lorentz invariance plus some other general requirements, which one would expect an adequate physical theory to fulfill, do entail one or more types of locality. The additional requirements that first come to mind are conservation principles.

In Newtonian theories, energy might be attributed to a field postulated to mediate the interactions of particles; however, it is otiose to attribute momentum to the field if, as is often the case, the total momentum of the particles is conserved. Relativity theory turns the tables. Relativistically, it is the combination of energy-momentum that is conserved; and if influences propagate with a finite speed, it would seem that the energy-momentum of the particles cannot be conserved without the help of an intervening field to carry it. This intuition is made precise in a simple and ingenious theorem of Van Dam and Wigner (1966) which shows that no nontrivial pure particle theory can satisfy both Lorentz invariance and a conservation law of energy-momentum of the form

$$P^i = \text{constant}, \quad i = 1, 2, 3, 4 \tag{1}$$

where P^i is the total linear energy-momentum of the system of particles. More specifically, if

$$P^i = \sum_{\alpha=1}^{N} P_\alpha^i$$

(the sum being taken over the individual particles labeled by the index α) is independent of t in every Lorentz frame (x, t), if the particles do not collide, and if asymptotically (as proper time $\tau_\alpha \to \pm \infty$) the particle orbits are straight lines, then for $N = 2, 3,$ or 4 the orbits are straight lines for the time.

Van Dam and Wigner show that in their theory of action-at-a-distance particle mechanics the conservation law can be restored in the form

$$P^i + V^i = \text{constant}, \tag{2}$$

where V^i is the "interaction momentum" of the system. But since V^i involves an integral over the actual orbits of the particles, equation 2 is seen by some commentators as a mathematical trick, for the integration in V^i is "precisely what we would regard as field momentum in a field theory (the integration over history arises from writing the field in terms of the past motions of all the particles that contributed to the field)" (Ohanian 1976, p. 80).

In response, the particle theorist may reply that we may simply have to live with the fact that conservation laws of the form of equation 1 cannot hold for each instant of time but can only hold asymptotically for particles that are widely separated both initially and finally. Also, while the introduction of fields as storehouses of energy and momentum may facilitate the maintenance of

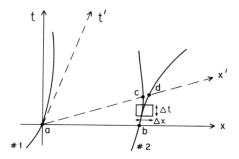

Figure 10

cherished forms of conservation principles, it may give rise to other problems. For instance, infinities can arise from particles conceived as creating fields which in turn act back on the particles. These infinities can sometimes be suppressed by clever subtraction procedures, but these procedures are no less artificial than the introduction of the interaction momentum in equation 2. It was precisely the avoidance of such infinities that motivated some of the action-at-a-distance formulations of classical relativistic electrodynamics—at least these theories have the virtue of consistency.

Nevertheless, it is hard to shake the feeling that in the context of STR pure particle theories are artificial and that various features of such theories indicate that fields are struggling to emerge from the formalism. For instance, in the Currie-Hill instantaneous-action-at-a-distance theories it is found that in multiparticle systems the total force acting on a particle is not the sum of two-body forces; and the presence of multibody forces can be read as the vestige of the fields suppressed in the pure particle description (Hill 1967a). Furthermore, a relativistic instantaneous-action-at-a-distance formalism is apparently inconsistent with the treatment of the measurement of particle position as a purely local interaction between the particle and a measurement apparatus (Hill 1967b). Figure 10 represents particle 2 passing through a measurement apparatus extending over the spatial interval Δx and sensitized for the time interval Δt. If the local interaction takes place, the world line of particle 2 is perturbed, say, from bc to bd. From the perspective of the unprimed (x, t) frame, this perturbation is not felt at event a by particle 1 (assumed to be interacting instantaneously at a distance with particle 2). But the opposite conclusion is drawn from the perspective of the (x', t') frame, since in this frame event a is simultaneous with c and d. In response, Thomas (1952) suggested that world lines of particles are non-invariant objects, a violation of L4. Hill (1967b) proposed to keep world-line invariance and postulate a compensating nonlocal interaction between the measuring apparatus and particle 1. The postulate remains a bare postulate.

More straightforward considerations of empirical adequacy can also be brought to bear against certain action-at-a-distance theories. Poincaré-type gravitational theories can be rigged so as to yield the correct result for the advance of the perihelion of Mercury. But there is no natural way to use such theories to explain the red shift and the bending of light (see Whitrow and Morduch 1965).

In sum, there are reasons to be uneasy about special-relativistic theories that flagrantly violate L6 and other locality principles by postulating particles that act at a distance without the help of an intervening field medium. The reasons are a mixture of considerations of empirical adequacy, the form of conservation principles, naturality, etc. These considerations may be persuasive to one degree or another, but they do not add up to an irrefutable proof that STR demands "normal causality." There remains room for maneuvering toward a détente, if not entirely peaceful coexistence, between special relativity and action at a distance.

The general theory of relativity (GTR) is a theory, not a meta-theory or a theory of theories. It is a theory of gravitation, and to the extent that we accept at face value the ontology and laws GTR postulates for gravitation, we have an argument for action by contact in gravitational interactions.

Reichenbach's Methodology of Explanation

Reichenbach seems to have taken it for granted that a synchronicity of distant events cannot be the direct result of basic physical laws but must be due, at least in part, to contingencies. The task is then to explain how the synchronization arose as a result of contingencies and how it is maintained as a result of the unfolding of events in accord with local laws displaying normal causality. A major obstacle to attempts to evaluate this methodology derives from the fact that Reichenbach gives us no general characterization of the types of synchronicity at issue. That my being in Minneapolis is perfectly synchronized with my being simultaneously absent from San Francisco is a kind of synchronicity of distant events,[18] but it does not seem to threaten any pernicious form of action at a distance. But then what forms of synchronicity do so threaten? For the moment we will simply have to acknowledge this hiatus and try to work around it.

Reichenbach's exhortation to look for an explanation in the common causal past is commonsensical enough, but it is worth noting that the global structure of some relativistic cosmologies blocks the search. Figure 11 is a schematic illustration of the type of null cone behavior found in the de Sitter cosmology. The perfect synchronicity of beeps emitted by the processes γ_1 and γ_2 cannot be explained by reference to their common causal past, since $C^-(\gamma_1) \cap C^-(\gamma_2) = \varnothing$.

In such a case Reichenbach would have us shift the search to a different

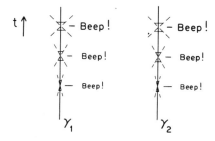

Figure 11

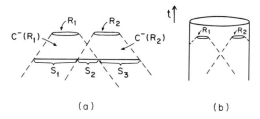

Figure 12

form of explanation, so let us assume that the events in question do have a common causal past. But only a moment's reflection is needed to show that the notion of a deterministic explanation by reference to the common causal past is in general incoherent. In the typical case illustrated in figure 12a,

$$D(C^-(R_1) \cap C^-(R_2)) = C^-(R_1) \cap C^-(R_2),$$

i.e., the domain of dependence of the common causal past of the relatively spacelike regions R_1 and R_2 is trivial. This common causal past can determine exactly nothing outside of itself. There are exceptions, one of which is illustrated in figure 12b, where $D(C^-(R_2)) \cap C^-(R_2)$ is the entire space-time.

Of course, if the localizability of determinism (L10) holds, the state on $S = S_1 \cup S_2 \cup \dot{S}_3$ (figure 12a) will determine the states on R_1 and R_2. And if the states on S_1 and S_3 are held fixed, the variations on R_1 and R_2 can be attributed to variations on the slice S_2 of the common causal past. But more is needed for Reichenbach's methodology to work—namely, that for a variey of fixed background conditions on S_1 and S_3 the variations on S_2 do in fact account for the presence or absence of synchronizations of events on R_1 and R_2. This is a large piece of contingent physics that may or may not be true in the real world; even if it is true, it does not commend itself as a keystone of the methodology of explanation. Moreover, while the size of the problem may have been reduced by assigning the responsibility to the S_2 portion of S, the problem itself has not been resolved if the events on S_2 themsevles show the types of synchronicity we were concerned to explain in the first place. Reichenbach and his followers tend to avoid this difficulty by concentrating on punctal events as common causes. But again it is a contingent piece of physics that punctal events will serve.

The patterns of explanation of synchronicities encountered in relativistic theories are too broad to be accommodated by Reichenbach's common-cause model. The explanation of synchronicities due to the instantaneous constraint equations in Maxwell's theory is a relevant example. The explanation of global conservation laws is another. That the total value of a quantity Q is constant through time implies a kind of sympathetic harmony between distant events, since the values Q_1 and Q_2 associated with spatially separated parts of a two-part system must contravary so as to maintain a constant sum $Q_1 + Q_2$. The most common relativistic explanation of a global conservation law, and hence of the sympathetic harmony, is to seek a timelike vector quantity Q^i obeying the local conservation law $Q^i_{,i} = 0$. Then for a spatially localized system, with Q^i vanishing outside some compact region of space between the times t_1 and t_2, and application of Gauss' theorem yields[19]

$$Q(t_1) \equiv \int Q^i d\sigma_i(t_1) = Q(t_2) \equiv \int Q^i d\sigma_i(t_2).$$

The fundamental defect in Reichenbach's approach can now be stated. Choose your favorite locality principles. Mine are L1–L6, L10, and possibly some variants of L7 and L8. For ease of discussion, suppose for the moment that your preferences agree with mine. Then if the laws of nature conform to these principles, there is no action at a distance. The notion that the synchronicity of distant events has to be explained away à la Reichenbach, or else action at a distance threatens, is simply a false dichotomy. From my perspective, Reichenbach and his followers are guilty of conflating epistemology and ontology. Failure to find a common-cause explanation of the synchronicity of distant events may serve as *prima facie* evidence that the operative laws are not local. But the common-cause principle is not constitutive of locality, and if further investigation confirms that the operative laws fulfill the stipulated locality conditions, then expressing frustration over the failure to find a Reichenbachian common-cause explanation by muttering about action at a distance or nonlocality only serves to sow confusion.

At this juncture you may decide to break the initial agreement you made to take L1–L6 etc. as explicating locality, and you may add to the meaning of locality a further principle demanding that Reichenbachian explanations prevail. That is your privilege. But I would note that your additional principle bears no natural connection to any of the other locality principles encountered so far, and I would remind you that you have still not specified a definite principle until you have said what kinds of synchronizations of what sorts of events are at issue.

I conclude this section with some remarks about the probabilistic form of the PCC. If A and B are relatively spacelike events that are probabilistically dependent, i.e., $\Pr(A \& B) \neq \Pr(A) \cdot \Pr(B)$, then the PCC exhorts us to search for factors C in $C^-(A) \cap C^-(B)$ that induce conditional stochastic independence[20]:

$$\Pr(A \& B/C) = \Pr(A/C) \cdot \Pr(B/C). \tag{F}$$

If $\Pr(A/B)$ and $\Pr(B/C)$ are nonzero, we can infer from factorization (F) that

$$\Pr(A/C \& B) = \Pr(A/C),$$
$$\Pr(B/C \& A) = \Pr(B/C), \tag{S}$$

which says that the probabilistic common cause C screens off A from B and vice versa.

Drawing on the above discussion, I do not believe that a failure of F or S is necessarily indicative of action at a distance.[21] It might be thought, however, that a variant of S is constitutive of locality in the stochastic setting. Supposing that $\Pr(\cdot)$ stands for objective propensity probability, the locality advocate might want to demand, in analogy with Einstein locality (L10), that the propensity probability $\Pr(X)$ for any event in region R be determined by the conditions in some appropriate portion of $C^-(R)$. But, the reasoning goes, this demand cannot be met if $\Pr(A/C \& B) = \Pr(A/C)$ fails to hold when C is not confined to $C^-(A) \cap C^-(B)$ but is allowed to range over all the conditions in the appropriate portion of $C^-(A)$.[22] The reasoning is fallacious. The most direct analogue of L10 would require that the probability $\Pr(X)$ for any X in R is locally determined in the sense that agreement on any two models of the theory on the appropriate region of $C^-(R)$ forces agreement on $\Pr(X)$. The fact that this form of stochastic locality does not entail factorization or screening off will be relevant to our discussion of quantum locality.

Locality and Nonlocality in Quantum Physics

One cannot penetrate very far into the voluminous literature on the Einstein-Rosen-Podolsky paradox and Bell's inequalities without encountering rather breathless claims about quantum nonlocality. The experiments to test the Bell inequalities have been likened to the Michelson-Morley experiment (see Popper et al. 1981), the intimation being that we are on the brink of a new conceptual revolution in which quantum action at a distance may force revisions in the relativistic concepts of space, time, and causation. While not wishing to denying the seriousness of the issues, I have to say that a lack of a clear characterization of the alleged quantum action at a distance has served as a self-exciting inducement to extravagant pronouncements. But if the problem is obvious, the remedy is elusive; the goal of providing an intelligible characterization of the alleged nonlocality seems like a shimmering mirage that recedes as we try to advance toward it. If nothing else, I hope to be able to show why this is so. I begin with a superficial review of quantum mechanics, looking to see what elements of locality and nonlocality can be read from the surface structure of the theory.

In the approach to locality questions taken above, it was assumed that the theories under scrutiny are space-time theories. Ordinary quantum mechanics is not such a theory, though in special cases it can be made to look a bit like one. Consider the L_C^2 realization of the Hilbert space for a single spinless particle. The state vector is realized as a wave function $\psi(x,t)$, which can, with some fiddling, be regarded as a geometric object field on Newtonian space-time, or more precisely as part of such an object field. $\psi(x,t)$ does not satisfy the definition of a geometric object, since the transformation $\psi(x,t) \mapsto \psi'(x',t')$ from one Galilean chart (x,t) to another (x',t') depends not only on the value of ψ in the old chart (x,t) and the relation of (x,t) to (x',t') but also on the mass of the particle. (This fact can be seen as the origin of Bargmann's (1954) superselection rule for mass in nonrelativistic quantum mechanics, forbidding the superposition of different mass states.) But the pair ψ,m, where m is the "scalar mass field," is a geometric object.

The Schrödinger equation for ψ,m reveals the combination of localities and nonlocalities expected for a Newtonian field; as a local partial-differential equation the Schrödinger equation exhibits contiguity, but as a parabolic equation it permits infinitely fast disturbances in the ψ field, violating L1 just as the classical heat equation did.

So far, so good. But ordinary quantum mechanics provides no representation for quantum magnitudes (or "observables," as they are usually called) as space-time quantities. Thus, talk such as "Consider a local measurement in region R..." has no precise meaning in theory. And worse, for two or more interacting particles there is no way to fiddle ψ so as to make it part of a geometric-object field on space-time. The two-particle wave function $\psi(x_1,x_2,t)$ can be regarded as a geometric-object field on a seven-dimensional space; however, only in the trivial cases where the function factors into $\varphi(x_1,t) \cdot \kappa(x_2,t)$ can it be reinterpreted as a pair of geometric objects on real four-dimensional space-time. These violations of prelocality seem to block any further discussion of physical locality if the discussion is to proceed along the lines suggested above.

The case for locality in one-particle relativistic quantum mechanics is both better and worse than in the nonrelativistic case. In the relativistic setting we will want to demand L2. But then, on pain of contradicting L2, we cannot have localized states L5. For supposing, for sake of contradiction, that the one-particle state has compact spatial support at any instant, it follows that for positive-mass particles the wave function at any later time is spread over all space (see Hergerfeldt 1974). The Dirac wave equation is hyperbolic, so it should not violate L2, and so we should be able to infer that it is impossible to construct localized states from positive energy solutions, which indeed proves to be the case.

For the case of two or more interacting particles we can say nothing in relativistic quantum mechanics; the theory is essentially a one-particle theory. To discuss interacting particles requires the apparatus of quantum field theory. From the perspective of the first main section of this essay, this main advantage of quantum field theory is that it goes much farther than ordinary wave mechanics is putting quantum phenomena into a space-time format and thus opens the way for a discussion of quantum locality in terms of quantum analogues of the conditions developed above.

The violation of prelocality in wave mechanics virtually demands a hidden-variable interpretation, not in the traditional sense of the introduction of subquantum parameters designed to restore determinism or the like but rather in the sense of the proper use of quantum variables to restore enough pre-locality that questions of locality can be meaningfully posed. If, for purposes of illustration, a Newtonian theory employed a non-prelocal object $F(x_1, x_2, t)$, where x_1 and x_2 both range over all space, then one would guess that pre-locality was to be restored by the introduction of additional object fields $f_1(x_1, t)$ and $f_2(x_2, t)$ such that F can be interpreted as a functional of f_1 and f_2. In a way, this is what quantum field theory does for the quantum case. Here is the way Dirac envisioned the program in 1948:

A *localized* dynamical variable is a quantity which describes physical conditions at one point of space-time. Examples are field quantities and derivatives of field quantities. . . . A dynamical system in quantum theory will be defined as localizable if a representation for the wave function can be set up in which all the dynamical variables are localizable. . . . Let S denote any three-dimensional spacelike surface extending to infinity. We suppose a state of our dynamical system can be fixed by a wave function $\psi(q)$ involving variables which are all localized on S. The variables q may consist of certain discrete variables, denoting positions of particles on S . . . and may also consist of three-fold infinities of variables, denoting field quantities on S. If any three-fold infinities occur, then ψ is to be understood as a functional. (Dirac 1948, pp. 1092–1093)

The program of quantum field theory as it unfolded in the decades following Dirac's article did not quite conform to his vision. My discussion will be confined to the "axiomatic" versions of quantum field theory.[23]

The "local fields" of quantum field theory are not fields in the usual sense, nor are they entirely local. They are operator valued, and they exist only in a distributional sense. In the Wightman approach, for example, it is meaningless to ask for the value of a field at a space-time point p. One can, however, compute the smeared value of the field in as small a neighborhood of p as one likes by letting it act on test functions whose supports are contained in the

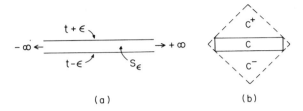

Figure 13

chosen neighborhood. Since the smearing has to be done in time as well as space, the Dirac at-a-given-time formulation is inappropriate.

Quantum field theory also implies the need to revise the classical concept of localized state. Let U and V be open regions of Minkowski space-time, and let $\mathscr{P}(U)$ and $\mathscr{P}(V)$ be the polynomials of Wightman fields smeared with test functions whose supports are contained respectively in U and V. If U and V are nonoverlapping, and especially if they are relatively spacelike, one might think that $P\psi_0$ and $P'\psi_0$, where ψ_0 is the vacuum state and $P\in\mathscr{P}(U)$ and $P'\in\mathscr{P}(V)$, are orthogonal, since the resulting states should be localized to U and V. But not so; in fact, for any open U, $\mathscr{P}(U)\psi_0$ spans the entire Hilbert space.[24]

Knight (1961) defines a state ψ of the quantum field A to be localized to a space-time region U just in case measurements made outside of U cannot distinguish ψ from the vacuum state, i.e.,

$$(\psi, A(f_1)A(f_2)\cdots A(f_n)\psi) = (\psi_0, A(f_1)A(f_2)\cdots A(f_n)\psi_0)$$

for any test functions f_i with no support on U. Knight shows that, in analogy to the result mentioned above for relativistic wave mechanics, localization for a free field is impossible with a finite number of particles. Licht (1963) showed that localized states need not form a linear space since the superposition of two localized states may not be localized.[25]

I turn now to various locality conditions that have been imposed on quantum fields. The Wightman axioms explicitly demand locality in the sense of commutativity; for boson fields A and B, this means that $[A(f), B(g)] = 0$ whenever the supports of f and g are relatively spacelike.[26] This axiom guarantees that measurements of field quantities at relatively spacelike positions can be made independently of one another.[27] The status of the commutativity axiom as a locality requirement can be further clarified by locating it in the network of other locality conditions.

Wightman's original axiom set did not imply any analogue of Laplacian determinism. Later writers took this to be a defect and proposed to remedy it by adding a "time-slice axiom" or a requirement of "primitive causality" (see Haag and Schroer 1962). Since the notion of fields at a given time is now suspect, determinism must be formulated by reference to a sandwich of events lying between two spacelike hypersurfaces. Thus, in figure 13a, S_ϵ is the infinitely extended open region sandwiched between $t_0 + \varepsilon$ and $t_0 - \varepsilon$, where ε is a positive number and t is some Lorentzian time coordinate. Primitive causality then demands that the fields $\mathscr{P}(S_\varepsilon)$ uniquely determine the fields over all space-time, or that the $\mathscr{P}(S_\varepsilon)$ form an irreducible set for the entire Hilbert space. The quantum-field-theoretic analogue of the classical principle of Einstein causality (L10) is discussed under such labels as the "diamond property." For instance, in figure 13b the fields $\mathscr{P}(C)$ on the slab C should uniquely determine the fields for the future C^+ and past C^- caps subtended by C. Haag and Schroer (1962) show that, as expected, primitive causality plus various technical assumptions entail the diamond property.

Other locality conditions in quantum field theory are usually discussed in terms of the *-algebra approach, which postulates that with each open set $U \subset \mathbb{R}^4$ with compact closure there is an associated algebra[28] $\mathscr{A}(U)$ of bounded operators and that the self-adjoint elements of $\mathscr{A}(U)$ correspond to local observables measureable on U. A statistical state ω for $\mathscr{A}(U)$ is an expectation-value assignment.[29] What is variously called "local independence," "causal independence," and "statistical independence" is the quantum-field-theoretic analogue of the classical requirement (L9) of independence for relatively spacelike regions. It requires that for any relatively spacelike U_1 and U_2 and any statistical states ω_1 and ω_2 for $\mathscr{A}(U_1)$ and $\mathscr{A}(U_2)$ respectively, there is a statistical state ω that extends the local states to the entire algebra $\mathscr{A}$ of observables, i.e., $\omega(X) = \omega_i(X)$ for each $X \in \mathscr{A}(U_i)$, where $i = 1, 2$.[30] This property implies what is called "extended locality": If U_1 and U_2 are relatively spacelike, then $\mathscr{A}(U_1) \cap \mathscr{A}(U_2)$ contains only multiples of the identity, so that, operationally, measurements in U_1 and U_2 must measure different things, apart from the identity.[31] It is known that statistical independence is neither necessary nor sufficient for commutativity (Ekstein 1969; De Facio 1974). But if $\mathscr{A}(U_1)$ and $\mathscr{A}(U_2)$ commute elementwise, then they are statistically independent iff $XY \neq 0$ for any nonzero $X \in \mathscr{A}(U_1)$ and $Y \in \mathscr{A}(U_2)$ (Roos 1970).

In sum, relativistic quantum field theory does pose some puzzles about locality, especially for the notion of a localized state. But if the program of local axiomatic quantum field theory (which admittedly, in its present state, boasts more axioms than quantitative results) succeeds in producing empirically adequate theories that conform to the various locality requirements reviewed above, then there will be a strong presumption in favor of no action at a distance in the quantum domain. This comforting presumption may have to give way in the face of the problem of quantum measurement and the EPR paradox.

Measurement, Miracles, and Quantum Action at a Distance

In the preceding subsection, some of the thorniest foundation problems in quantum mechanics were ignored. We must now confront these problems if we wish to understand recent discussions of quantum nonlocality.

According to the orthodox interpretation of quantum mechanics, a quantum observable A does not have a sharp or determinate value unless the state vector ψ is an eigenstate of the operator A representing A. A sharp value of A does emerge when an A measurement is made, and if prior to measurement ψ was not an eigenstate of A then the measurement process necessarily involves a collapse of ψ into an eigenstate of A. This collapse is, quite literally, a miracle in the sense of contravening the presumed laws of evolution. In ordinary quantum mechanics the point is that measurement collapse is inconsistent with the Schrödinger equation if measurement is treated as a physical interaction between an object system and a measurement apparatus.[32] Bringing in a conscious observer to accomplish the collapse[33] only trades one miracle for another.[34] In the relativistic setting, there is the additional difficulty of reconciling collapse (assumed to be instantaneous in ordinary quantum mechanics) with the relativity of simultaneity.

When the miracle of measurement is performed on correlated systems of particles, disturbing nonlocal effects can emerge, as is emphasized in the famous Einstein-Rosen-Podolsky paradox (Einstein et al. 1935). The Bohm (1951) version of the EPR construction will be discussed here, since it is tied more directly to the Bell inequalities. Two spin-$\frac{1}{2}$ particles, which may have interacted in the past, are now supposed to be widely separated in space. According to ordinary quantum mechanics it is possible to prepare such a composite system in a singlet state whose state vector is

$$\psi_s(1, 2) = \frac{1}{\sqrt{2}}(\psi_+(1)\psi_-(2) - \psi_-(1)\psi_+(2))$$

$$= \frac{1}{\sqrt{2}}(\psi'_+(1)\psi'_-(2) - \psi'_-(1)\psi'_+(2)),$$

where the unprimed ψ's are eigenvectors of spin up ($+$) along the x axis and spin down ($-$) along the x axis and the primed ψ's are eigenvectors of spin along the y axis. If a measurement of the x component of spin is made on particle 1, yielding the value up, the singlet state collapses to $\psi_+(1)\psi_-(2)$. If, alternatively, the y component of spin is measured on particle 1, giving a down result, the singlet state collapses to $\psi'_-(1)\psi'_+(2)$. Thus, on the orthodox interpretation, not only do different choices of measurement on particle 1 create different physical situations with regard to that particle, but different physical situations are thereby created with regard to particle 2. The latter creationism clashes with the presumption that the measurements performed on particle 1 are physical operations that affect only a limited part of space and have no direct influence on the physical reality in the distant part of space occupied by particle 2.[35] Just what form of action at a distance is in the offing is not readily apparent;[36] there is, for example, no possibility of using the correlations between the particles to send superluminal messages. But undeniably the upshot is, to use Einstein's word, spooky.

It is a curious business to try to draw conclusions about nonlocal actions from miracles, for even in an action-by-contact world it can be hard to localize the effects of even the most localized of miracles. Introduce a local miracle in the form of a local contravention of the laws of an otherwise local theory and dramatic nonlocal effects can result. Introduce a local creation of electric charge in Maxwell's theory of electromagnetism and the electromagnetic field must instantaneously adjust over all space. No firm conclusions can be drawn from miracles about action at a distance, or about anything for that matter.

The miraculous nature of quantum measurement, regardless of whether it is used to promote larger nonlocal miracles by means of composite systems, provides incentive enough for exploring alternatives to the orthodox interpretation. The alternative that comes most readily to mind is to view quantum mechanics as giving an incomplete description[37] of an essentially classical ontology of magnitudes with pre-existing determinate values and to view quantum measurement as a process of revealing the pre-existing values. The hope would be that these views can be filled in consistently with whatever cherished locality principles we choose to apply, and certainly with those locality principles that are required by relativity theory. Bell's theorems have been taken to dash this hope by showing that hidden-variable interpretations that are local in accord with the requirements of relativity theory cannot reproduce the verifiable statistics of quantum mechanics. I will argue that the moral of Bell's theorem is different but no less dark.

Bell's Inequalities and Locality

Suppose that, in a two-particle correlation experiment, particles 1 and 2 have associated with them respectively the operators A_a and B_b ($a, b = 1, 2$). The two A_a's are assumed to be noncommuting, as are the two B_b's, but the A_a's commute with the B_b's. Further, it is sufficient for present purposes to suppose that the four operators are bivalent, taking the values ± 1. A deterministic hidden-variable model[38] for the quantum state ψ and for the measurements of singleton observables and compatible pairs uses a total state space Λ; a normalized probability density $\rho(\lambda)$, where $\lambda \in \Lambda$; and response functions $A_a(\lambda)$ and $B_b(\lambda)$ giving the outcomes of the measurements as functions of the total state λ. The following constraints are imposed:

(i) *Spectrum condition.* $A_a(\lambda)$ and $B_b(\lambda)$ have the values ± 1 for all $\lambda \in \Lambda$.

(ii) *Condition of recovery of quantum-mechanical probabilities.* The hidden-variable probabilities (denoted by pr) for singleton observables, e.g.,

$$\mathrm{pr}(A_a = +1) = \int_\Lambda \chi[A_a^+](\lambda)\rho(\lambda)\,d\lambda,$$

and for pairs of compatible observables, e.g.,

$$\mathrm{pr}((A_a = +1) \& (B_b = -1)) = \int_\Lambda \chi[A_a^+](\lambda) \cdot \chi[B_b^-](\lambda)\rho(\lambda)\,d\lambda,$$

all agree with the quantum-mechanical probabilities for the state ψ.

Here $\chi[\ ]$ denotes the characteristic function of the relevant subspace of Λ; i.e., $\chi[A_a^+](\lambda) = 1$ if $A_a(\lambda) = +1$, and zero otherwise, etc. It follows from a purely number-theoretic inequality proved by Clauser and Horne (1974) that

$$-1 \leqslant \chi[A_1^+](\lambda) \cdot \chi[B_1^+](\lambda) + \chi[A_1^+](\lambda) \cdot \chi[B_2^+](\lambda)$$

$$+ \chi[A_2^+](\lambda) \cdot \chi[B_2^+](\lambda) - \chi[A_2^+](\lambda) \cdot \chi[B_1^+](\lambda)$$

$$- \chi[A_1^+](\lambda) - \chi[B_2^+](\lambda) \leqslant 0. \tag{3}$$

Then multiplying by $\rho(\lambda)$ and integrating over all Λ gives:

$$-1 \leqslant \mathrm{pr}((A_1 = +1) \& (B_1 = +1)) + \mathrm{pr}((A_1 = +1) \& (B_2 = +1))$$

$$+ \mathrm{pr}((A_2 = +1) \& (B_2 = +1)) - \mathrm{pr}((A_2 = +1) \& (B_1 = +1))$$

$$- \mathrm{pr}(A_1 = +1) - \mathrm{pr}(B_2 = +1) \leqslant 0. \tag{4}$$

Permutations produce other relations similar to (4), and the family of such relations is referred to collectively as the Bell-Clauser-Horne (BCH) inequalities. Fine (1982a) showed that the converse of the above result is also true: The BCH inequalities imply the existence of a deterministic hidden-variable model. The BCH inequalities are provably violated in some quantum states.

Nothing, in the above derivation of the BCH inequalities, was assumed about determinism in the proper sense of unique evolution through time; indeed, nothing about temporal evolution entered, and we are free to think of λ as characterizing the state of the system at the very instant of measurement. Nor does the derivation hinge on an assumption of physical locality in any sense identified in the first main section of this essay. We are free to imagine that the most viscous forms of Newtonian action at a distance are at work and still the BCH inequalities emerge. The only locality I can see to be involved is the semantic locality of treating the quantities represented by A_a and B_b as nonrelational. (Thus, it seems more accurate to call the model a determinate-value model rather than a deterministic model.) If, on the contrary, the quantities are relational, then some form of semantic action at a distance may be involved, but such a form of distant action may be no more shocking than the fact that I became a father at a distance when my son was born in Boston while I was in Washington.

The irrelevancies of determinism and physical locality strengthen the grounds for denying that quantum mechanics is incomplete in the straightforward sense in which classical statistical mechanics is incomplete. For if it were possible to interpret quantum mechanics as giving an incomplete description of a classical ontology of quantities with simultaneously determinate values, and if the quantum-mechanical probabilities were only reflections of our ignorance of these values, then it should be possible to construct a hidden-variable model of the above form. Bell's analysis has, however, foreclosed this possibility by demonstrating the clash between the BCH inequalities entailed by the model and quantum statistics, and the direct experimental verifications of quantum statistics (see Clauser and Shimony 1978) tend to foreclose the possibility that the blame lies with quantum mechanics rather than with the model.

Unable to force a strong form of incompleteness on quantum mechanics, we might still hope to enforce a weaker form according to which quantum probabilities to display values upon measurement are split into finer-grained probabilities conditioned on a hidden parameter. Thus, a stochastic hidden-variable model eschews deterministic response functions in favor of the more modest postulation of probability distributions $\text{pr}(A_a/\lambda)$, $\text{pr}(B_b/\lambda)$, and $\text{pr}(A_a \& B_b/\lambda)$ which return the correct quantum-mechanical probabilities when averaged over λ—e.g.,

$$\int_\Lambda \text{pr}(A_a = +1/\lambda)\rho(\lambda)\,d\lambda = \text{Pr}^\psi_{A_a}(\{+1\}).$$

The subclass of *factorizable* models are those that satisfy $\text{pr}(A_a \& B_b/\lambda) = \text{pr}(A_a/\lambda) \cdot \text{pr}(B_b/\lambda)$ for all $\lambda \in \Lambda$. The proof that factorizable stochastic models entail the BCH inequalities parallels the above derivation for deterministic models. For those who equate factorizability with locality,[39] the impossibility of factorizable stochastic models for quantum mechanics pinpoints a sense in which quantum mechanics is nonlocal.

I have seen no persuasive reason for the equation. I would resist an appeal to Reichenbach's PCC as a justification for taking factorizibility as a locality principle, for (as argued above) the PCC is not constitutive of local action. I would accept as a legitimate locality demand the requirement that, insofar as probabilities refer to objective propensities rather than epistemic uncertainties, the probabilities of outcomes made in a region R should be determined by the conditions in $C^-(R)$ (or, more stringently, by the conditions on a slice or slab across $C^-(R)$). But, as already noted, the way to express the relevant sense of 'determined' is not by factorizability but by the condition that the agreement of any two models of the theory on the appropriate region of $C^-(R)$ forces agreement on the propensity probabilities for measurement outcomes in R. Barring measurement collapse, this condition is one which any adequate local relativistic quantum field theory should satisfy—although, of course, without satisfying factorizability and the BCH inequalities.[40] The condition will fail for local quantum field theory if the models are broken by introducing miraculous measurement collapse, but that failure does not undercut the point that factorizability is not a consequence of the most natural locality requirement which the setting calls for.

Returning to the stochastic hidden-variable models: There remains the fact that nothing in the derivation of the BCH inequalities from factorizability requires that λ refer to conditions in the past light cones of the measurement events or that the hidden parameters propagate or interact in accord with any recognizable principle of physical locality. The impossibility result is that quantum mechanics admits of no factorizable model, local or nonlocal. The impossibility springs from the the same roots as the impossibility for a deterministic (or determinate value) model; since the BCH inequalities entail the existence of a deterministic model, factorizability guarantees the existence of a deterministic model.[41]

Thus far the Bell theorems have given us some negative information but very little positive information about quantum locality. The theorems show that some attempts to draw the spookiness of the correlation experiments are non-starters, but they have not helped to pinpoint the nature of the threatened action at a distance. If the Bell inequalities were indeed the key to understanding quantum nonlocality, there should be results of the form (locality + X) $\Rightarrow$ Bell inequalities. The weaker the auxiliary assumptions X, the stronger and more informative the result. The results examined so far are not of this form, since the inequalities emerge from the various X's regardless of whether there is any recognizable form of physical locality at work. In the next subsection I describe an attempt to derive a strong result of the desired form.

Other Bell Arguments

Using a construction due to Stapp (1971), Eberhard (1977) and Nordin (1979) argue that a form of Bell's inequalities can be derived directly from the demand that the outcome of a measurement on one wing of the correlation experiment be independent of the setting of the measurement device on the other wing.[42] Before I proceed to the argument itself, it is necessary to distinguish two forms of the independence demand.

It is supposed that on each wing there is a measuring device with a knob that admits of two settings. The settings correspond to internal states of the devices appropriate to measuring A_1 and A_2 on the left and B_1 and B_2 on the right. On the first form of the independence demand, the probability of an outcome on one wing is to be independent of the knob setting on the other wing. This condition has sometimes been called "signal locality," for it would seem that if it failed a message could be sent from one wing to the other by switching the knob one setting to the other and observing the difference in the frequency counts of measurement results at the other wing (see Eberhard 1978 and Jarrett 1984). However, "signal locality" is not sufficient for signal locality in sense of the first main section of this essay, for ordinary nonrelativistic quantum mechanics is "signal local" even though it permits infinitely fast propagation of disturbances. Ordinary quantum mechanics simply assumes that the operators A_a and B_b associated with different particles commute, and as a result quantum mechanics defines a joint probability $\text{Pr}^\psi_{A_a, B_b}(S, T)$ for joint outcomes lying in (S, T). It follows immediately and trivially that

$$\text{Pr}^\psi_{B_b}(T) = \text{Pr}^\psi_{A_1, B_b}(\{+1, -1\}, T)$$

$$= \text{Pr}^\psi_{A_2, B_b}(\{+1, -1\}, T),$$

which is a way of saying that the probability of an outcome of measuring B_b on the right is independent of whether A_1, or A_2, or nothing is measured on the left. Moreover, the fact that commutativity implies "signal locality" should make one leery of saying that a violation of "signal locality" means that messages can be sent from one wing to the other, for to take advantage of such a violation would involve simultaneous measurement of noncommuting operators—an idea that is suspect on other grounds. We have yet another illustration of the difficulty, if not the hopelessness, of expressing basic ideas about physical locality in terms of conditional probability statements.

The second form of the independence demand, and the one actually used in the Stapp-Eberhard-Nordin construction, requires that individual outcomes of measurement on one wing be independent of the setting of the measuring device on the other wing. The relevant sense of independence is called "counterfactual definiteness." To illustrate, suppose that in fact the pair A_2, B_1 was measured, yielding $+1$, -1. Counterfactual definiteness demands that, had A_1 been measured instead (or had no measurement been made on the left), the outcome of measuring B_1 on the right would still have been -1. The Stapp-Eberhard-Nordin proof shows that another form of the Bell inequalities, first derived by Clauser et al. (1969), follows from counterfactual definiteness and from the assumption that the frequency counts in a large ensemble of measurements approximate the true values. Eberhard and Nordin took this demonstration to show that quantum mechanics, independent of considerations of hidden variables and determinism, is nonlocal; Nordin took the nonlocality to be in potential conflict with relativity theory.

Counterfactuals are catsup of philosophy. Just as greasy fast food calls for catsup, so greasy issues about causal influence call for counterfactuals. My worry here is not simply a reflection of a general wariness about counterfactual talk, though I admit to that. To give the worry a more concrete form, recall from above that Maxwell's theory of electromagnetism entails lawlike constraints on instantaneous data at relatively spacelike positions. To the extent that laws entail counterfactuals, such constraints would seem to imply that it is false that the magnetic flux in one region would have been the same even if the flux in an appropriately chosen relatively spacelike region had been different. But the failure of counterfactual definiteness here warrants no fear of action at a distance. My point is that one cannot hope to read lessons about locality and action at a distance merely by staring hard at counterfactuals. If we know the details of the underlying dynamics and can apply the locality principles developed above, then we can decide issues about locality without ever having to resort to counterfactuals. In between, we are left to conjure with plausibility arguments.

In an irreducibly stochastic world it is implausible that counterfactual definiteness manages to capture any interesting sense of locality. Imagine a counterfactual scenario in which the experiment is run over again, this time with A_1, B_1 being measured instead of A_2, B_1. Since determinism is not at work, we are free to suppose that the starting state in the counterfactual scenario is the same as that in the actual experiment. But even so, there is no good reason to think that this state will eventuate in the same outcome (-1) for B_1—otherwise the assumption of irreducible stochasticity is not being taken seriously.[43] Thus, counterfactual definiteness will fail quite generally in stochastic worlds, whether or not there is something we wish to call action at a distance.

Under the alternative assumption of determinism, the fate of counterfactual definiteness is less apparent. If we assume complete determinism, with the settings on the measurement devices being determined along with everything else, then the starting state in the counterfactual scenario must be different from the actual starting state if the A-measurement setting is different but the laws are the same. And the different starting state may determine a different outcome for B_1. Counterfactual definiteness can, perhaps, be made plausible by delimiting the form of determinism and by specializing the hardware of the experiment.[44] Thus, suppose that determinism is of the local relativistic form, with the state on a region R determined by the state on a slice contained in $C^-(R)$. Suppose further that the A-measuring device is rigged so that the setting at the measurement event in question turns on causal chains that never enter the past light cone of the B-measurement event in question (e.g., the setting could be made to depend on the frequency of a source-free light ray from a direction chosen so that the space-time trajectory of the ray is disjoint from the past cone of the B-measurement). Then, plausibly, even if the A-setting had been different (because, contrary to fact, a light ray of a different frequency had been received), the state on any slice in the past cone of the B-measurement event would still have been the same, and thus by local determinism the outcome of the B_1 measurement would still have been -1. But then what we have is a plausibility argument against local relativistic determinism and not an argument against locality *per se*. And if in the face of such an argument we choose to abandon locality in order to save determinism, there are, as we saw above, nonlocal forms of determinism that are compatible with Lorentz invariance.

Conclusion

There is good news and bad. The good news is that we have found no justification for the alarmist pronouncements that the EPR-Bell experiments threaten the relativistic conception of space-time. Nor do the Bell theorems preclude the possibility of a local explanation of the correlations.[45] The bad news is that half a century after the formulation of the EPR paradox we have made little progress toward understanding the spookiness of which Einstein complained. The lack of progress is due to the fact that the spookiness traces largely if not entirely to the miracle of measurement collapse. Without a coherent theory to test for local and nonlocal features, we are reduced to hand-waving and speculation. Even if our speculations should lead to a coherent theory of quantum phenomena, we should be prepared for the fact that our questions about local and nonlocal actions in the quantum domain may be rendered moot. For unless the final theory can be put in a form that allows the application of principles like those developed in the first part of this essay, there is no firm basis for assertions about local versus distant action—at least if these assertions are to have meanings anything like those enjoyed in Newtonian and classical relativistic theories.[46]

Notes

1. For an antidote, see Stein 1970a, 1977.

2. For a critical assessment of Reichenbach's PCC and the uses to which it is put, see van Fraassen 1980, 1982a, 1982b.

3. For a sampling of views on this matter, see Bell 1965; Stapp 1971; Eberhard 1977, 1978; Shimony 1978, 1980; Nordin 1979; Hellman 1982a.

4. In a letter to Max Born dated March 3, 1947, Einstein began with a conciliatory remark: "I admit, of course, that there is a considerable amount of validity in the statistical approach which you were the first to recognize clearly as necessary given the framework of the existing formalism." But then he added: "I cannot seriously believe in it because the theory cannot be reconciled with the idea that physics should represent a reality in time and space, free from spooky actions at a distance." (Born 1971, p. 158)

5. These matters are discussed in detail in Earman 1986.

6. Familiar examples of geometric object fields are scalar, vector, and tensor fields. For a precise definition of the general concept, see Schouten 1954.

7. If space had an even number of dimensions, Huygens' principle would fail for the wave equation, and as a result sharp signals would be impossible—the disturbance would "ring on" long after the source was turned off.

8. The notions of 'field' and 'field theory' are discussed in the following subsection.

9. The source-free heat equation reads

$$\nabla^2 \phi = \kappa \frac{\partial \phi}{\partial t},$$

where κ is the coefficient of thermometric conductivity. The definitive discussion of this equation is to be found in Widder 1975.

10. For a sampling of various approaches and opinions, see Hesse 1961, 1967; Stein 1970b; Cantor and Hodge 1981.

11. See also Maxwell 1890b.

12. Thus, in reviewing pre-Faraday theories of the electric field, Heilbron (1981) states that "a field and its representation must act by local forces, by forces exerted between contiguous elements of space of ether, between a particle and its nearest neighbor" and that "mediation by a classical field takes time." Faraday himself used finite time of transmission to distinguish fields from action-at-a-distance forces; see Hesse 1961, 1967.

13. Faraday demanded that the transmission of action by a true field be independent of receptors; see Hesse 1961, 1967.

14. This definition assumes that we are working with time-orientable space-times and that the future orientation has been singled out.

15. It is usually assumed also that S is achronal, i.e., is not intersected more than once by the same future-directed causal curve.

16. See Earman 1986 for a discussion of this and other problems in formulating determinism in general relativity theory.

17. This issue is bound up with what constitutes an "absolute object"; see Anderson 1967; Jones 1981; Friedman 1983.

18. Examples of this type have been urged by Patrick Suppes.

19. Here $d\sigma_i(t)$ is the future-oriented surface element of the hypersurface $t = $ constant.

20. Reichenbach also required that $\Pr(A \, \& \, B/\bar{C}) = \Pr(A/\bar{C}) \, \Pr(B/\bar{C})$. See Reichenbach 1971, pp. 159 ff., for a full discussion of the requirements of common cause for a "conjunctive fork."

21. As I understand his latest position, Salmon agrees; see Salmon 1984, Chapter 6.

22. Bell (1976) seems to rely on such reasoning.

23. For reviews of this approach, see Streater and Wightman 1964; Streater 1975; Bogolubov et al. 1975.

24. In the technical jargon, $\mathscr{P}(U)$ is cyclic in the vacuum; this was first proved by Reeh and Schlieder (1962).

25. Licht's approach is based on Knight 1961 but is slightly different. Licht (1963) says that a state ψ is localized outside U just in case $(\psi, \hat{A}\psi) = (\psi_0, \hat{A}\psi_0)$, where $\hat{A}$ is any member of the ring of bounded operators generated by the projection operators associated with the field $A(x)$ in U.

26. Actually what is wanted is commutativity of bounded functions of $A(f)$ and $B(g)$, but this does not quite follow from the Wightman axioms. The *-algebra approaches mentioned below were motivated in part by this difficulty.

27. On this point, see the discussion below.

28. It is sometimes assumed that the algebra is a C^* algebra, sometimes that it is a von Neumann algebra. The difference lies in the topology in which the completion is taken. See De Facio 1974 for cases where the difference can have important consequences.

29. ω is a linear functional on $\mathscr{A}(U)$, assigning the value 1 to the unit element which is assumed to belong to each $\mathscr{A}(U)$.

30. For a review of these and other locality conditions in axiomatic quantum field theory, see Avishai and Ekstein 1971 and De Facio 1974.

31. Landau (1969) gives a sufficient condition for extended locality.

32. For some sharp results on this point, see Fine 1970 and Shimony 1974.

33. As suggested by Wigner (1961).

34. There are other ways to escape the problem; e.g., replace the Schrödinger equation by a nonlinear equation of motion (see Pearle 1969).

35. See Einstein 1948, p. 172. This reference provides a clearer and more accessible version of Einstein's argument than the more often cited paper, Einstein et al. 1935.

36. Here is the conception of locality applied by Einstein:

If one asks what, irrespective of quantum mechanics, is characteristic of the world of ideas of physics, one is first of all struck by the following: the concepts of physics relate to a real outside world, that is ideas are established relating to things such as bodies, fields, etc., which claim a 'real existence' that is independent of the perceiving subject—ideas which, on the other hand, have been brought into as secure a relationship as possible with the sense-data. It is further characteristic of these physical objects that they are thought of as arranged in a space-time continuum. An essential aspect of this arrangement of things in physics is that they lay claim, at a certain time, to an existence independent of one another, provided these objects 'are situated in different parts of space'. Unless one makes this kind of assumption about the independence of the existence (the 'being thus') of objects which are far apart from one another in space—which stems in the first place from everyday thinking—physical thinking in the familiar sense would not be possible. It is also hard to see any way of formulating and testing the laws of physics unless one makes a clear distinction of this kind. This principle has been carried to extremes in the field theory by localising the elementary objects on which it is based and which exist independently of each other, as well as the elementary laws which have been postulated for it, in the infinitely small (four-dimensional) elements of space.

The following idea characterises the relative independence of objects far apart in space (A and B): external influence on A has no direct influence on B; this is known as the 'principle of contiguity', which is used consistently only in the field theory. If this axiom were to be completely abolished, the idea of the existence of (quasi-)enclosed systems, and thereby the postulation of laws which can be checked empirically in the accepted sense, would become impossible. (Einstein 1948, pp. 170–171)

I trust that the reader of the first section of this essay can detect the multiple ambiguities in Einstein's discussion. Einstein did not attempt to say specifically which aspect of his conception of locality was threatened in the EPR paradox, and this tradition continued in much of the subsequent discussion.

37. Einstein (1948) argues that quantum mechanics either is incomplete or else involves action at a distance. In the context of the Bohm version of the EPR paradox, Einstein would take the assertion that ψ provides a "complete description" to mean that in the singlet state the particles do not have definite spins along the x and y axes and that the

sharply defined spin values, obtained by measuring spin, cannot be interpreted as values possessed prior to measurement (see Einstein 1948, p. 169). Thus, the assumption of completeness implies that different choices of measurement on particle 1 create different spin values for particle 2—a seeming violation of locality. Actually, Einstein puts the point in a slightly different manner. Completeness implies that "two ψ-functions which differ in more than trivialities always describe two different real situations" (p. 170). Thus, depending on the choice of measurement performed on particle 1, a different real situation is created in regard to particle 2 (p. 172). Einstein then fleshes out the argument in two steps. First, he asserts that locality entails the principle that every statement about particle 2 that is arrived at by means of a complete measurement on particle 1 has to be valid even if no measurement is made on particle 1. Second, he argues that this principle implies that all the statements that can be deduced from the settlement of $\psi_+ (2)$, $\psi'_- (2)$, etc. must be simultaneously valid for particle 2, which contradicts the completeness of ψ.

Einstein could have stated his dilemma of incompleteness or action at a distance in terms of a one-particle system; in fact, initially he did (see Einstein 1928). Consider what happens when a position is measured on a particle and the particle is found to be in the interval Δx. The initial wave function $\psi(x,t)$, which, we may assume, had support on all space, collapses to a $\psi'(x, t)$, which has support only on Δx. On the assumption that ψ provides a complete description, this collapse as a result of the measurement in Δx looks like action at a distance. However, the use of a composite system makes the argument more compelling. Einstein assumed that the two parts of the composite system can be localized and separated in space at a time t in that the composite wave function $\psi(x_1, x_2, t)$ is 0 unless x_1 belongs to a limited region R_1 of space and x_2 belongs to a limited region R_2 separated from R_1 (Einstein 1948, p. 171). Under these circumstances it is hard to deny that, whatever disturbance the measurement of particle 1 causes in R_1, the situation in R_2 is not thereby affected. It is also easier to apply to the composite system the style of argument outlined at the end of the last paragraph.

The Bell theorems described below are traditionally presented for two-particle systems. The theorems require certain kinds of operators and quantum states; I believe it should be possible to realize them for a one-particle system.

38. The terminology and the formalism used below follow Fine 1982a. For reasons which will soon become apparent, I would prefer to call this a determinate-value model rather than a deterministic model.

39. Proponents of equating factorizability with locality are Clauser and Horne (1974), Clauser and Shimony (1978), and Wessels (1985). Opponents include Suppes and Zannoti (1976) and Fine (1981).

40. See Hellman 1982a for a detailed discussion of the point that local stochastic determinism does not entail factorizability.

41. Proofs that factorizability is a necessary and sufficient condition for a deterministic (or, as I would prefer to say, a determinate-value) model can be found in Fine 1982a, Fine 1982b, and Suppes and Zannoti 1981.

42. In the paper in which the Bell inequalities were first derived, Bell concluded: "In a theory in which parameters are added to quantum mechanics to determine results of individual measurements without changing the statistical predictions, there must be a mechanism whereby the setting of one measuring device can influence the reading of another instrument, however remote. Moreover, the signal involved must propagate

instantaneously, so that the theory could not be Lorentz invariant." (Bell 1965, p. 199) The Stapp-Eberhard-Nordin argument is supposed to show that the same conclusion emerges even when hidden variables are not used.

43. See Hellman 1982a and Redhead 1983 for a detailed discussion of this point.

44. Hellman (1982b) argues that Einstein locality (L10) entails the Bell inequalities and, therefore, that the experimental tests confirming the quantum-mechanical statistical predictions provide direct empirical evidence against local relativistic determinism. I do not believe that this conclusion follows without specialized assumptions about the measurement set up.

45. See Fine 1982c for an example of a possible local explanation of the correlation experiments.

46. This paper was completed in the spring of 1985. Since then a number of authors have helped to clarify the problem of locality in quantum mechanics. I would especially call attention to the following papers: L. E. Balentine and J. P. Jarrett, Bell's Theorem: Does Quantum Mechanics Contradict Relativity? (unpublished); J. Hellman, EPR, Bell, and Collapse: A Route Around "Stochastic" Hidden Variables (unpublished); D. Howard, "Einstein on Locality and Separability," *Stud. Hist. Phil. Sci.* 16 (1985): 171–201; D. Howard, Locality, Separability, and the Physical Implications of the Bell Experiments: A New Interpretation (unpublished); H. Krips, Quantum Measurement and Bell's Theorem (unpublished); M. Redhead, Relativity and Quantum Mechanics—Conflict or Peaceful Coexistence (unpublished). Although I will have to modify the opinions expressed in the second section of this paper to take account of these contributions, I believe that the essay stands as a useful approach to the topic.

References

Anderson, J. L. 1967. *Principles of Relativity Physics.* New York: Academic.

Avishai, Y., and H. Ekstein. 1971. Causal Independence. *Found. Phys.* 2: 257–270.

Avishai, Y., H. Ekstein, and J. E. Moyal. 1972. Is the Maxwell Field Local? *J. Math. Phys.* 13: 1139–1145.

Bargmann, V. 1954. On Unitary Ray Representations of Continuous Groups. *Ann. Math.* 59: 1–46.

Bell, J. S. 1965. On the Einstein-Podolsky-Rosen Paradox. *Physics* 1: 195–200.

Bell, J. S., 1976. The Theory of Local Beables. *Epistemological Letters,* March: 11–24.

Bogolubov, N. M., A. A. Logonov, and I. T. Todorov. 1975. *Introduction to Axiomatic Quantum Field Theory.* Reading, Mass.: Benjamin.

Bohm, D. 1951. *Quantum Theory.* Englewood Cliffs., N.J.: Prentice-Hall.

Born, M. 1971. *The Born-Einstein Letters.* London: Macmillan.

Cantor, G. N., and M. J. S. Hodge, eds. 1981. *Conceptions of the Ether.* Cambridge University Press.

Clauser, J. F., and M. A. Horne. 1974. Experimental Consequences of Objective local theories. *Phys. Rev.* D 10: 526−535.

Clauser, J. F., and A. Shimony. 1978. Bell's Theorem: Experimental Tests and Consequences. *Rep. Prog. Phys.* 41: 1881−1927.

Clauser, J. F., M. A. Horne, A. Shimony, and R. A. Holt. 1969. Proposed Experiment to Test Local Hidden-Variable Theories. *Phys. Rev. Lett.* 23: 880−883.

Currie, D. G. 1966. Poincaré Invariant Equations of Motion for Classical Particles. *Phys. Rev.* 142: 817−826. Reprinted in Kerner 1972.

De Facio, B. 1974. Causal Independence in Algebraic Quantum Field Theory. *Found. Phys.* 5: 229−237.

Dirac, P. A. M. 1948. Quantum Theory of Localizable Dynamical Systems. *Phys. Rev.* 73: 1092−1103.

Driver, R. D. 1969. A 'Backwards' Two-Body Problem of Classical Relativistic Electrodynamics. *Phys. Rev.* 178: 2051−2067.

Driver, R. D. 1977. *Ordinary and Delay Differential Equations.* New York: Springer-Verlag.

Earman, J. 1986. *A Primer on Determinism.* Reidel.

Eberhard, P. H. 1977. Bell's Theorem Without Hidden Variables. *Nuo. Cim.* 38 B: 75−79.

Eberhard, P. H. 1978. Bell's Theorem and the Different Concepts of Locality. *Nuo. Cim.* B 46: 392−414.

Einstein, A. 1928. Causalité, Determinisme, Probabilité. In *Electrons et Photons.* Paris: Gauthier-Villars.

Einstein, A. 1948. Quanten-Mechanik und Wirklichkeit. *Dialectica* 2: 320−324. English translation: Born 1971, pp. 169−173.

Einstein, A., B. Podolsky, and N. Rosen. 1935. Can Quantum Mechanical Description of Reality Be Considered Complete? *Phys. Rev.* 47: 777−780.

Ekstein, H. 1969. Presymmetry. II. *Phys. Rev.* 184: 1315−1337.

Fine, A. 1970. Insolubility of the Quantum Measurement Problem. *Phys. Rev.* D 2: 2783−2787.

Fine, A. 1981. Correlations and Physical Locality. In P. Asquith and R. Giere (eds.), *PSA 1980*, vol. 2. East Lansing, Mich.: Philosophy of Science Association.

Fine, A. 1982a. Hidden Variables, Joint Probabilities, and the Bell Theorems. *Phys. Rev. Lett.* 48: 291−295.

Fine, A. 1982b. Joint Distributions, Quantum Mechanics, and Commuting Observables *J. Math. Phys.* 23: 1306−1310.

Fine, A., 1982c. Some Local Models for Correlation Experiments. *Synthese* 50: 279−294.

Friedman, M. 1983. *Foundations of Space-Time Theories*. Princeton University Press.

Geroch, R., and G. T. Horowitz. 1979. Global Structures of Space-Times. In *General Relativity*, S. W. Hawking and W. Israel, eds. Cambridge University Press.

Haag, R., and B. Schroer. 1962. Postulates of Quantum Field Theory. *J. Math. Phys.* 3: 248–261.

Hawking, S. W., and G. F. R. Ellis. 1973. *The Large Scale Structure of Space-Time*. Cambridge University Press.

Heilbron, J. 1981. The Electrical Field Before Faraday. In Cantor and Hodge 1981.

Hellman, G. 1982a, Stochastic Einstein Locality and the Bell Inequalities. *Synthese* 53: 461–504.

Hellman, G. 1982b. Einstein and Bell: Strengthening the Case for Microphysical Randomness. *Synthese* 53: 445–460.

Hergerfeldt, G. C. 1974. Remark on Causality and Particle Localization. *Phys. Rev.* D 10: 3320–3321.

Hesse, M. B. 1961. *Forces and Fields*. London: T. Nelson.

Hesse, M. B. 1967. Action at a Distance and Field Theory. In *The Encyclopedia of Philosophy*, vol. 1. New York: Macmillan and Free Press.

Hill, R. N., 1967a. Instantaneous Action-at-a-Distance in Classical Relativistic Mechanics. *J. Math. Phys.* 8: 201–220.

Hill, R. N., 1967b. Canonical Formulation of Relativistic Mechanics. *J. Math. Phys.* 8: 1756–1773.

Hill, R. N., 1982. The Origins of Predictive Relativistic Mechanics. In J. Llosa, ed., *Relativistic Action at a Distance: Classical and Quantum Aspects*. New York: Springer-Verlag.

Jarrett, J. P. 1984. On the Physical Significance of the Locality Conditions in the Bell Arguments. *Noûs* 18: 569–589.

Jones, R. 1981. Is General Relativity Generally Relativistic? In P. Asquith and R. Giere, eds., *PSA 1980*, vol. 2. East Lansing, Mich.: Philosophy of Science Association.

Kerner, E. H., ed. 1972. *The Theory of Action-at-a-Distance in Relativistic Particle Mechanics*. New York: Gordon and Breach.

Knight, J. L. 1961. Strict Localization in Quantum Field Theory. *J. Math. Phys.* 2: 459–471.

Landau, L. J. 1969. A Note on Extended Locality. *Comm. Math. Phys.* 13: 246–253.

Licht, A. L. 1963. Strict Localization. *J. Math. Phys.* 4: 1443–1447.

Maxwell, J. C. 1890a. Attraction. In D. V. Niven, ed., *The Scientific Papers of James Clerk Maxwell*, vol. 2. Cambridge University Press.

Maxwell, J. C. 1890b. On Action at a Distance. Ibid.

Nordin, I. 1979. Determinism and Locality in Quantum Mechanics. *Synthese* 42: 71–90.

Ohanian, H. 1976. *Gravitation and Space-Time*. New York: Norton.

Pearle, P. 1969. Reduction of the State Vector by a Non-Linear Schrödinger Equation. *Phys. Rev.* D 13: 857–864.

Popper, K., A. Garuccio, and J.-P. Vigier. 1981. An Experiment to Interpret EPR Action-at-a-Distance. *Epistemological Letters*, July: 21–29.

Redhead, M. 1983. Non Locality and Peaceful Coexistence. In R. Swinburne, ed., *Space, Time and Causality*. Dordrecht: Reidel.

Reeh, H., and S. Schlieder. 1962. Bemerkungen zur Unitäräquivalenz von Lorentzinvarienten Felden. *Nuo. Cim.* 22: 1051–1068.

Reichenbach, H. 1958. *Space and Time*. New York: Dover.

Reichenbach, H. 1959. *Modern Philosophy of Science*. London: Routledge and Kegan Paul.

Reichenbach, H. 1971. *The Direction of Time*. Berkeley: University of California Press.

Roos, H. 1970. Independence of Local Algebras in Quantum Field Theory. *Comm. Math. Phys.* 16: 238–246.

Salmon, W. 1975. Theoretical Explanation. In S. Körner, ed., *Explanation*. New Haven: Yale University Press.

Salmon, W. 1978. Why Ask Why? *Proc. Amer. Phil. Assoc.* 51: 638–705.

Salmon, W. 1984. *Scientific Explanation and the Causal Structure of the World*. Princeton University Press.

Schouten, J. A. 1954. *The Ricci-Calculus*, second edition. New York: Springer-Verlag.

Shimony, A. 1974. Approximate Measurement in Quantum Mechanics, II. *Phys. Rev.* D 9: 2321–2323.

Shimony, A. 1978. Metaphysical Problems in the Foundations of Quantum Mechanics. *Inter. Phil. Quart.* 18: 4–17.

Shimony, A. 1980. The Point We Have Reached. *Epistemological Letters*, June: 1–6.

Stapp, H. P. 1971. S-Matrix Interpretation of Quantum Theory. *Phys. Rev.* D 3: 1303–1320.

Stein, H. 1970a. Newtonian Space-Time. In R. Palter, ed., *The Annus Mirabilis of Sir Isaac Newton*. Cambridge, Mass.: MIT Press.

Stein, H. 1970b. "On the Notion of Field in Newton, Maxwell and Beyond. In R. Stuewer, ed., *Minnesota Studies in the Philosophy of Science*, vol. 5. Minneapolis: University of Minnesota Press.

Stein, H. 1977. Some Pre-History of General Relativity. In J. Earman, C. Glymour, and J. Stachel, eds., *Minnesota Studies in the Philosophy of Science*, vol. 8. Minneapolis: University of Minnesota Press.

Streater, R. F. 1975. Outline of Axiomatic Relativistic Quantum Field Theory. *Rep. Prog. Phys.* 38: 771–846.

Streater, R. F., and A. S. Wightman. 1964. *PCT, Spin, Statistics and All That*. New York: Benjamin.

Suppes, P., and M. Zanotti. 1976. On the Determinism of Hidden Variable Theories with Strict Correlation and Conditional Independence of Observables. In P. Suppes, ed., *Logic and Probability in Quantum Mechanics*. Dordrecht: Reidel.

Suppes, P., and M. Zanotti. 1981. When Are Probabilistic Explanations Possible? *Synthese* 48: 191–199.

Thomas, L. H. 1952. The Relativistic Dynamics of a System of Particles Interacting at a Distance. *Phys. Rev.* 85: 846–872.

Van Dam, H., and E. P. Wigner. 1965. Classical Relativistic Mechanics of Interacting Point Particles. *Phys. Rev.* 138: 1576–1582. Reprinted in Kerner 1972.

Van Dam, H., and E. P. Wigner. 1966. Instantaneous and Asymptotic Conservation Laws for Classical Relativistic Mechanics of Interacting Point Particles. *Phys. Rev.* 142: 838–843. Reprinted in Kerner 1972.

van Fraassen, B. C. 1980 *The Scientific Image*. Oxford: Clarendon.

van Fraassen, B. C. 1982a. The Charybdis of Realism. *Synthese* 52: 25–38.

van Fraassen, B. C. 1982b. Rational Belief and the Common Cause Principle. In R. McLaughlin, ed., *What? Where? When? Why?* Dordrecht: Reidel.

Wessels, L. 1985. Locality, Factorability and the Bell Inequalities. *Noûs* 19: 481–520.

Wheeler, J. A., and R. P. Feynman. 1945. Interaction with the Absorber as a Mechanism for Radiation. *Rev. Mod. Phys.* 17: 157–181. Reprinted in Kerner 1972.

Wheeler, J. A., and R. P. Feynman. 1949. Classical Electrodynamics in Terms of Direct Interparticle Action. *Rev. Mod. Phys.* 21: 425–433. Reprinted in Kerner 1972.

Whitrow, G. J., and G. E. Morduch. 1965. Relativistic Theories of Gravitation. In A. Beer, ed., *Vistas in Astronomy*, vol. 6. Elmsford, N.Y.: Pergamon.

Widder, D. V. 1975. *The Heat Equation*. New York: Academic.

Wigner, E. P. 1961. Remarks on the Mind-Body Problem. In I. J. Good, ed., *The Scientist Speculates*. London: Heinemann.

Arthur Fine

With Complacency or Concern: Solving the Quantum Measurement Problem

At the conclusion of a Stern-Gerlach experiment, after the beam of atoms passes through the inhomogeneous magnetic field, the beam is split. How does this come about? More generally, how does it ever come about that quantum-measurement procedures actually produce definite results? This is the quantum-measurement problem, whose central difficulty can be illustrated by an example as simple as the statement of the problem itself:

The system is a substance in chemically unstable equilibrium, perhaps a charge of gunpowder that, by means of intrinsic forces, can spontaneously combust, and where the average life span of the whole setup is a year. In principle this can quite easily be represented quantum-mechanically. In the beginning the ψ function characterizes a reasonably well-defined macroscopic state. But, according to your equation, after the course of a year this is no longer the case at all. Rather, the ψ function then describes a sort of blend of not-yet and already-exploded systems. Through no art of interpretation can this ψ function be turned into an adequate description of a real state of affairs; in reality there is just no intermediary between exploded and not-exploded.

These are Einstein's words written to Schrödinger on August 8, 1935.[1] The problem is how to account for the fact that after a year something definite has happened: the gunpowder has exploded, or it has not. The difficulty is that the Schrödinger evolution of the system produces a final state that is a superposition of "exploded" and "not exploded" states. This final superposed state has no natural interpretation; in particular, it does not entail either that the gunpowder has exploded or that it has not. Thus, the definite fact of the matter remains unaccounted for. The situation is not different in the Stern-Gerlach experiment, or in any other.

The standard approach to this problem involves invoking the "projection postulate" (also known as the "collapse of the wave packet"). In the picturesque words of Dirac (1958), "a measurement always causes the system to jump into an eigenstate of the dynamical variable that is measured...." As a resolution of the problem this approach is implausible—especially in the case of macroscopic observables (e.g., the explosion in Einstein's example), where the idea of causing by merely looking seems rather farfetched. In any event, to invoke the projection postulate is just to repeat in the language of quantum theory that measurements do, in fact, issue in results. The problem of accounting for this fact remains.

The basic idea of such an account, however, is straightforward. According to Bohr (1958, p. 73). "every atomic phenomenon is closed in the sense that its observation is based on registrations obtained by means of suitable amplification devices with irreversible functioning such as, for example, permanent marks on the photographic plate, caused by the penetration of electrons into the emulsion." Like most of Bohr's descriptions, this has a tangible and physical ring. Since it is an account of the "registrations" that concern us, it leads us to anticipate the possibility of a physical analysis explaining how "amplification devices with irreversible functioning" bring them about. Thus we would treat the measured object, the measuring instrument, and their interaction from a quantum-physical point of view and show, in terms of this application of the physics, how the instrument succeeds in registering something as a result of the interaction. This means that the end product of such an object-instrument interaction should be a state, or a mixture over states, where the instrument already displays some definite record. Is this possible? Can one show by means of an ordinary piece of physics that measurements produce results?

In the case of the exploding gunpowder, Einstein's intuitions were to the contrary. He seems to have thought that the state of the complex system (where the gunpowder's explosion would irreversibly amplify the functioning of some microchemical component) would not evolve into either an exploded state or a not-exploded one (or a mixture of these) according to standard principles of the quantum theory. He was right.

Quantum Measurement

The quantum theory of measurement is just the purely quantum-theoretical treatment of the measurement process to which Bohr's physical description of that process would lead one. In this section I want to sketch some of this treatment in order to state clearly the negative result concerning the measurement problem, as well as to establish a technical vocabulary for the remainder of this essay.

We associate a state space H_o with the object being measured and a state space H_a with the measuring apparatus, and treat their interaction in the tensor-product space H. Since the measurement problem concerns the transition from a pure case to a mixture, it is useful to adopt the formalism that represents "states" by normalized statistical operators W, distinguishing pure from mixed according to whether $W^2 = W$ (pure) or not. Recall that in this formalism the probability measure $P(W, Q)$ that quantum theory associates with the observable Q of a system in state W is given by

$$P(W, Q)(B) = \text{tr}[W\, C_B(Q)] \tag{1}$$

for any Borel set B, where C_B is the characteristic function of the set B. Using these measures we can introduce an equivalence relation that will play an important role in the discussion of measurement. For an observable Q, as above, we define states W and W' to be *Q-equivalent*, and write $W \underset{Q}{=} W'$, according to

$$W \underset{Q}{=} W' \text{ iff } P(W, Q) = P(W', Q). \tag{2}$$

For each Q, this relation partitions the states into equivalence classes, which I will denote by $[W]_Q$ (for the class of all states W' satisfying $W \underset{Q}{=} W'$). This relation lumps together as Q-equivalent pure states with mixed states, provided they yield the same probability distribution for Q.

Suppose we set out to measure the observable E on an object system by letting that system interact with an apparatus system having some "indicator" observable A that registers information about E. Assume that at the start of the measurement the apparatus is prepared in some specially "tuned" state W_a. If the object starts out in state W, then, according to the standard Schrödinger dynamics, the interaction corresponds to some unitary operator U on the tensor-product space **H** as follows:

$$W \otimes W_a \to U(W \otimes W_a)U^{-1}. \tag{3}$$

For fixed W_a and U, equation 3 defines a measurement map **m** from states on $\mathbf{H_o}$ to states on **H**:

$$\mathbf{m}(W) = U(W \otimes W_a)U^{-1}. \tag{4}$$

For the interaction to constitute a measurement of E by A, we should expect, at a minimum, that the statistics of the indicator observable A at the conclusion of measurement reflect the statistics of E contained in the initial state of the object. Thus, by repeatedly observing the indicator we find out about the object. This very minimal requirement on what counts as a measurement is just the idea that measurements effect a transfer of probability from the object to the apparatus. This is concisely expressed by requiring that

$$P(W, E) = P[\mathbf{m}(W), I \otimes A], \tag{5}$$

where I is the identity operator on $\mathbf{H_o}$ (or, equivocally, on $\mathbf{H_a}$).

In previous publications (Fine 1969, 1970) I have called measurement interactions satisfying equation 5 *filters*. They can be characterized, alternatively, by

$$P(W, E) = P(W', E) \tag{6a}$$

iff

$$P[\mathbf{m}(W), I \otimes A] = P[\mathbf{m}(W'), I \otimes A]. \tag{6b}$$

In terms of "equivalence" this amounts to

$$W \underset{E}{=} W' \text{ iff } \mathbf{m}(W) \underset{I \otimes A}{=} \mathbf{m}(W'), \tag{7}$$

or, treating $\mathbf{m}$ equivocally as a set function,

$$\mathbf{m}([W]_E) = [\mathbf{m}(W)]_{I \otimes A}. \tag{8}$$

Thus the filter requirement (equation 5) is just the requirement that $\mathbf{m}$ respects equivalence; i.e., that it mirrors initial E-equivalence by final $(I \otimes A)$-equivalence while (similarly) distinguishing inequivalent states.[2]

These filter requirements on the measurement map $\mathbf{m}$ restrict the types of unitary interactions U that can count as measurements. They are rather weak requirements; nevertheless, they are strong enough to rule out any strict solution to the measurement problem. Such a "solution" would involve measurement interactions terminating in object-apparatus states $\mathbf{m}(W)$ that can be represented as mixtures over states W_n where the indicator observable A displays a definite value, i.e.,

$$\mathbf{m}(W) = \sum c_n W_n, \tag{9}$$

where for every n there is some A value μ_n such that

$$P(W_n, I \otimes A)(\mu_n) = 1. \tag{10}$$

It turns out, however, that equations 9 and 10 are already inconsistent with equation 5 for the simplest case, namely that in which E is a bivalent observable having distinct eigenstates ψ_1 and ψ_2. (Think of "exploded" versus "not-exploded.") If we take for initial states W projections onto (respectively) ψ_1, ψ_2, and $(\psi_1 + \psi_2)2^{-1/2}$, then there is no single unitary motion U and initial apparatus state W_a, satisfying the conditions for a filter, for which all three measurements of E could produce definite results (as in equations 9 and 10). I shall refer to this as the *insolubility theorem*. Indeed, even in this simplest case the insolubility theorem holds for a condition weaker than equation 5, namely

$$\text{if } \mathbf{m}(W) \underset{I \otimes A}{=} \mathbf{m}(W'), \text{ then } W \underset{E}{=} W'. \tag{11}$$

Notice that condition 11 is the "only if" part of condition 7. Its violation would mean that one could not use the statistics accumulated at the end of an interaction to distinguish between initial object states differing in probability for some E value. In the simple case at hand, one could not distinguish between the two eigenstates of E and a superposition concentrated equally on both. Surely no interaction of this sort would count as a satisfactory measurement procedure for E.[3]

Of course, the inconsistency of conditions defining a measurement interaction with the further requirement that measurements have results would not be very interesting if it turned out that those measurement conditions were themselves inconsistent. Thus a "theory of measurement" requires an existence theorem demonstrating that there are indeed measurement interactions. For the weak theory of filters represented by equation 5 (or, weaker still, for theories represented by equation 11), such existence theorems are quite easy to come by. In particular, the classical "nondisturbing" measurements studied by von Neumann, as well as the "disturbing" ones studied by Landau, are all filters. These classical theories involve *eigenstate correlations*.[4] That is, they impose the special requirement that initial object-system eigenstates of E give rise to final apparatus eigenstates for A. If $\mathbf{H}_o^n$ are the E-eigenspaces (corresponding to eigenvalues λ_n) and $\mathbf{H}_a^n$ are the A-eigenspaces (corresponding to eigenvalues μ_n), then eigenstate correlations require the unitary evolution U on the product space to satisfy

$$U: \mathbf{H}_o^n \otimes \mathbf{H}_a \to \mathbf{H}_o \otimes \mathbf{H}_a^n. \tag{12}$$

If the initial object state W_n satisfies

$$P(W_n, E)(\lambda_n) = 1, \tag{13}$$

then W_n is a mixture over the λ_n-eigenstates ψ_{in}; i.e.,

$$W_n = \sum_i c_i P_{[\psi_{in}]}. \tag{14}$$

It follows from the evolution in equation 12 that

$$\mathbf{m}(W_n) = \sum_k w_k P_{[\phi_{kn}]}, \tag{15}$$

where, for each k, ϕ_{kn} belongs to $\mathbf{H}_o \otimes \mathbf{H}_a^n$. Hence equations 12 and 13 imply

$$P(\mathbf{m}(W_n), I \otimes A)(\mu_n) = 1. \tag{16}$$

We can paraphrase this by saying that, in measurements involving eigenstate correlations, if initially the object observable has a definite value then, when the measurement is over, the apparatus indicator also displays a definite value. So, for these special starting states, the problem of measurement is "solved." This solution extends to mixtures over such starting states. That is, if initially the object state W is a mixture

$$W = \sum c_n W_n, \tag{17}$$

where the W_n satisfy equation 13, then, since U is linear and the tensor product is bilinear,

$$\mathbf{m}(W) = \sum c_n \mathbf{m}(W_n). \tag{18}$$

Thus, at the conclusion of the measurement the state of the object-apparatus system is represented by a mixture over states in each of which, according to equation 16, the apparatus observable displays a definite value. By the insolubility theorem, the conclusion that measurements initiated in states W satisfying equation 17 "have results" does not extend to starting states that are pure states projecting onto superpositions of eigenstates of E.

The Source of the Problem

The insolubility theorem reviewed in the preceding section, as well as the partial solubility for eigenstate correlations discussed there, involves a tacit understanding of how to talk about measurements "having a definite result." More generally, I have been using a basic and standard interpretive principle for talking about an observable "having a value." I have called the simplest version of this principle the *eigenvalue-eigenstate link* (Fine 1973, p. 20). The principle says that a necessary and sufficient condition for an observable to have a value is that the state of the system be an eigenstate for that value. In the statistical-operator formalism we want a generalization of this, which I shall separate into two parts, beginning with a sufficient condition.

The Rule of Law. (a) If the state of a system is W_n, where $P(W_n, Q)(\lambda_n) = 1$, then observable Q has value λ_n in state W_n. (b) If the state W of a system is a mixture over states W_n, as in part a, then Q has a value in state W.

Part b of the Rule of Law involves the "ignorance interpretation" of mixtures, and one would probably want to supplement it by criteria for picking out certain preferred representations of mixed states. I shall assume this to have been done. Part a is relatively uncontroversial, unless one subscribes to a radically positivist line. Wheeler and Zurek (1983, p. 184) attribute such a radical view to Bohr, under the slogan "No elementary phenomenon is a phenomenon until it is a registered (observed) phenomenon." I think this may serve Wheeler's conception of the quantum theory better than it does Bohr's. In any case, I think we can live by the Rule of Law. What we would like is a decent necessary condition. The standard one is the following.

The Rule of Silence. If the state W of a system does not yield probability 1 for an observable Q to have a value (as in part a of the Rule of Law) and is not a mixture over such states (as in part b), then one is not to speak of "the value of Q" in state W.

In this formulation I have tried to adopt the point of view of Bohr, and to state the Rule of Silence as a meta-linguistic rule (i.e., one restricting how we are to talk about the theory). But one could put things in the object language and say that under the stated conditions the observable Q simply does not *have* any value. For example, it may seem paradoxical to say of an atom that right now it simply has no position, and less paradoxical to say instead that for now we should not talk about its position. I have no clear feeling about this issue of object language versus meta-language, and will comfortably flip-flop back and forth as seems appropriate. One way or another, however, I believe that the Rule of Silence is part of the standard interpretive lore of the quantum theory.

The insolubility of the measurement problem amounts to a demonstration that the Rule of Silence is inconsistent with the unitary dynamics of the theory, for, as we saw in the preceding section, the insolubility theorem shows that at the conclusion of a measurement interaction the state of the combined system is generally not one where the Rule of Silence permits talk about any definite result of the measurement (i.e., a value of the apparatus observable). Thus, Bohr's promise of an ordinary physical account of the measurement process cannot be kept. Standard techniques of quantum physics cannot be integrated with their standard interpretation so as to account for the fact that measurements do have results.

Bohr's own response to this dilemma was an explicitly philosophical one. He thought that epistemology would provide a way out, and proposed his well-known doctrine of the "cut." According to this doctrine, every complete application of the quantum theory involves an externally imposed distinction (or cut) between observer and observed. Bohr's idea was to enforce the restrictive rule that what falls on the observer side of the cut should not be treated at all by the physical principles of the quantum theory. Thus, when we come to the part of the measurement process that we call an observation, Bohr's philosophical doctrine directs us to hold the observer system exempt from the Rule of Silence. The difficulty with this way around the insolubility theorem is surely that it gives in too quickly, shifting abruptly from physical analysis to philosophical counsel.

Implicitly, at least, the main-line response in the scientific literature seems to recognize this difficulty with Bohr's philosophical doctrine of the cut, for that response attempts to justify the cut on physical grounds. In particular, it attempts to show how various practical considerations warrant ignoring the difference between the actual final state $\mathbf{m}(W)$ in a measurement interaction and the mixture M, over measurement results, that one would rather have. The strategy is to argue that, for all practical purposes, M "approximates" $\mathbf{m}(W)$ sufficiently well. This amounts to the claim that

$$\mathbf{m}(W) \underset{Q}{=} M \tag{19}$$

for all "genuinely observable" Q. Of course the insolubility theorem shows that not all observables Q can be considered in the "genuinely observable" class, and the literature does not yet show any way to resolve controversies over the structure and composition of this class.[5] A feature of this strategy not usually made explicit is that it involves amending the Rule of Silence; the strategy of approximation allows us to speak about the value registered by the instrument in situations where the Rule of Silence holds that, literally, there is no value. Thus these approximate solutions want to add on to the Rule of Law a clause that sanctions the attribution of a value to the apparatus observable provided that equation 19 holds for all "genuinely observable" Q, and that (correlatively) lifts the Rule of Silence for these cases. Roughly speaking, this amendment to the Rule of Silence associates values with certain observables in certain superposed states. In the foundational literature such associations are generally referred to as "hidden variables." Thus, attempts to solve the problem of measurement by approximation are reinterpretations of the quantum theory that introduce hidden variables into the end state of a measurement interaction, thereby amending the Rule of Silence. Because hidden-variable theories have a bad reputation among physicists, proponents of schemes of approximation will no doubt resist this description.

Be that as it may, the strategy of approximation is clear enough. It seeks to supplement Bohr's purely epistemological way around the measurement problem by arguing on physical grounds that we may treat the final object-apparatus state as though it were a mixed state in which the instrument actually shows a definite reading even though, strictly speaking, the final state is not such a mixture. The physical basis for thus lifting the Rule of Silence is the contention that the actual final state cannot be distinguished from the desired mixture by observations that are feasible—from a practical point of view. This contention of practical indistinguishability can be challenged. Even where it is admitted, one can wonder whether it constitutes good scientific practice to let such grossly practical considerations determine the shape of fundamental interpretive principles. Moreover, if one is going to do so, then precisely how does one proceed, and when? This is a scientific area for which there are no established ground rules; there are only judgment calls concerning which opinions may be expected to differ. John Bell's description of the situation seems to me apt, and fair:

The continuing dispute about quantum measurement theory is not between people who disagree on the results of simple mathematical manipulations. Nor is it between people with different ideas about the actual practicality of measuring arbitrarily complicated observables. It is between people who view with different degrees of concern or complacency the following fact: so long as the wave packet reduction is an essential component, and so long as we do not know exactly when and how it takes over from the Schrödinger equation, we do not have an exact and unambiguous formulation of our most fundamental physical theory.   (Bell 1975, p. 98)

The approach that moves from Bohr's doctrine of the cut to practical approximations tries to get around the problem of measurement by modifying the Rule of Silence. There is an alternative approach that focuses on the other horn of the dilemma, the Schrödinger evolution. This approach has a conservative and a radical wing. The conservative idea is to replace the Schrödinger evolution by a different dynamics in such a way that the nonunitary object-apparatus interaction dilates to a unitary motion on a larger Hilbert space. The classical development of this idea (due originally to von Neumann) is that of London and Bauer (1939), who introduce an "observer" system and show how tracing out over the observer space in an object-apparatus-observer unitary interaction (of a nondisturbing kind) produces just the right object-apparatus mixture. Zurek (1981, 1982) has taken up this approach, calling his third system the "environment" (rather than the "observer") in order to connect it with the idea of an open system. Among the motivations for Zurek in this "third system" approach is a puzzle he has noted in the (two-system) object-apparatus analysis: that unitary two-system interactions in general allow for many pairs of correlated observables between the systems. (Think of the Einstein-Podolsky-Rosen paradox, where both relative positions and conjugate linear momenta are correlated between the two particles.) The puzzle is how to tell which object variable is being measured by the interaction! The resolution in the three-systems approach is achieved by the trace over variables of the third system, selected so as to ensure that the final interaction is non-disturbing. More radical proposals for a different dynamics involve introducing stochastic processes that only approximate the partial traces of larger, unitary evolutions. The quantum theory of open systems provides a framework in which this idea can be developed.[6]

Although the program for introducing a non-unitary dynamics might be seen as an attempt to save the Rule of Silence, I believe that it actually exhibits a good deal of unresolved tension with regard to that interpretive principle. In the conservative approach the object-apparatus system is a component of a larger one at the conclusion of measurement. Therefore, in general, the object-apparatus system literally has no quantum state at that time, and the partial trace assigned to it is merely an improper mixture.[7] In these circumstance, talk about any definite result of the measurement violates the Rule of Silence. Hence the conservative approach must be supplemented by some strategy for modifying the Rule of Silence, and, just like the strategy of approximation, it is properly subject to the same variations of complacency or concern. The more radical approach involves physical processes that achieve a reduction of the wave packet (or an evolution into the right object-apparatus mixture) as an asymptotic result, usually in infinite time. It is then argued that, from the temporal perspective of the object system, the infinity can be truncated for all practical purposes. Depending on the details, these may well be sensible arguments. But they do involve a modification of the Rule of Silence, insofar as they are supposed to warrant attributing values before the state has, literally, been reduced.

The Solution

This brief survey of attempts to "solve" the problem of measurement emphasizes that the standard approaches actually involve some fundamental reinterpretations of the theory, at least insofar as they tamper with the Rule of Silence. These approaches also share the following tactic: They focus on the final state in the measurement process, and try (somehow) to "fix" it. I believe that if we shift the focus of our attention to the other end, to the initial rather than the final state, we can find a way of treating measurements that respects the Rule of Silence, employs only the standard unitary dynamics, and nevertheless allows measurements to have results. Here is the way.

Think of quantum measurement as a species of ordinary observation. Such an observation involves a selective (or "tuned") interaction. The selectivity is built into the way the measuring instrument physically interacts with the measured object: by filtering out from among all the actual physical features present in the object just those features relevant to the particular kind of observation being made. Thus, in ordinary circumstances a measuring apparatus interacts with only part of the physical object being measured, not with the whole object. For example, ordinary optical observations involve interactions only with features of the surfaces of objects, and x-ray observations basically ignore soft tissue surfaces. In the sense in which observation or measurement is a kind of seeing, on any particular occasion we only see (and *can* only see) part of what is there. In the quantum theory of measurement, however, this fact about observation has been neglected. I propose to begin the quantum analysis of a measurement interaction by attending to it.

The idea I want to explore is that, in interacting with an object to be measured, the measuring apparatus responds to only certain physical aspects of the object, not to all of them. In particular, if we have an apparatus for measuring an object observable E (call this an E *apparatus*), it will "see" only those aspects of the object that are relevant to E, and it will not be responsive to other features of the object. More concretely, we should expect that if the object initially has a value of E then the apparatus observable A will register it at the conclusion of measurement. But if we take the Rule of Silence seriously, and the initial state W of the object is neither an eigenstate of E nor a mixture over these, then what "aspect relevant to E" is there for the apparatus to respond to? The only sensible answer, I believe, is that the E apparatus responds to the probability distribution $P(W, E)$ associated with the object in state W. If we hold, further, that an E apparatus will not respond to other aspects of the object, then, statistically, the pattern of responses for E-equivalent states should be the same. This amounts to the idea of identifying the E aspect of an object in state W with the equivalence class $[W]_E$, and then requiring that an E apparatus not discriminate in its pattern of responses between objects having the same E aspect. This last requirement embodies our condition 5 (that the measurement interaction be a filter), but to deal with the problem of measurement it must include something more. What it must include is the idea of selectivity, the idea that when the E apparatus looks at the object it does not see *more* than the E aspect. Here is one way to express this formally.

Suppose that $E = \sum \lambda_n E_n$ is a discrete observable with finite multiplicities; i.e., $\mathrm{tr}(E_n) < \infty$. Let

$$W_n = \frac{E_n}{\mathrm{tr}(E_n)}. \tag{20}$$

Then each W_n is a normalized statistical operator satisfying equation 13, i.e.,

$$P(W_n, E)(\lambda_n) = 1,$$

and, according to the Rule of Law, we can say that in state W_n observable E has the value λ_n. If the object is in state W, then define $W(E)$ by

$$W(E) = \sum_n \text{tr}(W E_n) \cdot W_n. \tag{21}$$

Clearly $W(E)$ is also a normalized statistical operator, and it is in the equivalence class $[W]_E$; i.e.,

$$P(W(E), E) = P(W, E). \tag{22}$$

I shall call $W(E)$ the *standard representative* of its class $[W]_E$.

Recall that the idea being explored is that when an E apparatus looks at an object in state W, it does not see all aspects of the object, only the object's E aspect. This would be the case if, when looking at an object in state W, the E apparatus saw only $W(E)$; i.e., if the interaction with the object did not involve the entire state W, only the standard representative of the class $[W]_E$. Thus, instead of starting an E measurement interaction with

$$W \otimes Wa \rightarrow \tag{23}$$

we begin with

$$W(E) \otimes Wa \rightarrow . \tag{24}$$

If the arrow in equation 24 represents a unitary interaction that is an eigenstate correlation, then it follows from equations 17 and 18 that the object-apparatus system evolves into a mixture over states in each of which (according to the Rule of Law) the apparatus observable displays a definite value. That is, measurements effected by means of eigenstate correlations and starting out as in equation 24 always produce results. Hence, if it were correct to treat measurement interactions as interactions only with an aspect of the measured object, as represented by the starting position in equation 24, the measurement problem would be solved.

There are no strict rules for how a formalism applies to particular systems. There are only precedents and rules of thumb based on what has worked well enough in the past. This is as true of the interaction formalism of the quantum theory as of anything else. Standardly, the interaction of two systems is treated by starting the composite system in a state that is the product of the states of the components. But what should one do if this standard recipe fails to give the right (i.e., observed) answer in a certain class of cases? One might try to alter the dynamics of the formalism, or argue that the received answer is good enough if not exactly right. These are the ways with the measurement problem that we have already rehearsed. But one might also respond by examining the character of the difficult class of faulty cases to see whether it offers some sensible physical rationale for deploying the standard formalism differently. This is the response I have been trying to explore here.

My exploration starts out from the idea that some interactions are selective. They do not actually involve the whole system, only some physical subsystem. Thus, the interaction formalism ought not to be applied to the state of the whole system, only to a representative of the subsystem engaged in the interaction. To give a manageable model of this for measurement interactions, I have suggested, in effect, that the subsystem actually engaged in the measurement process has the standard representative for its state. The *-algebra formalism provides a rationale for this choice.[8] In that formalism a physical system is represented by an algebra of operators. This provides a natural way of talking about subsystems: as (closed) subalgebras. States, in this formalism, are linear functionals ("expectation values") on the operators. Distinct states on the full algebra of a system may have identical restrictions on some subalgebra. We can use these ideas to pick out the subsystem of the object that is involved in an E measurement; it is the one represented by the function algebra $\mathscr{F}(E)$ generated by the operator E. Restrictions to $\mathscr{F}(E)$ of distinct states W and W' are identical just in case $W \underset{E}{=} W'$. Thus, from the point of view of the subsystem engaged in a measurement, the states W and $W(E)$ are actually the same. Hence, if we take seriously the idea that the measuring apparatus interacts only with the physical subsystem represented by $\mathscr{F}(E)$, then our choice of starting the interaction with $W(E)$, as in equation 24, seems warranted.

One might, however, respond to this line of thought by suggesting the following puzzle. If E-equivalent states are actually identical relative to the subsystem represented by $\mathscr{F}(E)$, then how can we justify using any one of them, say $W(E)$, as opposed to any other, say W? I think the only answer can be that the E equivalence warrants using one or another state, indifferently, provided that the particular use yields satisfactory results. In the measurement case, starting with W, as in equation 23, fails to produce measurements with results. Starting with other E-equivalent states (as in equation 24) does yield results. Thus, in the analysis of those interactions that have results, we are free to choose from among E-equivalent starting states those that produce the results we find. A corollary to this way of approaching the analysis of interactions is that where results are not produced (as in a Stern-Gerlach setup where the beams are recombined) it would be a mistake to take equation 24 as the starting position; we need W, as in equation 23. Such an interaction involves more than the E aspect of the object. Correlatively, such an interaction is not an E measurement.[9]

This approach to the measurement problem puts no special interpretive weight on the concept of a measurement, nor on the distinction between micro and macro. The weight is on the idea of a selective interaction (an interaction with a subsystem only), and on how to treat that idea formally. Clearly one could imagine interactions involving subalgebras larger than $\mathscr{F}(E)$

—for instance, those involving operators commuting with E (that is, the subalgebra of operators "simultaneously measurable" with E). States with identical restrictions to this subalgebra form a natural equivalence class. One could select a representative from this class that would display the joint distribution function for all the commuting observables. One could then treat measurement interactions as starting from such a representative state, and develop a generalized theory of joint measurement interactions for their analysis in a straightforward way. (See, for example, de Muynck and van den Eijnde 1984, section 3). My intuition in this case, as in the case of a single observable, is that quantum probability distributions represent physical aspects of the object system, aspects that can be selectively interacted with and displayed by patterns of results of the interaction. This selective interaction could occur entirely at the micro level, and in contexts free of observers and their special arrangements.[10] However, if we want to know about it, the display will have to be amplified and registered in some way accessible to us.

Concluding Remarks

The customary responses to the quantum-measurement problem, by focusing on the final evolved state, have grabbed hold of the wrong end of the stick. If one retains a unitary dynamics, it is clear from the insolubility theorem that things will not come right in the end unless they start off right. In this sense one might say that the problem contains the seeds of its own solution. I have tried to prepare the ground for reexamining the starting states of the object system by emphasizing the pragmatic elements involved in the application of any formalism, and by developing a simple model for treating measurements as a species of selective interactions. This is an approach that deemphasizes the role of the observer, the peculiarities of "measurement," and the consolations of epistemology. It takes the standard formalism seriously and nonreductively, including the standard interpretive principles (the Rule of Law and the Rule of Silence). Thus, it supposes that the most salient features of the quantum theory, the probability distributions, pick out physical aspects that are sufficient to determine the patterns of outcomes of selective interactions. Although my simple model for selective interaction may not be good enough to cover all the varieties of cases that can arise, it seems that emphasizing flexibility in the way the interaction formalism is used to model physical processes is the right way around the measurement problem. Certainly such a flexible and pragmatic approach to problem solving is well within the recognized boundaries of scientific practice. On Bell's scale, which stretches from complacency over the measurement problem to concern, I would count this selective-interaction approach as justifiably complacent regarding the *feasibility* of a solution, and as appropriately concerned over the details in particular cases.

Notes

1. One is bound to recognize the similarity between this example and Schrödinger's cat paradox, which is usually cited as an instance of the measurement problem. Therefore, one might well conclude that Einstein and Schrödinger were discussing that cat paradox, or perhaps the measurement problem more generally, in their correspondence. In fact the topic of the correspondence was the Einstein-Podolsky-Rosen paradox. Schrödinger described the cat paradox only after receiving Einstein's "exploding gunpowder" letter, and not in connection with the measurement problem at all. See chapter 5 of Fine 1986.

2. So far it has not been necessary to restrict the spectra of the observables mentioned. For the remainder of this essay, however, I shall suppose that the object observable E and the apparatus observable A are discrete, although not necessarily maximal.

3. In Fine 1970, interactions satisfying equation 11 for an initial apparatus state W_a are called W_a-measurements and the insolubility theorem is stated and proved for them. The proof (section IV) uses the principle that the actual decomposition of the time evolute of a mixture, initially decomposed over orthogonal pure states, is a mixture over the time evolute of those pure states. This is called the RUE (Real Unitary Evolution) principle by Brown (1986), who defends its use. The insolubility theorem for W_a measurements is correct, however, even without assuming this principle, as is shown in Shimony 1974 and in d'Espagnat 1976 (exercise 3, p. 228). See Brown 1986 for a discussion of different lines of approach to such insolubility results.

4. Proposition 5 in Fine 1969, a general existence theorem for eigenstate correlations, is preceded by a discussion of the literature.

5. See section V of Fine 1970 for a discussion of the concept of an approximate solution and some of its difficulties. Peres 1980 is a recent example.

6. Gisin 1984 contains an interesting proposal of this radical kind, along with references to others. In an exchange between Gisin and Pearle (*Physical Review Letters* 53 [1984]: 1775–1776), Pearle criticizes Gisin, as I do below for this whole class, because his model involves the limit of infinite time. Bell's (1975) undoing of Hepp's (1972) model shows the danger here. On the other hand, Gisin points to difficulties over relativistic causality in the nonlinear models proposed by Pearle (1976). Among philosophers, Cartwright (1983) favors the abandonment of uniticity as a way out of the measurement problem.

7. See section 7.2 of d'Espagnat 1976 for the concept of an improper mixture.

8. Emch 1972 is a good reference for the algebraic approach.

9. A dramatic version of a nonselective interaction is the SQUID arrangement of Leggett and Garg (1985), who are careful to emphasize the unsuitability of their arrangement for measurements of magnetic flux.

10. Thus the negative-result aspect of a "negative-result measurement" (Renninger 1960) is easily accounted for on the selectivity approach, since the initial probability distribution being measured is over results both positive and negative. This approach also nicely solves Zurek's problem with regard to what observable is being measured— it is the one corresponding to the subsystem actively engaged in the measurement interaction.

References

Bell, J. S. 1975. On Wave Packet Reduction in the Coleman-Hepp Model. *Helv. Phys. Acta.* 48: 93–98.

Bohr, N. 1958. *Atomic Physics and Human Knowledge.* New York: Wiley.

Brown, H. 1986. The Insolubility Proof of the Quantum Measurement Problem. *Found. Phys.* 16: 857–870.

Cartwright, N. 1983. *How the Laws of Physics Lie.* Oxford: Clarendon.

de Muynck, W. M., and J. P. H. W. van den Eijnde. 1984. A Derivation of Local Commutativity from Macrocausility Using a Quantum Mechanical Theory of Measurement. *Found. Phys.* 14: 111–146.

d'Espagnat, B. 1976. *Conceptual Foundations of Quantum Mechanics,* second edition. Reading, Mass.: Benjamin.

Dirac, P. A. M. 1958. *Principles of Quantum Mechanics,* fourth edition. Oxford: Clarendon.

Emch, C. G. 1972. *Algebraic Methods in Statistical Mechanics and Quantum Field Theory.* New York: Wiley.

Fine, A. 1969. On the General Quantum Theory of Measurement. *Proc. Camb. Phil. Soc.* 65: 111–122.

Fine, A. 1970. Insolubility of the Quantum Measurement Problem. *Phys. Rev.* D 2: 2783–2787.

Fine, A. 1973. Probability and the Interpretation of Quantum Mechanics. *Brit. J. Phil. Sci.* 24: 1–37.

Fine, A. 1986. *The Shaky Game: Einstein, Realism and The Quantum Theory.* University of Chicago Press.

Gisin, N. 1984. Quantum Measurements and Stochastic Processes. *Phy. Rev. Lett.* 52: 1657–1660.

Hepp, K. 1972. Quantum Theory of Measurement and Macroscopic Observables. *Helv. Phys. Acta* 45: 237–248.

Leggett, A. J., and A. Garg. 1985. Quantum Mechanics Versus Macroscopic Reality: Is There Flux When Nobody Looks? *Phys. Rev. Lett.* 54: 857–860.

London, F. W., and F. Bauer. 1939. *La théorie de l'observation en mécanique quantique.* Paris: Hermann. Translated in Wheeler and Zurek 1983.

Pearle, P. 1976. Reduction of the State Vector by a Nonlinear Schrödinger Equation. *Phys. Rev.* D 13: 857–868.

Peres, A. 1980. Can We Undo Quantum Measurements? *Phys. Rev.* D 22: 879–883. Reprinted in Wheeler and Zurek 1983.

Renninger, M. 1960. Messungen ohne Störung des Messobjekts. *Z. Phys.* 158: 417–421.

Shimony, A. 1974. Approximate Measurement in Quantum Mechanics, II. *Phys. Rev.* D 9: 2321–2323.

Wheeler, J. A., W. H. Zurek 1983. *Quantum Theory and Measurement.* Princeton University Press.

Zurek, W. H. 1981. Pointer Basis of Quantum Apparatus: Into What Mixture Does the Wave Packet Collapse? *Phys. Rev.* D 24: 1516–1525.

Zurek, W. H. 1982. Environment-Induced Superselection Rules. *Phys. Rev.* D 26: 1862–1880.

Thomas Nickles

From Natural Philosophy to Metaphilosophy of Science

Time was when physical scientists regularly invoked philosophical considerations—both epistemological and metaphysical—to back their scientific claims and programs. Since Kelvin's day, however, the influence of philosophy on physics has dropped off as an exponentially decreasing function of time, even though philosophy is today more "scientific" than ever and there now exists a special branch of philosophy called "philosophy of science."

There are a multitude of reasons for the divergence of philosophy and physics over the past century. I shall explore three developments that combined to transform traditional discussions of scientific method (in which scientists played a leading part) into contemporary metamethodology (in which physicists are scarcely interested). These developments grew out of the deep changes in science and in methodological thinking that occurred in the nineteenth century—changes that were already well underway in William Thomson's youth. Thus, my story begins around 1800 rather than in 1884, the year of the Baltimore Lectures.

My first topic will be the Great Logical Inversion—the nearly total replacement of the old, constructive-generative methodologies associated with Bacon, Newton, and Descartes by the purely consequentialist methodologies associated with William Stanley Jevons, Karl Popper, and many of the logical positivists. (My special terminology will be explained below.) John Herschel and William Whewell were transitional figures here. If I am right, the philosophical inversion was more extreme than the changes in the working methodology of physical science itself. Greater use of the method of hypothesis by physicists did not signal their abandonment of generative methodology. On the contrary, as the physical sciences matured into modern mathematical physics, generative considerations became more important and more manageable than ever.

In the next section I identify another dimension along which nineteenth-century philosophy and working science drifted apart. Philosophical methodology became more global and abstract just as scientific methodology became increasingly local and concrete. With the rapid emergence of new specialty areas within science, the methods and techniques actually employed by scientists became more and more context-specific and content-specific; nevertheless (and partly for that very reason), philosophical discussions became ever more general and abstract and unified.

Next I address the emergence of metaphilosophy, which in much twentieth-century philosophy of science virtually came to replace descriptive and critical first-order inquiry into how science works. John Stuart Mill's role in this change is surprisingly large, given his coolness toward the method of hypothesis and given his lack of association with Kantian philosophy.

Finally, I attempt to weave the three strands of my story into a coherent but partial account of the divergence of philosophy and physics. The constellation of these factors in twentieth-century professional academic philosophy of science accelerated the divergence.

After Kant and Mill, philosophers and logicians of science came to concern themselves more with metamethodology and less with practical scientific methods, more with ultimate philosophical justification and less with the problem-solving process. For the philosophers and the logicians, methodology was derivative from epistemology. For scientists, meanwhile, heuristics and economy of research became increasingly important. Pedagogical obligations aside, research physicists tend to be interested in physical foundations and in methodological matters for heuristic reasons rather than for foundational reasons of a philosophical (epistemological) kind. For them, methodology directs the routine treatment of old problems and provides heuristic guidance to the solution of new ones; and epistemology is derivative from methodology. As physics has become ever more "progressive," forward-looking, heuristic, economy-minded, and opportunistic over the past century and a half, philosophy has become more epistemological, anti-heuristic, uneconomical, and backward-looking.

My emphasis throughout will be on methodology in Britain rather than on the Continent. Developments in France, Germany, and elsewhere receive slight treatment not because they are unimportant but because I am unable to discuss everything. My intentions are not purely historical in any case. In my view, not all of the historical developments recounted here represented unequivocal philosophical progress. I do not hesitate to draw critical conclusions.

The Nineteenth-Century Methodological Turn

William Thomson belonged to the first generation of fully professional physicists in Britain. As is well known, the term 'physicist' itself was only invented by William Whewell in 1840, as a counterpart to the French *physicien*, who, in the persons of Ampère, Fresnel, Fourier, et al., can be argued to have invented the subject of physics (Cannon 1978, chapter 4). Thomson, who read Fourier at an early age and helped transmit French science to Britain, had a major role in the later phases of these changes, which affected not only the content of the relevant fields but also the way in which research was conducted.[1]

Actually, major changes in British science and methodology were well underway by the time Thomson entered the University of Glasgow (1834) and Cambridge (1841) as a student. Around the turn of the century, Scottish methodologists, particularly Dugald Stewart, had begun to allow much wider use of hypotheses and analogies than did the official "Newtonian" inductivist methodology. Herschel's *Preliminary Discourse on the Study of Natural Philosophy* had appeared in 1830. Whewell published his *History of the Inductive Sciences* in 1837, and the first edition of *Philosophy of the Inductive Sciences, Founded upon their History* some three years later. The latter work was soon overshadowed by Mill's *System of Logic* (1843), despite the fact that Mill was much less a scientist than Whewell or Herschel.

The standard accounts of nineteenth-century changes in methodology make no sharp distinction between words and deeds, between scientific practice and methodological pronouncements in philosophical and semi-popular writings.[2] This practice is justified up to a point, for in the decades around the turn of the nineteenth century much of the more interesting methodological writing was by scientists, and their writing was intimately related to their scientific work. The main development here was the rejection of narrow inductivist methodology in favor of the less constraining "method of hypothesis." In Britain, the doctrinaire Newtonianism of the day had made hypotheses unwelcome. The use of hypotheses raised the specter of scepticism, which constituted a threat to good sense, religion, morality, and the social order (Strong 1978). Science was to proceed by induction from observation and experiment. On the other hand, speculative hypotheses seemed to bear fruit in the work of LeSage, Prevost, Priestley, and others (Laudan 1980). These men, partly through realizing the fertility of well-directed hypotheses and partly out of fear that their own work was being unfairly treated by orthodox methodologists, campaigned for toleration of hypotheses.

This call was soon taken up by Stewart and the second generation of Scottish methodologists (mostly philosophers, with an admixture of scientists), who stressed the great fertility of the controlled use of hypotheses and analogies and who contended (rightly) that the use of hypotheses in the early stages of inquiry had actually been Newton's method all along (Olson 1975, chapter 4; Cantor 1971; Laudan 1981, chapter 7). However, the Scottish methodologists and virtually all the major methodologists up to 1860 or so agreed that the method of hypotheses was only a method of *discovery* (albeit a more fertile one than induction). Something more—which some still termed "a complete induction"—was required to *justify* adequately any claim of positive science.[3] In effect, full justification meant showing, retrospectively, that investigators now knew enough to see how the claim could have been discovered "inductively," i.e., by being derived from observation and experiment in conjunction with established principles. Justification was a kind of *post hoc,* rational reconstruction of a logically plausible discovery path.

For these methodologists, the method of hypothesis was important because it contributed to the economy of research. It was a shortcut to discovery but not the final word on scientific justification. Although Herschel and Whewell gave novel prediction special probative force, Victorian methodology remained constructive and, indeed, generative. A *constructive* methodology concerns itself with the discovery process, with problem solving and heuristic fertility. It provides guidance for the construction of new claims and for the discovery of new effects. Most nineteenth-century British methodologists denied that there was a general logic of discovery in the strict sense of a set of rules for generating interesting theories, but they believed that methodology could furnish valuable advice concerning discovery. A *generative* methodology holds that adequate scientific justification requires more than successful prediction. It requires direct ("generative") support. Predictive confirmation of a hypothesis provides only indirect support; it reasons *from* the hypothesis (plus necessary auxiliary assumptions) to one or more of its logical consequences. Generative support is provided by reasoning *to* the hypothesis itself from phenomenal and theoretical claims already considered well established. This is one reason why physicists were not (and are not) satisfied with a predictively successful hypothesis unless they can derive it from first principles.

In short: Constructive methodology concerns discovery, whereas generative methodology concerns discoverability—justification construed as the derivability of a claim from suitable premises. No matter how a hypothesis was first thought up ("discovered"), to adequately justify it is to show how to derive it from what we now think we know about the world—that is, to show how it could have been discovered had we known then what we know now.[4]

British methodology was both constructive and generative. Those passages in Herschel and other writers about the importance of rigorously testing hypotheses over their full range—passages that lead whiggish readers to project Popperian views onto Victorian methodology—virtually always give way to a generative conception of justification as the ultimate aim of inquiry. For these early Victorian writers, the proposal and testing of hypotheses was a tool for learning enough about the problem, the empirical facts, and other constraints to eventually provide some sort of plausible derivation of the claim.

Although the Victorian methodologies were far from Popperian, they did contain some of the materials out of which a twentieth-century hypothetico-deductive (H-D) view of science such as Popper's can be fashioned. William Stanley Jevons (a late Victorian) came very close to the modern conception, although he remained far more interested in heuristic fertility and economy of research than do most twentieth-century H-D theorists. Jevons was a *pure consequentialist* in that he explicitly rejected any need for generative support (although he proceeded to qualify his view by requiring consistency with known laws). In *The Principles of Science* (1874) Jevons wrote: "*Agreement with fact is the sole and sufficient test of a true hypothesis*" (p. 510 of 1883 edition; Jevons's emphasis). As far as I am aware, Jevons was the first methodologist to state what was to become the Central Dogma of H-D methodology, which I formulate as follows: All empirical support = empirical evidence = empirical data = successful test results = successful predictions = true empirical consequences (not antecedents) of the claim. Many contemporary methodologists (including Popperians and Lakatosians) add to the Central Dogma the further stipulation that only novel predictions provide empirical support. According to these *novel consequentialists*, information known to a theorist, or at least information built into a theory, cannot count in its support.[5]

The sea change from the Newtonian-Scottish methodology of Stewart, Herschel, et al. to the pure H-D method of Jevons and Popper, half a century later, is the Great Logical Inversion that I spoke of above. Historically explaining and logically justifying this change are the two aspects of the "mystery" of the Great Logical Inversion.

The inversion was not total, since predictive success had been important since Newton and before. Moreover, the change took more than a century to reach complete fruition in the Popperian version of the Central Dogma. Nonetheless, the overall change in methodological perspective was immense.

The big choice in methodology is among pure generativism, pure consequentialism, and some consistent combination of the two. Most nineteenth-century methodologists reasonably chose the "combination" view, as did most physicists. Remarkably, many twentieth-century philosophers write as if they are unaware of this option. At least they hastily reject all generative ideas as throwbacks to the inductivism, discovery logics, and foundationist epistemologies of the seventeenth century.

In my view an adequate logical justification of the inversion is impossible; only a historical account can be given. The generativists were correct: a viable methodology must include a generative component (Nickles 1985, 1987).

I already have mentioned some good reasons for the revival of the method of hypothesis, viz., this revival was scientifically liberating and heuristically fruitful. A nineteenth-century outbreak of skepticism or fallibilism (Charles Peirce's later term) could in principle provide another good reason.[6] Why so?

First, on a fallibilistic view of scientific claims, one no longer pretends to have the absolute truth in hand. Our theories are always subject to future revision, which (we hope) will take them ever nearer the truth. It is easy to see that a healthy dose of fallibilism could shift methodology from a Truth Now mode to a progressive program of starting with crude hypotheses and then successively correcting and complicating them. A progressive view tends to blur the distinction between hypotheses and established theories. It is a step toward modern hypotheticalism. However, it is hardly a step away from discovery and heuristics; for on the progressive view, the proverbial "final" justification becomes less important (indeed, nonexistent), and there is a premium on heuristic appraisal and on heuristic guidance to ever-better hypotheses. This is just what we find in the Scottish methodological tradition, as we have seen. The Scots and their many heirs did not conceive the method of hypothesis as antidiscovery, as twentieth-century philosophers from Popper to Laudan have. On the contrary, it was a method of discovery, although not a strict logic of discovery.

Second, to suppose that the Great Logical Inversion was a response to an outbreak of scientific skepticism makes good sense in view of the following logical considerations. Infallibilist methodological aspirations dictate that one must be a strong generativist; whereas the truth of the premises of a valid argument is a sufficient condition for the truth of the conclusion, true testable consequences furnish only necessary conditions for the truth of the hypothesis from which they are derived. The well-known fallacy of affirming the consequent lurks here.[7] Hence, no amount of successful testing guarantees truth. Logically speaking, then, one can be a total, epistemological optimist about science only if one holds that there exist generative methods for constructing true theories. Infallibilism entails generativism. On this view, hypotheses and consequential testing are reduced to inferior, heuristic, temporary aids to research—the status they had for Bacon, Newton, and the Scots.

But suppose that we embrace fallibilism instead of infallibilistic optimism. Now everything is turned upside down, it seems. No longer can we trust implicitly the premises of our arguments, so generative justification becomes suspect. And since inductive arguments from different information bases can yield quite different and even quite opposite results, the best test of a good idea would seem to be how it survives predictive criticism. The most rigorous tests are furnished by startling, "novel" predictions. We have exchanged generative for consequential justification. And since we no longer hope to obtain the absolute truth, at least not immediately, hypotheses become the primary vehicle of scientific communication. What was a heuristic device (the H-D method) is now the chief basis of scientific justification, and what used to be that chief basis (generative method) is now reduced to a heuristic device or worse. What used to be the main business of science (fact gathering) is now demoted to a low grade of inquiry, and what was low (speculative hypothesizing) is now exalted. What used to count for almost everything in scientific justification now counts for little or nothing, and vice versa. The very concept of evidence has flip-flopped. Previously, evidence in the primary sense was the empirical basis from which the theory somehow could be derived. On the more extreme H-D views, old evidence counts for nothing. For novel consequentialists, all genuine evidence is logically (and for some even temporally) posterior to the theory. Previously, pedigree was important. Now, "ye shall know them by their fruits."

How good is this neat logical solution to the inversion mystery? Not very.

First, the fallibilism argument does not show that generative justification is either impossible or useless. It only shows that consequential testing is important and suggests that on many occasions science may have to "wing it" on the basis of consequential justification alone. We may accept the argument and still insist on the importance of generative justification, when it can be had. In short, the argument does not establish pure consequentialism. The "combination" option remains available. Infallibilism entails generativism, but rejecting infallibilism does not entail rejecting generative justification.

Second, the fallibilism argument not only fails as a convincing logical solution to the mystery; it fails as a historical solution also. The argument is not even consistent with history, let alone explanatory of it. Although fallibilism is certainly relevant to the developments (especially around the turn of the nineteenth century), there is a fatal flaw in using it as the key to the mystery. In this case history fails to recapitulate logic.

The method of hypothesis came into flower precisely when the threat that skepticism posed to truth and the social order had been turned aside. Most nineteenth-century philosophers believed that Kant and the Scottish Common Sense philosophers had adequately "handled" Hartley, Hume, Priestley, and other materialists, skeptics, and atheists (Strong 1978). It was only because Stewart, Herschel, Whewell, and others were so confident that the controlled use of hypotheses posed no threat that they could speak so boldly of hypotheses. Although these writers paid lip service to a minimal sort of fallibilism, they were bursting with confidence that the principal results of physical science were true.

This confidence grows as we move deeper into the nineteenth century and further away from the failures of Newtonian chemistry and (especially) Newtonian optics. As the various sciences chalked up one impressive success after another, Whewell imposed fewer restrictions on hypotheses than did the Scots and Herschel. Jevons imposed fewer still. By Jevons's time we hear little about postulated entities having to be *verae causae* (true, Newtonian, observable causes, or at least analogical causes), for example. Thus it is misleading indeed to suggest that a deepening fallibilism explains the victory of the H-D methodology and the Great Logical Inversion.[8]

What killed generative methodology more than anything—besides its facile identification with inductivism, which is only one variety of generativism—was the increasing importance attached to predictive confirmation and especially to novel prediction. Once one takes this view to the limit and says that predictive justification is justification enough, one becomes a pure consequentialist and embraces the Central Dogma, as Jevons and others eventually did. And in the limit in which novel prediction carries all the epistemic weight, information which a theory was designed to explain or which was built into the theory can carry no weight at all.

Forty years before Jevons, Whewell had been enchanted by novel prediction and the related phenomenon of "consilience of inductions" (the case in which a theory advanced to handle one body of facts also explains a previously independent domain of information). In making novel prediction *cum* consilience a hallmark of truth, Whewell, too, in a strange sort of way, became a pure consequentialist (Laudan 1981). However, Whewell's methodology was too complex to be labeled simply an H-D methodology.

An interesting historical question is this: Precisely why did Herschel and others (especially Whewell) place so much emphasis on novel prediction, raising it to such a high methodological status? Historical studies suggest that scientists themselves did not attach special importance to genuinely novel predictions, such as the "spot" of diffracted light that Poisson deduced from Fresnel's wave theory of light (Worrall 1987). Did the emphasis on novel prediction constitute a methodological weapon with which to attack philosophical opponents?

I have no neat solution to this problem, but about Whewell's idea of consilience of inductions (and, to some extent, novel prediction) I can safely make three points, the first a normative one. First, consilience does fit our intuitions about strong confirmation, for it represents a kind of robustness of results. If two or more different lines of investigation yield the same result, this is surely confirmation of the value of that work, although Whewell went too far by seeing in consilience and novel prediction the hallmark of truth. Second, as to why it was Whewell who "discovered" and emphasized consilience, it was not until the early nineteenth century that the physical sciences had matured to the point that several clear examples of consilience were available for the history-and-methodology-minded student of scientific development. (To a lesser extent, this was also true of novel prediction.) Third, Whewell's historical methodology needed something more than ordinary predictive confirmation. Although not so much a metamethodologist as Mill, Whewell was enough of a logician to see that routine predictive confirmations were insufficient to yield a consequentialist methodology as optimistic as the generative methodology it was designed to replace.

The striking thing is that, so far as I know, not one nineteenth-century (pure) consequentialist directly faced the issue of showing that generative support is impossible or useless, that all genuine support derives from consequential testing. Despite the fact that Stewart had clearly distinguished Newton's actual method of research from inductivism, most later methodologists wrongly assimilated Newton's method, which was in fact generative but not ardently inductive, to inductivism on the one hand or to pure consequentialism on the other. These later methododologists simply took for granted that in attacking inductivism they were attacking generative methodology in general. Thus, pure consequentialists quietly abandoned noninductive versions of generative methodology, without serious argument. This is the more surprising since Mill had lodged well-known protests against the doctrine of novel prediction and to the hypotheticalism for which it stands. That Mill's opposition was largely unheeded, despite his immense reputation, is evidence that in mid-Victorian Britain Mill stood practically alone in taking metamethodological questions utterly seriously.

While many philosophers eventually abandoned generative methodology for a purely consequentialist position, the physicists did not go so far in their practice. To be sure, many of them did employ hypotheses right and left, and they recognized the importance of predictive adequacy and rigorous testing. I do not question the historical fact that there was an outbreak of hypothesizing in many quarters, and I agree that this was a legitimate and healthy development. My point is that we can cite any number of cases in which physicists in the nineteenth century and later were (and are) not content to have confirmed hypotheses.[9] As the physical sciences accumulated a large body of accepted results and sophisticated techniques for dealing with them, the possibilities for generative justification increased rather than decreased. The mathematical deductions which the H-D method made so important were not restricted in their employment to the deducing of testable predictions from hypotheses.

If we set aside the more phenomenological research traditions and look at the more hypotheticalist programs, we still find that physicists typically consider even predictively confirmed hypotheses to be conjectural in a more or less pejorative sense unless they have been derived from "established" principles. (Of course, a well-tested hypothesis is far better than a rank conjecture.) There are several motives for wanting such a derivation (theoretical unity, completeness, simplicity, and so on), but one powerful motive is surely epistemic: A well-founded hypothesis is considered more reliable as a basis for further research than one that is merely well tested.

The Great Logical Inversion amounted to a revolution in methodology, a dramatic switch in aims, methods, and basic units of communication and analysis. But if I am right, once the surface features of the inductivism-versus-hypotheticalism debate are stripped away, no logical inversion is to be found deeply embedded in scientific practice. There is no sweeping change in overall methodological strategy, only a tactical change in emphasis. The inversion is largely a philosophical fiction. Despite the fact that the mystery concerns a philosophical revolution, a revolution in the "logic" of science, it has no logical or methodological solution.[10] The mystery has only a historical explanation, the outlines of which I have attempted to sketch.

Methodology and Scientific Change

Having criticized the standard treatment of the nineteenth-century methodological turn, let me propose an alternative account. I am painfully aware that decades of rapid historical development cannot be adequately captured in a few pages. Nevertheless, it is possible to identify some important methodological changes that resulted from changes within science itself.[11]

My overall claim is that as the various branches of science (including the numerous specialty areas of physical science) matured in the nineteenth century, their research methods and techniques became highly domain-specific, problem-specific, and context-specific—that is, specific to the scientific field in question, to the problem addressed, and to the facet of the problem under investigation. Powerful modes of investigation became available, including quasi-logics and strong-heuristic methods of discovery in a few cases, but the more powerful methods tended to be locally rather than globally applicable. However, the philosophers and logicians writing about science either missed this shift toward locality in methodology or deliberately abstracted away all local differences in scientific work, on the ground that philosophers are supposed to write about the sciences in general.[12] The result was that as working scientific methodology became more content- and context-specific, philosophical accounts (partly for that very reason) became more and more general and abstract. If I am right that these divergent tendencies characterize methodology over the past century and a half, it is no wonder that scientists today feel that philosophical accounts do not really grasp what they do.

As Lord Rutherford once characteristically quipped, "There are two kinds of science: physics and stamp collecting." In the early nineteenth century, as everyone knows, many varieties of stamp collecting began (or continued) a remarkable transformation into mature, mathematical and/or theoretical sciences. A multitude of formerly "Baconian" or "Humboldtian" sciences—including optics, electricity, magnetism, heat theory, and chemical subjects—began to acquire interesting mathematical form and, eventually, deep theoretical structure. Even such still largely natural-historical fields as geology, paleontology, botany, and zoology underwent important theoretical development.

As the century progressed, so did these sciences. Ampère, Fourier, Fresnel, and others showed how to apply sophisticated mathematical methods to nonmechanical phenomena, and several previously independent or merely analogous branches of physical science became theoretically unified. William Thomson had a good deal to do with the emergence of electromagnetic theory and mathematical thermodynamics from the 1840s on.

These developments naturally had a major impact on methodology and were to some extent the result of changes in methodology. Mathematical methods, for example, provided an immense infusion of new material into the heuristic and justificatory components of methodology. Powerful formal analogies could now be added to physical analogies and could even "inform" those physical analogies. Once some basic assumptions were given mathematical form, an elaborate mathematical theory sometimes could be worked out relatively quickly and testable implications could be deduced—now by complex, quantitative calculations rather than by the old series of qualitative syllogisms. Moreover, such mathematical formulation typically added new empirical content to a theory.[14] And once formulated mathematically, many scientific problems could be solved routinely.

A mathematized method of hypothesis replaced "induction" as the main method of maturing disciplines. The primary virtue of induction had been that it was supposedly both general and powerful. Assiduous application of the method was purportedly sufficient to ascertain the scientific truth in any domain. But by this time, inductive methods appeared weak in comparison with the method of mathematical hypothesis. Induction even appeared to be limited in generality, since, as Stewart, Herschel, and many other methodologists noted, once one gets beyond the first level of empirical generalizations to higher-order "inductions" it is far less clear how one is to proceed. Science is not "data driven" at this heady level. The method of hypothesis sparkled just where induction faded.

Yet it would be wrong to say that mathematics furnished a method of science that was both general and powerful enough to replace induction as an account of scientific inquiry. There was, and could have been, no general yet detailed, content-neutral replacement for inductivist methodology. Most relevant here is the fact that mathematics itself exploded into a dozen special fields during this period, especially in France, and sometimes in response to previously intractable problems of "physics" and engineering. The more powerful mathematical methods had to be specially tailored to their particular subject matters. "Pure" mathematics is content neutral. To fruitfully apply specific mathematical techniques to physical problem solving requires that those techniques become physically "embodied" (one of Thomson's favorite terms). The philosophers' later fascination with the pure mathematics and pure logics that were then emerging made them lose sight of this fact.[14]

The abandonment by scientists of classical, inductivist methodology has a significance that has gone unrecognized by recent philosophers: In effect, it was the abandonment of the idea that there exists a single, content-neutral method of science (whether a method of discovery or a method of justification) that is both powerful and completely general. Scientists implicitly discovered that powerful methods are local, knowledge-based, "expert" methods rather than global, purely formal methods.[15] Powerful methods are not neutral logical or conventional rules; they are heavily laden with specific theoretical and experimental content. If I am right, then it is a mistake to describe the nineteenth-century change in methodology as the abandonment of logic of discovery, for it was equally the abandonment of general, content-neutral logic of justification. Local research contexts, however, continued to evidence "logics" of both kinds.

Since historical purity is not my aim in this essay, permit me to draw an analogy to a contemporary field: artificial intelligence. It is today generally conceded that Herbert Simon, Allen Newell, and their associates failed in their attempts of the late 1950s and the early 1960s to develop a General Problem Solver that would be both powerful and globally applicable. Since then, artificial intelligence has evolved from this content-neutral, general-purpose approach to a "knowledge-based" or "expert systems" approach. Instead of trying to get by with general, context-independent heuristics (which are now considered weak), artificial-intelligencers today recognize that nontrivial problem solving and problem-solving efficiency both demand "strong," context-specific heuristics. Efficient chess-playing programs, for example, must include a great deal of specific information about how to play good chess. The superiority of knowledge-based problem-solving systems is bought at a price, however, for such programs are only locally applicable. As E. A. Feigenbaum once put it, "There is a kind of 'law of nature' operating that relates problem solving generality (breadth of applicability) inversely to power (solution successes, efficiency, etc.) and power directly to specificity (task-specific information)." [16]

I want to tell a similar story about classical methodology, including the classical discovery program. In the infancy of modern science, Bacon, Descartes, Newton, and their associates believed that general-purpose procedures (e.g., eliminative induction) would suffice to solve all or most scientific problems. At that time, relatively few laws and still fewer high theoretical results had been established. Many investigators conceived the remaining problems to be difficult but conceptually shallow problems of finding the empirical laws governing the various domains of observable phenomena. Insofar as they thought that all the necessary concepts were already in hand or close to hand, it was not unrealistic for them to think that they could generate all possible (or plausible) law claims for a given field and then eliminate all but the correct one. This was Bacon's idea of eliminative induction. Newton's optical experiments more or less fitted an eliminative pattern, although his own methodological pronouncements favored a more direct form of reasoning to a law or theory (direct discoverability). In any case, since few established results were already in hand, it seemed appropriate that discovery methods be data driven rather than theory driven. And it was not unreasonable to think that the same kind of procedure by which a law claim was generated could justify it.

By the 1830s, however, and increasingly as the century progressed, science had outrun the classical inductivist program, not to mention the Cartesian program. There clearly existed no general, content-neutral method for routinely generating deep and interesting theories—or for justifying them. Whatever powerful methods existed were content- and context-specific. Since inductive methods (despite Bacon's protests to the contrary) were in fundamental respects just an updating of Aristotle's universal, biological-taxonomic method of science, it can be said that the rejection of inductive methodology by nineteenth-century physicists amounted to the final rejection of the Aristotelian conception of science as plainly inadequate to physical research.

Now suppose that a late-nineteenth-century or a twentieth-century philosopher stands back to interpret what happened in the early decades of the nineteenth century. If such a philosopher makes the standard assumption that the philosopher's task is to abstract from the multifaceted complexity of scientists' daily activities a general account of a scientific method common to all the branches of science at all stages of their development, then it looks to all the world as if the classical constructive-generative, discovery-*cum*-justification program was simply abandoned, to be replaced by nothing at all on the side of discovery and by the H-D method of testing on the side of justification. The H-D view is only the "least common denominator" of scientific method. It omits all the discipline-specific richness and, accordingly, tells us very little about scientific practice.

The conception that only those methods common to all the sciences are the proper concern of philosophers—that philosophers should confine their interest to the form of science and not its content—has effectively eliminated methodology in the sense of a descriptive-critical account of what scientists do in favor of formal logical and/or epistemological puzzles that are of no concern to working scientists (Gale 1984). The latter fact does not worry most philosophers, because their interest is not in methods but in metamethodology. Metamethodology certainly does address some important problems. However, by directing most of their energy to meta-issues, modern, "general" philosophers of science (as opposed to those studying the "physical foundations" of scientific theories) have virtually changed the subject. Philosophy no longer attempts to see how the sciences work. Philosophy is no longer the interpreter of science.

How did the subject come to be changed? Who changed it, and why?

The Emergence of Metaphilosophy of Science

The Kantian conception of philosophy was dominant when university reform initiated modern, professional, academic life in Germany. Indeed, Kantian philosophy influenced the reform movement. Partly for this reason and because of the intellectual power of Kant's thought, Kantian and post-Kantian philosophy became the official academic philosophy, and philosophy's proper task came to be seen as one of demarcation and legitimation or the "grounding" of the various disciplines (Merz 1904–12; Haines 1969; Rorty 1979). However, the German model did not take root in Britain for some time, even among pro-Germans such as Whewell. In early-nineteenth-century Britain, philosophy was still regarded as a general theoretical enterprise that encompassed the theory of the mind and society and was not far removed from theoretical concerns about the physical world (which, however, Newton had begun to separate out as special disciplines for mathematicians and experimentalists). Everyone knows that science was called 'natural philosophy' and 'experimental philosophy", but far more than traditional labels was involved.

In Whewell, philosophy of science, history of science, and several branches of science itself were all wrapped up in one. Whewell claimed to derive his philosophy of science from his history of science; he clearly made the descriptive enterprise as important as the normative; and he saw the descriptive-critical task as more important than inquiry into ultimate epistemological grounding. Although scientists could of course be faulted for logical blunders, Whewell held more strongly than anyone until the present generation that agreement with the history of science was a condition of adequacy for any philosophy of science. Whewell claimed to advance "a new method of pursuing the philosophy of human knowledge." This method was the study of history of science. He wrote that "we are most likely to learn the best methods of discovering truth by examining how truths, now universally recognized, have really been discovered." [17] However, he attached no special importance to recent and contemporary science. The lessons must be culled from a systematic examination of the entire history of the physical sciences.

I have already noted the importance of consilience of inductions and novel prediction in Whewell's account of scientific justification. Despite his anticipation of later developments, Whewell remains a transitional figure. For him as for the Scottish methodological tradition and for Herschel, the ultimate justification of scientific method still resided in theology (Strong 1978). The fact that God had created the human mind and nature in harmony with one another made an elaborate metamethodology unnecessary. Ultimately, our methods worked because nature, like human thought, followed regular, nonarbitrary patterns. In effect, God had preselected our world for us, or us for the world; at the very least, God preserved a stable world order.

This short-circuiting of metamethodological investigation may be surprising in Whewell's case, since Whewell is often thought to have been some sort of Kantian. He did begin with a Kantian-looking question: Given that we have by now achieved certain, virtually self-evident knowledge in several branches of physical science, how is that possible? Whewell's un-Kantian answer suggests that he was not asking Kant's question at all. Whewell proceeded to examine the history of science for an answer. What he wanted to know was the "how?" in the sense of the process of inquiry that had brought us to see the self-evidence of dynamical principles, not the "how possible?" in Kant's sense of an ultimate, philosophical grounding. Whewell, like British writers generally, was not as sensitive as Kant to questions of "philosophical" justification and legitimation. Whewell did take a big step in the direction of metamethodological justification (consilience and novel prediction), but he, like most of his Victorian contemporaries, was prepared to appeal to religion, whereas later philosophers would be metamethodological "all the way down."

Although it short-circuited what was to become the major topic of philosophy of science, such an appeal was easy to make, since there was still a pretty comfortable consensus about correct religious belief in early Victorian Britain.[18] Actually, the appeal to religion is not as important as the fact that problems of ultimate legitimation were not yet burning issues for these thinkers. It is nevertheless interesting that such appeals to theology and/or metaphysics carry a broad implication for scientific methodology, though one that was not pursued by the dominant methodologists. Anyone making such an appeal recognizes, at least implicity, that methods are not content neutral but depend for their success on the way the world is. This means that a given methodology will not be equally good in all possible worlds; rather, we must match the choice of methodology, as closely as we can, to the world in which we live. In other words, as a latter generation might put the point, methodology is not a purely logical subject but depends on what we think we know about the world. At this level, the content dependence of method was still compatible with the unity of science, for it could be assumed that the world was equally "connected" or regular in all its phenomena. And, of course, no appeal to religion could tell scientists which method worked best. They had to find that out for themselves.

Mill's overall orientation was in some ways the reverse of Whewell's and more in line with the conception of methodology that has dominated much of the twentieth century. This is no accident, for Mill had something to do with the formation and reception of this conception. Despite the fact that Mill's knowledge of the natural sciences was shallow, his *System of Logic* (1843), which went through eight revised editions by 1881, became by far the most respected Victorian logical and methodological work. Mill outlived Whewell by seven years, and his reputation remained strong while Whewell's rapidly faded. (Whewell had no significant followers who outlived him.) Although Mill himself was not an academic, his *Logic* eventually became the official academic textbook of logic and methodology in most British Commonwealth and American universities. Alexander Bain, in the preface to his famous text *Logic: Deductive and Inductive* (1870) apologized for appearing to compete with his mentor, the incomparable Mr. Mill.

Whewell's "new method" had placed methodology logically prior to epistemology: any adequate epistemology must square with philosophy of science as derived from the study of the history of science. Mill reversed these priorities. For him, scientific methodology must square with general epistemology, as discussed by philosophers, and it amounted to applied logic. Mill bucked the trend (DeMorgan, Venn, later Jevons) to found inductive reasoning on probability theory by insisting that probability theory itself be founded on the logic of induction. For Whewell, induction was a process; for Mill, induction was proof.[19] Mill criticized Whewell's methodology as insufficiently logical and as epistemologically thin (history was one thing, logic quite another), while Whewell considered Mill's logic well nigh irrelevant to science as practiced.

Mill did, of course, have his famous four (or five) methods—which he considered uniform methods that could be applied to almost any subject matter.[20] But these methods are the exception that proves the rule. To Whewell's criticism that the methods of agreement, difference, concomitant variation, and residues beg all the interesting questions about discovery and that no actual scientific discoveries had been made by applying these methods, Mill (1843, book III, chapter ix) responded that they provide a goal for discovery just as syllogisms do for deductive reasoning. The four methods amount to rules of proof. For Mill, induction concerns not the inferential processes of research but the form the ultimate product of inquiry must have it if is to have logical standing.

Mill is modern in another respect. Always suspected of atheism or agnosticism, he was certainly not willing to follow Herschel, Whewell, and others in giving a religious or metaphysical justification for methodology. The question was a logical one. More than any of his British contemporaries, Mill made metamethodology a serious subject for philosophical inquiry. In this respect, he pushed philosophy in roughly the same direction as Kant, albeit from the opposite side of the philosophical fence. Mill attempted to tease out the presuppositions of inductive practice. His principles of uniformity and limited variety were his substitutes for the harmony of God's creation.[21] (Again, however, Mill's principles imply that methodology depends on the way the world is and does not reduce to pure logic.) Nor was Mill content to accept logical principles as simply *a priori* true. He sharply formulated the questions about the justification of induction and of deductive reasoning that would absorb the attention of philosophers, but of few scientists, up to the present. However, Mill's metamethodology was broadly empirical, whereas those of Popper and the positivists would stick more closely to the Kantian *a priori*.

Although Mill criticized "empiricists," his approach to scientific methodology was, in our terms, strongly empiricist. He favored claims susceptible of solid, inductive "proof" over speculative, postulatory hypotheses that yielded successful novel predictions. (In respect of his rejection of Whewell's ardent hypotheticalism, Mill does not sound modern.) Hence, although Mill did not mention Hume in this connection, his conception of ampliative inference in science was, like Hume's, thin. Mill did clearly recognize the problem of underdetermination of hypothesis by data, whereas Whewell had scarcely been struck by it. However, for Mill the essential *logical* problem of the underdetermination of law claims by data arose as much for the law of refraction as for atomic theories of matter.[22]

Since scientists worry far more about theoretical underdetermination than about simple, Hume-Mill type inductions, and since many philosophers of science were to follow Hume and Mill on induction, we have here a specific illustration of the divergence of interests and the perceived lack of relevance of philosophical discussions to science as practiced. Logically speaking, ampliative inference is ampliative inference. If anything, the later formulation of the problem in formal logical terms (how get from a number of singular claims to a universally quantified conclusion) was to reinforce the position that there was no real, methodological difference in these underdetermination questions.

Retrospectively, we can see the transition from Herschel and Whewell to Mill as "progress" toward the dominant modern view that methodology is not (much) concerned with the process by which scientists reason or otherwise make their way to new problems and new solutions, but only with evaluating the final products and the philosophical legitimation of the whole scientific enterprise. Methodology had traditionally been the domain of logicians. As Whewell's historical methodology lost its philosophical following and as the subject of logic narrowed to the study of proofs and came to exclude the study of problem-solving methods, logic tended to take methodology with it.

During the mid and late nineteenth century, important developments took place within mathematics and logic. Four related developments were the program to improve the rigor of analysis and ultimately (with Hilbert) the rigor of geometry, the discovery of alternative geometries, the emergence of the idea of "pure" mathematics and logic, and the development by Gottlob Frege and others of modern symbolic logic.[23] The eventual upshot of these developments was a narrower but more rigorous conception of "pure" mathematics and logic as being expressible by formal and abstract symbolic systems which, although susceptible of various interpretations, had no more to do with physics, with the psychology of human thinking, or even with numbers than with mining or rugby. William Thomson was among the physicists who frowned on this purification movement. For these physicists, mathematics must retain its physical meaning, its embodiment, to remain a fruitful, properly motivated, intellectual pursuit.

I conclude this section with brief remarks on some German developments that influenced twentieth-century philosophy of science.

In the latter third of the nineteenth century, many German scientists and their sympathizers reacted strongly against post-Kantian idealism, which had virtually become the official philosophy of the universities and the government. The idealistic strain in German thought had helped to make high theory respectable, but it had also brought with it much questionable metaphysics, which many scientists came to perceive as a threat to scientific progress. It is not surprising, then, that there were anti-theoretical, positivistic programs among the jousting methodologies.

During this period philosophy lost a great deal of intellectual respect in Germany. This was, for many intellectuals, the time in which science knocked philosophy from her privileged place in German intellectual life. Scientific materialists launched a direct attack on philosophy, which they wished to replace by their materialist-scientific world picture. Rudolf Virchow and others explicitly proclaimed the new sovereignty of science and declared philosophy superfluous (Gregory 1977, chapter 7). The neo-Kantians, under heavy attack from the materialists and reductionists, themselves assaulted woolly Hegelian metaphysics and what was left of *Naturphilosophie.*

Neo-Kantian philosophy was to be a major influence on the twentieth-century logical positivists—for a time, a greater influence than the positivism of Ernst Mach. Actually, Mach's work itself grew out of a tradition of drawing epistemological consequences from research on the physiology of the human senses. Johannes Müller, Hermann von Helmholtz, and others had pursued a research program that might be termed "naturalistic Kantianism." Their idea was to give empirical, physiological content, in terms of specific nerve energies and the like, to Kant's forms of intuition (Moulines 1981; Richards 1977).

As usual, these various methodologies and philosophies of science were reactions to some perceived threat. They were weapons with which to attack opponents. Despite its "up"-sounding name, positivism too was basically a negative doctrine, as Einstein eventually recognized. In a 1917 letter to his friend Michele Besso, Einstein wrote: "I do not inveigh against Mach's little horse; but you know what I think about it. It cannot give birth to anything living, it can only exterminate harmful vermin." (quoted in Holton 1984, p. 1233) With that happy thought, I pass to the twentieth century.

Twentieth-Century Philosophy of Science

In the twentieth century, philosophy of science, like so many other fields, became a full-fledged, professional academic discipline. Now ideas that are (or become) dominant during the period in which a field becomes a professional discipline are of special importance to that field. One reason for the intellectual success of these particular ideas may be their utility in defining the field as something that could gain recognition as a professional discipline in the first place. However, the action is mutual: Because they are intellectually success-ful, the ideas also help determine the institutional structure of the discipline. In either case, the leading ideas actively "present at the creation" gain a privileged place. Since the field is more or less defined in their terms, it is difficult to criticize these ideas effectively, for any such criticism can be viewed as an attack on the profession itself. Professionalization thus tends to be a conservative process in the sense of attenuating criticism of the central constellation of ideas. The very process of becoming a distinct discipline of specialists tends to insulate the practitioners from external criticism.[24] And internal criticism is dulled by the fact that practitioners who launch radical critiques jeopardize their standing within their community.

These conservative tendencies are weaker in the sciences than in less critical disciplines, and perhaps weaker still in philosophy, with its premium on mutual criticism and disagreement. But the very fact of pervasive disagreement has done much to shape professional philosophy, by pushing philosophers into metamethodology in the hope of finding a neutral ground on which to resolve first-order methodological disputes. "Methodological ascent" is a standard response to any unresolved disagreement about substantive matters, whether in science, in law, or in politics. When the disagreement is over methodology itself, then the discussion ascends to metamethodology.

The three related ideas whose emergence we have followed in the previous three sections came together to form much of the constellation of ideas characterizing professional, academic philosophy. Or, rather, these and other ideas were ingeniously fused into a coherent whole by early positivist thinkers (Frank 1949, chapter 1) and, somewhat differently, by Popper. These ideas are the abandonment of generative methodologies for consequentialist, H-D methodologies; the replacement of general inductive methodology by a view of science that is equally general and unitary and more abstract; and the conception of philosophy of science as an *a priori*, Kantian-Millian meta-inquiry into ultimate epistemological foundations rather than a descriptive-critical inquiry into the actual problem-solving work of scientific research.

In this section I shall assemble some reminders about how these themes were developed by Moritz Schlick, Rudolf Carnap, Otto Neurath, Karl Popper, and more recent writers over the past sixty years or so (see note 5). It can be argued that their unification in what became professional philosophy of science accelerated the dropoff of physicists' interest in what the philosophers were doing.

It may appear that my position is doomed from the start by its failure to take into account both the well-known influence of philosophy on relativity theory and quantum mechanics and the less well-known but equally strong influence of these theories on philosophy of science. Did not modern philosophy of science and twentieth-century physics somehow mutually assist one another at birth?

Indeed, a few scientists, including Max Born, P. W. Bridgman, and Werner Heisenberg, have credited positivism with stimulating the twentieth-century revolutions in physics. I shall not enter into the debate over the merits of these claims.[25] My own view is that, although the claims contain a grain of truth, philosophy of science owes more to modern physics than modern physics owes to philosophy of science. Hence I shall turn at once to the other side of the interaction: the influence of physics on philosophy. And here there is an irony. Careful study of physical theories led philosophers to views (or to more extreme versions of views already entertained) that ultimately pushed the two disciplines further apart. Attraction is not the only form of interaction!

Before they became logical positivists, the leaders of the Vienna Circle and the Berlin Circle—Schlick, Carnap, and Reichenbach—were Kantians. (Popper, too, was heavily influenced by Kant.) Despite pronouncements to the contrary, the logical positivists and like-minded scientists were deeply indebted to Kant by way of the neo-Kantian movements that flourished in Germany around the turn of the century. The Kantianism of Hans Reichenbach's position in *The Philosophy of Space and Time* (1928) is plain to see in his doctrine of coordinative definitions, according to which we try to analyze theories into two distinct components, one representing nature's purely factual contribution and one representing our human intellectual contribution.[26] For Kant the contribution of the human mind was *a priori* true, but this idea had to be transformed by the positivists and their precursors because Kant's theory had been undermined by the discovery of non-Euclidean geometries and by Hilbert's purely discursive, axiomatic geometry, which required no Kantian spatial intuition. Einstein's relativity theory also had a prominent role in the transformation of neo-Kantian ideas into logical-positivists ideas.

Michael Friedman (1983b) has illuminated the influence of relativity theory on the development of positivism, particularly in pushing Schlick, Carnap, and Reichenbach away from a neo-Kantian position opposed to Mach to a pro-Machian position.[27] Friedman identifies relativity theory as second only to symbolic logic as a formative influence on positivism. (However, he argues that relativity theory ultimately supports a "realist" rather than a positivist or a conventionalist account of theories—despite the young Einstein's avowed positivistic, Machian, and Leibnizean motivations.) For example, the fundamental positivist distinction between observational language and theoretical language grew out of Kant's distinction between content and form by way of a distinction between the factual and the merely conventional. Non-Euclidean geometry had shown that Euclidean geometry is not an unalterable form of human intuition and an *a priori* true theory of physical space. Contrary to Kant, alternative forms of intuition and alternative categories of thought are therefore possible. In place of Kant's Euclidean, classical physical paradigm of the formal element of knowledge contributed by the human mind, the positivists substituted the physical, space-time coordinate system. But according to relativity theory, there are any number of these that are observationally equivalent, so the choice of one of them is an arbitrary convention. Only what is invariant across coordinate systems represents nature's contribution to our theories.

A curious reversal has occurred here, as Friedman notes. Whereas Kant's formal elements were the most "objective" and the most stable components of our knowledge, the positivists reduced them to arbitrary conventions, free choices, subjective preferences. Schlick, Reichenbach, et al. claimed a happy congruence between general relativity and epistemology: What was objective according to relativity was precisely what was observable according to epistemology. By now, Kant's form-content distinction had become the theoretical-observational distinction.

The positivists (wrongly, Friedman argues) interpreted special and general relativity as reducing to a minimum the amount of theoretical structure necessary to formulate physical statements, and as successively eliminating from physics ever more excess baggage—unobservable, metaphysical entities such as absolute space. Clearly, on their nonrealistic view of theories, theories are not deep-structural attempts to describe the reality underlying our observations. Nature's contributions—empirical facts and invariances—are the only real elements. Popper, too, spoke of theories as Kantian impositions of the human mind on nature, albeit with a "realist" interpretation of theories. We can now see how the old problem of the underdetermination of theory by facts reemerged as the fundamental problem of twentieth-century philosophy of science, with a peculiar neo-Kantian-*cum*-positivist stamp.[28]

It would be unfair to the early positivists to say that they ignored the *process* of physical reasoning. After all, Einstein's path to relativity and Heisenberg's path to quantum mechanics were of great interest to them—although they relied too heavily on Einstein's and Heisenberg's misleading remarks about what they had done. And, in general, the better work on the "physical foundations" of various theories has taken into account discovery paths and their rational reconstructions as "discoverability" paths. As long as it has not attempted to force physical theories into oversimplified formal languages, careful work (such as Friedman's) on physical foundations has constituted the best work in philosophy of physics.

General methodology (as opposed to physical foundations research) is a different story. The philosophers most interested in theorizing about the process of inquiry, the American Pragmatists (especially Peirce and Dewey), were eventually swamped by the logical positivists, who in their maturity aimed chiefly to disclose the logical structure of the finished products of research (Rorty 1979). Rather than broaden logic to include theory of inquiry (as Dewey tried to do in *Logic: The Theory of Inquiry* [1938] and elsewhere), the positivists narrowed general philosophy of science to an application of formal deductive and inductive logics. A quotation from the article by Rudolf Carnap that opened the very first issue of *Philosophy of Science* conveys the flavor of the the enterprise:

Philosophy deals with science only from the *logical* viewpoint. *Philosophy is the logic of science*, i.e., the logical analysis of the concepts, propositions, proofs, theories of science, as well as ... possible methods of constructing concepts, proofs, hypotheses, theories.... [Philosophy is the] logical syntax of the language of science. (Carnap 1934, pp. 6, 9)

For Carnap, formal philosophy of science was philosophy enough. The constructive side of this program came to little more than discussions of the construction of concepts in isolation from any theoretical context, e.g., how to form operational definitions. Hardly anything was said abut theory construction or problem solving. Gradually the realization came, as Carl Hempel (1950, p. 113) later emphasized, that "theory formation and concept formation go hand in hand; neither can be carried on successfully in isolation from the other." Yet this realization, clearly stated by Hempel in 1950, did not lead positivists to take up "discovery" topics such as theory construction. In general, the logical positivists' attempt to translate scientific results into an all-purpose, permanent, neutral, formal language, now physically interpreted, did not promise to capture the more subtle physical content of these claims. Constrast Dewey's historicist challenge:

> ... all logical forms ... arise within the operation of inquiry and are concerned with control of inquiry so that it may yield warranted assertions. This conception implies much more than that logical forms are disclosed or come to light when we reflect upon processes of inquiry that are in use. Of course it means that; but it also means that the forms *originate* in operations of inquiry. To employ a convenient expression, it means that while inquiry into inquiry is the *causa cognoscendi* of logical forms, primary inquiry is itself *causa essendi* of the forms which inquiry into inquiry discloses. (Dewey 1938, pp. 3 ff.)

> ... it is cause for surprise that writers who energetically reject the intervention of the supernatural or the non-natural in every other scientific field feel no hesitancy in invoking Reason and a priori Intuition in the domain of logical theory. It would seem to be more incumbent upon logicians than upon others to make their position in logic coherent with their [naturalistic] beliefs about other matters. (ibid., pp. 24 ff.)

For Dewey, logical and methodological frameworks themselves are the products of particular inquiries, including empirical inquiries. A permanent, neutral framework, independent of inquiry, is not simply given in the positivist way anymore than in the Kantian manner.

It is true that the pragmatists jumped on the consequentialist bandwagon and repeatedly emphasized that it was consequences, not antecedents, that counted in science as well as in democratic society. And like the logical positivists, the pragmatists pushed operational and instrumental thinking too far. However, despite his formal logical interests, Peirce appreciated the content specificity of methods more than anyone else of his day (see, e.g., Peirce 1877). And Peirce was, if anything, more sensitive even than Mach to problems of economy of research—a concern that practically dropped out of formal logical positivism and was never central to Popper's methodology.[29]

I have not yet spoken directly about the positivists' conception of science as a unitary enterprise. A passage from Otto Neurath about his "central conviction" will serve to introduce this topic:

> ... the elaboration of the differences between the various sciences is an inessential task ... on the contrary, it was especially important to develop an account of all the sciences using only one kind of a scientific 'style'. (Neurath 1937, p. 178)

Although Neurath's intentions were admirable, it is fortunate that this project, which recreated all sciences in the same image, had no serious impact on the natural sciences; attempts to produce a uniform scientific style, even in a single field, ultimately stifle research.[30] Of course, Neurath did not dream of imposing anything on anyone. For him, the converging of the sciences into one unified language and methodology was a historical fact.

I turn now to Popper, who introduced a strongly fallibilistic (nonjustificationist) element into methodology, partly in response to the revolutionary developments in the physics of his youth. Whereas most scientists optimistically came to regard relativity and quantum theory as great successes, Popper pessimistically saw them as great failures—failures of the best classical theories and of classical methodologies. Popper concluded that there was no method of science in the sense of a set of rules that carried even a probabilistic guarantee of success. Even the Scottish conception of methodology was too strong. Popper was correct that there is no *general* method of this sort.

Popper's Humean, Kantian, deductivist, and anti-discovery strains are all apparent in a famous passage:

> The initial stage, the act of conceiving or inventing a theory, seems to me neither to call for logical analysis nor to be susceptible of it. The question how it happens that a new idea occurs to a man—whether it is a musical theme, a dramatic conflict, or a scientific theory—may be of great interest to empirical psychology; but it is irrelevant to the logical analysis of scientific knowledge. This latter is concerned not with *questions of fact* (Kant's *quid facti?*), but only with questions of *justification or validity* (Kant's *quid juris?*). . . .
>
> . . . the method of critically testing theories, and selecting them according to the results of tests, always proceeds on the following lines. From a new idea, put up tentatively, and not yet justified in any way . . . conclusions are drawn by means of logical deduction. These conclusions are then compared with one another and with other relevant statements, so as to find what logical relations (such as equivalence, derivability, compatibility, or incompatibility) exist between them. (Popper 1959, pp. 31–32)

As the final sentence hints, Popper's reduction of scientific reasoning to simple logical relations is still more restrictive than Carnap's. Such a simple, deductivist view is not calculated to grasp the richness of scientific inference, although Popper is more successful than Carnap in capturing the spirit of scientific inquiry. Indeed, one social function of Popper's constant harping on creative inspiration and his adamant rejection of inductive or any other discovery methods is to present scientists in a favorable light—as more like poets and musical composers than fact-grubbing drudges.[31] This theme is not original with Popper; one can find the idea that inductive research is drudgery in nearly all the major nineteenth-century methodologists.

Popper acknowledges that his view of method as strongly normative and as content neutral derives from Hume. Popper credits Hume with clearly distinguishing questions of validity (answers to the question *quid juris?*) from questions of fact (answers to the question *quid facti?*).[32] Having drawn the distinction, Popper follows Kant in identifying philosophical questions with the former rather than the latter. Although Popper's methodology is anti-epistemological in the sense that he attacks foundationist tendencies wherever he finds them, he is so concerned with Humean-Kantian problems and so unconcerned with the problem-solving process that he, too, makes methodology derivative from epistemology, rather than the other way around.

Popper thinks that any constructive methodology is generative and that any generative methodology must be "justificationist" (foundationist). Hence, his adamant anti-justificationism leads him to conclude that methodology must be entirely critical and not at all constructive. Only a negative sort of justification is possible: A claim is temporarily "justified" (a term Popper does not like to use) only in the sense that it has survived rigorous tests that all its known competitors have failed. A claim is "supported" only by answering its harshest critics and by criticizing its competitors.

While Popper's methodology is interesting for its fallibilism and its emphasis on criticism, its main attraction comes from its power in combatting opposing positions, such as inductivism and justificationism. As a positive account of science, Popper's method of conjectures and refutations does not tell us very much. Popper would reply that there is not much to tell, but others see the large domain of heuristic problem-solving methods—which lies between algorithmic methods and the divine madness of poets—as a worthy field for methodological inquiry (Nickles 1986).

In telling us that all sciences at all times follow (or should follow) the same basic methodological rules, Popperians reinforce the grand methodological myth that one must choose only one basic methodology, no matter what one's task is. One must be either an inductivist, a conventionalist, or a purely consequential, H-D theorist regardless of one's current problem situation. In my experience, most scientists and many beginning philosophy-of-science students find this suggestion silly. Is it not obvious that which methods are best employed depends heavily on the particular field of investigation, the stage of maturity the field has reached, and the specific problem at hand?

The great importance the classical Popper attached to the "problem of demarcation"—the problem of distinguishing science from pseudoscience, metaphysics, and other nonscientific pursuits—pressed his philosophical thinking toward a unitary, neutral conception of method, independent of particular scientific content. Popper himself emphasized how much of his methodology grew out of his early concern with the demarcation problem (Popper 1962, chapter 1). Yet despite Popper's repeated attacks on essentialism (the doctrine that things have real natures or essences, which it is the job of science or philosophy to discover), it is hard to dispel the impression that, in making demarcation problems so central, Popper presupposed that there is some key feature that captures the very nature of all scientific research. A suitable criterion of demarcation would express this feature. For Popper, of course, this key feature was empirical falsifiability—an effective weapon against the inductivist criterion of demarcation (according to which a theory is scientific provided that it is inferred from the data) but not otherwise very informative about scientific problem solving.

General methodologies as developed by inductivists, positivists, and Popperians are all examples of what I call *whig methodologies*. As the historian Herbert Butterfield observed in *The Whig Interpretation of History* (1931), general history, in contrast with research monographs, can hardly fail to provide a distorted picture of historical developments. General history reduces the vagaries and vicissitudes of the historical process to a few simple patterns and makes it appear that all history was aiming at the present as its goal, its *terminus ad quem*. Historical agents may be judged as progressive or not according as they facilitated or obstructed progress toward the goal. For Butterfield, such "whig" history is so far from being genuine history that the term 'general history' is practically an oxymoron. The same can be said for the term 'general methodology', in my opinion. Whig methodology reduces all varieties of research to their least common denominator; it forces scientific work into a few basic patterns. By assuring us that what really matters to research is revealed by the simple patterns disclosed by the philosophers of science, it discourages inquiry into the details of scientific inquiry and misses the open-future standpoint of the working scientist. Scientific results appear to be the natural and expected products of the method applied to the data of nature rather than the result of a good bit of human construction and human judgment. It becomes tempting to describe and explain pieces of physical research only in terms of the ways in which nature constrained scientific thought, when in fact this conception of nature was precisely what the scientists were in the process of hammering out (see Pickering 1984, chapter 1).

In equating methodology (philosophy of science) with the study of inferential and developmental patterns which all sciences at all times possess in common, whig methodology places a premium on demarcating science from pseudoscience, metaphysics, and nonscience generally instead of, say, demarcating good scientific work in a particular area from bad work or differentiating work within one specialty or on one type of problem from other specialized work; on establishing the unity of science as a more or less homogeneous field for philosophical study; and on finding completely general principles, rules, or "laws" for this domain. For positivistic writers, these rules were largely *a priori* laws of logic and probability theory. For Popper they were community conventions motivated by logical principles. For a more recent generation of history-oriented philosophers, they are general models of rational, scientific change. Whether logical or historical, any general methodology or model of science—one that aims to capture all sciences, or even one science at all stages of its historical development—will be formal in the sense that it cannot link methodology with the specific content of any particular speciality at a particular time.

But what of the recent historical revolution in philosophy of science, the reader may ask. Doesn't this show that many philosophers are joining historians and sociologists in studying ever smaller groups and pieces of work? Yes, in many instances it does. However, ironically, one genre of historical philosophy of science is actually more abstract and vapid than ever. Surprisingly, perhaps, a good deal of historical methodology is still whig methodology. A little historical knowledge can contribute to a most unhistorical result. How does this happen?

Philosophers appeal to history to suggest and to test methodological theses. They find a single case (their one case study?) in which the methodological principle in question was not followed. Hence, they reject the principle as mistaken and seek something more general, more abstract, more highly qualified—a rule that will survive this case and fit all the others as well. The trouble, of course, is that practitioners of this genre leave in place the crucial assumption that there exists a general methodology that unifies all sciences. Left unchecked, this mode of inquiry can only result in more and more abstract conceptions of methodology which, despite their historical motivation, stray further and further from the concrete problems of the historical agents whose work is studied.

Fortunately, not everyone is moving in this direction. We find in Thomas Kuhn's idea of a paradigm and its offspring—Imre Lakatos's research programs and Larry Laudan's research traditions—the realization that method and theory are every bit as intertwined as observation and theory (see Kuhn 1962; Lakatos 1970; Laudan 1977). Methods are not content neutral. Not only are they specific in their applicability, but they carry with them theoretical presuppositions. Recently, some philosophers have begun to seek and use empirical information about science more seriously than did the history-oriented philosophers of the 1960s and the 1970s. They study micro-history, they attend to sociological studies of ongoing research, and they even engage in direct observation of scientific communities at work (Giere 1985). It is now relatively safe for philosophers to practice empiricism as well as to preach it! If philosophers do not become empirical, it appears, historians, sociologists, and psychologists of science will take over completely the job of interpreting science.

Why have philosophers, scientists, and nearly everyone else automatically equated philosophy of science with the study of inferential and developmental patterns which all sciences have in common? Strong intellectual traditions are a main part of the answer, but there are also, I suggest, complementary social factors that make this conception of philosophy of science professionally self-serving. The constellation of ideas that we have been discussing was peculiarly effective in solving for academic philosophy of science the two fundamental problems faced by any aspiring new discipline: the problem of integrity or disciplinary unity and the problem of autonomy (Nickles 1976b). The first order of business for proponents of a new (or reformed) discipline is to convince their colleagues and the powers that be that there exists a unified subject matter to be studied. This domain must be sufficiently homogeneous to imply the existence of general rules or laws governing the domain, which it will be the discipline's task to discover. The second order of business is to show that such an enterprise is sufficiently distinct from existing fields to form a separate subject.

In order to come into existence and to flourish, academic philosophy of science had to emphasize both the unity of its subject matter (methods of scientific inquiry) and its distinctness from what working scientists themselves were already doing (the empirical and pragmatic mutual adjustment of methodological means and research ends). It is not difficult to see how emphasis on general, *a priori*, logical, content-free methodologies helped solve both the integrity problem and the autonomy problem for philosophy of science. In this essay I have focused on the intellectual side of this story. Elaboration of the "sociological" account must await another occasion.

To conclude the present story, let us notice how different are the current interests of philosophers and scientists. Their interests need not be identical, of course; however, the philosophers are supposed to be studying the doings of scientists. Nor is the approval of scientists a condition of adequacy for good philosophical work, although sharp disagreement or complete lack of interest should be cause for concern. Insofar as they address scientists at all, philosophers too often try to speak to all of them at once and end up addressing no one in particular. As professionals must, they spend most of their time talking to one another. Since philosophical disagreements are many and run deep, there is always much to argue about. Unfortunately, many of the twentieth-century issues have concerned spurious problems generated by the various philosophical programs rather than problems of interpreting the sciences themselves. Not surprisingly, the more metamethodological the discussion, the less interesting it is to scientists and other "outsiders." [33]

Given the philosophers' emphasis on consequentialism, fallibilism, and anti-foundationism, it is striking how epistemologically conservative, how backward-looking, philosophers are in comparison with physicists. Philosophers have, until quite recently, ignored the interesting problems of heuristic appraisal—of judging the future problem-solving ability or "heuristic potential" of ideas, theories, and research programs. [34] Philosophical theories of justification have been backward-looking in evaluating theories entirely on the basis of their "track record" of empirical success. Yet for working scientists, the justification of research decisions rests as much on heuristic appraisal as on epistemic appraisal. Problems which philosophers of physics consider central to contemporary physics are often ignored as "old hat" by the more creative physicists. For example, philosophers attach far more significance to the issue of hidden variables and quantum locality and to Einstein-Podolsky-Rosen-type experimental tests of quantum mechanics than physicists do. [35]

In contrast with the conservative philosophers, highly creative scientists appear to be light-footed, quick-handed opportunists who will gladly explore even ideas they know to be defective in order to produce some new results or develop new techniques which may be valuable to future work. Even when they concern themselves with foundational questions, scientists usually have one eye on the future. Their hope is that settling an old foundational question will provide heuristic guidance to the solution of outstanding problems. Why would anyone today write a book intended to clear up old problems of classical field and particle theory, for example? F. Rohrlich, who did just that in his well-known textbook *Classical Charged Particles* (1965), stated that heuristic illumination was the primary purpose of his investigations. He hoped that seeing his way through the old problems would shed light on the new ones.

Although John Dewey was far from being a physical scientist, physicists might well adopt his words as their motto: "We live forward." [36]

Acknowledgments

I am grateful to the National Science Foundation for summer research support. Helpful conversations with William Scott and Geoffrey Cantor do not commit them to my excesses.

Notes

1. See Buchwald 1977 for Fourier's influence on Thomson's mathematization of Faraday's ideas about electrostatic fields.

2. The best general treatments of methodology during this period are those of Larry Laudan (1981) and his student John Strong (1978), to which I am greatly indebted throughout. There are a number of excellent scholarly articles on individual people and developments.

3. Whewell is an exception, since his view of induction was peculiar. But Whewell held that novel prediction and consilience of inductions could establish the truth of scientific claims with virtual certainty. Established principles were necessary truths, he said, so they clearly did not have the status of mere confirmed hypotheses, in the modern sense.

4. I discuss these matters in more detail in Nickles 1984b and Nickles 1985.

5. See, e.g., Worrall 1978 and Zahar 1983. For a critique, see Nickles 1987. By no means are all modern theories of confirmation purely consequentialist. For example, Hempel's (1945) early confirmation theory was not, nor was Reichenbach's (1938) theory of induction. A few Bayesians recognize the generative possibilities of the Bayesian framework (see, e.g., Salmon 1966; Dorling 1979; Howson 1984). Glymour's (1980) confirmation theory has a generative side. For simplicity, I shall make H-D methodology the "modern" methodology in this essay and shall ignore Bayesian methodologies.

6. Fallibilism is a central feature of Laudan's (1980) explanation, but he agrees that the argument does not establish what I term pure consequentialism.

7. The fallacy of affirming the consequent concludes that P is true from the premises "If P then Q" and "Q."

8. On the H-D movement in Germany, see Caneva 1978.

9. I do not have space here to discuss examples. One particularly revealing case is Wilhelm Wien's 1896 "derivation" of his blackbody energy-distribution law, discussed in Nickles 1984b. Nickles 1976a contains examples from Ehrenfest's work, although my purpose there was different. See also Dorling 1973.

10. For an entry into the discussion of novel prediction versus old evidence, see Worrall 1978 and various essys in Earman 1983.

11. However much I may differ over what the changes were, I agree with Laudan's general thesis that nineteenth-century changes in methodology were due mainly to the changes within science itself.

12. I do not have space to discuss the scientific countertendency toward unity of science in Germany. The ideal of *Wissenschaft*, implying broad unity of the sciences, was quite different from the specialization occurring in America. In some respects, it resembled the Scottish view that the unity of science rested on unity of method, which resided in the fact that the same human mind was involved in all scientific work. The *Wissenschaft* ideal was backed by a Kantian conception of mind. See Haines 1967, p. 111, and Turner 1971.

13. See Zahar 1980. On the case of William Thomson, see Wise 1979 and Buchwald 1977.

14. This point is emphasized in the postscript to the 1970 edition of Kuhn 1962.

15. Given a fallibilistic orientation, the move to content-dependent methods means that method itself becomes hypothetical. But we must not equate this new hypotheticalism in methodology with H-D methodology of the usual sort. The H-D "method" is every bit as "formal" (content-free) and general as the inductive method.

16. Quoted in Feigenbaum et al. 1977, p. 84. See also Duda and Shortliffe 1983. For an account of the General Problem Solver and many other developments, see Newell and Simon 1972.

17. Both quotations are from the 1847 edition of Whewell 1840 (introduction to chapter 1 of book I).

18. On William Thomson's use of theological arguments to further support his scientific positions and arguments, see Smith 1976.

19. See aphorism XIII and other aphorisms in the 1847 edition of Whewell 1840. Compare chapters ii, iii, ix and other sections of book III of Mill 1843.

20. In the introductory essay to the *International Encyclopedia of Unified Science*, Otto Neurath praised Mill as follows: "Modern scientific empiricism attained very late in its development a comprehensive work which analyzes empirical procedure in all scientific fields: John Stuart Mill's *A System of Logic*.... Mill does not question the fact that astronomy and social science, physics and biology, are sciences of the same type." (Neurath 1938, p. 9)

21. I am indebted to John Strong (1978, p. 230) for this way of putting it. See Strong 1976 for an important development which I omit.

22. On Hume and the problem of induction, see Laudan 1981, chapter 6. Bertrand Russell (1912) was to solidly reconnect the problem of scientific induction with Hume's problem. Together, Russell, Popper, and the positivists made Hume into one of the major heroes of the history of philosophy.

23. Poincaré (1908) reacted strongly against the non-intuitive, purely discursive view of mathematics advanced by Hilbert, Russell, and others. He countered with an unabashedly romantic account of mathematical discovery that has plagued attempts to deal with scientific discovery ever since.

24. William Thomson discovered this when he used heat-theoretic arguments to attack the geologists' conclusions about the age of the earth.

25. See Zahar 1977 for references to the debate. Zahar rejects these claims. Friedman (1983a, pp. 24 ff. and passim) also rejects them, on various grounds.

26. See Kamlah 1984 for the neo-Kantian roots of Reichenbach's theory of induction.

27. The next several paragraphs are based on chapter 1 of Friedman 1983a.

28. Kant's idea of a conceptual framework made possible the conception of alternative conceptual frameworks, which has contributed to the skepticism that characterizes post-Kuhnian, historical philosophy of science. Against the Kantian legacy, see chapter 1 of Rorty 1979 and the introduction and the first chapter of Rorty 1982.

29. On the neglect of economy of research in recent philosophy, see Rescher 1976. See also Nickles 1986, where I apply Herbert Simon's distinction between maximizing and satisficing. Most scientists are satisficers, whereas philosophers tend to be optimizers.

30. For the extreme case of Soviet physics in the 1930s, see Vucinich 1980.

31. The biologist Peter Medawar (1982) praises Popper on this score.

32. In chapter 2 of Popper 1962 (originally a lecture given in 1953), Popper states that "it was Hume's great achievement to break this uncritical identification of the question of fact—*quid facti*—and the question of justification or validity—*quid juris*" (p. 45). Later in the chapter Popper explains that in important respects he is a Kantian and that Kant's correct idea that the human mind imposes its ideas on nature rather than vice versa (as in the Baconian image of the mind as a mirror to nature) amounts to rejecting the idea that our theories are copies or representations of nature. Popper also discusses Kant and nonrepresentationalism in his autobiography (Popper 1974). However, Popper remains a physical realist.

33. See Gale 1984. The drastic dropoff in the leading scientists' interest in philosophy is indicated by the dramatic change in the editorial board of *Philosophy of Science*, which celebrated its fiftieth anniversary during the Kelvin centennial year. The original board of 1934 was composed almost entirely of leading scientists from several fields, with a very few historians of ideas added for leavening. In 1984, every member of the board was a card-carrying philosopher of science, although a few had also published in scientific journals.

34. For preliminary work on heuristic appraisal, see Lakatos 1970; Urbach 1978; Laudan 1977; Wimsatt 1980; Nickles 1981.

35. For one point of view on Holt's failure to confirm quantum theory, see Harvey 1980, 1981.

36. Dewey 1917, p. 64.

References

Bain, A. 1870. *Logic: Deductive and Inductive*. London.

Buchwald, J. 1977. "William Thomson and the Mathematization of Faraday's Electrostatics." *Historical Studies in the Physical Sciences* 8: 101–136.

Butterfield, H. 1931. *The Whig Interpretation of History*. London.

Caneva, K. 1978. "From Galvanism to Electrodynamics: The Transformation of German Physics and Its Social Context." *Historical Studies in the Physical Sciences* 9: 63–159.

Cannon, S. F. 1978. *Science in Culture: The Early Victorian Period*. New York: Science History Publications.

Cantor, G. N. 1971. "Henry Brougham and the Scottish Methodological Tradition." *Studies in History and Philosophy of Science* 2: 69–89.

Carnap, R. 1934. "On the Character of Philosophic Problems." *Philosophy of Science* 1: 5–19. Reprinted in *Philosophy of Science* 51 (1984): 5–19.

Dewey, J. 1917. "The Need for a Recovery of Philosophy." Reprinted in J. McDermott, ed., *The Philosophy of John Dewey*. University of Chicago Press, 1981.

Dewey, J. 1938. *Logic, The Theory of Inquiry*. New York: Henry Holt.

Dorling, J. 1973. "Demonstrative Induction: Its Significant Role in the History of Physics." *Philosophy of Science* 40: 360–372.

Dorling, J. 1979. "Bayesian Personalism, the Methodology of Scientific Research Programmes, and Duhem's Problem." *Studies in History and Philosophy of Science* 10: 177–187.

Duda, R. O., and E. H. Shortliffe. 1983. "Expert Systems Research." *Science* 220 (April 15): 261–268.

Earman, J., ed. 1983. *Testing Scientific Theories* (Minnesota Studies in the Philosophy of Science, vol. 10). Minneapolis: University of Minnesota Press.

Feigenbaum, E. A., B. G. Buchanan, and J. Lederberg. 1971. "On Generality and Problem Solving: A Case Study Using the DENDRAL Program." *Machine Intelligence* 7: 165–190.

Frank, P. 1949. *Modern Science and Its Philosophy*. New York: Braziller.

Friedman, M. 1983a. *Foundations of Space-Time Theories*. Princeton University Press.

Friedman, M. 1983b. "Moritz Schlick, Philosophical Papers." *Philosophy of Science* 50: 498–514.

Gale, G. 1984. "Science and the Philosophers." *Nature* 312 (December 6): 491–495.

Giere, R. 1985. "Philosophy of Science Naturalized." *Philosophy of Science* 52: 331–356.

Glymour, C. 1980. *Theory and Evidence.* Princeton University Press.

Gregory, F. 1977. *Scientific Materialism in Nineteenth-Century Germany.* Dordrecht: Reidel.

Haines, G. 1969. *Essays on German Influence upon English Education and Science, 1850–1919* (Connecticut College Monograph No. 9). Hamden: Archon.

Harvey, B. 1980. "The Effects of Social Context on the Process of Scientific Investigation: Experimental Tests of Quantum Mechanics." In K. Knorr, R. Krohn, and R. Whitley, eds., *The Social Process of Scientific Investigation.* Dordrecht: Reidel.

Harvey, B. 1981. "Plausibility and the Evaluation of Knowledge: A Case-Study of Experimental Quantum Mechanics." *Social Studies of Science* 11: 95–130.

Hempel, C. G. 1945. "Studies in the Logic of Confirmation." Reprinted in *Aspects of Scientific Explanation.* New York: Free Press, 1965.

Hempel, C. G. 1950. "Problems and Changes in the Empiricist Criterion of Meaning." Reprinted, with changes, in *Aspects of Scientific Explanation.* New York: Free Press, 1965.

Herschel, J. F. W. 1830. *Preliminary Discourse on the Study of Natural Philosophy.* London.

Holton, G. 1984. "Do Scientists Need a Philosophy?" *Times Literary Supplement* (November 2): 1231–1234.

Howson, C. 1984. "Bayesianism and Support by Novel Facts." *British Journal for the Philosophy of Science* 35: 245–251.

Jahnke, H. N., and M. Otte, eds. 1981. *Epistemological and Social Problems of the Sciences in the Early Nineteenth Century.* Dordrecht: Reidel.

Jevons, W. S. 1874. *The Principles of Science.* Second edition: London, 1883.

Kamlah, A. 1984. The Neo-Kantian Origins of Hans Reichenbach's Principle of Induction. Read at the University of Pittsburgh.

Kuhn, T. 1962. *The Structure of Scientific Revolutions.* Second, expanded edition: University of Chicago Press, 1970.

Lakatos, I. 1970. "Falsification and the Methodology of Scientific Research Programmes." In I. Lakatos and A. Musgrave, eds., *Criticism and the Growth of Knowledge.* Cambridge University Press.

Laudan, L. 1977. *Progress and Its Problems.* Berkeley: University of California Press.

Laudan, L. 1980. "Why Was the Logic of Discovery Abandoned?" In Nickles 1980; revised version in Laudan 1981.

Laudan, L. 1981. *Science and Hypothesis*. Dordrecht: Reidel.

Medawar, P. 1982. *Pluto's Republic*. Oxford University Press.

Merz, J. T. 1904–1912. *A History of European Thought in the Nineteenth Century*. Four volumes. London: Blackwood & Sons. Reprinted New York: Dover, 1965.

Mill, J. S. 1843. *A System of Logic*. London. Eighth edition: New York, 1881.

Moulines, C.-U. 1981. "Hermann von Helmholtz: A Physiological Approach to the Theory of Knowledge." In Jahnke and Otte 1981.

Neurath, O. 1937. "Unified Science and Its Encyclopedia." Reprinted in Neurath's *Philosophical Papers*, ed. R. S. Cohen and M. Neurath. Dordrecht: Reidel.

Neurath, O. 1938. "Unified Science as Encyclopedia Integration." *International Encyclopedia of Unified Science*, vol. 1, combined edition. University of Chicago Press, 1955.

Newell, A., and H. A. Simon. 1972. *Human Problem Solving*. Englewood Cliffs: Prentice-Hall.

Nickles, T. 1976a. "Theory Generalization, Problem Reduction, and the Unity of Science." In A. Michalos and R. S. Cohen, eds., *PSA 1974*. Dordrecht: Reidel.

Nickles, T. 1976b. "On Some Autonomy Arguments in Social Science." In Suppe and Asquith 1976.

Nickles, T., ed. 1980. *Scientific Discovery, Logic, and Rationality*. Dordrecht: Reidel.

Nickles, T. 1981. "What Is a Problem that We May Solve It?" *Synthese* 47: 85–118.

Nickles, T. 1984a. "Scoperta e Mutamento Scientifico" ("Discovery and Scientific Change"). *Materiali Filosofici* 10: 7–27.

Nickles, T. 1984b. "Positive Science and Discoverability." In P. Asquith and P. Kitcher, eds., *PSA 1984*, vol. 1. East Lansing: Philosophy of Science Association.

Nickles, T. 1985. "Beyond Divorce: Current Status of the Discovery Debate." *Philosophy of Science* 52: 177–206.

Nickles, T. 1986. "'Twixt Method and Madness." In N. Nersessian, ed., *The Process of Science*. The Hague: Martinus Nijhoff.

Nickles, T. 1987. "Lakatosian Heuristics and Epistemic Support." *British Journal for the Philosophy of Science* 38 (in press).

Olson, R. 1975. *Scottish Philosophy and British Physics 1750–1880*. Princeton University Press.

Peirce, C. S. 1877. "The Fixation of Belief." *Popular Science Monthly* 12: 1–15.

Pickering, A. 1984. *Constructing Quarks*. University of Chicago Press.

Poincaré, H. 1908. *Science et Méthode*. Paris. English translation: London, 1914. Reissued New York: Dover, n.d.

Popper, K. 1959. *The Logic of Scientific Discovery*. London: Hutchinson. Expanded translation of *Logik der Forschung* (Vienna: Julius Springer, 1934).

Popper, K. 1962. *Conjectures and Refutations*. New York: Basic Books.

Popper, K. 1974. "Autobiography: Unended Quest." In P. A. Schilpp, ed., *The Philosophy of Karl R. Popper*. LaSalle Ill.: Open Court.

Radnitzky, G., and G. Andersson, eds. 1978. *Progress and Rationality in Science*. Dordrecht: Reidel.

Reichenbach, H. 1928. *Philosophie der Raum-Zeit-Lehre*. Berlin. Translated as *The Philosophy of Space and Time* (New York: Dover, 1957).

Reichenbach, H. 1938. *Experience and Prediction*. University of Chicago Press.

Rescher, N. 1976. "Peirce and the Economy of Research." *Philosophy of Science* 43: 71–98. Reprinted with revisions in *Peirce's Philosophy of Science* (University of Notre Dame Press, 1977).

Richards, J. 1977. "The Evolution of Empiricism: Hermann von Helmholtz and the Foundations of Geometry." *British Journal for the Philosophy of Science* 28: 235–253.

Rohrlich, F. 1965. *Classical Charged Particles*. Reading, Mass.: Addison-Wesley.

Rorty, R. 1979. *Philosophy and the Mirror of Nature*. Princeton University Press.

Rorty, R. 1982. *Consequences of Pragmatism*. Minneapolis: University of Minnesota Press.

Russell, B. 1912. *The Problems of Philosophy*. Oxford University Press.

Salmon, W. 1966. *Foundations of Scientific Inference*. University of Pittsburgh Press.

Smith, C. 1976. "Natural Philosophy and Thermodynamics: William Thomson and 'The Dynamical Theory of Heat.'" *British Journal for the History of Science* 9: 293–319.

Strong, J. V. 1976. "The Infinite Ballot Box of Nature: DeMorgan, Boole, and Jevons on Probability and the Logic of Induction." In Suppe and Asquith 1976.

Strong, J. V. 1978. Studies in the Logic of Theory Assessment in Early Victorian Britain, 1830–1860. Ph.D. dissertation, Department of History and Philosophy of Science, University of Pittsburgh.

Suppe, F., and P. Asquith, eds. 1976. *PSA 1976*, volume I. East Lansing: Philosophy of Science Association.

Turner, S. 1971. "The Growth of Professorial Research in Prussia, 1818 to 1848—Causes and Context." *Historical Studies in the Physical Sciences* 3: 137–182.

Urbach, P. 1978. "The Objective Promise of a Research Programme." In Radnitzky and Andersson 1978.

Vucinich, A. 1980. "Soviet Physicists and Philosophers in the 1930's: Dynamics of a Conflict." *ISIS* 71: 236–250.

Whewell, W. 1837. *History of the Inductive Sciences*. London.

Whewell, W. 1840. *Philosophy of the Inductive Sciences*. London.

Wimsatt, W. 1980. "Reductionistic Research Strategies and Their Biases in the Units of Selection Controversy." In T. Nickles, ed., *Scientific Discovery: Case Studies*. Dordrecht: Reidel.

Wise, M. N. 1979. "William Thomson's Mathematical Route to Energy Conservation: A Case Study of the Role of Mathematics in Concept Formation." *Historical Studies in the Physical Sciences* 10: 49–83.

Worrall, J. 1978. "The Ways in Which the Methodology of Scientific Research Programmes Improves on Popper's Methodology." In Radnitzky and Andersson 1978.

Worrall, J. 1987. "The Role of Successful Experimental Predictions in Theory Change." In D. Gooding, T. Pinch, and S. Schaffer, eds., *The Uses of Experiment*. Cambridge University Press.

Zahar, E. 1977. "Mach, Einstein and the Rise of Modern Science." *British Journal for the Philosophy of Science* 28: 195–213.

Zahar, E. 1980. "Einstein, Meyerson and the Role of Mathematics in Physical Discovery." *British Journal for the Philosophy of Science* 31: 1–43.

Zahar, E. 1983. "Logic of Discovery or Psychology of Invention?" *British Journal for the Philosophy of Science* 34: 243–261.

Name Index for Essays

DATE DUE

DEMCO 38-297